International Federation of Automatic Control

LOW COST AUTOMATION 1992
Techniques,
Components and Instruments,
Applications

IFAC Symposia Series, 1993. Number 13

IFAC SYMPOSIA SERIES

Janos Gertler, *Editor-in-Chief*, George Mason University, School of Information Technology and Engineering, Fairfax, VA 22030-4444, USA

DHURJATI & STEPHANOPOULOS: On-line Fault Detection and Supervision in the Chemical Process Industries *(1993, No.1)*
BALCHEN *et al*: Dynamics and Control of Chemical Reactors, Distillation Columns and Batch Processes *(1993, No.2)*
OLLERO & CAMACHO: Intelligent Components and Instruments for Control Applications *(1993, No.3)*
ZAREMBA: Information Control Problems in Manufacturing Technology *(1993, No.4)*
STASSEN: Analysis, Design and Evaluation of Man-Machine Systems *(1993, No.5)*
VERBRUGGEN & RODD: Artificial Intelligence in Real-Time Control *(1993, No.6)*
FLIESS: Nonlinear Control Systems Design *(1993, No.7)*
DUGARD, M'SAAD & LANDAU: Adaptive Systems in Control and Signal Processing *(1993, No.8)*
TU XUYAN: Modelling and Control of National Economies *(1993, No.9)*
LIU, CHEN & ZHENG: Large Scale Systems: Theory and Applications *(1993, No.10)*
GU YAN & CHEN ZHEN-YU: Automation in Mining, Mineral and Metal Processing *(1993, No.11)*
DEBRA & GOTTZEIN: Automatic Control in Aerospace *(1993, No.12)*
KOPACEK & ALBERTOS: Low Cost Automation *(1993, No.13)*
HARVEY & EMSPAK: Automated Systems Based on Human Skill (and Intelligence) *(1993, No.14)*

BARKER: Computer Aided Design in Control Systems *(1992, No.1)*
KHEIR *et al*: Advances in Control Education *(1992, No.2)*
BANYASZ & KEVICZKY: Identification and System Parameter Estimation *(1992, No.3)*
LEVIS & STEPHANOU: Distributed Intelligence Systems *(1992, No.4)*
FRANKE & KRAUS: Design Methods of Control Systems *(1992, No.5)*
ISERMANN & FREYERMUTH: Fault Detection, Supervision and Safety for Technical Processes *(1992, No.6)*
TROCH *et al*: Robot Control *(1992, No.7)*
NAJIM & DUFOUR: Advanced Control of Chemical Processes *(1992, No.8)*
WELFONDER, LAUSTERER & WEBER: Control of Power Plants and Power Systems *(1992, No.9)*
KARIM & STEPHANOPOULOS: Modeling and Control of Biotechnical Processes *(1992, No.10)*
FREY: Safety of Computer Control Systems 1992

NOTICE TO READERS

If your library is not already a standing/continuation order customer or subscriber to this series, may we recommend that you place a standing/continuation or subscription order to receive immediately upon publication all new volumes. Should you find that these volumes no longer serve your needs your order can be cancelled at any time without notice.

Copies of all previously published volumes are available. A fully descriptive catalogue will be gladly sent on request.

AUTOMATICA and *CONTROL ENGINEERING PRACTICE*

The editors of the IFAC journals *Automatica* and *Control Engineering Practice* always welcome papers for publication. Manuscript requirements will be found in the journals. Manuscripts should be sent to:

Automatica

Professor H A Kwakernaak
Deputy Editor-in-Chief
AUTOMATICA
Department of Applied Mathematics
University of Twente
P O Box 217, 7500 AE Enschede
The Netherlands

Control Engineering Practice

Professor M G Rodd
Editor-in-Chief, CEP
Institute for Industrial Information Technology Ltd
Innovation Centre
Singleton Park
Swansea SA2 8PP
UK

For a free sample copy of either journal please write to:

Pergamon Press Ltd
Headington Hill Hall
Oxford OX3 0BW, UK

Pergamon Press Inc
660 White Plains Road
Tarrytown, NY 10591-5153, USA

Full list of IFAC publications appears at the end of this volume

LOW COST AUTOMATION 1992
Techniques, Components and Instruments, Applications

Selected Papers from the 3rd IFAC Symposium, Vienna, Austria, 9 - 11 September 1992

Edited by

P. KOPACEK
University of Technology, Vienna, Austria

and

P. ALBERTOS
Technical University of Valencia, Spain

Published for the

INTERNATIONAL FEDERATION OF AUTOMATIC CONTROL

by

PERGAMON PRESS

OXFORD • NEW YORK • SEOUL • TOKYO

UK	Pergamon Press Ltd, Headington Hill Hall, Oxford OX3 0BW, England
USA	Pergamon Press, Inc., 660 White Plains Road, Tarrytown, New York 10591-5153, USA
KOREA	Pergamon Press Korea, KPO Box 315, Seoul 110-603, Korea
JAPAN	Pergamon Press Japan, Tsunashima Building Annex, 3-20-12 Yushima, Bunkyo-ku, Tokyo 113, Japan

First edition 1993
Transferred to digital 2007

Library of Congress Cataloging in Publication Data

IFAC Symposium on Low Cost Automation (1992: Vienna, Austria) Low cost automation 1992: techniques, components and instruments, applications: selected papers from the 3rd IFAC symposium, Vienna, Austria, 9-11 September 1992/edited by P. Kopacek and P. Albertos.
p. cm. — (IFAC symposia series: 1993, no. 13)
Includes index.
1. Automatic control—Congresses. 2. Automation—Congresses. I. Kopacek, Peter.
II. Albertos, Perez, P. III. Title. IV. Series.
TJ212.2.I3389 1992 629.8—dc20 93-36687

British Library Cataloguing in Publication Data

A catalogue record for this book is available from the British Library

ISBN: 9780080417165

Printed and bound by CPI Antony Rowe, Eastbourne

These proceedings were reproduced by means of the photo-offset process using the manuscripts supplied by the authors of the different papers. The manuscripts have been typed using different typewriters and typefaces. The lay-out, figures and tables of some papers did not agree completely with the standard requirements: consequently the reproduction does not display complete uniformity. To ensure rapid publication this discrepancy could not be changed: nor could the English be checked completely. Therefore, the readers are asked to excuse any deficiencies of this publication which may be due to the above mentioned reasons.

The Editors

IFAC SYMPOSIUM ON LOW COST AUTOMATION 1992

Techniques, Components and Instruments, Applications

Sponsored by
International Federation of Automatic Control (IFAC)
Technical Committees on
- Components and Instruments (COMPON)
- Education (EDCOM)
- Applications (APCOM)
- Development Countries (DECOM)

Supported by
University of Technology, Vienna

Organized by
Austrian Centre for Productivity and Efficiency
Institute of Robotics, University of Technology, Vienna

International Programme Committee
P. Albertos (E) (Chairman)
A. Aguado (RCH)
Z. Binder (F)
T. Boromissza (H)
A. Casucci (RA)
T.Y. Chai (PRC)
A. Cipriano (RCH)
A. de Carli (I)
A. Gomide (BR)
F. Harashima (J)
H. Hinssen (B)
U. Jaaksoo (SU)
K.B. Kim (ROK)
P. Kopacek (A)
T. Kuusisto (SF)
H.J. Leskiewicz (PL)
D. Mee (AUS)
P. Nikiforuk (CDN)
G. Schmidt (D)
J. Tornero (E)

National Organizing Committee
P. Kopacek (Chairman)
N. Girsule
J. Hähnel
R. Probst

PREFACE

After Valencia and Milano this third symposium was organized by the Austrian NMO, the "Austrian Centre for Productivity and Efficiency" in cooperation with the Institute of "Handling Devices and Robotics" at the Technical University of Vienna. The International Program Committee, chaired by P. Albertos (Spain) had the difficult task to select 94 papers from 155 submitted abstracts. These 94 technical papers were arranged in 21 sessions. In addition, five plenary papers were given and one round table session was held.

According to the scope the main reason of the symposium was to bring together end-users and control system specialists to evaluate the possibilities of technique, design procedures, components and instrument to get a Low Cost Automation not only considering all economic aspects but also improvements in productivity, reliability, flexibility and easy to apply. Special emphasis should be placed on small and medium sized enterprises. Therefore the plenary papers should give an overview on this field. These were

⊗ New Techniques to Improve Performance of Simple Control Loops by R. Ortega (France)

⊗ Low-Cost PLC Design and Application by H. P. Jörgl and G. Höld (Austria)

⊗ Neuro-Fuzzy Controllers by F. Gomide, A. Rocha (Brazil) and P. Albertos (Spain)

⊗ Low-Cost Factory Automation by P. Kopacek (Austria)

⊗ Using Low Cost PLC Electric Ultrasonic Sensors for 3-D-Measurement in robots by G. Lindstedt and G. Olson (Sweden)

The technical sessions dealt with topics from process automation as well as from manufacturing automation. From the side of automation of continuous processes, two sessions on adaptive controllers, two sessions on process control applications, two sessions on drives and one session on power engineering application, and signal processing were given. In the field of the automation on processes - manufacturing automation, two sessions on modelling and CAD, robotics, computer systems and one session on intelligent control were given. A round table discussion organized by P. Albertos (Spain) delt with problems on Low Cost Automation in former East-European countries.

135 participants attended this symposium. All accepted papers were published in a preprints volume and selected papers are included in this proceedings volume. According to the high interest and the high quality of most of the papers it was decided to continue this symposium series in three years. Argentina was suggested as the country to host the fourth edition of this Symposia Series.

P. Kopacek
(NOC chairman)

P. Albertos
(IPC chairman)

CONTENTS

PLENARY PAPERS

THEORY

SIGNAL PROCESSING

MODELLING AND CAD

SENSORS

ADAPTIVE CONTROLLERS

ROBOTICS

PROCESS CONTROL APPLICATIONS

INTELLIGENT CONTROLLERS

PLC DESIGN

COMPUTER SYSTEMS

DRIVERS

APPLICATIONS (MISCELLANEOUS)

NEW TECHNIQUES TO IMPROVE PERFORMANCE OF SIMPLE CONTROL LOOPS

R. Ortega

Université de Technologie de Compiègne, Heudiasyc URA C.N.R.S. 817, Centre de Recherches de Royallieu, BP 233, 60206 Compiègne Cedex, France

1 Introduction

The subject of feedback control theory has witnessed some striking developments in the last 15 years. Interestingly enough, it has become both more rigorous and potentially more applicable. The rigor, that has helped to clarify the problems and provide methodological solutions, stemmed from the incorporation of powerful analytic tools to the field. On the other hand, the wider applicability is a consequence of new formulations, that best capture the practical issues of control problems, and new closed form mathematical solutions. Furthermore, the readily availability of computers and software to implement and test the control algorithms has had a profound impact in control engineering design. These developments have already had significant consequences in high-tech applications. However, at the *low-cost automation* edge little if no impact has yet been detected.

One of the reasons for this state of affairs is the lack of *simple*, yet rigourous, presentations of the recent theoretical developments. These results are usually reported in highly specialized journals written in cryptic languages essentially devoted for their narrow circle of initiated people. Textbooks that should help to bridge this gap are unfortunately also lacking. Very few authors are willing to spend several years of their "productive" scientific life to write a clear pedagogical textbook, basically because the reward system very poorly acknowledges these efforts. In consequence, there is a strong trend among the potential book authors to, at best, write hurriedly produced research monographs (most often than not collectively written) more concerned with the novelty and rigour of the contents than with the clarity of exposition and the putting in perspective of the results. In other words, in control theory the transfer of knowledge between the academic circles and the users in industry is far from satisfactory. This is particularly true for the small and medium size industries which cannot afford the expense of a high-tech research centre or the services of high-charging consulting companies. It is our belief that these potential users could considerably benefit from the significant achievements accomplished in the theory provided they had easy access to their developments, in a language which is familiar to them. Furthermore, and perhaps more importantly, the underlying principles for these developments is very simple and intuitive, and can be presented appealing to concepts with which the engineers in practice are accustomed and confortable. In particular, some of them rely on basic physical principles of *energy dissipation*, which are captured in terms of generalized loop-gain and loop-shift notions or *linearization-based* designs.

The purpose of this paper is to present some recent developments in control theory that we believe can have an inmediate impact in low-cost automation applications. In particular, we concentrate on *synthesis techniques for single-loop controllers*. We review first the practically important area of controller tuning. We concentrate here on *gain-scheduling* techniques where some new insightful research has been reported. Then, we describe some of the results on *robust linear* controller design. Here, we divide the material into designs which are robust *vis a vis* structured (parametric) or unstructured (dynamic) uncertainty.

In closing this introduction, we would like to remark that even though the full transition toward practical implementation of the theoretical results is far from complete, and most of the current work is apparently not aimed at bridging this gap, we are confident that the new lines of research pursued at various institutions throughout the world will reverse this tide and aim instead at a fresh reconsideration of the subject, one that exploits the new theoretical developments while emphasizing their connection with engineering practice.

2 Controller Tuning.

Following [1] we distinguish between three different techniques for controller tuning: *i) Automatic tuning* by which we mean a one-shot method where the controller is tuned automatically on demand from the user. It usually involves the injection of a probing signal and it is typically carried out by sending a command to the controller. *ii) Gain scheduling* is a table look-up (thus, open-loop) procedure where the control law is changed depending on measured operating conditions according to some *a priori* determined rule. *iii) Adaptive control* where the controller depends on some adjustable parameters which are continuously adjusted on-line as functions of the systems evolution. That is, adaptive control is a truly on-line method that searches continuously for a "good" parametrization of the given controler structure.

An excellent updated survey of automatic tuners may be found in [1]. On the other hand, adaptive techniques have been recently surveyed by the author in [7]. Therefore, we restrict our discussion here to *gain scheduling* techniques and in particular present the recent interesting work of [9] and [8].

2.1 Classical Gain Scheduling

The basic idea in gain scheduling techniques is to design local linear controllers based on linearizations of the plant at several operating points. Then, a global nonlinear controller is obtained by "glueing" the linear controllers, e.g. by "scheduling" their gains. The procedure is an *ad hoc* methodology where two basic heuristic rules of thumb have emerged to guide the succesful designs. Namely, that the scheduling variable should *vary slowly* and should *capture the plants nonlinearities.*

In [9] it is argued that since gain scheduled designs are based on linearizations, the limitation of capturing the nonlinearities can be addressed in the selection of the scheduling variables. Thus, any theory developed for this problem will simply verify the intuition obtained from an understanding of the physical system. In contrast with this, the restriction to slow variations on the scheduling variable is most likely due to the nature of the scheduling algorithms. In other words, that even though one can find good controller gains for all *fix* interpolating condition, performance will be degraded in the presence of fast variations. This is an interesting, though not surprising, fundamental limitation of gain scheduling as currently applied. In the paper the potential hazards of scheduling on a slow variable are illustrated via examples of a linear time varying system. It is also proposed to abandon the idea of scheduling on slow variables and it is suggested to schedule the fix operating gains in a manner which explicitely addresses the possibility of rapid variations. Of course, without the knowledge of "future time variations" the existing theory on control of time-varying systems provides very little help to carry out this modification.

2.2 New Approach

In a series of interesting papers [8] and coauthors have proposed an alternative practically appealing local approach to gain scheduling. The approach is a natural extension of the classical idea of linearizing the plant at a nominal *constant operating point* (equilibrium), and apply linear control theory, to the case where the linearization is done *pointwise to all constant operating points* in an open neighborhood of the nominal. Then the key consideration is to show that the parametrized linear control law provided by this extension satisfies the conditions for existence of a nonlinear control law with the desired properties. Interestingly enough, in many practical problems it can be shown that this is the case provided the initial states are close to the constant operating point manifold and the external inputs vary slowly enough.

To fix the ideas let us consider the problem of *noninteracting control* with stability [1]. The procedure starts from a state-space nonlinear model of the plant $\dot{x}(t) = f(x(t), u(t))$, $y(t) = h(x(t))$, with the classical definitions of state $x \epsilon \mathcal{R}^n$, input $u \epsilon \mathcal{R}^m$ and output $y \epsilon \mathcal{R}^m$. We consider dynamic nonlinear control laws of the form $u(t) = k(x(t), r(t), z(t))$, $\dot{z}(t) = g(x(t), z(t))$ where $r(t)$ is a reference input that we want the output $y(t)$ to track asymptotically. There are two basic requirements that have to be satisfied in an open set $\Gamma \epsilon \mathcal{R}^n$ containing the origin. First, the definition of a local family of closed-loop constant operating points, corresponding to a constant value of $r(t)$, say $\mathbf{r}$, and which satisfies $y(t) = r$. Second, that the closed loop linearized system is asymptotically stable, and noninteracting in the sense of linear systems *for each* $r \epsilon \Gamma$. Under these conditions, it can be shown that there exists a nonlinear controller that achieves the objectives *if and only if* the corresponding pointwise linearized noninteracting control solves the problem.

3 Robust Linear Control

It is well known that the *raison d'etre* of feedback is to compensate for the adverse effects of modeling errors on the performance of a control system. The first step to systematically carry out this task is to characterize the modeling errors. The basic technique is to model the plant as belonging to a set $\mathcal{P}$. Such a set can be either *structured* if the uncertainty is captured by unknown parameters, or *unstructured* which is obtained considering a fully known linearized model in a frequency domain ball. That is, characterizing the plant by the Nyquist locus of a nominal model and a band, inside of which lies the frequency response of the unknown plant. In this section we present two recent interesting results pertaining to systems with structured and unstructured uncertainties respectively.

3.1 $\mathcal{H}_\infty$-control: An Energy Dissipation Perspective

We consider here the problem of designing an LTI controller for a given LTI plant such as to attain some suitable disturbance rejection properties [2]. The problem set up is shown in Fig. 1, where r, u, y and z represent exogenous inputs (disturbances), controlled inputs, regulated outputs and measured outputs respectively. The transfer matrix $P(s)$ of the plant is partitioned conformally with the dimensions of r, u, y and z as

$$\begin{bmatrix} z \\ y \end{bmatrix} = \begin{bmatrix} P_{11} & P_{12} \\ P_{21} & P_{22} \end{bmatrix} \begin{bmatrix} r \\ u \end{bmatrix}$$

where $u = Ky$, P_{ii} are scalar transfer functions, and the closed-loop transfer matrix assumes the form $z = Tr$ with $T = P_{11} + P_{12}K[I - P_{22}K]^{-1}P_{21}$ [3]. The disturbance rejection objective is to —em minimize the energy of the output signal z due to exogenous inputs r of finite energy, and of course, we want this to be attained with internal stability. In other words, we pose the following minimization problem

$$inf_{K(s)\epsilon\mathcal{K}} sup\{\| z \|_2 : \| r \|_2 \leq 1\}$$

where $\| \cdot \|_2 = \int_0^t (\cdot) d\tau$ is the signal energy (its $\mathcal{L}_2$ norm) and $\mathcal{K}$ denotes the set of *all stabilizing controllers* for P_{22}. It can be shown [4] that this minimization problem is equivalent to

$$inf_{K(s)\epsilon\mathcal{K}} \| T \|_\infty$$

where $\| T \|_\infty := sup_\omega \mid T(j\omega) \mid$ is the $\mathcal{H}_\infty$ norm of T. In other words, we want to find the stabilizing controller that gives the closed loop frequency response $T(j\omega)$ with the "smallest resonant peak". In practice we usually look for a simpler suboptimal solution where we specify the disturbance rejection level, say $\gamma > 0$, in this case we look for $\| T \|_\infty \leq \gamma$. Dividing T, P_{11}, P_{12} by γ reduces the problem to the case when $\gamma = 1$. Thus, we are interested in finding K which yield internal stability and makes $\| T \|_\infty \leq 1$.

The problem described above is a paradigm of the celebrated $\mathcal{H}_\infty$-control theory that has attracted the attention of many researchers in the last several years. Many monographs are devoted to the problem (see, e.g. [4] and

[1] This problem has been discussed in detail in [5] for the case of constant operating point.

[2] It can be shown, e.g. [4], that the robust stabilization and robust performance problem can be recast in a form similar to our disturbance rejection problem

[3] Notice that the classical problem of additive output disturbance with the same regulated and measured output is obtained from this formulation by setting $P_{21} = P_{11} = 1$, and $P_{22} = P_{12}$ is the plant transfer function

references therein) and several software packages are already commercially available, for instance [3]. We refer the reader to these references for further details.

Here we present the interesting energy dissipation interpretation of the $\mathcal{H}_\infty$-control problem solution recently derived by [10]. To this end, we interpret the configuration of Fig.1 in the context of classical circuit theory as shown in Fig.2. Within this framework the input signals r, u are viewed as *incident waves* to a two-port, z, y are viewed as reflected waves and $\tilde{P}$ is the *inverse scattering matrix* defined as

$$\begin{bmatrix} z \\ r \end{bmatrix} = \begin{bmatrix} \tilde{P}_{11} & \tilde{P}_{12} \\ \tilde{P}_{21} & \tilde{P}_{22} \end{bmatrix} \begin{bmatrix} u \\ y \end{bmatrix}$$

The feedback K is viewed as a termination of the two-port and T is the reflection coefficient from r to z.

The characterization of $\mathcal{K}$ in terms of two-ports has a very appealing interpretation. Namely, *all* $K \epsilon \mathcal{K}$ are obtained by plugging-in *any* stable transfer function Q as a termination to the two-port of Fig. 3, where

$$M_1^{-1} = \begin{bmatrix} D_R & -U_L \\ N_R & V_L \end{bmatrix}$$

and D_R, U_L, N_R, V_L are transfer functions univocally defined from P_{22} [4]. A further interesting fact is that the cascade connection of the two-ports $\tilde{P}$ and M_1^{-1} yields

$$\begin{bmatrix} z \\ w \end{bmatrix} = \begin{bmatrix} T_{11} & T_{12} \\ T_{21} & 0 \end{bmatrix} \begin{bmatrix} r \\ v \end{bmatrix}$$

indicating that the reflection coefficient from v to w is *zero*. This means that M_1^{-1} acts as a matching network for $\tilde{P}$, absorbing the energy reflected by the plant two-port.

Once the set of all stabilizing controllers has been characterized we can recast our disturbance rejection problem as one of finding a stable transfer function Q such that $\| T \|_\infty \leq 1$ where $T = T_{11} - T_{12}QT_{21}$. Once again the solution of this problem in terms of two-ports is very illuminating. It is shown in [10] that it can be represented as shown in Fig. 4. That is: *i)* The plant is factored as $\tilde{P} = LM_2M_1$, where L is a *lossless* two-port, i.e., its inputs u_1, u_2 and outputs y_1, y_2 satisfy the energy conservation equation

$$\| u_1 \|_2 + \| u_2 \|_2 = \| y_1 \|_2 + \| y_2 \|_2$$

[4] They form a doubly coprime factorization of P_{22}

and $M = M_2M_1$ is invertible, (it contains all the "stable" zeros of the plant); *ii)* The controller is factored as $K = M_1^{-1}M_2^{-1}E$ where $\| E \|_\infty \leq 1$. In summary, the $\mathcal{H}_\infty$ suboptimal feedback laws are characterized by an invertible two-port which, when cascaded with the plant two-port, makes it lossless, and the load (termination) on the resulting lossless two-port has $\mathcal{H}_\infty$ norm smaller than one.

3.2 Systems with Structured Uncertainties

In a celebrated paper [6] it was established that the stability of the family of polynomials $D(s, r) = s^n + r_{n-1}s^{n-1} + \cdots + r_1 s + r_0$, where $r_i^+ \leq r_i \leq r_i^-, i = 0, 1, \cdots, n-1$ could be established by checking stability of only *4 polynomials*. The latter, called Kharitonov polynomials are obtained by taking the extreme values of the coefficients in some specified ordering. This seminal work has sparked a whole new line of research in robust linear control and several important problems have already been solved (see [2] for some key references of the literature).

A problem of particular interest is the design of stabilizing controllers for the so-called *interval plant*, i.e. the uncertainty in the plant is manifested via *a priori* interval bounds for each numerator and denominator coefficients, that is

$$P(s, r, q) = \frac{N(s, q)}{D(s, r)}$$

where $N(s, r) = q^m s^m + q_{m-1}s^{m-1} + \cdots + q_1 s + q_0$ and $q_i^+ \leq q_i \leq q_i^-, i = 0, 1, \cdots, m$. A question of great practical relevance is to give conditions under which stabilization of some distinguished subset of the plants implies stabilization of the entire interval family. Since we want this results to be useful in design then the issue of "computability" of the robust stabilizing controller becomes paramount.

In [2] it was shown that the first order controller

$$C(s) = K\frac{s - z}{s - p}$$

with K, z, p real numbers, stabilizes the whole interval plant *if and only if* it stabilizes *sixteen* well defined plants. The latter are obtained by taking all combinations of the Karitonov polynomials for the numerator and denominator of the interval plant.

As a corollary of these result we dispose of a simple technique to solve the following problem: Given an interval plant (that is, known bounds on the numerator and denominator polynomial coefficients of a transfer function) and a PI controller $C(s) = K_1 + \frac{K_2}{s}$ find *all controller tunings* that will insure stability of the whole family. The solution is obtained by setting up 16 Routh tables, whose first colums are left as functions of K_1, K_2. The positivity requirement for stability of the Routh test leads to a set of inequalities in the plane $K_1 - K_2$ whose intersection provides the desired answer.

4 Conclusions

We have presented four results from the recent literature in control theory that, in our opinion, can have an impact on *low-cost automation* applications. The first two results pertain to gain-scheduling and gives new insights and a rigourous analytical framework to design linearization-based controllers. The third result shows that the celebrated $\mathcal{H}_\infty$ suboptimal controllers can be interpreted in terms of simple well-known ideas of circuit theory. Finally, the result on stabilization of plants with structured uncertainties provides a powerfull Routh table based tool to design robust PI controllers.

References

[1] K. Astrom, T. Hagglund, C. Hang and W. Ho, Automatic Tuning and Adaptation for PID Controllers- A Survey, *Proc. IFAC Symp. ACASP*, Grenoble, France, July 1-3, 1992.

[2] B. Barmish, C. Hollot, F. Kraus and R. Tempo, Extreme Point Results for Robust Stabilization of Interval Plants with First order Compensators, *IEEE Trans. Aut. Control*, AC-37, June 1992.

[3] R. Chiang and M. Safonov, **Robust Control Toolbox**, The Math Works Inc., 1988.

[4] J. C. Doyle, B. Francis and A. Tannenbaum, **Feedback Control Theory**, MacMillan Publ. Co., N.Y., 1992.

[5] L. Gras and H. Nijmeijer, Decoupling in Nonlinear Systems: From Linearity to Nonlinearity, *IEE Proc*, Vol. 136, Pt. D, 1989.

[6] V. Kharitonov, Asymptotic Stability of an Equilibrium Position of a Family of Systems of Linear Differential Equations, *Differentsyalne Uravnenyia*, Vol. 14, 1978.

[7] R. Ortega and T. Yu, Robustness of Adaptive Controllers: A Survey, *Automatica*, Vol. 25, No. 5, Sept. 1989, pp. 651-677.

[8] W. Rugh, Analytical Framework for Gain Scheduling, *IEEE Control Syst. Magazine*, January 1991.

[9] J. Shamma and M. Athans, Gain Scheduling: Potential Hazards and Possible Remedies, *IEEE Control Syst. Magazine*, June 1992.

[10] M. Verma and G. Zames, An Energy and Two-Port Framework for $\mathcal{H}_\infty$-control, *McGill University Int. Rep.*, June 1992.

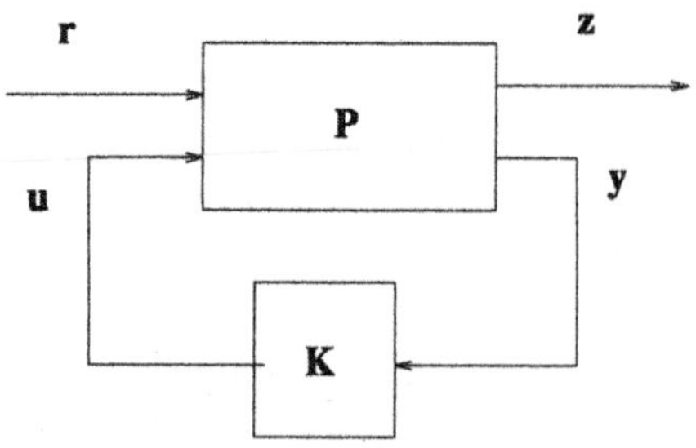

Fig. 1

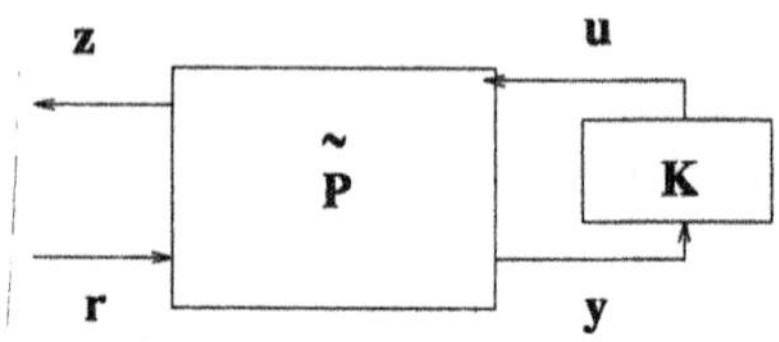

Fig. 2

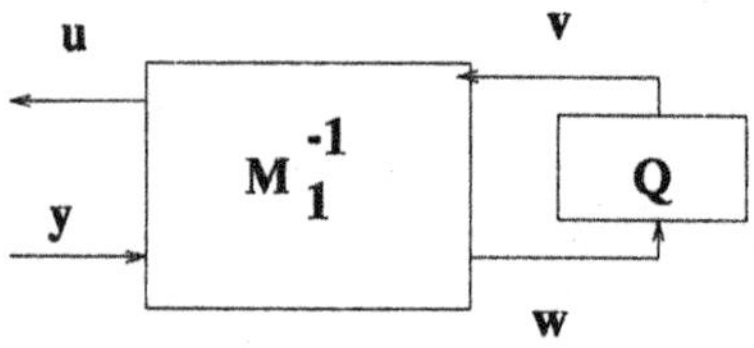

Fig. 3

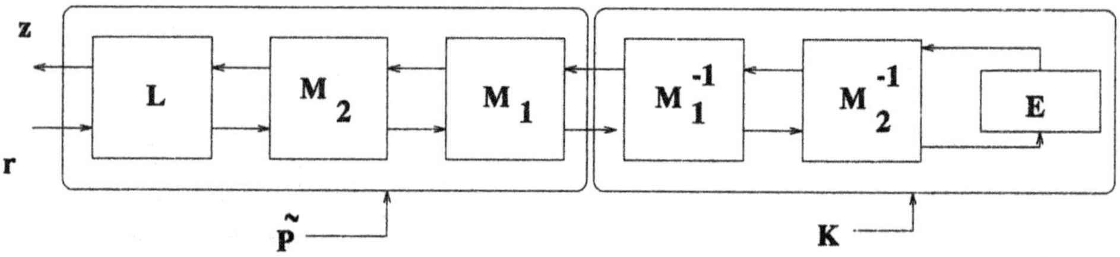

Fig. 4

LOW COST PLC DESIGN AND APPLICATION

H.P. Jörgl and G. Höld

Institute of Machine and Process Automation, Technical University of Vienna, Gußhausstraße 27-29, A-1040 Vienna, Austria

Abstract. In recent years small PLCs have gained importance as true low cost systems for industrial automation. Equipped with up to 100 digital I/Os, programmable timer and counter functions, PID-controller modules, analogue I/Os and communication ports, they are widely used for open and closed loop control of small machines and processes. In this paper, a short survey of the functionality of micro-PLCs with PID-control capabilities is given and an application of a PLC in the field of small hydropower plant control is presented.

Keywords. Low cost controller, PLC, PID-control, small hydropower plant control, frequency control.

INTRODUCTION

In recent years, the technical properties of true low cost PLCs have changed dramatically. Originally designed for pure open-loop binary logic control tasks, they are nowadays able to perform additional and intelligent functions. In most cases low cost PLCs are equipped with standard microprocessor CPU's, some with special bit processors as co-processors. Most low cost PLCs are able to communicate via a series interface. Although the operating systems remain manufacturer specific, standard PCs (MS-DOS, OS/2) can be used for the graphic-oriented programming. Nevertheless, inexpensive terminals for manual programming remain the most often used tool for programming the PLC in form of an instruction list. Low cost PLCs have also become increasingly more efficient in handling closed loop control tasks. Besides hardware analogue I/Os and temperature sensor inputs, many manufacturers offer PID-control algorithms in form of hardware units as well as software modules. The means for visualisation of control processes is rather limited. In the majority of applications the man-machine communication will have to be realised using a simple terminal or some other peripheral equipment.

When the decision is made to use a low cost PLC, the costs that arise while planning the project and putting it into operation, as well as the documentation costs have to be considered as important cost factors. For this reason, efficient programming (e.g. with a standard PC), allowing to comment the program and to use different programming languages, becomes an important factor as well. The user ought to take these aspects into full consideration when choosing his system.

In the first part of the paper, a somewhat detailed discussion of the functionality and programmability of low cost PLCs with PID-control capabilities will be presented. The main part of the paper will be devoted to the design of a PLC for frequency and headwater level closed loop control as well as several open loop control tasks in a small hydropower plant.

LOW COST PLC FUNCTIONALITY

The CPU

In recent years the functionality of small PLCs has been increasing continually. For the *CPU*, predominantly 8 bit processors have been used until recently. Since today small PLCs are more and more taking over tasks that were originally reserved for "true" process computers, small PLCs with 16 bit processors and special co-processors dedicated to arithmetic are already on the market. These arithmetic co-processors are able to carry out arithmetic tasks like subtraction, addition, multiplication and division, trigonometric functions, etc,. This is of great importance if software PID controllers are to be implemented. Figure 1 shows a representative blockdiagram of a CPU [2].

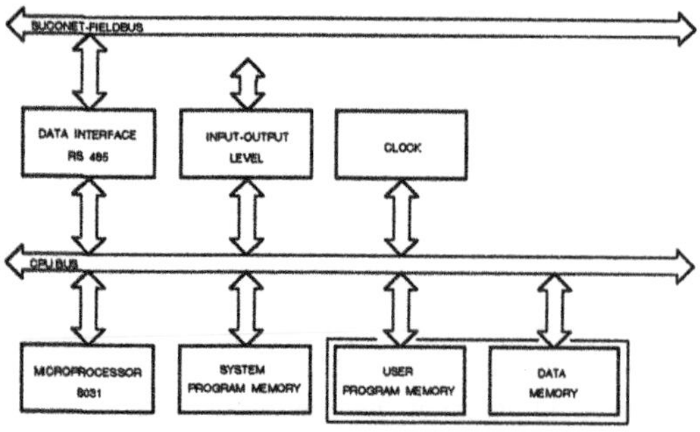

Fig. 1. Blockdiagram of a CPU

In the classical field of binary control, a 16 bit processor has more disadvantages than advantages, since one is only interested in bit processing. In case of analogue data processing 16 bit processors are advantageous if one uses analogue/digital and digital/analogue converters with a resolution greater than 8 bit which, due to higher quality requirements, is very often the case. If a 12 bit converted analogue value is to be processed using an 8 bit CPU, two code instructions are necessary, whereas a 16 bit processor only needs one instruction to process the value.

Memory

Concerning the user memory in which the program is stored, various different technologies are implemented. Very common are battery buffered RAMs, EEPROMs and in rare cases EPROMs. Battery buffered RAMs have the disadvantage that the battery has to be replaced from time to time. A program development environment using hard discs or floppy discs is not provided for in small PLCs.

Time response

As a characteristic value for the time response of a PLC the time needed to process 1000 instructions is commonly used. Here, one can observe tremendous differences, the range being 0.4 ms - 75 ms. Average values lie around 5 ms.

Programming a PLC

Programming a PLC can be done using a manual programming device or in many cases on standard PCs under MS-DOS. Several manufacturers offer especially designed programming terminals. Five different methods for programming a PLC are commonly used: *instruction list*, *contact plan*, *logic plan*, *function plan* and *assembler language*. The fact that, on some PLCs, one can switch from one method to another during the process of programming can be of great advantage. This feature permits the development of the binary control part of a program e.g. in form of an contact plan, whereas the PID-control part can be programmed using the instruction list method. Recently the possibility to use a higher programming language like the C-language is offered.

Documentation and debugging

For documentation purposes a print-out of the program can be made, where one can again choose which form of representation is is wanted. For debugging purposes so-called *cross reference lists* are provided. These lists show, where in the program a certain internal or external subject of an operation (operand) is used. Furthermore, the *status display* mode can be used to display internal and external operands on the programming terminal, thus allowing a check of their functionality.

Other functions and features

Small (or micro-) PLCs are built in two different versions: *compact* and *modular* devices. The modular version of a PLC makes extensions with e.g. a positioning module, a hardware PID module, a communication module, a graphics module as well as with analogue inputs and outputs easily possible. With compact devices, on the other hand, the master-slave principle can be used to link several PLCs together, thus increasing the number of inputs and outputs. In addition, a few compact PLCs offer software PID-algorithms.

SMALL HYDROPOWER PLANT CONTROL

The majority of existing small hydropower plants is not equipped with state of the art control systems, many of them still operating with, e.g., outdated hydraulic controllers. Therefore, there is a great demand to equip these plants with modern control technology, although under the constraint that the plant owners are usually not willing to make large financial investments. In these kind of situations, when a modernisation is necessary for technical reasons, low cost PLCs constitute an ideal solution to the problem. One has to make sure though, that low cost does not mean low performance, i.e. the performance requirements have to be satisfied.

Small hydropower plants have two possible modes of operation. They are either connected to the public electric power supply system (parallel mode) or are disconnected from the net and operate on their own (single mode). In the first mode, the generator frequency and thus the turbine speed are fixed by the net, leaving the headwater level as the variable to be controlled. In the second mode of operation the variable to be controlled is the generator frequency (i.e. the turbine speed), for which a specification of 50 ± 0.5 Hz has to be guaranteed.

A further distinction has to be made concerning the type of turbine being used. Small plants operate with either Francis-, Kaplan-, or Pelton-turbines. Consequently, the control tasks will also have differ significantly. In this paper, however, only the control of a plant with a Francis-turbine is discussed.

Description of the plant and control requirements

Plant description. The hydropower plant for which the control systems are to be designed consists of a Francis-turbine coupled via a belt drive to a synchronous generator with a rated power of 80 kW. The headwater channel is approximately 1.5 km long, the nominal headwater level being 1.8 m.

Control system requirements. Due to the fact that two modes of operation have to be considered, the turbine controller has to handle two different closed loop tasks depending on the mode of operation:

- Turbine speed control (frequency control) in the single mode.
- Headwater level control in the parallel mode.

For economical reasons these two tasks have to be achieved by a single controller, a condition which strongly influenced the final choice of the PLC.

Besides the main closed loop control tasks described above, the PLC has to handle the following additional tasks:

- Net synchronisation when changing from the single mode to the parallel mode.
- Control of the start-up and shut-down procedures of the plant.
- Monitoring, and treatment of disturbances.
- Emergency shut-off in case of excessive speed variations (too high or too low turbine speed).

HARDWARE SELECTION

The PLC

An extensive and comprehensive market analysis was conducted in order to find the best suitable hardware for the control objectives stated above. The final decision was to use a compact micro-PLC with a software PID-module (SUCOS PS3-DC by Klöckner-Moeller) with the following configuration:

- CPU: 3.6 k Byte RAM (usable for typically 1 k binary instructions)
  cycle time: < 5 ms/1 k instructions
  real time clock
- 1 fast counter, input frequency $\geq$ 10 kHz
- 16 binary inputs, 16 binary outputs
- 4 analogue inputs, 1 analogue output

The reasons for choosing this product in this particular configuration were:

- It best solves the control problems stated above, in particular the implementation of two controllers is easily accomplished with the help of a PID software module.
- An inexpensive manual programming device is available.
- A number of inexpensive display terminals can be used with it.
- It is a true low cost PLC
- A more powerful compatible product is expected on the market in the immediate future.

The disadvantages of this configuration are:

- 8 bit resolution provided for the analogue inputs
- If a second analogue output is required, it can only be provided using a second PLC.

In the practical application the 8 bit resolution does not play a role at all if the sensor range is chosen appropriately. The disadvantage of having only one analogue output available may cause problems when controlling a Kaplan- or Pelton-turbine due to the more involved control strategies there, but did not matter in the application for the Francis-turbine.

The sensors

Three variables have to be measured in this application. Sensors for the turbine speed respectively the generator frequency considered were: impulse counters, tachometer generators and commercially available frequency sensors with an analogue output. The latter one was chosen for this application. For the measurement of the headwater level a simple pressure meter was chosen. During the synchronisation procedure the phase angle difference between the generator output and the net signal is needed. For this purpose a commercial $\Delta\phi$ sensor was used.

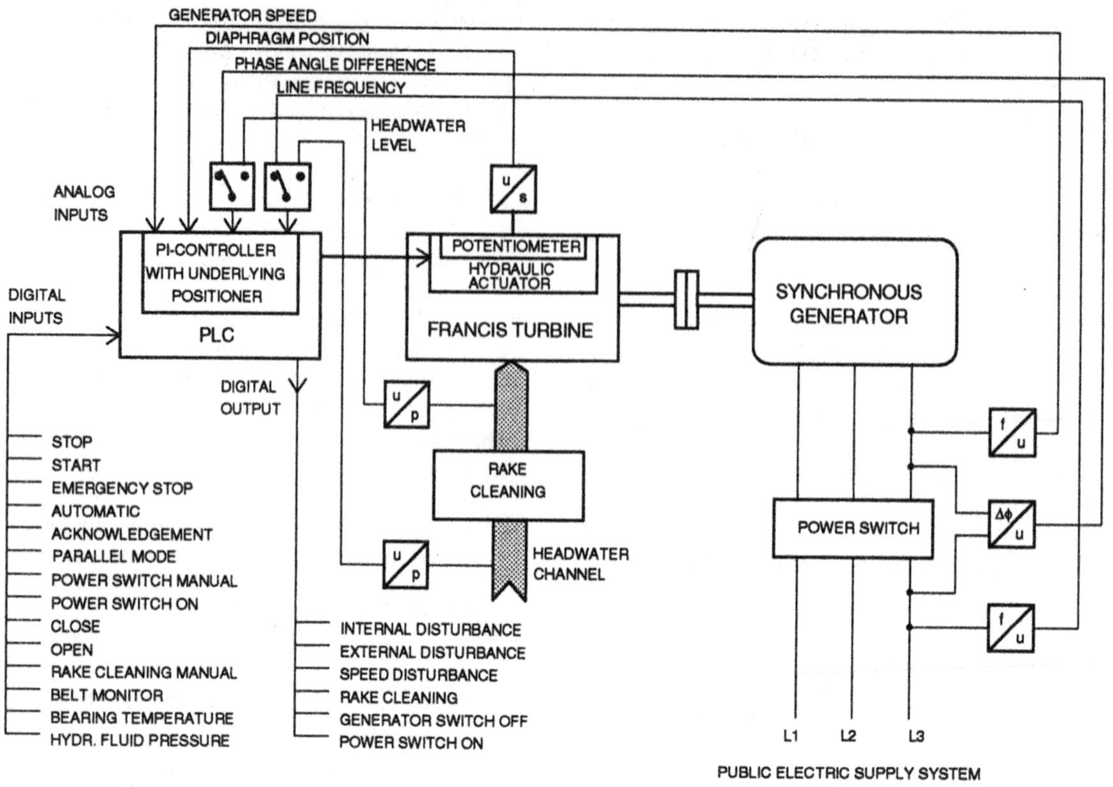

Fig. 2. Scheme of the controlled hydropower plant

The actuator

The flow through the turbine is controlled by changing the diaphragm position. This can either be accomplished using a hydraulic or an electric actuator. In the first case an analogue electric control signal (output voltage of the controller) acts as the input to a proportional valve in the hydraulic fluid flow. In the second case an electric elevating spindle motor is supplied with a digital signal determining its direction of motion. In both cases a lever mechanism moves the diaphragm. An underlying positioning control system increases the actuator performance.

CONTROLLER IMPLEMENTATION

Scheme of the controlled plant

In figure 2, a scheme of the controlled power plant consisting of the turbine including the hydaulic actuator, the generator, the sensors and the PLC is depicted. Furthermore, all the analogue and digital inputs and outputs are shown.

Program description

A modular program structure was chosen in order to guarantee convenient adaptability to other similar plants. Also, extensions and/or modifications can be performed quite easily. A further advantage of this modular program structure is that the documentation can be updated conveniently.

In order to make sure that all parameters can be easily adjusted, a special parameterisation module was designed. In this module parameters like sampling time, switching pulse lengths, frequency and phase angle ranges and the delay times for causing the alarms to go off in case of excessive speed deviations (e.g. over speed). Once a plant is operating with "optimally" adjusted parameters and a new program version is implemented, the old parameters are automatically transferred to it.

The program structure

The program is organised following a stepping mechanism. This allows to determine the system state at any given time and is of great help when installing the program. Figure 3 illustrates the individual steps of the program in a flow diagram. Digital inputs, logically connected by AND-gates permit the next step. They can either be set manually (e.g. the signal START) or are set automatically by the program (e.g. the signal REFERENCE SPEED ATTAINED). It is important to point out that *only* the immediately following step is activated, if all logical signals necessary take on the value 1 (true).

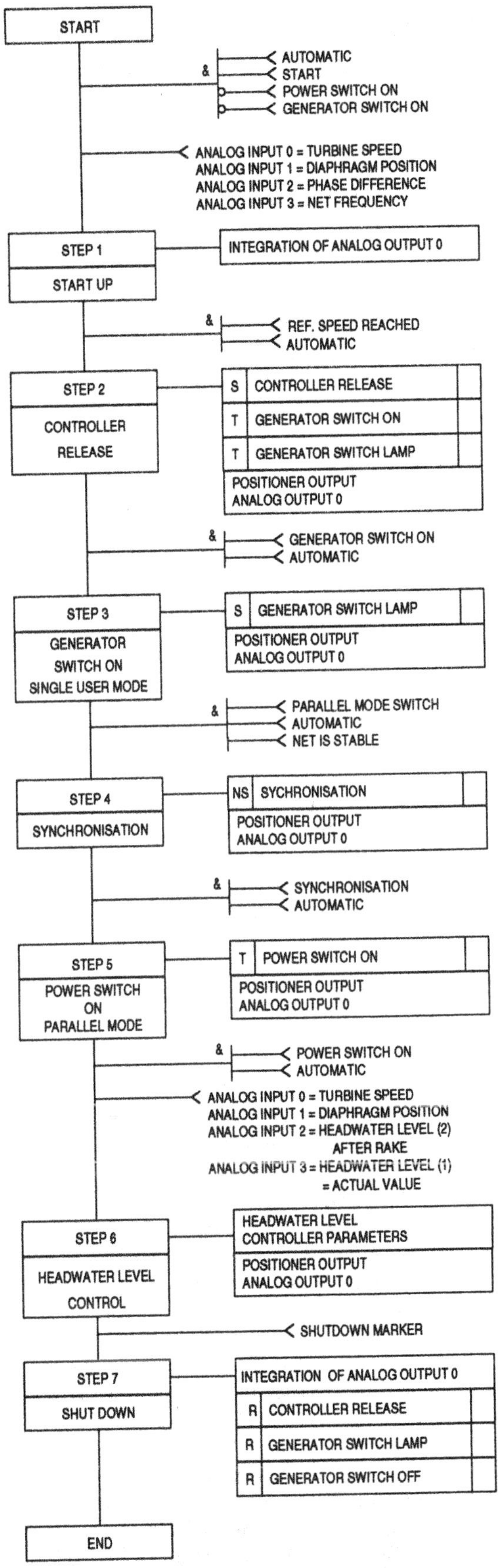

Fig. 3. Program structure

RESULTS

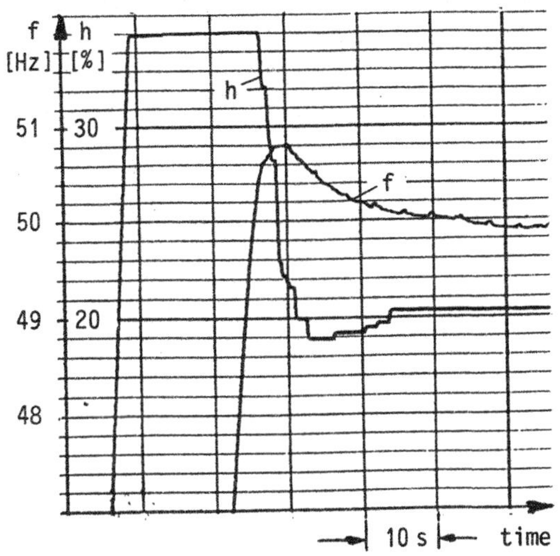

Fig. 4. Start-up of the plant, single user mode

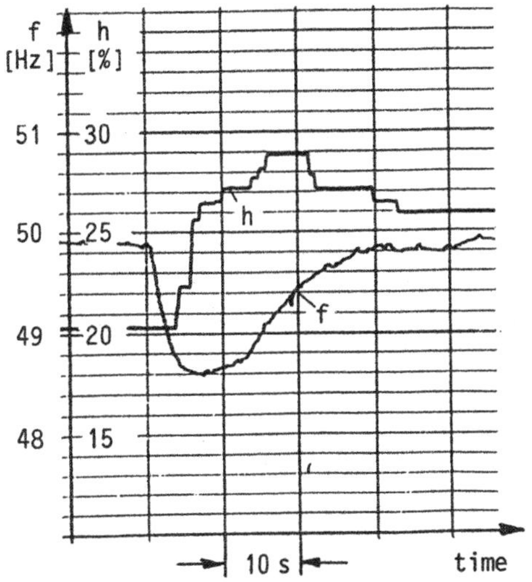

Fig. 5. Disturbance response, load increase,single user mode

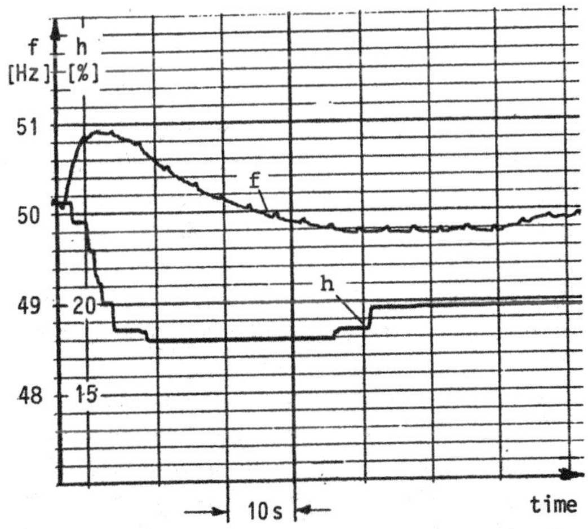

Fig. 6. Disturbance response, load decrease, single user mode

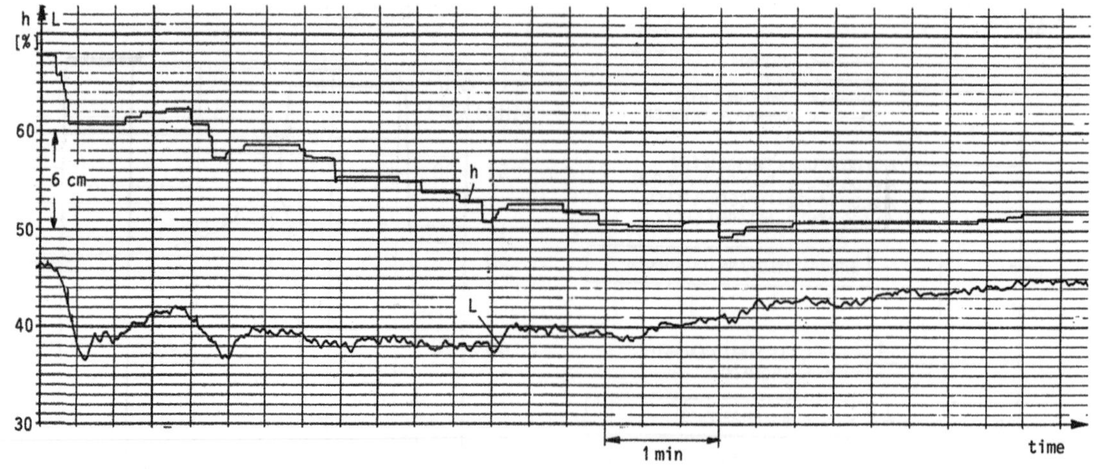

Fig. 7. Headwater level control, parallel mode

After the implementation of the controller in the plant, the exemplary results depicted in figure 4-7 were obtained. It should be mentioned that due to the comparatively low water inflow, the plant was operating at only approximately 60 kW, i.e. at 75 % of the rated power.

Figure 4 shows the generator frequency and the actuator position when the plant is started from standstill in single user mode with no external load. In figures 5 and 6, the response of the system to a stepwise load increase and load decrease of approximately 13 kW is depicted. Due to the rather large and sudden load change which is unlikely to occur in reality, the overshoot exceeds the specified ± 0.5 Hz.

Figure 7 shows the headwater level L and the actuator position h with the plant in parallel mode. As a disturbance, part of the water is branched off in front of the turbine, causing the level to drop off temporarily, before being taken back to its set value by the control system.

CONCLUSIONS

Due to their extended functionality, and here in particular through the possibility to implement PID-controllers, micro-PLCs have generally gained importance for the automation of small machines and processes. This fact was illustrated by the application presented in this paper, namely the control of a small hydropower plant.

The developed control software has been implemented in the plant and is functioning satisfactorily ever since. In the meantime a variant of the first application, where the hydraulic actuator has been replaced by an electric, is also in operation. Due to the modular structure of the program, the necessary modifications of the control software were easily realised.

In this application the "optimal" controller parameter settings were determined experimentally (i.e. in the plant itself). A simulation of the controlled plant in order to increase the system performance was not carried out in this case. For practical reasons this would not be a good approach, since for this purpose each plant would have to be modelled, and the low cost property would be lost.

In future applications the following options are possible: The PLC could be connected to a man-machine interface (MMI). this would allow for a clearer display of the system state, error messages and measurements.

REFERENCES

Borebach, K.H., G. Kraemer, W. Mock, E. Nows (Eds.) (1990). *Speicherprogrammierte Steuerungen mit der SUCOS PS3*. Verlag Europa-Lehrmittel.

Heyde, J. (1988). SUCOS PS3 System Manual, Klöckner-Moeller, Bonn, (in German).

Klinker, W. (1991). Small Programmable Controllers with increasing functionality, *atp - Automatisierungstechnische Praxis,* 3-20, (in German).

Klinker, W. (1991). Medium PLCs are comprehensively prepared. *atp - Automatisierungstechnische Praxis,* 22-37, (in German).

Klinker, W. (1991). Complex Programmable Controllers include features of computer technology. *atp - Automatisierungstechnische Praxis,* 41-57, (in German).

Kucera, G. (1989). *Automatisieren mit SPS,* (2. ed.). Markt & Technik Verlag AG, Haar bei München.

NEUROFUZZY CONTROLLERS

F. Gomide*, A. Rocha* and P. Albertos**

**UNICAMP, FEE/DCA, CP 6101, CEP 13081, Campinas-SP, Brazil*
***Universidad Politécnica de Valencia, E-46071, Valencia, Spain*

Abstract: Fuzzy modeling and control is a technique for handling qualitative information in a formal way. The greater simplicity of implementing fuzzy control systems may reduce design complexity and solve classes of previously intractable problems. Neural nets have come to mean architetures that have massively parallel interconnections of single, neuronlike processors. In control systems, they where first introduced to learn input-output mappings. This paper reviews the underlying ideas and applications of fuzzy and neural control systems. Neurofuzzy control and decision systems that are being developed are particularly emphasized. The key ideas behind these systems are outlined, and currently avaiable hardware and support tools described. Finally, it is suggested how neurofuzzy systems may be used to construct control systems with improved capatibilities.

Keywords: Fuzzy Control, Neural Nets, Computer Control.

Introduction

In modern and classical control theory applications, the first step is to derive a mathematical model to describe the controlled process. This requires a detailed understanding of all variables in the process, a requirement not always feasible if the process is complicated. Existing control theories can deal with a variety of control problems in areas where the system is well defined. Techniques for linear multivariable control (Doyle and Skin, 1981), state estimation from noisy measurements (Anderson and Moore, 1979), optimal control problems (Sage and White, 1977), linear stochastic systems (Bertsekas, 1976), and certain classes of deterministic nonlinear problems (Holtzman, 1970) have been developed and sucessfully applied in numerous well posed problems. However, all of these techniques have been unable to cope with the potential provided by a large body of mathematical knowledge (Tong, 1977). For instance, in many situations, a considerable amount of essential a priopri information is available only in qualitative form, and performance criteria are only specified linguistically. These features introduce a source of inexactness or imprecision which precludes most of the current theories of being used.

Fuzzy modeling and control (Lee, 1990) is a technique for handling qualitative information in a rigorous way. It makes assumptions about the way in which inexactness is described and, in doing so, is powerful enough to handle the many forms in which this arises. Its implementation in real time process control computers or microcontrollers is very convenient since it does not, in general, involve any severe computational problems. Fuzzy modeling and control theory deals with the relationship of the output to the input, aggregating many process and controlled parameters together. This affords the inclusion of high-order variables so that the resulting control systems is of a high order and often provides a more accurate, stable, and robust performance. The greater simplicity of implementing fuzzy control systems may reduce design complexity to a point where previously intractable problems may now be solved.

Artificial neural networks have come to mean architectures that have massively parallel interconnections of simple, neuronlike processors. They were first introduced to learn a mapping between an input space and an output space. Given a set of inputs and the corresponding classes, the problem is to determine a network which yields the desired output class when given an input, and generalizes when given a new input. Depending on the input/output pair, the input-output mapping may be associated with a process model, a disturbance model or with a controller (Hetch-Nielsen, 1990). Thus, neural networks provide a generic class of nonlinear models or nonlinear controllers which are determined by training procedures. Part of a control system implementation effort is left to learning which, somehow, is an automatic design procedure.

Neural networks and fuzzy logic systems are both numerical model-free estimators. They share the common ability to deal with difficulties arising from uncertainty, imprecision, and noise. Both systems and associated techniques have been succesfully applied to several fields to improve intelligent behavior (Self, 1990; Sugeno, 1985; Zadeh, 1988). Neural systems numerically process unstructured knowledge (to structure it), whereas fuzzy systems numerically process structured knowledge. Moreover, is has been shown recently (Rocha, 1992) that neural systems are also capable to symbolically process structured knowledge in a way similar to knowledge-based systems. A promising approach to get both the benefits of neural networks and fuzzy logic systems and to solve their respective problems is to combine them into an integrated system such that we can bring the learning and computational power of neural networks into the fuzzy logic systems, and the representation and reasoning of fuzzy logic systems into the neural networks. For system modeling and control purposes their combination should provide an approach where structured knowledge of complex ill-defined systems is processed in a qualitative way, allowing reasoning and consideration of essential a priori information and performance criteria. Learning features should provide training procedures for controller synthesis, design, and implementation. Systems that combine neural network with fuzzy logic are called here neurofuzzy systems.

This paper reviews the underlying ideas and applications of fuzzy and neural control systems. Neurofuzzy control and decision systems that are being developed are particularly emphasized. The key ideas behind these systems are outlined. Currently available hardware and support tools are also described. Finally, it is suggested how neurofuzzy systems may be used to construct control systems with improved capabilities.

Fuzzy Control Systems

In this section the basic ideas of fuzzy logic control systems are presented. Although there is a great deal of theoretical foundation underlying fuzzy control, only that which is required to understand the basic fuzzy control theory is included (Pedrycz, 1989; Yager et al., 1987; Lee, 1990; Albertos, 1992).

In standard set theory, an object is either a member of a set or it is not a member at all. Given an universe of objects $\mathcal{U}$ and a particular object $x \in \mathcal{U}$, the degree of membership $\mu_A(x)$ with respect to a set $A \subseteq \mathcal{U}$ is:

$$\mu_A(x) = \begin{cases} 1 \text{ if } x \in A \\ 0 \text{ if } x \notin A. \end{cases}$$

The function $\mu_A : \mathcal{U} \mapsto \{0, 1\}$ is called characteristic function in standard set theory. Often, a generalization of this idea is used, for instance, to handle data with error bounds. A degree of membership of one is assigned to all the mumbers within some percent error, and all the numbers outside that interval a degree of membership of zero, figure 1.a. For the precise case, the membership degree is one at the exact number and zero everywhere else, figure 1.b.

Zadeh (1965) proposed a further generalization in which some

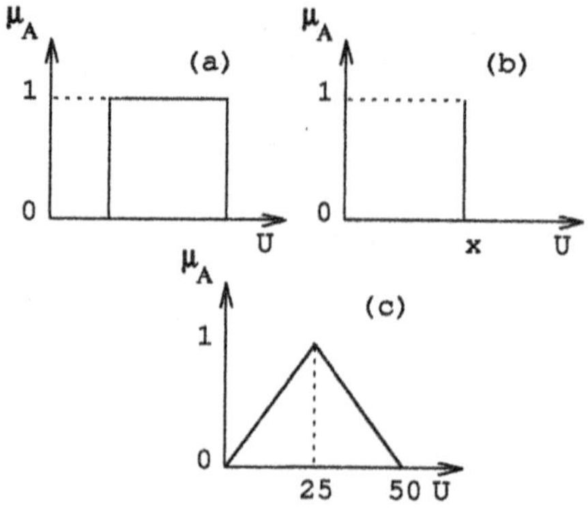

Figure 1: Membership functions for crisp and fuzzy data.

objects are more member of a set than others. The degree of membership takes on various values between zero and one, where a zero value indicates complete exclusion and a value of one indicates complete membership. This generalization expands the expressiveness power. For instance, to express the idea that a temperature is around 25 we way use a triangular membership function (figure 1.c) with its peak at 25 to express the idea that the closer a number is to 25 the better it qualifies.

Formally, let $\mathcal{U}$ be a collection of objects denoted generally by $\{u\}$. $\mathcal{U}$ is called the universe of discourse and may be continuous or discrete. A fuzzy set A in an universe $\mathcal{U}$ is defined by a membership function μ_A which takes values in the interval $[0, 1]$:

$$\mu_A : \mathcal{U} \mapsto [0, 1].$$

The *support* of a fuzzy set A is the crisp set of all points u in $\mathcal{U}$ such that $\mu_A(u) > 0$. A fuzzy set whose support is a single point in $\mathcal{U}$ with $\mu_A = 1$ is called a fuzzy singleton.

Let A and B be two fuzzy sets in $\mathcal{U}$ with membership functions μ_A and μ_B, respectively. The set theoretic operations of union ($A \cup B$), intersection ($A \cap B$) and complement ($\overline{A}$) for fuzzy sets are defined as follows:

$$\begin{aligned} \mu_{A\cup B}(u) &= \mu_A(u)\mathcal{S}\mu_B(u) \\ \mu_{A\cap B}(u) &= \mu_A(u)\mathcal{T}\mu_B(u) \\ \mu_{\overline{A}}(u) &= 1 - \mu_A(u) \end{aligned}$$

where $\mathcal{T}$ is a triangular norm ($\mathcal{T}$-norm) and $\mathcal{S}$ a triangular co-norm ($\mathcal{S}$-norm). A triangular norm is a two-place function $\mathcal{T} : [0, 1] \times [0, 1] \mapsto [0, 1]$ with

$$\begin{array}{lll} \text{(i)} & x\mathcal{T}w \le y\mathcal{T}z, & \text{if } x \le y,\ w \le z \\ \text{(ii)} & x\mathcal{T}y = y\mathcal{T}x & \\ \text{(iii)} & (x\mathcal{T}y)\mathcal{T}z = x\mathcal{T}(y\mathcal{T}z) & \\ \text{(iv)} & x\mathcal{T}0 = 0;\ x\mathcal{T}1 = x & \\ & & x, y, z, w \in [0, 1]. \end{array}$$

A triangular co-norm is such that $\mathcal{S} : [0, 1] \times [0, 1] \mapsto [0, 1]$ and satisfies the properties (i)–(iii) above and

$$\text{(iv)} \quad x\mathcal{S}0 = x;\ x\mathcal{S}1 = 1.$$

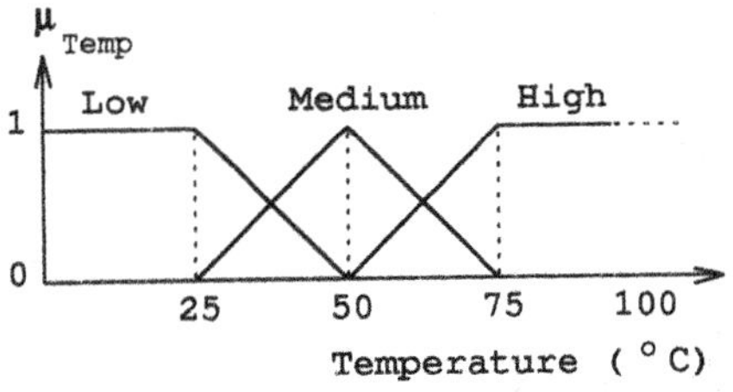

Figure 2: Linguistic variable *Temperature*.

Examples of $\mathcal{T}$-norms include min ($\vee$), algebraic product ($\cdot$), and of $\mathcal{S}$-norms max ($\wedge$), bounded-sum ($\oplus$) [Pedrycz, 1989].

It is quite common to use qualitative phrases instead of quantitative values for words that denote measures and counts, such as ***fairly high***, ***quite a few***, ***not many***. This idea is captured in the concept of a linguistic variable. A linguistic variable takes on one of a set of labels, i.e., words or phrases, as its value. For example, the linguistic variable *Temperature* could take one of the members of the set { *low, medium, high* } for its value. The labels are given meaning by associating with each one a fuzzy subset of some universe of discourse (see figure 2).

If $A_1, \ldots, A_n$ are fuzzy sets in $\mathcal{U}_1, \ldots, \mathcal{U}_n$, respectively, an n-ary fuzzy relaction is a fuzzy set in $\mathcal{U}_1 \times \cdots \times \mathcal{U}_n$ and is expressed as:

$$\mathcal{R} = \{[(u_1, \ldots, u_n), \mu_{\mathcal{R}}(u_1, \ldots, u_n)] \mid (u_1, \ldots, u_n) \in \mathcal{U}_1 \times \cdots \times \mathcal{U}_n\}.$$

If $\mathcal{R}$ and $\mathcal{P}$ are fuzzy relations in $\mathcal{U} \times \mathcal{V}$ and $\mathcal{V} \times \mathcal{W}$, respectively, the composition of $\mathcal{R}$ and $\mathcal{P}$ in a fuzzy relation denoted by $\mathcal{R} \circ \mathcal{P}$, defined by

$$\mathcal{R} \circ \mathcal{P} = \{[(u, w), \sup_v(\mu_{\mathcal{R}}(u, v)\mathcal{T}\mu_{\mathcal{P}}(v, w))]; u \in \mathcal{U}, v \in \mathcal{V}, w \in \mathcal{W}\}.$$

The basic idea of fuzzy control is to model the actions based on expert knowledge, rather than modeling the process itself. This take us outside the realm of standard approaches to process control, in which control algoritms developed via mathematical models of the process are used to derive control action as a function of process state. The motivation for this approach is to accomodate the situation in which control expertise is available from either operators or designers, and mathematical models are either not cost effective or too complex to develop.

The structure of a process controlled via a fuzzy controller is shown in figure 3, which emphasizes the basic components of a fuzzy controller: a fuzzification interface, a knowledge base, a data base, inference procedure, and a defuzzification interface.

The ***fuzzification interface*** gets the values of input variables, performs a scale mapping to transfer the range of values of input variables into corresponding universes of discurse, and performs the function of fuzzification to convert input data into linguistics values. The ***knowledge base*** comprises a rule

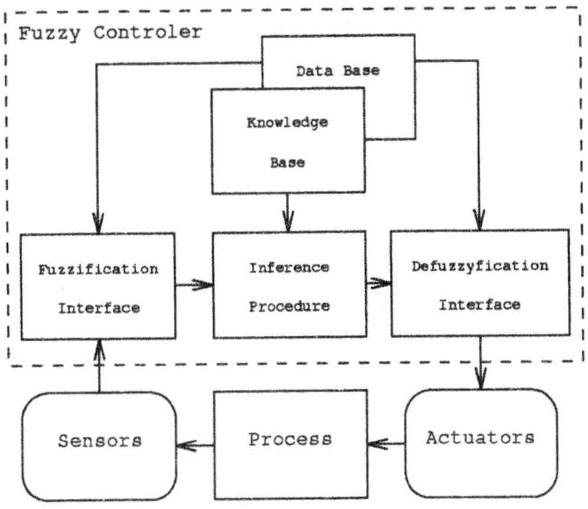

Figure 3: Basic structure of fuzzy logic controller.

base which characterizes the control policy and goals. The ***data base*** provides the necessary definitions about discretization and normalization of universes, fuzzy partition of input and output spaces, membership function definitions. The ***inference procedure*** process fuzzy input data and rules to inferr fuzzy control actions employing fuzzy implication and the rules of inference in fuzzy logic. The *defuzzification* interface performs a scale mapping to convert the range of values of universes into corresponding output variables, and transformation of an fuzzy control action inferred into a nonfuzzy control action.

The most commom form for expressing expert knowledge is the *condition-action* rule, in which a set of conditions describing an observable portion of the process output is associated with a control action that will keep or return the process to the desired operating condition. Typically a condition is a linguistic statement about the value of some output variable, such a ***the error is large positive***. Similarly, a typical control action is a linguistic action description, such as ***increase flow a little***. The idea is to capture the expert control knowledge in the form of a set of rules in which the conditions are expressed using a set of linguistic labels for each observed process output variable, and the action are expressed using a set of similar labels for the value of each control variable. The ***if-then*** rules are often called ***fuzzy conditional statements*** or a ***fuzzy control rule***.

A fuzzy control rule, such as

$$\text{If } (x \text{ is } A_i \text{ and } y \text{ is } B_i) \text{ then } (z \text{ is } C_i)$$

is implemented by a fuzzy implication (fuzzy relation) $\mathcal{R}_i$ and is defined as follows:

$$\mu_{\mathcal{R}_i} \triangleq \mu_{(A_i \text{ and } B_i) \to C_i(u,v,w)} = [\mu_{A_i}(u) \text{ and } \mu_{B_i}(v)] \to \mu_{C_i}(w)$$

where (A_i and B_i) is a fuzzy set $A_i \times B_i$ in $\mathcal{U} \times \mathcal{V}$; $\mathcal{R}_i \triangleq (A_i \text{ and } B_i) \to C_i$ is a fuzzy implication in $\mathcal{U} \times \mathcal{V} \times \mathcal{W}$; and $\to$ denotes a fuzzy implication function.

Examples of fuzzy implication functions include (Lee, 1990)

$$\mu_{\mathcal{R}_c}(u,v) = \mu_A(u) \wedge \mu_B(v)$$
$$\mu_{\mathcal{R}_p}(u,v) = \mu_A(u) \cdot \mu_B(v)$$

In a fuzzy controller, each fuzzy control rule is represented by a fuzzy relation. The overall behavior of a fuzzy system can be characterized by these fuzzy relations. The system can be characterized by a single fuzzy relation which is the combination of the fuzzy relations in the rule set. The combination involve a rule connective *also*, simbolically:

$$\mathcal{R} = \text{also}(\mathcal{R}_1, \mathcal{R}_2, \ldots, \mathcal{R}_i, \ldots, \mathcal{R}_n).$$

Usualy *also* is interpreted as union operator (under the *max* operation).

In fuzzy logic and approximate reasoning, an important inference rule, namely generalized modus ponens, is a follows:

Fact:	x is A'
Rule:	if (x is A) then (y is B)
Consequence:	y is B'.

Usualy, the fuzzy inference is based on the compositional rule of inference suggested by Zadeh (1973). A computational inference procedure proposed by Zadeh (1973), a fuzzy rule if (x is A) then (y is B) ($A \rightarrow B$, for short) is first transformed into a fuzzy relation $\mathcal{R}_{A\rightarrow B}$. For instance

$$\mu_{\mathcal{R}_{A\rightarrow B}}(u,v) = \min(\mu_A(u), \mu_B(v));$$
$$u \in \mathcal{U}, v \in \mathcal{V}$$

where *min* is the implication function. Given a fact x is A' (A' for short), and a rule $A \rightarrow B$, Zadeh's compositional inference rule says

$$B' = A' \circ \mathcal{R}_{A\rightarrow B}$$
$$\mu_{B'}(v) = \max \min(\mu_{A'}(u), \mu_{\mathcal{R}_{A\rightarrow B}}(u,v)).$$

When more than one rule is enabled, the consequents of all actived rules are combined by the *else* operator. For instance, supposing that $B'_1, B'_2, \ldots, B'_n$ are the derived results, the combined result B' can be, for example, $B' = \bigcup_i B'_i$. Figure 4 illustrates the inference process when there are two rules, $A_i \rightarrow B_i$ and $A_j \rightarrow B_j$. A' is the input fact represented as a fuzzy subset.

After a fuzzy control action is inferred, a nonfuzzy control action that best represents the decision is needed. Although there is no systematic procedure to choose a defuzzification strategy, the most commom include: the max criterion method (MAX), which produce the point where membership function of the control action reaches a maximum value; the mean of maximum method (MOM) which represents the mean value of maximum when they are not unique; and the center of area method (COA) which generates the center of area of the membership function of a control action (see figure 5).

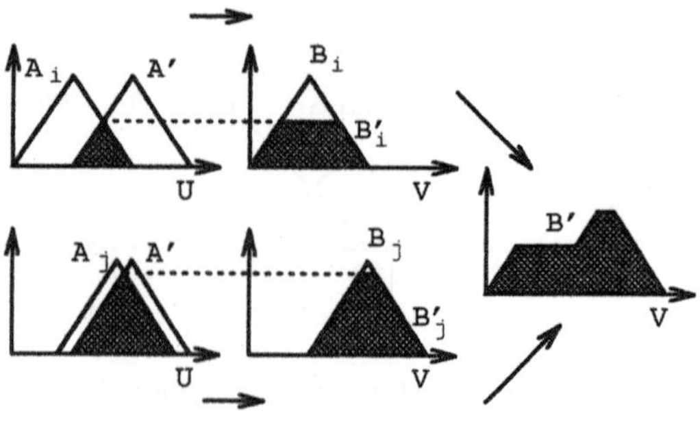

Figure 4: Fuzzy inference mechanism.

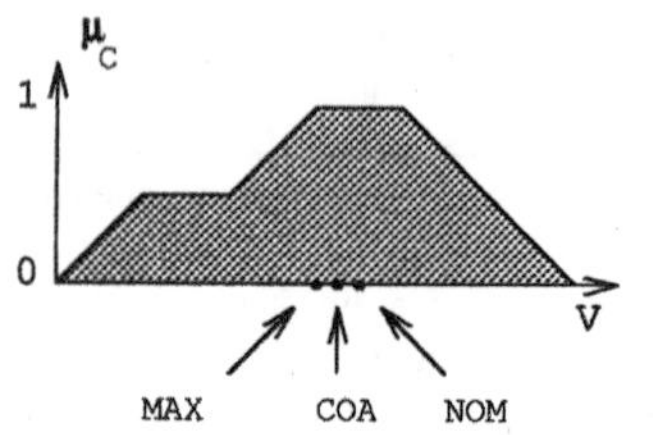

Figure 5: Defuzzification strategies.

Neural Networks in Control Systems

A neural network is a system composed of many simple processing elements operating in parallel whose function is determined by network structure, connection strenghts, and the processing performed at processing elements or neurons. Figure 6 depicts the structure of a generic processing element.

The input signals come from either the environment or outputs of other neurons and form an input vector $x = (x_1, \ldots, x_i, \ldots, x_n)$, where x_i is the activity level of the i-th neuron or input. Associated with each connected pair of neurons is an adjustable value called weight, forming a vector $w_j = (w_{1j}, \ldots, w_{ij}, \ldots, w_{nj})$ where w_{ij} represents the connection strenght from neuron n_i to the neuron n_j. Typically, neuron n_j computes the scalar product of x and w_j, subtracts the threshold, and pass the result throught an activation function $f(\cdot)$. Four common activations functions are the linear, ramp, step, and sigmoid functions.

An n-layer neural network with input $x \in R^n$ and output $y \in R^m$ can be described by the equation (Narendra et al., 1991)

$$\mathcal{N}[W_n \mathcal{N}[W_{n-1} \cdots \mathcal{N}[W_1 x + \Theta_1] +$$
$$+ \cdots + \Theta_{n-1}] + \Theta_n] = y$$

where W_i is the weight matrix associated with the i-th layer. The vectors Θ_i, $i = 1, 2, \ldots, n$, represent the threshold values for each node in the i-th layer and $\mathcal{N}[\cdot]$ is a nonlinear operator with $\mathcal{N}[u] = [f(u_1), f(u_2), \ldots, f(u_n)]^T$ where $f(u)$ is a smooth function. When the function $f(\cdot)$ is the threshold function, the multilayer network is called a multilayer perceptron.

Figure 7 shows a feedforward three layer network with an input vector $x \in R^n$ and output vector $y \in R^m$. The outputs of the intermediate layers are $v \in R^p$ and $z \in R^q$, respectively.

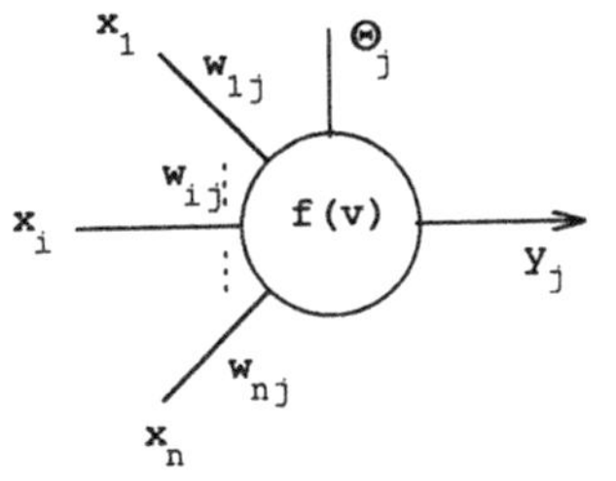

$$y_j = f(v)$$
$$v = \sum_{i=1}^{n} w_{ij} x_i - \Theta_j$$
$$f(\cdot) = \text{activation function}$$
$$\Theta_j = \text{threshold}$$

Figure 6: Typical artificial neuron model.

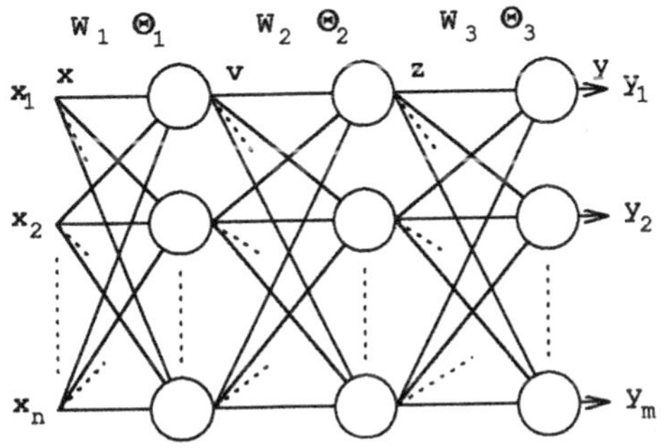

Figure 7: Feedforward architecture.

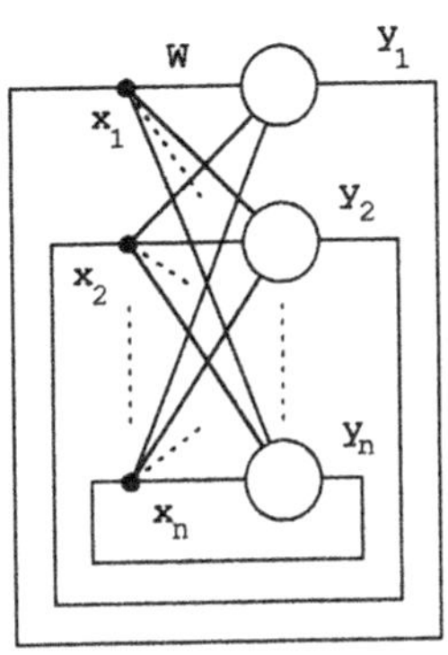

Figure 8: Feedback architecture.

The elements of matrices W_i as well as those of the vectors Θ_i represent the adjustable weights or parameters of the neural network. The total number of parameters is $p \cdot (n+1) + q \cdot (p+1) + m \cdot (q+1)$. Once the number of layers in the network and the number of nodes in each layer are designed, only the weights have to be determined to completely specify the network. The weights are generally found using input-output data by learning or training procedures.

In contrast with the feedforward architeture of the multilayer networks, when a collection of neuron like elements have their outputs fed back to inputs, a recurrent or feedback architecture is obtained, figure 8.

Artificial neural networks are useful for control purposes because they are able to learn input-output mappings. Given a set of inputs which belong to a compact set $\mathcal{X} \subset R^n$ and the corresponding outputs which belong to a set $\mathcal{Y} \subset R^m$, the problem is to determine a network which yields the desired output class when given an input and generalizes when a new input is given. An useful framework for a proper statement of this problem is approximation theory, which concerns the interpolation of a continuous multivariable functions from input-output data. In particular, it has been shown (Hornik et al., 1989) that multilayer feedforward networks with one hidden layer are capable of approximating any continuous function in $C[\mathcal{U}]$, the set of continuous functions mapping a compact set $\mathcal{U} \subseteq R^n$ to the real line. This result provides a sound basis for the use of multilayer neural networks. The nonlinear maps can be used for the approximation of arbitrary functions which arise in the representation of nonlinear control systems.

Neural networks can be used in many model based control schemes. They can be used as forward models (or part of forward models), as controllers, to minimize control action, to map from online to offline measurements, to model disturbances. For instance, when the neural networks is trained with an input-output set being plant input and plant output, respectively, a forward model is obtained, as shown in figure 9.

When the neural network is trained with the input-output set being the (desired) plant output and the (control) plant input, an inverse model may be obtained, as shown in figure 10. Figure 11 showns an inverse model controling a plant in

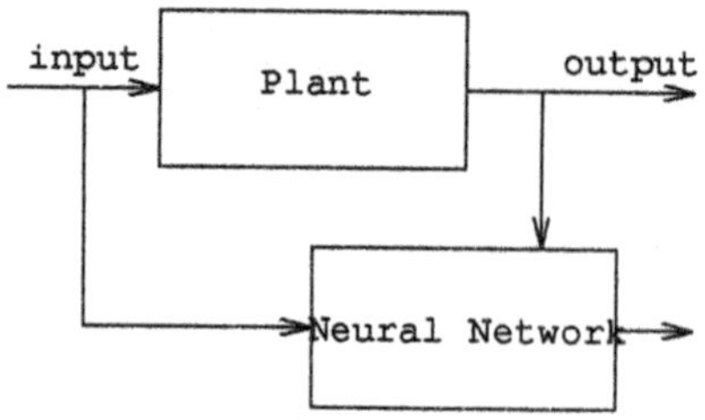

Figure 9: Neural network as a forward model.

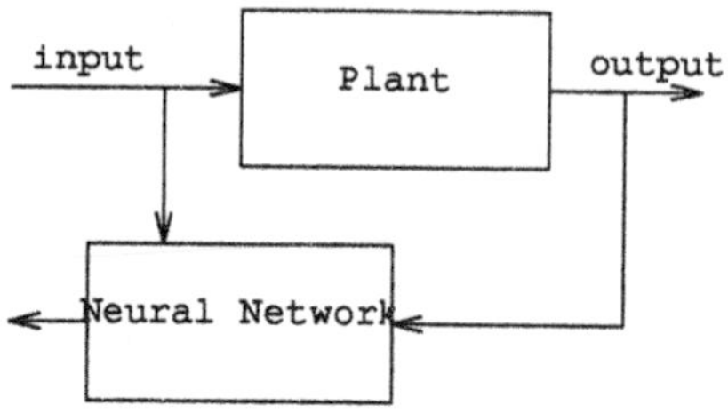

Figure 10: Neural network as an inverse model.

a internal model control scheme (Narendra et al., 1991).

Model based control schemes have shown to be very useful in several practical applications due to its ability to improve performance because of available information about the process. However modeling can be expensive and dificult. In this situation neural networks are very convenient because of their ability to code input-output mappings, to work with incomplete state information and to process information in parallel.

A tipical feedback control scheme is shown in figure 12.

Assuming the process being open loop stable, if the controller is a neural network which represents the inverse process model, an internal model control structure is obtained. When both, the controler and process model are inverse model and forward models, respectively, implemented via neural nets, a dual neural network controler is derived. Dual network controller may, however, fail if an inacurate inverse of the process model is used. When properly designed, these classes of controllers provide very good control and they can be successfuly used in nonlinear control when models are not known and only partial state information is available. Their main advantages include accurate nonlinear models relatively fast and easy to develop, avoid inverting nonlinear models, and estimate hard-to-measure parameters. The most important drawbacks include training time, limits due to data quality, need for verification to avoid overgeneralization.

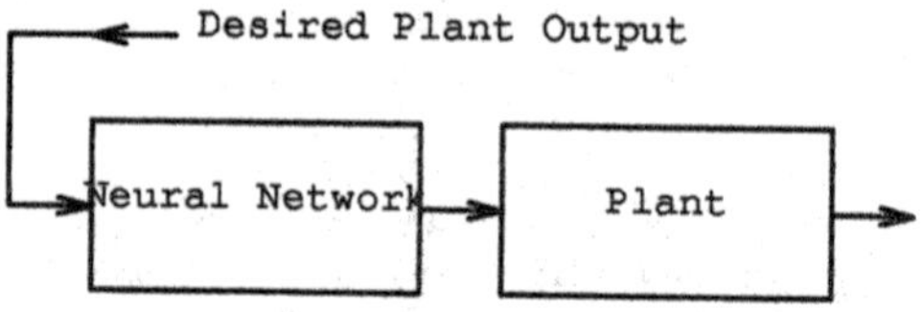

Figure 11: Internal model control.

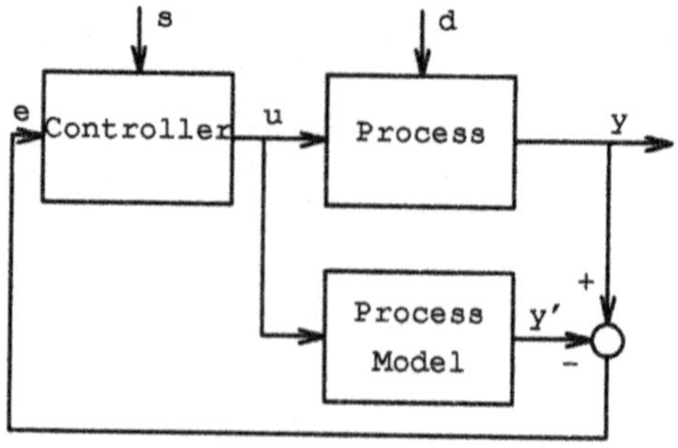

Figure 12: Model Based Control.

Neurofuzzy Systems

The introduction of fuzzy logic in neural network area was, in its early stages, concerned with the developments of multi-input/multi-output neuron models (Lee and Lee, 1974). As opposed to the McCullock and Pits model, they generalized the binary neuron model to provide intermediate values within a range. The correlation between neural physiology and fuzzy set theory has evolved since then to analyse nerve systems employing concepts such as fuzzy language, fuzzy entropy, and fuzzy automata (Rocha at al., 1980).

One of the mot important feature of fuzzy logic is the separation of its logic and fuzzines into a rule and a membership function, respectively. This feature allows, for instance, the quantitative expression and processing of fuzziness in knowledge processing. However, the methods for designing membership functions rely on experience and inevitably become a bottleneck in system design. Neural networks may, due its learning feature, be used in membership functions design. For instance, Takagi and Hayashi (1988) have proposed a neural network for fuzzy reasoning where a membership function can be found by feedforward neural nets. In this case, training data is first clustered to determine the number of fuzzy rules and to form each rule boundary in the neural net. This neural net encodes the membership functions of each rule.

An alternative scheme was derived by Furuya et al. (1988) also employing a feedforward neural network. Here, the degree of compatibylity between a pattern, previously learned by a neural network, and an input vector is used as membership function values. An architeture for knowledge processing were also proposed constituted by, at least, an associative memory neural net, a feedforward net as an operating manager, and a neural network as a sequence controller.

Learning vector quantization, which is a technique similar to fuzzy clustering (Takagi, 1990), maps an input vector into a new space which is fuzzy partitioned. After going through the mapping step, a non-linear fuzzy partition of the input space can be accomplished. Based on this learning technique, Yamaguchi et al.(1989) have developed a method to derive membership functions. Also, a learning fuzzy neural network has been proposed to control the fuzziness in the composition of membership functions by introducing bidirectional associative memories (Kosko, 1987) in the reasoning procedure.

A simplified method to derive membership functions has been developed in (Morita et al., 1988) where the gain of membership functions of fixed form are adjusted by neural networks. On the other hand, in (Ishibuchi et al., 1990), a fuzzy language is derived from a neural network. They further derived interval-based membership functions.

The following are the main advantages reported in using neural networks to determine membership functions: shorter design time because it is algorithmically determined without requiring manual procedures; design of nonlinear membership function due to the inherent nonlinearity of neural networks; automatic rule acquisition from experts using learning features of neural nets; dinamic adaptation to inference environment by neural net learning. For instance in (Takagi, 1990) an example of membership function adjustement, carried out between thirty minutes–one hour in a personal computer, as opposed to thirty–fourty hours, is reported. Nonlinear membership functions discovery as well as automatic rule acquisition are also reported in (Takagi, 1990). Dinamic adaptation to inference environment seems to be an open question.

The symbiosis between fuzzy logic and neural networks has also propagated within areas other than membership function determination. There are, for instance, many cases where approximate reasoning is combined with expert systems. When a neural network is incorporated therein, it takes a form of knowledge acquisition and knowledge expression (Machado, 1991).

Fuzzy cognite maps is another interesting area. A cognitive map is an oriented graph showing a causal relationship among different factors, wherein the causal relationship is expressed by either positive or negative sign for knowledge expressions. Fuzzy cognitive maps (Kosko, 1986) expresses this relationship. The problem of how to determine the degree of causal relationship is proposed to be solved by a differential Hebbian learning.

Fuzzy clustering and pattern recognition as well as cascade connections of neural nets and fuzzy processing are further examples of areas where neural networks and fuzzy logic share their features. For instance, one generally practiced strategy is where a neural net performs pattern recognition first and later process, by fuzzy logic techniques, a knowledge base.

Future directions, as far as neurofuzzy systems in general are concerned, should address: automatic aquisition of fuzzy inference rules using neural nets; adaptation of fuzzy inference rules to the environment; fast approximation by introducing neural networks structs in inference rules; development of fast learning schemes based on fuzzy theory (Takagi, 1990). In these directions, a strategy for learning of fuzzy inference criteria with neural network (Uehara et al., 1990), a learning method of fuzzy inference rules with neural nets (Yamaoka et al., 1991), a learning algorithm for max-min neural networks (Saito et al. 1991), composition methods of fuzzy neural nets (Horikawa et al., 1990), and truth-valued flow inference network (Zhuang et al., 1990) have been proposed.

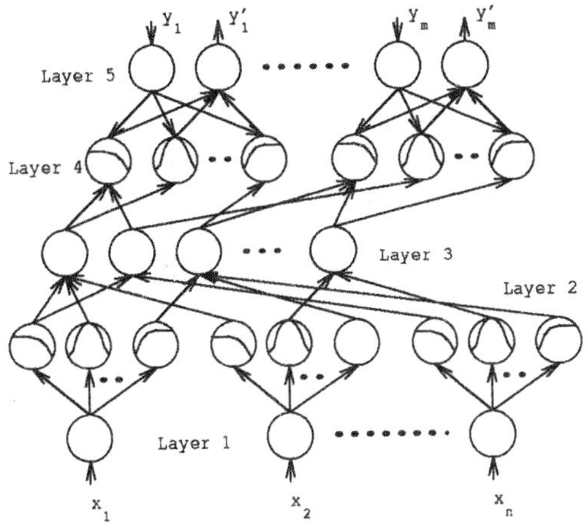

Layer 1 input linguistic nodes

Layer 2 input term nodes

Layer 3 rule nodes

Layer 4 output term nodes

Layer 5 output linguistic nodes

Figure 13: Neural fuzzy logic controller.

Neurofuzzy Control and Decision

Fuzzy theory and neural networks gave been combined to form a class of intelligent control systems. For instance, in (Lee et al., 1989), reinforcement learning is used to adjust the consequent of fuzzy logic rules for the cart-jole balancing problem. Fuzzy cognitive maps proposed by Kosko is another scheme to integrate neural networks and fuzzy logic. In these approaches, the membership function or fuzzy logic rules are chosen subjectively.

A general neural network has been recently proposed (Lin et al., 1991) for fuzzy logic control and decision systems. In this model, which has a feedforward multilayer topology, the idea of fuzzy logic controller and neural network structure and learning abilities are integrated into a neural network based fuzzy logic control and decision system. A network is constructed automatically by learning the training examples itself. The input and output nodes represent the input states and control signals, respectively, and in the hidden layers there are nodes that code membership functions and rules. The learning algorithm for this network combines unsupervised learning and supervised gradient descent learning algorithm to build the rule nodes and train the membership functions. Figure 13 shows the proposed neurofuzzy system.

In figure 13, the first layer transmit input values to the next layer directly. The link weight at layer one is unity. The second layer, if a single node is used to perform a single membership function, then the activation function of this node should be this membership function. If a set of nodes is used to perform a membership function, then the function of each node can be just in the standard form, and the whole subnetwork is trained off-line to perform the desired membership

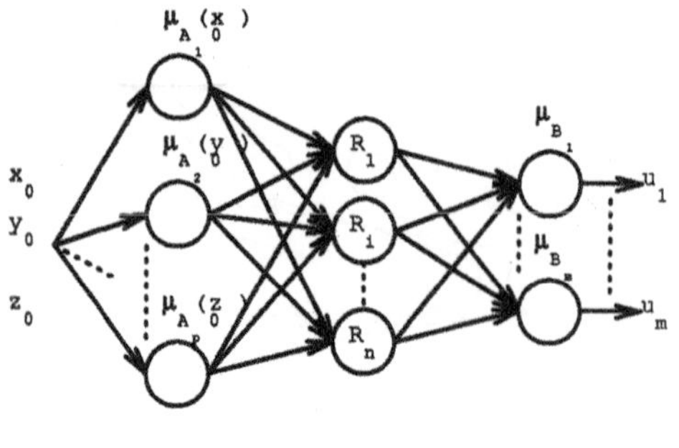

Figure 14: Neurofuzzy network structure.

function. The connections of the third layer are used to perform precondition matching of the fuzzy logic rules. The rule nodes should perform the fuzzy *and* operation. The weight in layer three is unity. The nodes of layer four have two operation modes: down-up transmission and up-down transmission modes. In the down-up transmission mode, the links at layer four perform the fuzzy *or* operation to integrate the fired rules that have the same consequent. The weights of layer four are also unity. The up-down transmission mode, the nodes behave exactly as those of layer two except that only a single node is used to perform a membership function for rule consequent linguistic variables. Finally, layer five also has two types of nodes. The first type performs the up-down transmission for the training data. The second type of node performs the down-up transmission for the control signal output. These nodes and the layer five connections attached to them act as the deffuzification interface. The hybrid learning algorithm developed is based on Kohonen's feature-map algorithm for the unsupervised phase, and on the backpropagation idea for the supervised phase.

More recently, a neurofuzzy controller designed for implementation on a neural chip has been derived (Glorennec, 1992). In this case, the used neuron model is adapted to model membership functions and fuzzy $\mathcal{T}$-norms. The learning scheme is based on backpropagation and because of this, membership functions are assumed to be differentiable. The inference procedure used is the Sugeno inference method (Lee, 1990) in which the terms in the rule consequent are nonfuzzy values. For a rule R_i of the form

$$\text{If } x \text{ is } A_{i1} \text{ and } y \text{ is } A_{i2} \text{ and } \ldots \text{ is } A_{in} \text{ then } u \text{ is } B_i$$

where $x, y, \ldots, z$ are controller inputs; A_{ij}, $j = 1, \ldots, n$ are fuzzy sets on their respective universes; u is the control action; and B_i, $i =, \ldots, n$ are either fuzzy sets on fuzzy singletons. In this simplest form, B_i is considered crisp and the weighted average method is used as a defuzzification method. Then the following is computed

$$\alpha_i = \min\{\mu_{A_{i1}}(x_0), \mu_{A_{i2}}(y_0), \ldots, \mu_{A_{in}}(z_0)\}$$
$$u(x_0, y_0, \ldots, z_0) = \frac{\sum \alpha_i \beta_i}{\sum \alpha_i}.$$

For a controler with n rules and m outputs, the neurofuzzy network is as shown in figure 14.

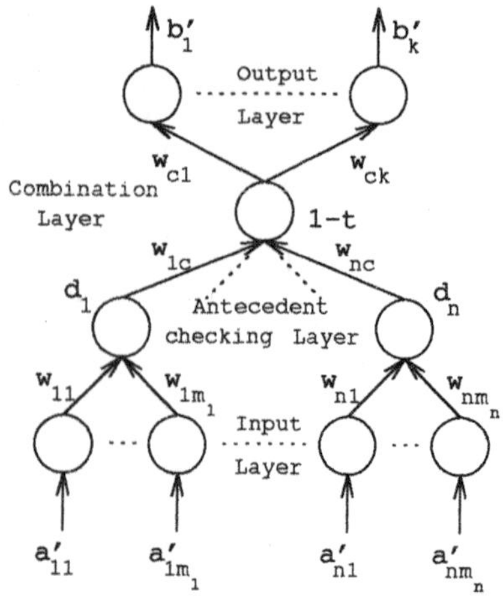

Figure 15: Neural network for fuzzy logic inference.

A neurofuzzy architeture which is used for fuzy logic inference in general and neurofuzzy control in particular has been developed in (Keller et al., 1992). Each basic network structure implements a single control rule in the rule base of the form

$$\text{If } (x_1 \text{ is } A_1 \text{ and } \cdots \text{ and } x_n \text{ is } A_n) \text{ then } y \text{ is } B$$

the fuzzy sets which characterize the facts x_i is $A'_1, \ldots, x_n$ is A'_n are presented to the input layer of the network, The fuzzy set A'_i is denoted by $A'_i = [a'_{i1}, \ldots, a'_{in}]^T$; these values being the membership grades of the fuzzy set A'_i at sampled points over its universe of discourse. As long as the finite universe is a totally ordered set, it is sufficient to describe a fuzzy set by only specifying the membership grades — the values of the domain being implicit in the order of the elements. Figure 15 depicts the neural network configuration for this rule.

In the antecedent checking layer, the weights w_{ij} are the fuzzy set complement of the fuzzy rule antecedent, i.e., for the i-th rule $w_{ij} = \bar{a}_{ij} = 1 - a_{ij}$. The weights are chosen this way because the first layer of the neural network will generate a measure of disagreement between the input and the rule antecedent. This is done so that as the input moves away from the antecedent, the amount of disagreement will rise to one. Thus, if each node computes the similarity (intersections) between the input and the complement of the rule antecedent, it will produce a local measure of disagreement. The amount of disagreement present between the rule antecedent and the input fact is, for the j-th node

$$d_j = \max_i\{w_{ij} \mathcal{T} a'_{ij}\} = \max_i\{(1 - a_{ij}) \mathcal{T} a'_i j\}.$$

Other forms of antecedent checking are also possible.

The disagreement values for each node are combined at the next level to produce an overall level od disagreement. The disagreement values provide inhibiting signals for the firing of the rule. The weights w_{ic} on these conections correspond to the importance of the various rule antecedents. These weights can be either supplied based on experience or obtained via

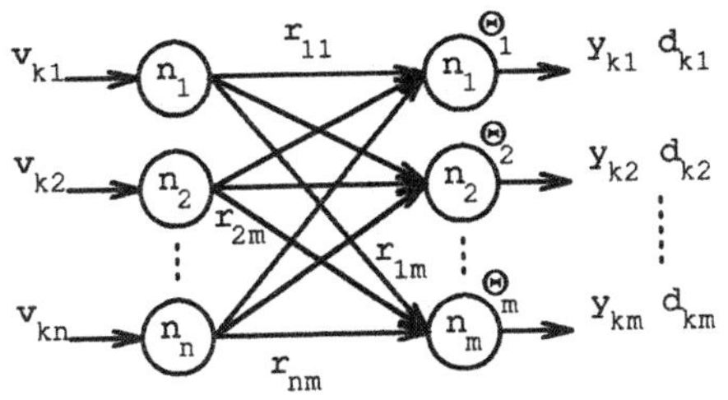

Figure 16: Max-min neurofuzzy controller.

training mechanisms. The combination nodes then computes

$$1 - t = 1 - \max_i \{w_{ic} \cdot d_i\}.$$

The weights w_{ci} on the output nodes carry the information from the rule consequent. If the proposition y is B is associated with the fuzzy set B, then

$$w_{ci} = \bar{b}_i = 1 - b_i.$$

Each output node forms the value

$$b'_i = b_i + t - b_i\, t.$$

From this equation, it is clear that if $t = 0$ then the rule fires with conclusion y is B exactly. If the overall disagreement is one, then the conclusion of the firing rule is composed entirely of ones. This network extends classical logic to fuzzy logic.

A max-min neurofuzzy controller has been developed in (Hayashi et al., 1992). The neural net is shown in figure 16.

There are n inputs neurons n_j, $j = 1, \ldots, n$, and m output neurons n_i, $i = 1, \ldots, m$. The learning set is $v_k \in R^n$ and the desired output $d_k \in R^m$, $1 \le k \le K$. The connection weights r_{ij} and a threshold term given by Θ_i, $i = 1, \ldots, m$ are associated with each output neuron n_i. For an given input v_R, the outputs are computed as

$$\begin{aligned} z_{ki} &= (v_{k1} \wedge r_{1i}) \wedge \cdots \wedge (v_{kn} \wedge r_{ni}) \\ y_{ki} &= z_{ki} \vee \Theta_i, \; i = 1, \ldots, n. \end{aligned}$$

The learning procedure is based on the modified delta rule for max-min nets due to (Pedrycz, 1991). The error measure to be minimized is

$$E = \frac{1}{2} \sum_{k=1}^{K} \sum_{i=1}^{m} (y_{ki} - d_{ki})^2$$

with the corresponding learning algorithm below

$$\begin{aligned} r_{ij}(l+1) &= r_{ij}(l) + \eta_1 \Delta r_{ij}(l+1) + \alpha_1 \Delta r_{ij}(l) \\ \Theta_i(l+1) &= \Theta_i(l) + \eta_2 \Delta\Theta_i(l+1) + \alpha_2 \Delta\Theta_i(l) \end{aligned}$$

where $\Delta r_{ij} = \partial E / \partial r_{ij}$, $\Delta\Theta_i = \partial E / \partial \Theta_i$, and η_i, α_i constants.

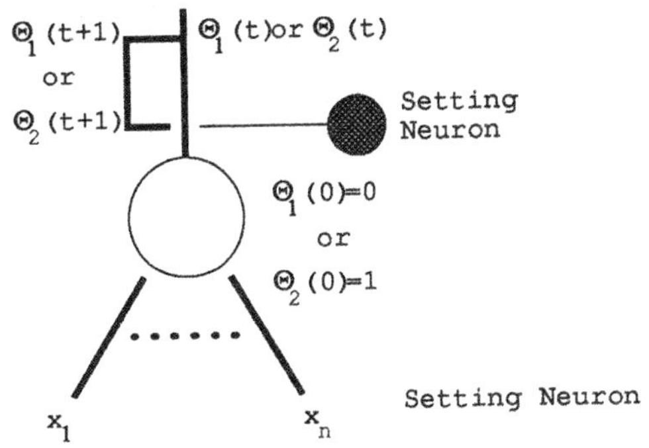

Figure 17: Threshold controlled neuron.

A Neurofuzzy Controller

Ordinary mathematical neuron models assume that the neuron fires when the weighted sum of the inputs exceeds a threshold level. The axonic activation y of a neuron is given by

$$y = \begin{cases} g(v) & \text{if } v \ge \Theta \\ 0 & \text{otherwise} \end{cases}$$

where Θ is the (axonic) threshold, $g(\cdot)$ the encoding function, and for a set of n inputs x_i and weights w_i, $i = 1, \ldots, n$, the pre-sinaptic activity v is:

$$v = \sum_{i=1}^{n} x_i w_i.$$

If $g : V \mapsto [0, 1]$, the neuron is named here fuzzy neuron. In this case y is modified by defining two axonic thresholds:

$$y = \begin{cases} 1 & \text{if } v \ge \Theta_1 \\ g(v) & \text{if } \Theta_1 \le v \le \Theta_2 \\ 0 & \text{otherwise} \end{cases}.$$

The fuzzy neuron model can be generalized by first redefining the pre-sinaptic activity and second by defining a threshold control function related with the axonic threshold. The proposal is to generalize the weighted sum by assuming that it is supported by any triangular norm $\mathcal{T}$:

$$v = \sum_{i=1}^{n} x_i \,\mathcal{T}\, w_i.$$

Note that this is equivalent to the previous case if $\mathcal{T}$ is the algebraic product.

Next, let us consider the n neuron inputs x_i such that they are each individually activated, and x_i is activated before x_j if $i < j$. This means that a pre-sinaptic neuron n_i always fires before another neuron n_j if $i < j$, and that the activity of neuron n_j is null for all $j \neq i$. This firing scheme is determined by a controlling setting neuron as shown in figure 17.

If the axionic threshold $\Theta_1(i)$ for each input x_i is updated according to

$$\Theta_1(i) \;=\; \Theta(i-1)$$

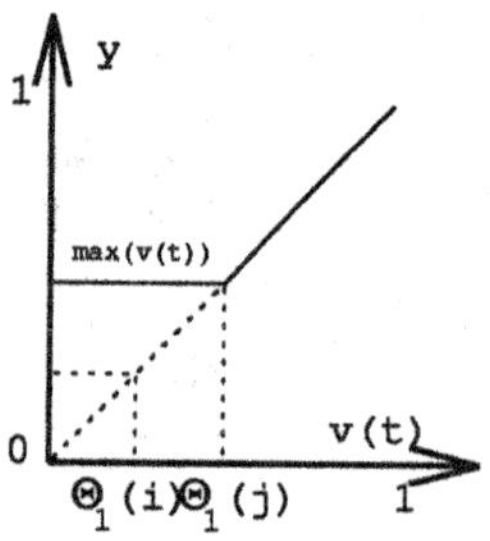

Figure 18: Max Neuron.

$$\Theta(i) = y(i)$$
$$\Theta(0) = y(0) = 0$$

where $y(i)$ is the neuron axonic activation due to input x_i, then

$$\Theta_1(i) = y(i-1)$$

and

$$y(i) = \begin{cases} \Theta_1(i) & \text{if } v(i) \leq \Theta_1(i) \\ v(i) & \text{otherwise} \end{cases}$$

where

$$v(i) = x_i \, \mathcal{T} \, w_i, \; i = 1, \ldots, n.$$

Hence, the ouytput $y(i)$ of the neuron encodes:

$$y(i) = \bigvee_{k=1}^{i} (x_k \, \mathcal{T} \, w_k)$$

and, if $w_k = 1$, $\forall k$, then:

$$y(i) = \bigvee_{k=1}^{i} x_k$$

where $\vee$ is the maximum operator. In this situation, threshold controlled neuron computes the maximum value of its n inputs and is called a max-neuron, figure 18. Min-neuron can also be defined if Θ_2 is initially set to zero, updated accordingly, and the inequality reversed.

Fuzzy control algorithms are described by a set of fuzzy conditional rules of the form:

$$\text{If } x \text{ is } A \text{ then } y \text{ is } B \text{ or } A \to B$$

where A and B are fuzzy sets defined is the universe of discurse $\mathcal{U}$ and $\mathcal{V}$, respectively. In this paper we consider discrete (or, otherwise, discretized) fuzzy sets. Thus, the fuzzy sets can be represented by vectors where each component is associated to the number of an element of the universe and its corresponding membership value. Hence, the fuzzy sets A and B are represented by

$$A = [x_1 \cdots x_n]^T \quad B = [y1 \cdots y_m]^T$$

where $x_i = \mu_A(u_i)$, $y_i = \mu_B(v_i)$; $u_i \in \mathcal{U}$, $v_i \in \mathcal{V}$ and $\mu_A : \mathcal{U} \mapsto [0,1]$, $\mu_B : \mathcal{V} \mapsto [0,1]$.

The implication $A \to B$ is defined by the following fuzzy relation:

$$R : A \times B \mapsto [0,1]$$
$$\mu_R(x,y) = \mu(u) \, \mathcal{T} \, \mu_B(v)$$

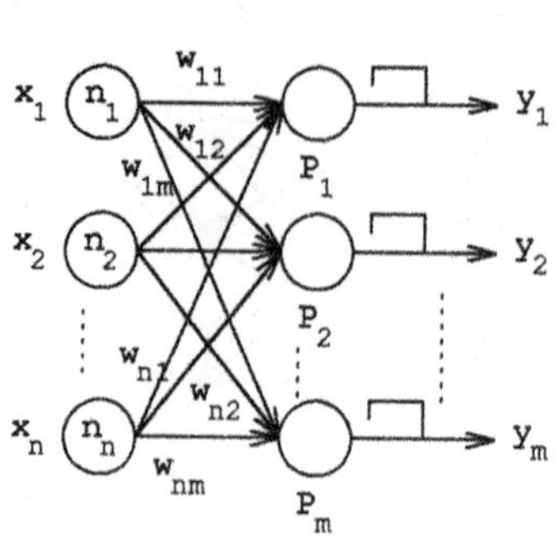

Figure 19: Neurofuzzy Controller.

and for discrete fuzzy sets, R is a $(n \times m)$ matrix $R = [r_{ij}]$ where $r_{ij} = x_i \, \mathcal{T} \, y_i$, and $\mathcal{T}$ triangular norm.

The reasoning scheme in fuzzy control is supported by the Generalized Modus Ponens whose most popular implementation is though the compositional rule of inference. That is, given A' and $A \to B$, B' is infered from

$$B' = A' \circ R$$

where $\circ$ is usually to be the max-$\mathcal{T}$ operator, and $\mathcal{T}$ a triangular norm.

The neurofuzzy controller may now be constructed, if the following are defined (see figure 19). The $n_1, \ldots, n_n$ are sensory neurons measuring the corresponding values of $x'_1, \ldots, x'_n$ of μ_A. The output neurons $P_1, \ldots, P_m$ are max-neurons. The weight w_ij between input neurons n_i and output neuron P_j is set as:

$$w_{ij} = x_i \mathcal{T} y_j.$$

Thus, the recording v_{ij} at the synapses between n_i and P_j becomes

$$v_{ij} = x'_i \mathcal{T} w_{ij}.$$

If $n_1, \ldots, n_n$ are synchronized such that n_i fires before n_j if $i < j$, we get

$$B' = A' \circ R$$

because $y'_i = \vee_{k=1}^{i}[x'_k \mathcal{T}(x_k \mathcal{T} y_j)]$, $j = 1, \ldots, m$. By definition $\mathcal{T}$ is a triangular norm.

Given a training data set

$$\mathcal{D} = \{(x^1, y^1), \ldots, (x^N, y^N)\},$$
$$x^k = [x_1^k, \ldots, x_n^k]^T,$$
$$y^k = [y_1^k, \ldots, y_m^k]^T,$$

and a performance index

$$P = \sum_{k=1}^{N} \sum_{j=1}^{m} \left[\bigvee_{i=1}^{n} (x_i^k \wedge w_{ij}) - y_j^k \right]^2$$

a learning scheme can be derived using the method proposed in (Pedrycz, 1991) which provides

$$w_{ij}(l+1) = w_{ij}(l) - \frac{\alpha}{2} \frac{\partial P}{\partial w_{ij}}$$

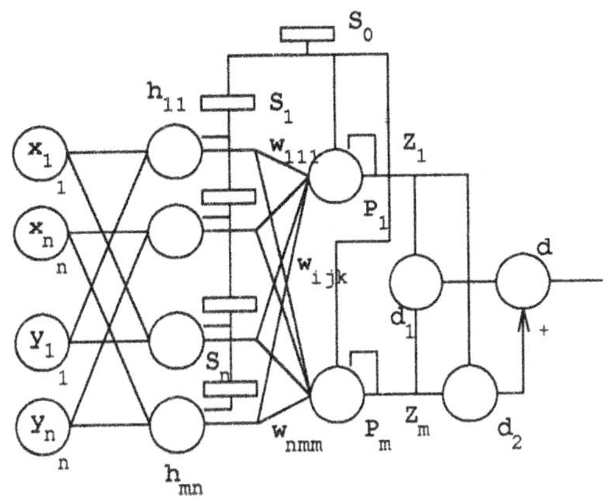

Figure 20: Neurofuzzy Controller and Defuzzification.

where α is a constant and $\partial P/\partial w_{ij}$ computed accordingly.

Note that, for a set of N fuzzy control rules of the form If x is A_i then y is B_i, $i = 1, \ldots, N$, connected by the Else Connective, the overall relation R associated with the N rules can be obtained by

$$R = \biguplus_{i=1}^{N} R_i$$

where R_i is the i-th implication relation, and $\uplus$ either a triangular norm, if the Else is interpreted as conjunctive operator, or a triangular co-norm, if Else is defined as disjunctive operator. The complexity of the neural network is not affected by the number of rules, but is obviously dependent on the degree of precision required by the input/output discretization.

A more general framework is to consider fuzzy control rules of the form

If x is A then y is B ... then z is C

where $A, B, \ldots$, and C are fuzzy sets in the corresponding universes $\mathcal{U}, \mathcal{V}, \ldots$, and $\mathcal{W}$. Then given A', B' we have

$$C' = (A' \wedge B' \wedge \cdots) \circ R$$
$$R : A \times B \times \cdots \times C \rightarrow [0,1]$$
$$\mu_R(u, v, \ldots, w) = [\mu_A(u) \wedge \mu_B(v) \wedge \cdots]$$
$$\mathcal{T} \mu_C(w)$$

where $\wedge$ and $\mathcal{T}$ are triangular norms. The neural network to compute C' can be obtained by including, in the previous single antecedent case, an associative layer to compute $(A' \wedge B' \wedge \cdots)$, where $\wedge$ defines the encoding function of the associative layer. See (Gomide et al., 1992-a) for further details, including the necessary synchronization mechanisms, and figure 20 for the simplest case with n and m discretization levels and two antecedents rule.

The output neurons P_i can be defuzzified by adding a layer of d_i neurons as shown in the figure 20 example. The defuzzification layer complexity and topology depends on the defuzzification method chosen. In figure 20, the center of area (COA) is considered. Here, the weights between the output neuron P_i and defuzzification neuron d_i are set as:

$$w_{P_i d_1} = \delta_i \epsilon_i$$

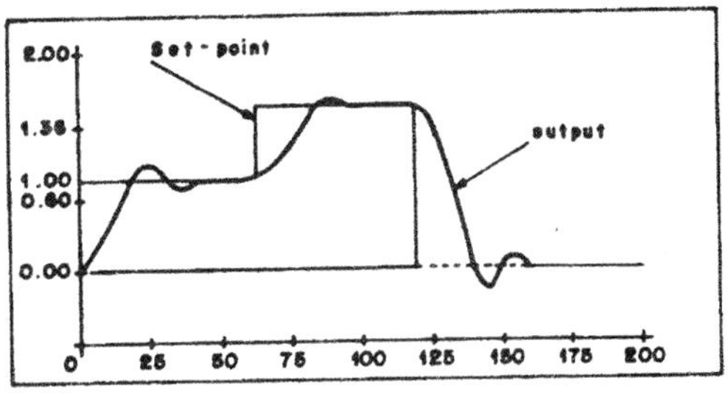

Figure 21: Process output.

where $\epsilon_i = \delta_{i+1} - \delta_i$, and δ_i is the discretication granule;

$$w_{P_i d_2} = \epsilon_i.$$

In this condition, the d_1 and d_2 neurons compute:

$$a_{d_1} = \sum_{i=1}^{m} z_i \epsilon_i \delta_i, \quad a_{d_2} = \sum_{i=1}^{m} z_i \epsilon_i$$

such that the output of neuron d becomes:

$$a_d = a_{d_1} / a_{d_2}$$

which corresponds to the COA method.

To illustrate the behavior of the neurofuzzy controllers, the control of a simple process described by the transfer function

$$G(s) = \frac{K}{s + \tau}; \; k = 1, \; \tau = 20$$

is considered. The fuzzy control rules that compose the controler are the same as considered by Mizumoto (1988). Here, the membership function of the fuzzy sets associated with the rules antecedents and consequents have discretized into five discretization granules. Hence, the fuzzy relations are $5 \times 5 \times 5$ matrices. The system response for a serie of set-point changes is shown by figure 21.

As it easily seen, the neurofuzzy controller is able to effectively control the process, without any a priori knowledge about process model and control rules.

Hardware and Development Tools

The first fuzzy logic chip was designed by Togai and Watanabe at AT&T Bell Laboratories in 1985. The fuzzy inference chip, which can process 16 rules in parallel, consists of four major parts: a rule-set memory, an inference-processing unit, a controller, and a input-output circuitry. Recently, the rule-set memory has been implemented by a static random access memory (SRAM) to realize a capability for dinamic changes in the rule set. The inference-processing unit is based on the sup-min compositional rule of inference. Timing tests indicate that the chip can perform approximately 250,000 FLIPS at 16-MHz clock. A fuzzy logic accelerator (FLA) based on this chip has also been developed. In March 1989 the Microeletronics Center of North Carolina successfully completed the fabrication of fuzzy logic chip, designed by Watanabe.

The full-custom chip comprises 688,000 transistors and is capable of making 580,000 FLIPS.

A high-speed FLC hardware system employing fuzzy reasoning has been proposed by Yamakawa (1986). It is composed of 15 control rule boards and an action interface (a defuzzifier based on COA). It can handle fuzzy linguistic rules labeled as NL, NM, NS, ZR, PS, PM, PL. The operacional speed is approximately 10 mega fuzzy logical inferences per second (MFLIPS).

The FLC hardware system has been tested by an application to the stabilization of inverted pendulums mounted on a vehicle. Two pendulums with different parameters were controlled. Each control rule board and action interface has been integrated to a 40-pin chip.

Yamakawa and Miki have also implemented nine basics fuzzy logic functions by the standard CMOS process in current-mode circuit systems. Later, a rudimentary concept of a fuzzy computer was proposed by Yamakawa and built by OMROM Tateishi Eletric Co. Ltd. The Yamakawa-OMROM computer comprises a fuzzy memory, a set of inference engines, a MAX block, a defuzzifier, and a control unit. The fuzzy memory stores linguistic fuzzy information in the form of membership functions. It has a binary RAM, a register, and a membership function generator. A membership function generator (MFG) consists of a PROM, a pass transistor array, and a decoder. Every term in a term set is represented by a binary code and stored in a binary RAM. The corresponding membership function are generated by the MFG via these binary codes. The inference engine employs MAX and MIN operations, which are implemented by the emitter coupled fuzzy logic gates (ECFL gates) in voltage-mode circuit systems. The linguistic inputs, which are represented by analog voltages distributed on data buses, are fed into each inference engine in parallel. The results inferred from the rules are aggregated by a MAX block, which implements the function of the connective *also* as a union operation, yielding a consequence which is a set of analog voltages distributed on output lines. In the FLC applications, a crisp control implementation, a fuzzy computer is capable of processing fuzzy information at the very high speed of approximately 10 mega-FLIPS. It is indeed an important step not only in industrial applications but also in common-sense knowledge processing.

More recently, Neuralogix (1991) has announced a fuzzy microcontroller NLX230 which is a configurable VLSI fuzzy logic engine based on the *max-min* inference scheme. The chip has sixteen fuzzifiers, a neural network for minimum comparison, a maximum comparator, rule memory and the associated timing and control circuits and registers. The chip is capable to process 30 M rules/sec when properly programmed.

In (Gomide et al., 1992-b) a low cost fuzzy-controller and development environment are decribed. This controller is implemented in conventional hardware platform and handles eight analog and digital input variables, eight analog and digital output variables, and several timers and internal variables for user conveniency. The controller is also able to communicate in a network system. There are several options to define the desired inference scheme, and *else* operator definition. It is able to process more than 600 rules.

Oki has also developed a VLSI chip for fuzzy logic inference in 132 pin ceramic PGA. The circuit architecture is based on pipelining structure and single instruction multi data stream. The chip is able to process at 7.5 MFLIPS rate and 960 rules. It contains all of the functional circuits to carry out fuzzy logic inference: input buffer, membership function memory, rule memory, fuzzifier, max-min circuit, defuzzifier and output register.

Omron has developed one of the first world fuzzy controller. Omron has also announced a new generation of fuzzy processors: the FP3000 digital fuzzy processor family. High-speed inference processing at 650μs using 24MHz clock (per 20 rules with 5 antecedents and 2 consequents) and several features for interfaces are provided.

From the software support point of view, currently a number of development environments are available. Most of them have functions for membership function editing supported by graphical resources, problem oriented languages for rule description, pre-compilers for several target processors, code generation in a high level linguage such as C. Examples include Togai Til shell, the CubiCalc, the Manifold Editor, fuzzyTECH environment, Aptronix Omron development tools, and Meiden Fuzzy Control System. Some of then also provide several reasoning procedures, defuzzification methods and learning capabilities.

Neurocomputers are being developed, and several architectures have been proposed (Treleaven, 1989). The main development efforts are the following: in the United States, AT&T have develop several prototype neural network chips. Developed chip designs include: hibryd digital/analog programmable connection matrix with adjustable connection strengths, and a digital pipelined best-match chip. Bell Communications Research (Bellcore Labs.) have built a first analog neural network with adaptative model synapses. At the California Institute of Technology (Caltech), a team led by Carver Mead, has implemented neural network chips. Their initial prototype contained 22 active elements and a full interconnection matrix of 462 elements, implemented in 4 microns NMOS technology. One of the latest was a 289 neuron chip in CMOS. Mead is also investigating neural chips for the early vision and auditory processing functions, using the intrinsic nonlinearities of MOS transistors in a subthershold regime. Synaptic Inc., a startup founded in 1986 is dedicated to the commercialisation of a new set of VLSI building blocks, neural network chips for sensory, motor and information processing applications. Lastly, the Jet Propulsion Laboratory are investigating associative memory structures using VLSI. Their chip uses a novel analog-digital hybrid architecture based on the utilization of high density digital RAMs for the storage of the synaptic weights of a network, and high speed analog hardware to perform neural computation.

In Japan numerous organizations have developed silicon neural networks. Nakano of Tokyo University has completed a number of neurocomputers, some dating from 1970. The Association, his best known neural network system, is a hardware, correlation-matrix type of associative memory. Other leading research centres include: Fujitsu, Nippon Telegraph

Telephone Company NTT and the Electrotechnical Laboratory. Fujitsu has recently announced work on neural network chips for tasks like character recognition and robot control.

In Europe, also, many organisation are engaged in implementing special-purpose hardware for neural networks. Kohonen of Helsinki University of Technology, a pioneer in associative memory neural networks, has experimented with several neural network distributed memories. He has recently completed a commercial-level neurocomputer based on signal processor modules and working memories to define a set of virtual processing elements. This neurocomputer, optimised for speech recognition allows 1000 virtual processing elements with 60 interconnections. It can perform a complete analysis and classification in phonemes every 10ms. Aleksander at Imperial College London has developed a series of systems called WISARD. WISARD II is organised as hierarchical network of processing elements, with each processing element being constructed from commercial RAM. The RAMs addresses inputs are used to detect binary patterns, with the input field for an image (comprising 512x512 binary pixels) being conected to the first layer of processing elements. Many other centres are working towards VLSI implementations of networks. Well known centres are Ecole Polytechnique Paris, University College London, Edinburg University, and the University of Dortmund. The Ecole Polytechnique are developing a fully digital integrated CMOS Hopfield network which includes the learning algorithm. University College London are producing a silicon compiler that will generate dedicated CMOS chips for a given neural network. Edinburgh University are building a bit-serial, hybrid analog/digital VLSI neural network with 64 neurons on a board operating at 20 MHz.

Conclusion

Despite of the sucess achieved in several applications such, as water quality control, automatic train operation, elevator control, automotive transmission control, consumer electronics products and appliances, just to name a few, fuzzy logic control still demands more fundamental results in, for instance, system stability, qualitative properties (controlability, observability), systematic analysis and design procedures. Stability and robustness issues concerning fuzzy expert control have been recently addressed in (Ollero et al., 1991). A considerable amount of hardware devices and software tools are available to help engineers in system development.

Neural networks are also being succesful in applications areas where conventional control theory is not effective. Although fundamental results have been found, there are many questions (e.g., network design procedures, more effective learning schemes, feedback network considerations, etc.) which still remain unanswered. However, recent results concerning learning strategies for neural controllers for control have been developed (Padilla, 1992). A number of tools is also available for neural network applications development.

Neurofuzzy systems and control still is in its infancy and, based on the recent advances in the area, we may expect significant results in the next years. However, there are currently several appliance manufacters which have announced their products being controlled by neurofuzzy controllers to adapt them to the user needs. A prototype application in controlling an automatically guided vehicle has also sucessfully been carried out (Souza et al., 1992).

In this paper we have reviewed the fundamentals of fuzzy control and neural networks, focussing on the recent developments in the fusion of neural nets with fuzzy logic; the neurofuzzy systems in general and neurofuzzy controllers in particular. Currently available software and hardware have also been discussed. However, it is clear that there remains much to be done before the field of neurofuzzy control becomes filled with a sound theoretical and methodological basis.

References

Albertos, P. (1992). Fuzzy controllers. *Notes of Erasmus course on AI techniques in Control*, Pergamon Press.

Anderson, B.D.O. and J.B. Moore (1979). *Optimal Filtering.* Prentice-Hall, N.J.

Bertsekas, D.P. (1976). Dynamic Programming and Stochastic Control in *Mathematics in Science and Engineering.* Vol. 125, Academic Press, N.Y.

Dougle, J.C. and G. Skin (1981). Multivariable feedback design: concept for a classical/modern synthesis. *IEEE Trans. on Automatic Control*, vol. AC-26, pp. 4–16.

Furuya, T.; A. Kokubu and T. Sakamoto (1988). NFS: Neurofuzzy inference system. *Iizuka-88*, pp. 219–230.

Glorennec, P.T. (1992). A neuro-fuzzy controller designed for implementation on a neural chip. *Iizuka-92*.

Gomide, F. and A. Rocha (1992-a). Neurofuzzy controllers. *Iizuka-92*.

Gomide, F.; R. Gudwin; M.A. Silva; H. Jacques and I. Ribeiro (1992-b). Fuzzy Control Supervision Workstation. IFAC-LCA'92, Austria.

Hayashi, Y.; E. Czogala and J. Buckey (1992). Fuzzy neural controller. *IEEE Int. Conf. on Fuzzy Systems*, pp. 197–202.

Hetch-Nielsen, R. (1990). *Neurocomputing.* Addison-Wesley, N.Y.

Holtzman, J.M. (1970) *Nonlinear System Theory.* Prentice-Hall, N.J.

Horikawa, S.; T. Furukashi; S. Okuma and Y. Uchikawa (1990). Composition methods of fuzzy neural networks. *IEEE* publication n° 087942-600-4/90/1100-1253, pp. 1253–1258.

Ishibushi, H.; H. Tanaka; R. Fujioka and R. Tamura (1990). Identification of membership functions by neural networks. *Trans. EICE Part D-II*, vol. I73, n° 8.

Keller, J.M.; R. Yager and H. Tahani (1992). Neural network implementation of fuzzy logic. Fuzzy Sets and Systems, n° 45, pp. 1–12.

Kosko, B. (1987). Bidirectional associative memories. IEEE Trans. Sust. Man and Cybernetics, vol. 18, n° 1, pp. 49–60.

Kosko, B (1986). Fuzzy cognitive maps. *Int. I. Man Machine Studies*, vol. 24, pp.65–75.

Lee, S.C. (1974). Fuzzy sets and neural networks. *J. Cybern.*, vol. 4, n° 2, pp. 83–103.

Lee, C.C. and H.R. Berengi (1989). An intelligent controller based on approximate reasoning and reinforcement learning. *Proc. IEEE Intelligent Machine*, pp. 200–205.

Lee, C.C. (1990). Fuzzy Logic in Control Systems: Fuzzy Logic Controller, part I and II. *IEEE Trans. on Systems, Man and Cybernetics*, vol. 20, pp. 404–435.

Lin, C.T., C.S.G. Lee (1991). Neural-network-based fuzzy logic control and decision system. *IEEE Trans. on Computers*, vol. 40, n° 12, pp. 1320–1336.

Machado, R. and F.A.M. Denis (1991). O Modelo Conexionista Evolutivo (in portuguese). IBM Rio Research Center Report, CCR-12.

Mizumoto, M. (1988). Fuzzy controls under various reasoning methods. *Information Sciences*, vol. 45, pp. 129–151.

Morita, A.; Y. Imai and M. Takegaki (1988). A method to refine fuzzy knowledge model of neural network type. *SICE* Joint Symp., pp. 343–348.

Narendra, K.S.; E.G. Kraft and L.H. Ungar (1991). Neural networks in control systems, *1991 ACC*, Boston, USA.

Ollero, A.; A. Garcia-Cerezo and J. Araci (1991). Design of rule-based expert controllers. *ECC91 European Control Conference*, Grenoble, France, pp. 578–583.

Padilla, F. (1992). Intelligent Control Algorithms Implemented by ANN. *PhD Thesis*, Valencia (in Spanish).

Pedrycz, W. (1989). *Fuzzy Control and Fuzzy Systems*. John Wiley and Sons Inc., N.Y.

Pedrycz, W. (1991). Neurocomputations in relational systems. *IEEE Trans. on Pattern Analysis and Machine Intelligence*. vol. 13, pp. 289–297.

Rocha, A.F.; E. Françozo; M. Handler and M. Balduino. Neural Languages. *Fuzzy Sets and Systems*, vol. 3, n° 1, pp. 11-53.

Rocha, A.F. (1992). *Neural Networks: A Theory for Brain and Machines*, in Lecture Notes in Artificial Intelligence, Spring-Verlag.

Sage, A. and C. White (1977). *Optimum Systems Control*. Prentice-Hall, N.J.

Saito, T. and M. Mukaidomo (1991). A learning algorithm for max-min network and its application to solve fuzzy relation equation. *IFSA'91 Brussels*, pp. 184–187.

Self, K. (1990). Designing with fuzzy logic. *IEEE Spectrum*, vol. 27, pp. 42–43.

Souza, E. and F. Gomide (1992). Automatically Guided Vehicles Control Using a Neurofuzzy Controller. *Internal report, Unicamp*, RT-DCA-1992.

Sugeno, M. (1985). An Introduction survey of fuzzy control. *Information Sciences*, vol. 36, pp. 59–83.

Takagi, H. and I. Hayashi (1988). Artificial neural network driven fuzzy reasoning. *Iizuka-88*, pp. 183–184.

Takagi, H. (1990). Fusion technology of fuzzy theory and neural networks: survey and future directions. *Iizuka-90*, pp. 13-26.

Tong, R.M. (1977). A Control Engineering Review of Fuzzy Systems. *Automatica*, vol. 13, pp. 559–569.

Treleaven, P. (1989). Neurocomputers. *Research note 89/8*, Depart. of Computer Science, University College London.

Yamaoka, M. and M. Mukaidono (1991). A learning method of fuzzy inference rules with a neural network. *IFSA'91, Brussels*, pp. 222–225.

Yamaguchi, T.; M. Tanabe and J. Murakami (1989). Fuzzy control using LVQ unsurpervised learning. *SICE* Joint Symp., pp. 179–189.

Yager, R.; S. Ovchinnikov, R.M. Tong and H.T. Nguyen (1987). *Fuzzy Sets and Applications*. Wiley Interscience, N.Y.

Zadeh, L. (1965). Fuzzy sets. *Information and Control*, vol. 8, pp. 338–353.

Zadeh, L. (1973). Outline of a new approach to the analysis of complex systems and decision processes. *IEEE Trans. on Systems Man and Cybernetics*, vol. SMC-3, pp. 28-44.

Zadeh, L. (1988). Fuzzy Logic. *IEEE Computer*, April.

Zhuang, W.P.; W.Z. Qiao and T.H. Heng (1990). The truth-valued flow inference network. *Iizuka-90*, pp. 267–281.

Acknowledgments: The first author acknowledgs the CNPq for grant #300729/86-3. The authors also thanks Mr. Luiz E. R. Cordeiro for his editorial help.

"LOW-COST" FACTORY AUTOMATION

P. Kopacek

Department of Mechanical Engineering, Technical University of Vienna, Austria

Abstract: "Low cost" and intelligent automation is an efficient tool especially for small and medium sized companies today. While "low cost" devices and concepts for automation of continuous processes are commercially available today concepts for discontinuous processes especially for manufacturing automation are in developing. Therefore in this paper a modular and open "low cost" concept will be presented and shortly discussed. It consists of a network of PC's superimposed by a host computer. This concept is illustrated by application examples.

Keywords: Production control; manufacturing processes; assembling; robots.

1. INTRODUCTION

In this paper some ideas and experiences are collected dealing with "intelligent" automation especially in small and medium sized companies. Such companies are very important for the productivity in various mainly smaller countries. For example in Austria approximately 50% of the industrial production is carried out in companies with less than 500 employees. Compared with larger companies one of the main advantages is the flexibility. Small companies are easier in the position to produce special products according to the demands of the market in small lots. Doing this in an efficient way automation is absolutely necessary now and in the future.

The facts indicated above are valid mainly for manufacturing companies (manufacturing automation) as well as for companies working with continuous processes (process automation). In the field of process automation, microprocessors offer various possibilities for "intelligent" devices like sensors, controllers, actuators and whole process control systems. Some of these devices e.g. microprocessor equipped ("intelligent") sensors and actuators, self tuning or adaptive microprocessor controllers are commercially available today. Problems for small and medium sized enterprises are the costs of such devices or complete automation solutions with those.

But today there is also a strong demand of companies for manufacturing automation. Modern production requires digital computers for various purposes, especially for design, planning and testing of products. These applications are summarised under the headline CIM. There are a lot of CIM packages commercially available today but only few of them are suitable for the demands of small and medium sized enterprises. Therefore a CIM concept for those was developed.

2. LOW COST CIM CONCEPT

The automation of discrete processes is one of the classical application fields of microcomputers today. Especially in production or manufacturing automation, NC, CNC or DNC controlled production machines have been in use for many years.

These machines might be the first step towards the so called "factory of the future", which will be totally computer controlled. The elements of such a concept are

- CAD (computer aided design)
- CAP (computer aided production planning)
- CAM (computer aided manufacturing)
- CAQ/CAT (computer aided quality control, computer aided testing).

For commercial purposes a PPS (production planning system) should also be implemented.

These elements together form a complete CIM concept. It allows a fully "computerized" production. Such systems fill the gap between inflexible high-production transfer lines and low production - but flexible - NC machines. Computer integrated manufacturing systems are flexible enough to handle a variety of part designs as well as having a sufficiently high production rate. Therefore they are suitable for small and medium sized factories.

The efficiency of the functional integration of CIM-components and CIM-modules depends on the existing integration of data. In order to achieve the economic objectives within the realisation of a CIM-concept, it is important to keep the data centrally organised.

This aim has already been accomplished in a "modular CIM-concept for small and medium sized companies". In addition to the central and relational CIM-database an "engineering-" and quality database is provided. The functional integration of the CIM-components by means of the client/server-concept is co-ordinated ideal with the existing infrastructure and the working places.

The base of the this concept is the database server (UNIX V), which is connected with the PC's (AT-286, AT-386, MS-DOS) with Ethernet. The NC-machine tools are integrated with the V24 interface.

Apart from the use of significant CIM-components like CAD or PPS, this CIM-concept supports the modules of a hierarchical group control system (GCS), which consists of GCSs and process control systems (PCS), the CAD/NC-File server, operating characteristics acquisition (OCA), alteration system as well as the CAD-construction- and tools management. Special attention has also been given to the fact, that a CAQ-system will be integrated into the CIM-landscape at a later date. With this stepwise introduction of this CIM-concept the employees of the company are confronted with less complexity, whereas higher acceptance is being supported.

This concept has following advantages for small and medium sized companies:

- "low cost"
- possibility of stepwise realization
- easy adjustment of the software
- possibility to include some methods of AI

- This concept is now in the developing stage and will be illustrated by some examples

3. APPLICATION EXAMPLES

3.1. Production of welding transformers

One of the first applications was the automation of the production of welding transformers in a medium sized company typical for Austria. This company produces six different types of welding transformers. Because of the market conditions the trend goes towards smaller lots and therefore more variations of the six basic types are necessary. Today approximately 1300 variations with an average lot size of 10 are produced.

The idea was to develop a special basic CIM concept for the whole production of these transformers taking into account the following boundary conditions:

- production of six different types of welding transformers with various dimensions and variations
- implementation of the machine tools available today in the CIM concept
- implementation of three different CAD systems running on PC's
- implementation of an existing factory data acquisition system
- network of low cost control computers (PCs or similar ones)

The features of this concept should be

- recording and administration of development data (drawings...)
- recording and administration of manufacturing data (NC programs...)
- implementation of factory data in the planning and manufacturing process
- automatic preparation of manufacturing tools (NC programs...)
- control of the whole manufacturing process

The basic concept of the computer network is shown in Fig.1

Central part of the concept is a host computer - a PC with the operation system UNIX - serving as a database as well as a network node. Manufacturing orders from the production planning system - connected by Ethernet with the host computer - are given to the manufacturing control station - a PC under the operating system MS-DOS. At the same time the necessary data from the database are submitted to the control station too. On the manufacturing control station the orders are assigned to distinct machine control stations - PC's under the operating system MS-DOS. One machine control station is responsible for the supervision of 4 machines or assembly cells. The factory data acquisition system transfer distinct data to the host computer.

Main problems were the different interfaces of the components. Commercially available CIM Software packages weren't suitable for this purpose because they haven't included important features absolutely necessary. The software was implemented one year ago. The modularity offers the possibility of a step by step realization. In the first stage the automatic assembling of the primary part of the welding transformer was chosen. This will be described in the following.

3.2. Assembly cell for primary parts of welding transformers.

The assembling of these devices is carried out by hand today in 10 working stations and 3 test stations. Parts of these transformers are manufactured by NC-, CNC- or DNC- machine tools. Unfortunately from the technical point of view it is impossible to carry out all assembling operations automatically. Some of the parts to be assembled are of a large size, some of them are "flexible" (cables, ...). Furthermore the necessary assembling operations include screwing, gluing and soldering and some of the parts are not constructed for automatic assembling. Therefore only a part of the necessary assembling operations can be carried out automatically today from the technical as well as the commercial point of view.

The necessary assembling operations are described for the smallest of the six different types. The primary part consists of a heat sink whose dimensions are determined by the attached electronic parts as well as two printed circuits. Today assembling is carried out by hand at three assembling stations. The time consumption amounts approximately 3 to 5 minutes depending on the type of the welding transformer.

The necessary operations in the automatic assembling station are:

- applying of a heat conductivity paste for the electronic parts onto distinct areas of the heat sink

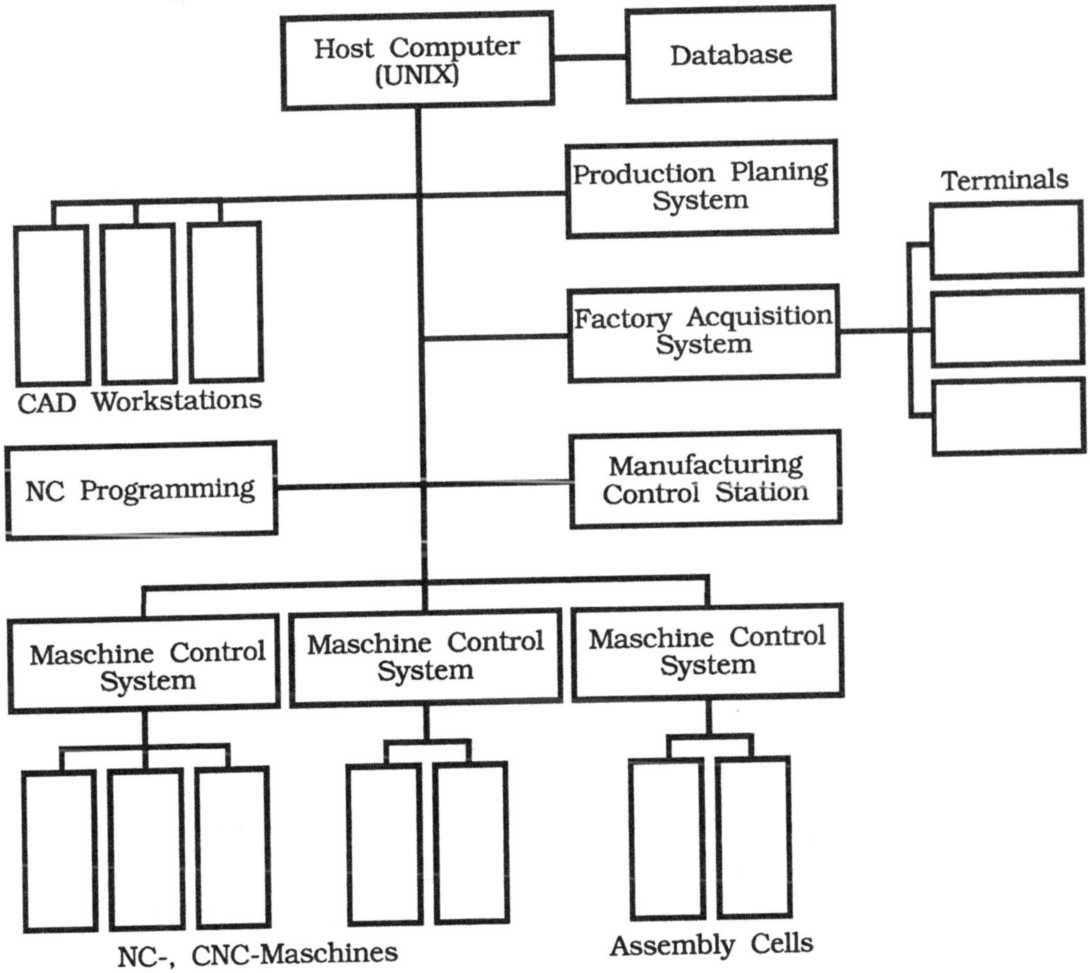

Fig. 1: Basis Concept of the Computer Network

- attaching of two transistors, 4 diodes, 2 resistors and 1 thermocouple
- screwing on of these electronic parts
- attaching of the printed circuit 1 - the 8 pins of the diodes have to be inserted in the holes of the printed circuit 1
- screwing of the printed circuit 1 onto the diodes
- screwing of 3 power cables to the printed circuit 1
- soldering of some cables and connections onto printed circuit 1
- attaching and soldering of a capacitor onto print 1
- attaching of a pressboard plate and the printed circuit 2
- screwing on of these two parts

The primary part is now ready for further assembling in the welding transformer. These further assembling operations are too complicated for automation today.

The layout of the assembling device for these tasks is shown in Figure 2. It consists of two robots, the storage units for the parts to be assembled, the screen printing machine, the necessary transportation devices, and the input-output station. First the head sink is fixed on a pallet. Together with the pallet it reaches the input storage unit. If the next part is required for assembling the controller will give the command for feeding the screen printing machine. After finishing this process the head sink reaches the assembling station. In the way described earlier the part is assembled completely by the two robots. A lift - marked "L" in Figure 2 - transports the part to the lower level of the transportation system and under the assembling station back to the output storage unit. The primary part of the welding transformer is now ready for further assembling while faulty assembled parts are removed by a locking out device. One of the robots - Type RT 280 (IGM) with 6 rotational degrees of freedom - is responsible mainly for tool handling operations the other - Type IRB 1000 with 5 rotational and one translatorial degree of freedom - is mainly responsible for part handling. Both robots are equipped with various grippers changed automatically by a tool changing system.

3.3. Expert System for Diagnosis in Assembly Cells

For this assembly cell mentioned above an expert system for diagnosis was developed and installed.

The plant can be supervised by one person, who is responsible for handling errors, feeding all necessary units including loading and unloading of heat sinks. The necessary assembly-sequence - according to the type of the welding transformer - works fully automatically. This diagnosis-system including a small expert system, is implemented on the PC in addition to the supervising system. The system allows less skilled operators to choose appropriate actions in order to recover errors. In

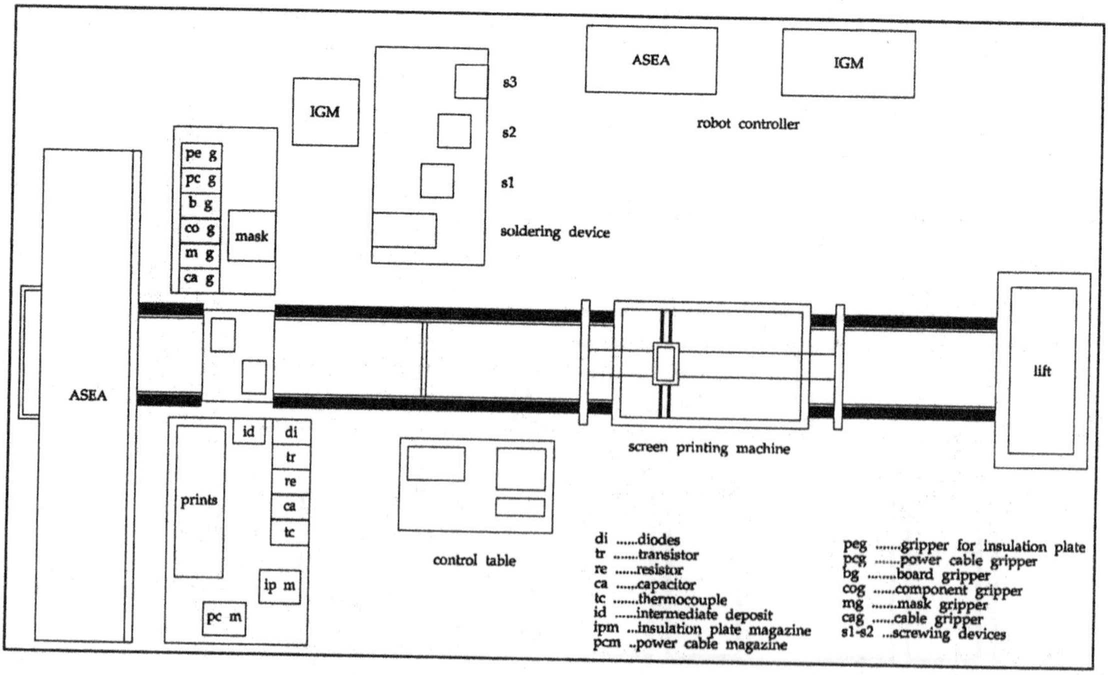

Fig. 2: Layout of the assembly cell

addition this knowledge based system serves for education of operators by simulation of different conditions of the assembly plant.

Knowledge and experience from experts was accumulated by interviews, by accompanying forms and extracted from error-protocols. All knowledge is transformed into logical rules. The knowledge base is maintained by the supervising person. Therefore it is possible to adapt the diagnose-system for assembly. The interface of the system is quite simple. If the supervising controller reports an error, it is possible to start the diagnosis system. After interpreting the state of the assembly plant and asking questions to the user, the error reasons are determined. Subsequently a way to fix the problem is suggested. First tests in practice turned out in a good conformity between the diagnoses of the knowledge based system and the experts treatment.

With this knowledge in a next step a computer aided system for planning flexible assembly cells was installed.

3.4. Computer aided planning of assembly cells

The development of the planning system is based on the assumption that flexible assembly can be described by a group of modules, such as the robots, grippers, delivery and orientation devices, various assembly tools, sensor and control, to work together simultaneously for certain assembly operations. An assembly cell consist at minimum of one robot with one tool or gripper equipped. It should perform a sub assembly or product assembly of one or various products in flexible assembly cell. Several cells are connected to the assembly lines or assembly systems trough a transportation system. One should also mention the supposition that the flexible cell is planned for given assembly operations.

Different effective planning methods are known in the assembly systems planning and the planners dispose of various support tools. In the development of the planning system those methods were briefly surveyed and taken into consideration.

In the development of the planning system the following tasks are taken into account:

- creation of an database and a database management system,
- determination of the planning system goals,
- the planning flaw or planning system building,
- collection of necessary data and knowledge,
- testing of the planning system.

A successful computer aided system for planning flexible assembly cells is based on the assembly planning oriented integrated database. The complexity of assembly planning process does not allow the use commercial database systems automatically. Different data were collected for planning reasons and stored into the data base as follows:

- information about assembled product, parts and assembly process, the *customer database*,
- information about assembly cells modules, *component database*,
- information about planning results, *assembly planning process database*,
- information and knowledge to support the planning decisions, *knowledge database*.

The planning process starts with the products and their part description, the goals definition, and customer data base creation. The assembly operations determination and their sequencing is done manually. As the result of the processing, in the following planning step, a rough proposal of the assembly cell is made. It consists of the required assembly operations, the assembly sequencing, the assembly time and the list of the chosen component with the assembly cell price and delivery time. The lay-out and the moving simulation of the proposed cell are the task of following steps in the planning system.

The assembly operations are described by standardized assembly operations, such are: part inserting, the part screwing, the assembled product testing, product manipulating. The standardized assembly operations for a robotised assembly cell are stored in the knowledge data base. Each assembly operations is by a symbol presented and the assembly minimum and maximum time is given, as well as the component group which can perform the assembly operations. The steps needed to perform the standardized assembly operation are also given.

The assembly operations, performed by robot, are executed by one or more grippers and one or more assembly tools. The selection of grippers is based on the assembly operations, part grasp and dimension demand. For a distinct part dimension one group of grippers is available. The final selection is carried out by the planner. The planner's task is to choose minimum number of grippers and assembly tools considering the lowest price and the shortest assembly time. The planning system supports the planner with stored data and graphical presentations. The changing system is connected with the number of grippers and tools an their sequencing. There are also some different changing systems available, the two parallel, the revolving gripper or the single changing system.

The standardized assembly operations are composed by different assembly steps. The assembly operation can be carried out with the following steps: tools or grippers changing, gripper setting, tools or grippers feeding, part preparation, part feeding and assembly operations execution. The selection of the robot and part preparation equipment and time calculation are based on the determination of assembly operations steps.

Corresponding to the selected parameters, such are, load, number of axes, accuracy etc. some robots are automatically chosen from the component data base and the planner has the opportunity to make the final decisions.

There are some different feeding and orientation equipment available. The selection is based on the relevant parameters.

For assembly time calculation the demands are the assembly operations steps time calculated and stored in the planning database. The assembly operations data are collected from manuals, catalogues and planner's experiences and some times they are also measured. The feeding time is calculated related to the robot moving speed and the path length. Since the lay-out is not present at the moment, the path length is defined by the planner. The calculated time and the cell price are compared to the given goals at the beginning of planning. The planner has the possibility to optimise the non acceptable results in the final plan.

The planning system allowed also the extension of the knowledge base in order that the system can take over more planner's tasks and connection with the layout presentation and simulation system

3.5. Low Cost CAQ Concept

Quality demands under the pressure of international competition can only be satisfied specific for small and medium enterprizes with respect to the standards ISO 9000ff and EN 29000ff.

The described quality ensurance system in the frame of quality control consists of 20 modules. Its implementation started with the most important modules maintenance, goods income, repair/complaint, and test routines including dynamic generation of test orders and test programs. Interfaces to the components PPS, purchase, update system, operational data acquisition, manufacturing control, commercial areas, and many others exist. Implementation and update of the integration of this system in the existing CIM-environment require considerable reduced expenditure because of the use of a logically central data base.

This CAQ-system has been considered at creation time of the existing CIM-system. Therefore the relational database of the CIM-system is a part of the CAQ-database and only had to be extended by special CAQ-relations. The PPS-database is the second part of the CAQ-database. The modular software is arranged in layers and supports the client/server-principle. Therefore introduction step by step is possible.

Optimal use of hardware components determines the use of one central CIM-database-server, of the PPS-database (4 Philips P9000 under UNIX V) , as well as 100% compatible DOS-PCs (AT-286, AT-386). All computers are connected by Ethernet with respect to the ISO/OSI standards. Front-end units like data acquisition terminals and test units communicate via RS485 or RS232C interfaces. Quality data loop back is done by operational data acquisition terminals (production area) and direct manual input (goods income).

3.6. Time acquisition in a low cost CIM/OCA solution

The company under investigation is divided into three strategical units: "Bank-, sawmill-, and industry automation". On the other hand there exist the sections marketing, finance, and technology. It is intended to find a CIM/OCA-solution, which will be realised at first in two steps, because most of the cash-flow and deployment of working power is being achieved in domains of development, construction and software design.

At the first level the introduction of a time acquisition system, and the time acquisition as related to the orders in the sections before the production will take place.

The concept of this OCA-solution has to take into account several frame conditions. On the one hand the working places are spread about three different buildings, which are connected by a LAN. On the other hand the data management will be realised by means of a relational database management system under SQL, in order to obtain an open interface to ensure an expandability of the database for the second level and the integration of the order-related data acquisition in the non production areas.

The OCA contains the connection of heterogeneous hardware components like OCA terminals, compatible PCs (AT-486), HP9000/730 (database server) and one IBM-AS400 platform.

The complexity of the project should be reduced by using standard software in separated modules.

4. SUMMARY

"Low cost" automation is absolutely necessary especially for small and medium sized companies today. Until now only low cost devices for process automation were described and discussed. Because of a strong demand of such companies working in the field of manufacturing, low cost concepts for manufacturing automation - low cost CIM concepts - were shortly discussed. Such a concept could consist on "open" software modules running on simple PC's (e.g. 386 compatibles).These PC's connected by a LAN are superimposed by a host computer. Such a concept is

- easy to adapt for different companies
- a relatively cheap solution
- stepwise realizable
- suitable for small companies.

This concept has been illustrated on three examples.

This work is supported by the Austrian "Forschungsförderungsfonds für die gewerbliche Wirtschaft - FFF."

5. LITERATURE

Kopacek, P. and K.Fronius (1989). *CIM concept for the production of welding transformers*. In Preprints of "INCOM'89", Vol.2, p. 737-740, Madrid 1989.

Kopacek, P. (1989). *CIM for small and medium sized companies*. Research Report, Austrian Ministery for Science and Technology, Vienna (in German).

Albertos, P. and J.Tornero (1989). *Low cost in CIM*. In Preprints of "LCA'89", Vol.I, T123-T127.

Kopacek, P., and N. Girsule: *A "Low cost" Modular CIM Concept for Small Companies*. Preprints of "INCOM'92", Vol.I, p. 188-192, Toronto 1992.

Kopacek, P., A. Frotschnig, and M. Zauner: *CIM for Small Companies*. Preprints of the IFAC Workshop on "Automated Control for Quality and Productivity ACQP'92", Vol.I, p. 35-41, Istanbul 1992.

USING LOW COST PIEZOELECTRIC ULTRASONIC SENSORS FOR 3D MEASUREMENT IN ROBOTS

G. Lindstedt and G. Olsson

Department of Industrial Electrical Engineering and Automation (IEA), Lund Institute of Technology, P.O. Box 118, S-22100 Lund, Sweden

Abstract. The problem of measuring the distance between a robot gripper and an object has been treated. A design, experimental verification and experience of alternate solutions using pulsed ultrasonics are reported. Ultrasonics is often superior to optical methods in a rough robot environment and has a relatively low cost. The developed sensor and methodology offers not only distance measurement capabilities but also space orientation and an object identification among a limited number of possible forms. The feedback potential is apparent.
Contrary to liquid environment, ultrasound in free air is significantly absorbed. This is part of the reasons why several frequencies have been used, 40 kHz for distances up to about one meter and 200 kHz for small distances.
A sensor device has been designed with four 40 kHz transducers mounted around one 200 kHz transducer.
Simple distance measurements are made by a pulse echo method. More advanced measurements use the responses from all the microphones. Also, the angle of a surface can be determined. By changing the frequency between 40 and 200 kHz for the ultrasonic excitation the range of the distance measurements is significantly increased while the accuracy is maintained.
It is not trivial to estimate the front of the reflected echo of the ultrasonic transducers, since the echo is a harmonic sinusoidal oscillation with growing amplitude, corrupted by noise. Particular algorithms have been developed to detect the front. The ultrasonic device requires fast sampling, and a special computer interface has been designed. The system has been built around a laboratory robot platform. Experimental experiences of both object identification and sensor based feedback are described.
Keywords. Ultrasonic sensors, distance measurement, pulsed ultrasonics, sensor feedback, robot sensor.

1. Ultrasonic feedback in robotics

1.1 Motivation and potential

The use of industrial robots in the manufacturing industry is a common technique today and has been so for several years. Nevertheless, robots with more "intelligent" sensor equipment than the built-in joint angle transducers are still rare. This has focused the use of robots on applications such as welding and spray-painting while areas like assembly operations or operations on individually shaped objects (e.g. agricultural products) are seldom robotized. The force-torque sensor is becoming more and more common but its capability is quite limited since it makes the robot act as a blind human working with a pair of pliers. Still many assembly operations can be carried out in this way.

Attempts to make the robot able to see have been rather discouraging. Different types of vision systems have been used in industrial applications but a lot of problems are often encountered. Lenses have to be kept clean, illumination has to be constant for the system to work, the data processing requires much computer power and the loop time in a feedback system becomes long. Still, it is an important step towards robot skillfullness to give the robot an overview of the working environment, both in 2 and in 3 dimensions.

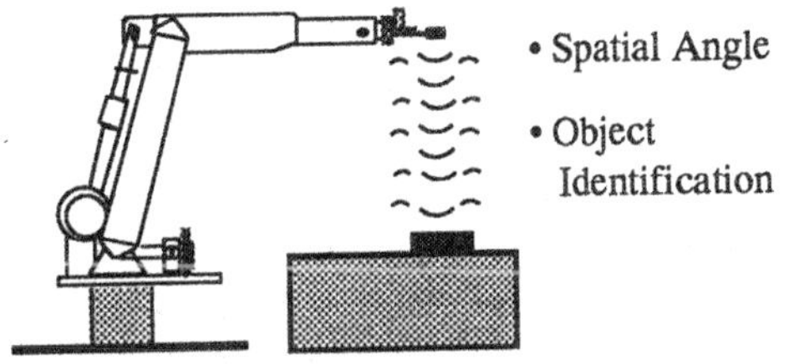

Fig. 1.1 The measurement task

Ultrasonics has proven to be a powerful technique for robot sensing. The pulse-echo method with sound (sonar) have much in common with radar technology. A pulse is transmitted and the received echo as a function of time describes the objects that reflects the wave. This can be used e.g. for simple distance measurements. Ultrasound in air is very robust since it is not influenced by light, smoke, electromagnetic interference or most other disturbance sources. The sensors are inexpensive (5-40 US$) and the total system cost can be kept low.

The reflected pulse echo actually contains a lot more information about the reflecting object than the distance to it. In Fig. 1.2 the echoes from two

objects are shown. A simple comparison reveals that an echo contains a signature from the object shape. If the number of receiving points is increased the angle of the reflecting surface can be measured; two receivers can measure the angle in one plane while three receivers measure a 3D spatial angle.
In summary:

- a lot of relevant information can be extracted;
- the sensors are robust in industrial environments;
- the transducers are inexpensive and cost effective.

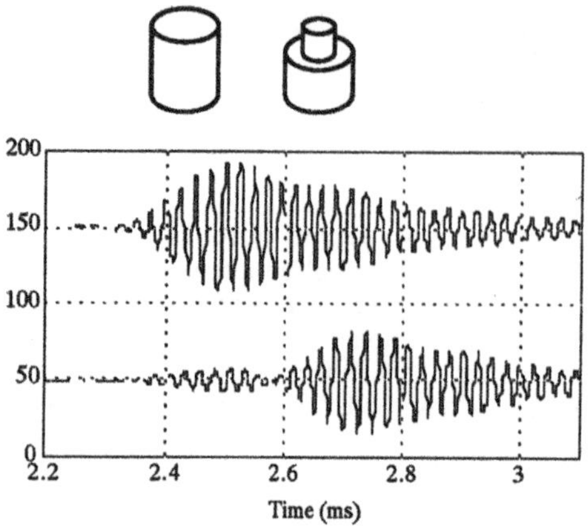

Fig 1.2 Echoes (signatures) from two objects.

1.2 Physical limitations

Ultrasound in liquids and in solid materials has been used for several years in medicine and in mechanics. In liquids and solid materials very little of the ultrasound is normally absorbed. Unfortunately this is not the case with air. For example, a 40 kHz signal can be used realistically to measure at a distance up to a couple of meters, while a 200 kHz signal is more appropriate for small distances (i.e. less than half a meter). The wavelength of a particular signal limits its accuracy. Consequently it has a major influence on the lower limit of the achievable distance.

The temperature influence on the speed of sound in air is the only severe physical factor that has to be considered. In most cases this can be compensated by a simple temperature transducer in the working area. If, however, the sound propagates through zones with large differences in temperature this problem can be difficult to solve.

Many sources of noise in an industrial environment also transmits noise at ultrasonic frequencies. In this case the high absorption of ultrasound in air reduces the influence. If this is not enough some additional shielding with damping materials can be needed. The noise is usually uncorrelated so that repeated measurements and statistical methods can be used. Correlation based methods can also be used as shown in (Holmberg 1992) where the ultrasonic signal is modulated with pink noise.

1.3 Applications in our project

An industrial application presents a number of severe demands on a sensory system. The information has to be accurate within the working area in terms of distances, angles, and topology. It has to be robust to environmental disturbances. The cost must be low and the sensor so small that it can be mounted in or above the robot gripper.

Taking all these items into consideration the following conclusions were made concerning a suitable experimental setup:

- the sensor layout is an array so that 3D measurements can be made;
- more than one frequency is used in order to make wide range measurements;
- only inexpensive and replaceable components may be used;
- piezoelectric transducers are preferred instead of electrostatic ones due to their physical size.

The basic measurements in our project concern distance and angle, but the ultimate goal is to obtain a more complete description of the working area. A major task is to formalize the information from the sensors by using them to their limit. A sensor system for a robot application must not deliver complex information that needs a lot of interpretation. Nevertheless, information for object localisation, object identification and obstacle avoidance is of high importance. The methods for these measurements must accordingly be developed so that the output is available as condensed and usable information. Methods for the basic measurements and object identification have been developed and are presented in this paper.

2. Distance Measurements

A piezo-electric transmitter element can be pulsed in different ways. Since the piezo-electric crystal is a highly resonant system a high amplitude can be achieved by pulsing the crystal with the resonance frequency for a number of periods. This is the method that mainly has been used.

The ideal echo of the sonar pulse reflected from a plane surface looks like the upper curve in in Fig. 1.2. It is apparent that the front of the oscillating signal has to be detected. This is however not a trivial task. Already before the true reflecting echo appears the signal is contaminated with noise due to stochastic disturbances as well as inherent resonant oscillations. The latter are initiated by the cross-talk from the sending transducer, as illustrated by Figure.2.1.

A couple of algorithms for the detection of the echo front have been developed and evaluated. Since the disturbances have the same oscillation frequency as the main echo it is not possible to separate them with any conventional filtering method. These disturbances can be considered having two components, one periodic with the dominating frequency from the transducer and the other a stochastic one. A filter is designed that uses the knowledge of the transducer properties. This makes it possible to subtract the periodic disturbance from the total echo. Figure.2.2 shows how the undesired oscillations have been subtracted from Fig.2.1. The stochastic part is further reduced by additional low-pass filtering.

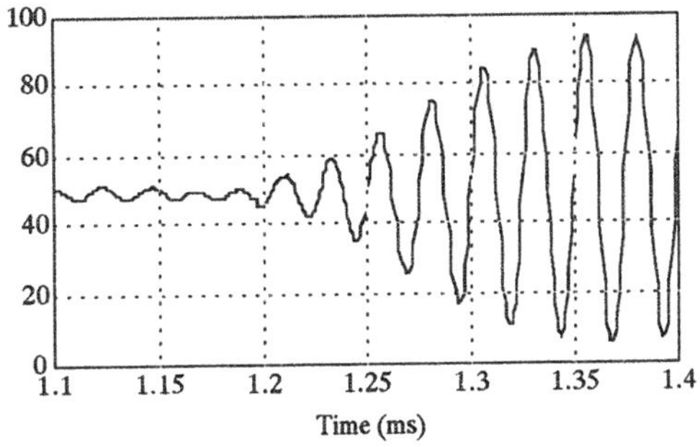

Fig. 2.1 Contaminated signal.

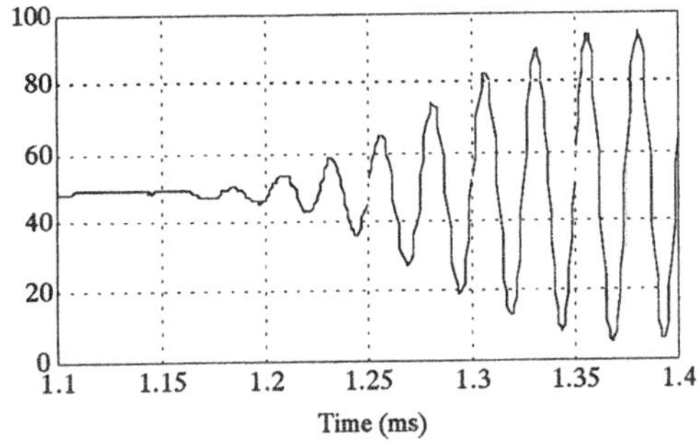

Fig. 2.2 Filtered signal.

The approximate front of the echo can now be detected. The data from the complete echo is extrapolated in order to find the front. However, this extrapolation uses the knowledge of the echo phases and frequencies. This can further increase the accuracy of the wavefront reconstruction.

3. Angular measurements

This measurement is based on the fact that an ultrasonic wave that is reflected at a sufficiently large plane surface proceeds as a wave parallel with this plane. Consequently, a configuration as the one shown in Fig.3.1 with one transmitting element and two receiving ones can be used to measure the inclination of a surface. By measuring the time from transmission of a pulse to reception at receivers 1 and 2 the distance r_0 and the angle ß can be calculated.

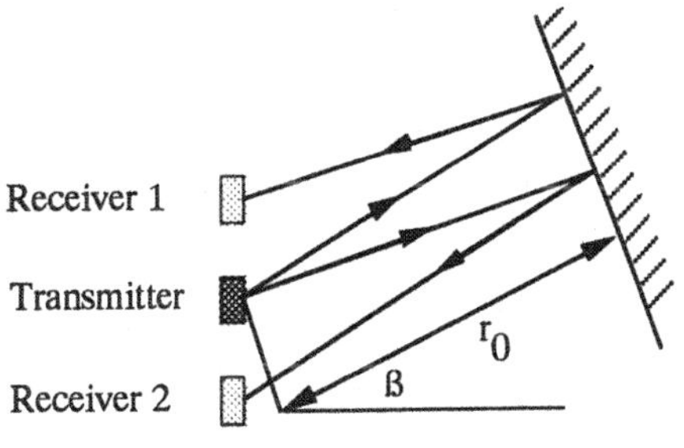

Fig. 3.1 Angle measurement.

If the phase angle between signals from two transducers are measured when the sensor plane is parallel with a base plane surface a calibration value can be stored. Using this value small deviations can be measured with high accuracy since the phase angle difference is easily detected. This measurement can be made with a constant ultrasonic signal. An application of this technique in assembly of fog lamps for cars is found in Klöör, Jonsson and Wernersson (1988).

We have combined this sensitive method with a measurement of the total number of periods , which is possible in the pulsed ultrasonic case, and obtained a high precision angle measurement.

Even if the previous discussion has been referring to a two-dimensional case, it is possible to expand the method to the measurement of spatial angles by placing receivers in another direction, see further section 4. Expressions for the angles in different receiver configurations are derived in Lindstedt (1992).

4. Object identification

Since an object identification usually is of interest at shorter distances this method has been developed and implemented for the 200 kHz sensor only. A short distance also makes the method more dependable. The initial conditions for the identification are that the sensor array is positioned over the object that is to be identified and that the height over the working area is constant during the data acquisition phase. This means that the position of the sensor vs. the object should be about the same in the learning phase and in the identification phase.

To start a measurement the 200 kHz piezoelectric sensor is pulsed. After a short delay the sampling of the incoming echo is started. The very same sensor is used both for sending and receiving. All the signal amplitudes, $y(kh)$, are sampled and stored. For an identification analysis it is unsuitable to use the whole data set. If a subset of the samples is selected by defining a time window, the number of sample points will be reduced. This time window can alternatively be expressed as a distance window.

Let us study a simple case where the working surface is smooth and equal to the xy-plane, the maximum object height z_0 and the sensor position z_s. We can define a distance window W as those points between positions slightly above the object and below the xy-plane, i.e.

$$W = \left[0.9\cdot(z_s - z_o)\ ;\ 1.1\cdot z_s\right]$$

Further reduction of the data amount can of course be done by grouping the samples. There are physically natural ways of grouping the data. Because of this we first study the following steps in the data treatment.

An ultrasonic echo is never 100% reproducible. Spikes and disturbances interfere with the signal during the sampling. The position of the sensor vs. the object is never exactly the same if the robot has moved between the measurements. It is obvious that statistical methods have to be used for the data comparison.

Since a sample-by-sample comparison is not meaningful due to the problems described above we have chosen to make a shape comparison. An envelope detection could be a possibility but since a noise eliminating peak detection requires interpolation this is computationally inefficient.

The chosen method is an energy oriented one. An rms-value of the incoming samples can easily be calculated. This means that the influence of large signal amplitudes (real echoes) is greater than small ones (often noise). The number of samples in the mean, n_m, must be chosen with care if low sensitivity for the phase angle in the ultrasonic signal is required. An n_m close to the number of samples in one half period in the signal or a multiple of this gives a low sensitivity. This is in fact a quadratic moving average (MA) filter.

The MA filter can, by advancing one sample at a time, be used to generate a smooth vector of the same length as the original data. It is, however, an advantage to reduce the amount of data. In our method we have been advancing n_m (where n_m now denotes the number of samples in *one period* of the signal) steps at a time with good results. These considerations gives us a new vector:

$$S = [r_0, r_1, r_2, \ldots\ r_{n-1}]$$

where

$$r_p = \sqrt{\sum_{k=0}^{n_m-1} y^2(n_m \cdot p + k)}$$

and *y(kh)* is the sample at time *kh*.

The *S* vector is a compact description of the object signature in terms of energy vs. time in the ultrasonic echo. An illustration of this energy function is found in Fig.4.1. For a window of about 10 cm using the 200 kHz signal a vector of about 60 values is produced.

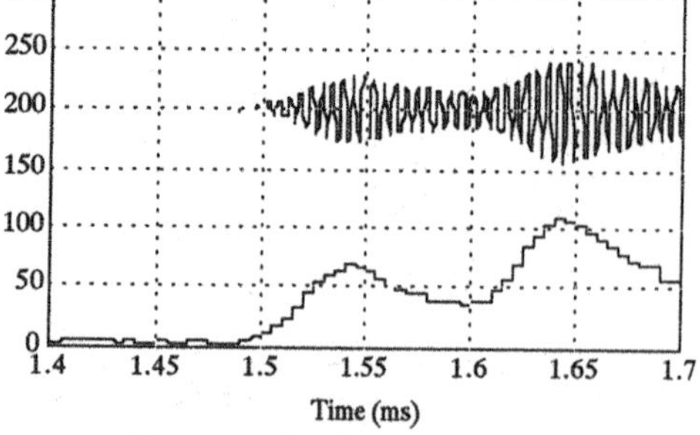

Fig. 4.1 Creation of signature function S.

A comparison of two echoes, S_a and S_b can now be made by calculating the sum:

$$Q_{ab} = \sum_{p=0}^{n-1} (r_{pa} - r_{pb})^2$$

where r_{pa} is the element r_p in vector S_a and r_{pb} is the element r_p in vector S_b. The lower the sum the better the fitting. It is now possible to let the interpreting computer store the signature vectors for a limited number of objects. From this data-base an unidentified object can be identified as the one with the smallest Q-sum. After some experience an absolute limit for acceptance can be drawn. In Fig.4.2 the energy functions from an identification experiment with 3 object in the data-base are shown.

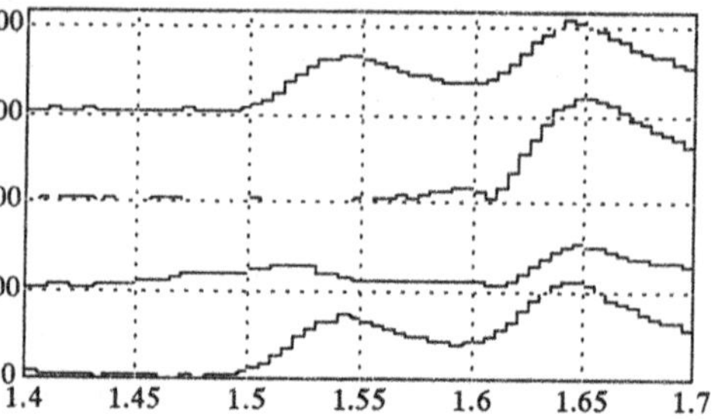

Fig 4.2 Object identification. (Offset added to make the curves separable.) Upper curve: unknown signature. Following: database for identification.

5. Sensor configuration

To be able to make the measurements described previously and still pay respect to the constraints mentioned in section 1 a special sensor unit has been designed. The unit includes the following items:

- ultrasonic transducer elements;
- transmission pulse generator;
- sensor amplifier;
- fast A/D-conversion;
- intermediate data storage.

The layout of the sensor matrix must be made with great care to make it possible to perform all measurements from one unit.

5.1 Sensor matrix layout

When working with low cost commercial sensors in the lower frequency range (about 40kHz) it soon becomes evident that the cross-talk from the transmitter to the receiver is a problem. Receiving with the transmitting sensor after a short time delay is usually not possible. These elements are easy to activate but once oscillating their low damping gives every echo a long tail. This requires a long distance between the transmitting and receiving sensor to avoid large pulses originating in cross-talk from the transmitted pulse. These false echoes can otherwise make it impossible to measure at short distances. The accuracy in angular measurements also requires that the sensor elements are separated. Several sensor matrix layouts have been tested within the project. The chosen one is shown in Fig.5.1.

There are several major advantages of this layout. Redundant information can be obtained when measuring spatial angles through alternation of the sending element (4 possibilities). There is a long distance between the elements while the overall dimensions are kept small. Furthermore, the short distance sensor, 200 kHz, is well centred and its position does not increase the overall sensor unit dimensions.

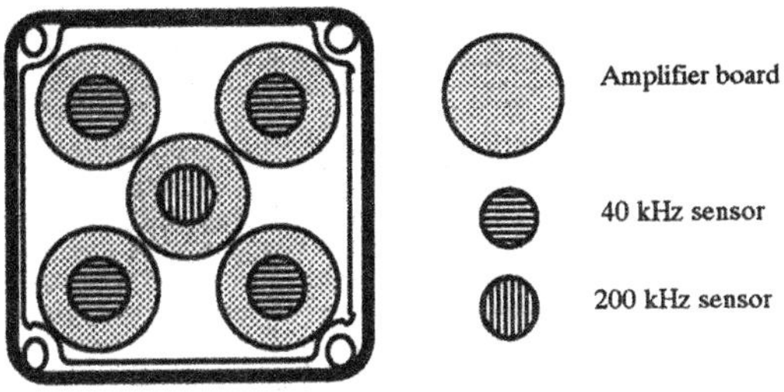

Fig. 5.1 Sensor layout

A more symmetrical layout of the sensors involved in the spatial angle measurement gives simpler expressions in the calculations but this is considered to be of less importance. No explicit method has been encountered that can calculate the two angles and the distance. Instead a numerical, fast converging method based on matrix Newton-Raphson iterations is used. Combined with explicit formulas that generate good initial values a six digit accuracy is always reached within 3 iterations (typically 1 or 2). Further details about this method is found in Lindstedt (1992).

5.2 Sensor unit design

The fact that one 40 kHz transducer at a time is used as a transmitter makes it necessary to supply the sensor system with a relay multiplexer.

The key decision when designing the hardware configuration is where to locate the different sensor sub-units. The unit that is to be mounted on the robot, the sensor unit, must be as small as possible. This is achieved by mounting only the transducers, the preamplifiers and the relay multiplexer there.

Since the signals from the sensors have got a very low amplitude they have to be amplified about 1000-5000 times. The original sensor signal is of course easily disturbed and ought to be amplified before it proceeds through any long cable. Because of this preamplifers are mounted directly on the sensor elements. These amplifiers have to be small and this is achieved by using SMD (surface mount device) techniques for the electronics. The amplification is about 1000 times. By using third order low pass filters as part of the amplifiers, the noise above the sensor frequency is highly reduced.

By using SMD technique the relay multiplexer can also be mounted within the sensor unit. Since the wire carrying the activating pulse is a large disturbance-source it has to be separately shielded. Using a relay multiplexer in the sensor unit makes it possible to use only two cables to connect the unit to the other equipment. One cable contains the activating pulse while the other contains both the power supply, the control signals and the measured signals. The external dimensions of the sensor unit are 100x100x90 mm.

5.3 Autonomous sampling unit[1]

To connect the sensor unit to the controlling computer a rack mounted unit, have been designed. This unit contains adjustable input amplifiers for the sensor signals, fast A/D-converters, RAM-memory, sampling control logic and the transmission pulse generator.

The sampling unit is started from the computer at the same time as the pulse generator is activated. To be able to meet the demands for fast sampling this unit is designed to work autonomously without influence from the computer. Once the sampling is started, the computer only waits for a *ready* signal.

Sampling is made using 4 channels in parallel with a programmable speed of max 2.5 Megasamples per second in all channels. The resolution is 8 bits and the data sequence, of programmable length, is stored in an 8 kbyte memory for each channel.

The sampling equipment only needs 4 channels since measurements can't be made at both 40 kHz and 200 kHz at the same time.

5.4 Computer equipment

The computer platform for the experiments, (please observe footnote), is VME-bus based and made by PEP Modular Computers®. The used processor is a Motorola® 68040. The programs are written in C and assembler and are running under the operating system OS–9.

This computer platform makes it possible to study the experimental results comfortably on a graphic screen and to save data for interchange with other evaluation environments e.g. the analysis package MATLAB®.

The transfer of data from the sampling unit is done via a fast parallel bus and this architecture has even made it possible to perform closed loop experiments where the computer controls the robot movements.

6. Experimental experience

Experiments have been done with the methods described in the sections 2-4. Some of these results have previously been described in Lindstedt and Kowalew (1991), where also experiences from transducer mounting is described.

<u>Distance measurement:</u> In the system both 40 kHz and 200 kHz transducers have been used. For the former sensors it has been possible to achieve an absolute precision better than 1-2 mm. The corresponding values for the 200 kHz device is about 0.5 mm. The relative accuracy is so good that we have not been able to detect corresponding errors.

---

1 Note that this part of the experimental setup is, together with the computer, not intended to be a model for an industrially applicable product. It rather serves as a flexible experimental platform for principal solutions and development of algorithms.

A closed loop experiment where the robot follows a moving surface in one dimension has been made. The total distance measuring range using both frequencies are from 6 cm up to more than one meter.

Angular measurements: These measurements are only possible to make with the 40 kHz sensors with the current sensor array design. The accuracy of the measurements towards large surfaces is better than 1°. Angles up to about 20°-30° can be measured depending of the distance.

Object identification: The 200 kHz transducer is used for the identification. About 20 arbitrarily chosen objects, mainly normal machine parts, have been used. Some of these objects have geometrical similarities like surfaces at the same height, or surfaces of the same size at different heights. Single measurements have been made both when learning a new object and at the identification stage. Repeated measurements could of course improve the security. Under normal conditions the probability for correct identification is about 24 out of 25 or 96%. The errors only occur for objects with apparent geometrical similarities. A slight sideway displacement of about 2-4 mm of the object still give a very good identification.

All presented measurements are made in a laboratory environment with little disturbances. However, additional disturbances have been purposefully added in order to test the method. The results have been encouraging.

7. Industrial significance

The equipment used for the presented measurements is, as earlier mentioned, not directly applicable in an industrial measurement system. This is not, and have not been, the goal for the project. The goal is to develop methods that together with a feasible hardware can be used in industrial robots or other flexible manufacturing systems. The sensor unit however is developed so that it can be a part of an industrial system.

To convert the measuring system into a usable product another important component is needed, namely a communication language to present the information received from the sensors. This language must be able to describe as much as possible of the information collected by the sensors. Still the language must be easy to use. This challenge leads to several problems and possibilities.

A suggestion for a possible hardware architecture for the final product is shown in Fig.7.1. The sampling is made in the same way as in the experimental unit. A dual ported memory gives the slave processors access to the data as soon as one sampling is completed. If it is desirable that the processors can work while a new echo is sampled, two alternating memories can be used for each processor. The slave processors probably do not have to be of extremely high performance type since most signal processing can be made in parallel in all channels. No floating point arithmetics are needed either.

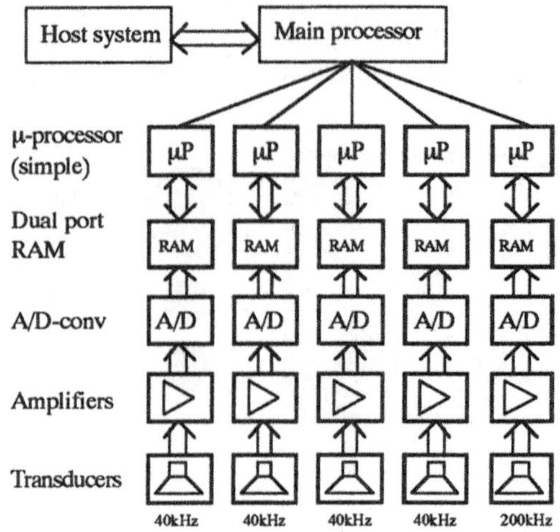

Fig 7.1 Suggested hardware architecture.

The master processor orders the slaves to do specific processing on the incoming data. The specified results are then collected at the slaves. A high communication speed between the master and the slaves is needed. Floating point operations will be executed by the master and a hardware support for this might be necessary. The master also communicates with the superior machine control-system.

The system cost for a final product is difficult to estimate at this stage. The number of units produced highly influences the price, and the development costs are unpredictable. What we can state is that no expensive hardware has to be included. The transducers cost less than 100 US$, the A/D-converters less than 30 US$, the memories less than 20 US$ and the microprocessors including their peripherals can be estimated to less than 200 US$. This gives us a grand total of less than 500 US$ for the complete hardware. Compared to other sensor systems this seems quite feasible and the economics seem to be in favour for a system of this type.

References

Holmberg, P. (1992). Applied signal processing in measuring applications - towards sensor controlled robots. (Licenciate thesis.) Dept. of Physics and Measurement technology. Linköping University. Sweden. Paper I.

Klöör, P., T. Jonsson and Å. Wernersson (1988). Monitoring the assembly sequence using pulsed coherent sonar. 9th Int. Conf. Assembly Automation, London, March 1988.

Lindstedt, G., K. Kowalew (1991). Distance Measurement in Robotics Using Ultrasonic Sensors. Report TEIE-7032, IEA, Lund Institute of Technology. Sweden.

Lindstedt, G. (1992). Computation of spatial angles from ultrasonic measurements. Report TEIE-7055, IEA, Lund Institute of Technology. Sweden.

A SIMPLE LINK BETWEEN POLE SHIFTING AND PHASE MARGIN

A. Rachid

Ecole Centrale de Lyon, BP 163, 69131 Ecully Cedex, France

Abstract : In this paper, we give simple relations relating the phase margin and pole shifting for single input systems. Some results are also presented in the context of optimal quadratic pole shifting.

Keywords : single input system ; continuous system ; pole shifting ; phase margin ; LQ Regulator

I) Introduction

This note deals with pole shifting through state feedback control for single input continuous systems. In particular, we present simple relations giving the phase margin and the cut-off pulsation in the case of one single pole shifting. The proposed results provide a very easy way to control SISO systems and to ensure some robustness properties.

When the feedback control is an LQR (Linear Quadratic Regulator), a new necessary condition relating closed loop poles and open loop ones is given and used to recapture the well known phase margin robustness property [1].

Besides their understanding simplicity, the developments addressed herein cope with industriel requirements and provide interesting insight into LQ and pole placement approaches.

In the following, we deal with a continuous time controllable system described by :

$$\begin{cases} \mathring{x} = Ax + bu \\ y = Cx \end{cases} \quad (1)$$

where x is the $n \times 1$ state vector and u the scalar input variable, under the state feedback control

$$u = -kx \quad (2)$$

so that the closed loop system is

$$\mathring{x} = (A - bk)x \quad (3)$$

Next, we define the following TFs (Transfer Function) :

$$H(s) = C(sI - A)^{-1}b$$

is the TF of the open loop system (1);

$$F(s) = C(sI - A + bk)^{-1}b$$

is the TF of closed loop system;

$$T(s) = k(sI - A)^{-1}b$$

is the TF such that $1 + T(s)$ is the return difference.

The open-loop and closed-loop characteristic polynomials h(s) and f(s) are given by :

$$h(s) = \det(sI - A) = \prod_{i=1}^{n} (s - \lambda_i) \quad (4)$$

$$f(s) = \det(sI - A + bk) = \prod_{i=1}^{n} (s - \sigma_i) \quad (5)$$

The λ_i's are the open-loop poles and the σ_i's are the closed-loop ones.

II/ Main results

With the above notations, we have the well known relation :

$$1 + T(s) = \frac{f(s)}{h(s)} \quad (6)$$

If the feedback law $u = -kx$ is designed in order to shift one single pole λ to a new stable position σ ($\neq \lambda$), the remaining poles being unchanged, then (6) gives :

$$1 + T(s) = \frac{s - \sigma}{s - \lambda} \qquad (7)$$

Of course, it is implicitly assumed that this regulator stabilizes the system (1) or in other words, λ is the only possible unstable pole of system (1) and also that both λ and σ are (obviously) real. An interesting question is to know how is this simple pole placement related to stability margins. In what follows, we give quantitative results to answer this.

A necessary condition for T(s) to present a finite phase margin φ is the existence of a gain crosssover pulsation ω_c such that $|T(j\omega_c)| = 1$ which gives using (7) :

$$\omega_c^2 = \sigma(\sigma - 2\lambda) \qquad (8)$$

The phase margin φ can be introduced using the equality :

$$(1 + T(j\omega_c))\,(1 + T(-j\omega_c)) = 2(1 - \cos\varphi).$$

In the case of a single pole shifting, (8) gives using (7) :

$$\frac{\omega_c^2 + \sigma^2}{\omega_c^2 + \lambda^2} = 2(1 - \cos\varphi) \qquad (9)$$

which yields by (8) :

$$\cos\varphi = \frac{\lambda}{\lambda - \sigma} \qquad (10)$$

To summarize, we state the following :

Proposition 1

When shifting one real single pole λ to a stable real position σ ($\sigma < 0$) and preserving other poles using any state feedback law, we have :

i) If $\sigma \geq 2\lambda$ then $\varphi \geq 180°$, the equality $\varphi = 180°$ occuring for $\sigma = 2\lambda$.

ii) If $\sigma < 2\lambda$ then φ is finite and given by (10); the corresponding gain crosssover pulsation is obtained from (8).

As a consequence of this proposition and with the same notations, one can derive the following :

Proposition 2

a) If λ is unstable then $\varphi \in (0, 90°]$, $\forall\ \sigma < 0$.

b) If λ is stable then $\varphi > 90°$, $\forall\ \sigma < 0$.

Proof

The result a) follows from proposition 1 and the facts that $\lambda > 0$ and $\sigma < 0$.

To prove b), we consider the two possible cases $\sigma \geq 2\lambda$, $\sigma < 2\lambda$. If $\sigma \geq 2\lambda$, the result is obvious from the proposition 1 (part i).

If $\sigma < 2\lambda$, then $\sigma < \lambda$ as $\lambda < 0$. Therefore and using (10) one gets $\cos\varphi < 0$, hence the result.

Proposition 3

- For a stable λ , the phase margin decreases as long as the modulus of σ increases.
- For an unstable λ, the phase margin increases with the modulus of σ.

Proof

This is a consequence of (10) and of propositions 1 and 2.

Detailed calculations are left to the reader.

This last conclusion contradicts the popular fact that robustness can be enhanced by speeding the system dynamics ([4]).

Remark 1

In the special case where A is *diagonalizable*, the above propositions can be recaptured by direct calculations on a diagonal canonical representation of the given system. In fact, if only one pole is to be moved, the diagonal form allows the interpretation of the state feedback regulator as a simple proportional feedback applied to a first order system. This remark cannot be made for a non diagonalizable form such as Jordan forms.

LQ optimal pole shifting case

It is possible to use the LQR approach to realize the above pole shifting regulator in which case one gets $\varphi \geq 60°$ [1]. In the following, the focus is made on the LQR for which we derive a necessary condition relating closed loop poles to open loop ones. When specializing to single pole placement, it is shown that this condition implies the well known $\varphi \geq 60°$.

So let us consider system (1) under an LQR minimizing the performance index :

$$J = \int_0^{\infty} \left(x^T Q x + u^2\right) dt \qquad (11)$$

where Q is a non-negative definite matrix and which yields to the control law :

$$u = -b^T P x \qquad (12)$$

P being the unique positive definite solution of the ARE (Algebraic Ricatti Equation) :

$$A^T P + PA - Pbb^T P + Q = 0 \qquad (13)$$

The frequency interpretation of the LQR yields to the following [1] :

$$(1 + T(s))\,(1 + T(-s)) = 1 + G^*(s)\,Q\,G(s) \qquad (14)$$

where

$$G(s) = (sI - A)^{-1} b \tag{15}$$

Combining (6) and (14), one obtains :

$$\frac{f(s)\ f(-s)}{h(s)\ h(-s)} = 1 + G^*(s)\ Q\ G(s) \tag{16}$$

This relation can be rewritten for $s = j\omega$:

$$\frac{f(s)\ f(-s) - h(s)\ h(-s)}{h(s)\ h(-s)} = G^*(s)\ Q\ G(s).$$

Multiplying each side of this equality by ω^2 and taking the limit when ω tends towards infinity yields :

$$\sum_{i=1}^{n} \sigma_i^2 - \sum_{i=1}^{n} \lambda_i^2 = b^T Q b \tag{17}$$

Remark 2

a) Relation (17) is a new formula which is valid for a general controllable single input system {A, b}. Notice that A doesn't appear explicitely in this relation.

b) Condition (17) is a generalization of the result proposed in [2], in the case of single input system.

When only one real pole λ is shifted to σ, (17) becomes :

$$\sigma^2 - \lambda^2 = b^T Q b \tag{19}$$

The pole shifting in the context of LQR can be achieved using, for instance, the algorithm in [3].

It can be shown that the necessary condition (19) implies the well known property that the LQR provides a phase margin of at least 60° [1]. In fact (19) with (9) gives :

$$\frac{\omega_c^2 + \lambda^2 + b^T Q b}{\omega_c^2 + \lambda^2} = 2(1 - \cos\varphi) \tag{20}$$

hence $2(1 - \cos\varphi) \geq 1$ equivalent to $\varphi \geq 60°$.

Remark 3

This result can be recaptured without using (17) (which is interesting since (17) is not standard). In fact, from proposition 2, the only case to be considered is with an unstable λ *i.e.* $\lambda > 0$. Next, as we are dealing with the LQR, one has $|\sigma| > |\lambda|$ which yields $\lambda - \sigma > 2\lambda$ or $\frac{\lambda}{\lambda - \sigma} < \frac{1}{2}$; with (10) one can see that $\varphi \geq 60°$.

The above obtained propositions can be reformulated in the context of an LQR pole shifting as follows :

Proposition 4

When shifting one real single pole λ to a new position σ ($\sigma < 0$) and preserving other poles using LQR state feedback law, we have :

i) If $\sigma \geq 2\lambda$, then $\varphi \geq 180°$.

ii) If $\sigma < 2\lambda$, then φ is finite and given by (20).

iii) If λ is unstable then $\varphi \in [60°, 90°]$, $\forall\ \sigma < 0$.

iv) If λ is stable then $\varphi > 90°$, $\forall\ \sigma < 0$.

Remark 4

a) The above developments can also be considered for a pole shifting which moves two poles of the given system while keeping the others unchanged. This case is a little more difficult since one has to consider the shifting of complexe conjugate or real poles.

b) Similar results can be obtained for discrete systems.

III/ CONCLUSION

Necessary conditions relating closed loop poles to open loop ones have been given for a single input continuous system controlled by an LQR. Simple relations involving the regulator phase margin and the cut-off pulsation have been obtained in the case of one single pole shifting. It is believed that the above results provide valuable a priori knowledge for regultors synthesis due to their simplicity. Notice that the above developments can be similarly performed for the gain margin.

REFERENCES

[1] ANDERSON B.D.O & J.B. MOORE, 1971: Linear Optimal Control. Englewood Cliffs, New Jersey.

[2] AMIN M.H., 1984 : Further comments on ' A necessary condition for optimization in the frequency domain' and on ' optimization and pole placement for a single input controllable system' Correspondance. Intern. Journal of Control, Vol. 40, N°4, pp. 863-865.

[3] AMIN M.H., 1985 : Optimal Pole shifting for continuous multivariable linear systems. Intern. Journal of Control, Vol. 41, N°1.

[4] LUNZE J., 1989 : Robust Multivariable Robust Control. Prentice Hall Intern.

A DESIGN OF TWO STEPWISE OPTIMAL CONTROLLER FOLLOWING THE REFERENCE MODEL OUTPUT

Y. Yamane* and P.N. Nikiforuk**

**Ashikaga Institute of Technology, Ashikaga, Tochigi Pref., Japan*
***University of Saskatchewan, Saskatoon, Saskatchewan, Canada*

Abstract
This paper exhibits some following features.
(1)Generalized model matching controller has feedback and feedforward gains in its structure, which keeps the resultant transfer function of the plant equal one of the reference model
(2)Two different cost functions are chosen to switch gains of the controller at a time when the deadbeat observer finishes the tasks. In the first transient period a choice of the cost function with a weighting coefficient concerning the control input being positive eventually suppresses the plant input to maintain a pure feedforward controller as well as output error regulating. In the last stable period assignment of the cost function with a zero coeffcient of input term causes the output error to approach to zero in the fastest time.

Keywords
feedback control, feedforward control, optimal control ,deadbeat control,model matching, reference model

1. Introduction

In a class of optimal tracking control problem the configuration of controller structure needs to solve a Riccati equation which contains reference input signals (Athans, 1966). It is not easy to express explicitly the controller structure of how the controller relates to the reference input. By the way the model matching technique is an useful and a simple approach to design a linear plant control system so that the transfer function of the resultant control system coincides with those of the reference model (Wolovich, 1974). Here once the desired reference model is assigned, the controller can be implemented in a explicit form including reference model inputs and stability of output error between the plant and the reference model is dependent not only on the poles of controller but also on the poles of reference model. Then this fact means that the reference model faces a severe limit on its choice of stability when it has to have slow decaying poles from a principle of design. Because inherently the output error must be regulated to zero as quickly as possible. As to the arguments of improving stability of error equation, the conventional controller with additional feedforward compensator in use of the reference model states named a generalized controller provide a striking tool because the error equation is able to have arbitrary poles independent on the assigned poles of the reference model (Yamane, 1991). That is, the controller composed of state feedback and feedforward gains has one free parameter vector to prescribe any location of stable region. By any suitable choice of the parameters several different kinds of controller can be constructed not only holding the model matching condition but also allowing any assignment of stable poles of the output error response. At the same time it is necessary and important to avoid magnitude of plant control input as well as one of output error from increasing urgently . From the viewpoint the optimal controller was designed in a framework of a kind of optimal regulation problem (Yamane,1992).

In this paper the both of weighting coefficients of cost function concerning plant output difference and plant input one are considered to be changable at a time when the plant observer finishes its convergence. In the first half part of operation it is natural to place the both of weighting coefficients of cost function positive numbers. Eventually in the transient period the plant control input is forced to the pure feedforward controller signal as well as the plant output is regulated to the reference model output. In the last half part of operation a complete achievement of the control effects is preferable to assigning the zero weighting coefficient of cost function concerning the plant input difference term. Resultantly the deadbeat control input fulfils the control objective.

2. Controller configuration with two free parameters

A n-th order plant with nonminimum phase, Σ_1 and a n-th order reference model, Σ_2, are written in controllable forms as follows,

Σ_J (J=1,2):

$$x_J(i+1)=A_J x_J(i)+bu_J(i)$$

$$y_J(i)=c_J^T x_J(i) \qquad i=0,1,2,3, \tag{1}$$

where $A_J=\begin{bmatrix} 0 & I \\ & a_J^T \end{bmatrix}$, $b=\begin{bmatrix} 0 \\ 1 \end{bmatrix}$

The parameters a_J and c_J are n dimensional vectors and $m_1 c_1=c_2$. It is assumed that the plant system is observable.
The z-transforms of equations (1) are expressed by

$$y_J(z)=\alpha_J r_J(z) p_J^{-1}(z) u_J(z) \qquad (2)$$
$$J=1,2$$

where α_J is a nonzero number, $r_1=r_2$ is unstable monic and p_2 is a stable monic polynominal of n-th degree.

The problem considered here is to design a controller so that the output difference between $y_1(i)$ and $y_2(i)$ is regulated to go to zero as quickly as possible.

The controller is given in a form,

$$u_1(i)=-k_1^T\hat{x}_1(i)+k_2^T x_2(i)+mu_2(i) \tag{3}$$

The plant state estimates can be constructed through a full order observer

$$\hat{x}_1(i+1)=A_1\hat{x}_1(i)+bu_1(i)+h\{y_1(i)-\hat{y}_1(i)\}$$
$$\hat{y}_1(i)=c_1^T\hat{x}_1(i) \qquad ,i=0,1,2,3 \tag{4}$$

An augmented plant equation combined with the reference model applied by the controller is given by

$$\eta(i+1)=\Lambda\eta(i)+\gamma u_2(i)$$
$$\xi(i)=\pi^T\eta(i) \tag{5}$$

where $\eta(i)=\{x_1^T(i), x_2^T(i), e^T(i)\}^T$

$$e(i)=\hat{x}_1(i)-x_1(i)$$

$$\xi(i)=y_2(i)-y_1(i)$$

$$\Lambda=\begin{bmatrix} A_1-bk_1^T & bk_2^T & -bk_1^T \\ 0 & A_2 & 0 \\ 0 & 0 & A_1-hc_1^T \end{bmatrix}$$

$$\pi=[\ -c_1^T\ ,\ c_2^T\ ,\ 0^T\ ]^T$$

$$\gamma=[\ mb^T\ ,\ b^T\ ,\ 0^T\ ]^T$$

A necessary and sufficient condition that $\xi(i)$ always converges to zero for any $u_2(i)$ means that the zero state response of Eq(5) is identical to be zero as well as the zero input response of Eq(5) is stable. From this statement we have

$$\pi^T(zI-\Lambda)^{-1}\gamma=0 \tag{6}$$

Putting the largest coefficient in z powers into zero, we get

$$m=\alpha_2/\alpha_1=m_1 \tag{7}$$

Substituting Eq(7) into Eq(6) we have a relation between k_1 and k_2,

$$a_1-k_1-a_2+k_2/m=0 \tag{8}$$

Eqs (7) and (8) mean the model matching condition of generalized controller with two free parameters.
Then the output of Eq(5) with regard of Eqs (7)and (8) is written by

$$\xi(z)=c_2^T(zI-A_1+bk_1^T)^{-1}z\{x_2(0)-x_1(0)/m$$
$$+bk_1^T(zI-A_1+hc_1^T)^{-1}e(0)/m\} \tag{9}$$

From Eq(9) the best choice of h is given by the deadbeat type observer gain to hold

$$(A_1-hc_1^T)^n=0 \tag{10}$$

Regarding of selection of k_1 we can propose different type gain.

1. When $k_1=0$ the controller means a pure feedforward controller.

$$u_1^{①}(i)=m(a_2-a_1)^T x_2(i)+mu_2(i) \tag{11}$$

However then the closed loop system of Eq(5) does not guarantee its stability.

2. When $k_2=0$ the controller results in to be a pure feedback controller (Yamane, 1990)

$$u_1^{②}(i)=(a_1-a_2)^T\hat{x}_1(i)+mu_2(i) \tag{12}$$

This form can be changed into well known conventional one in frequency domain such that $-k_1^T\hat{x}_1(i)$ is composed of plant input signals and plant output signals with a solution of Diophantine equation. Then we have $k_1=a_1-a_2$.

3. When $k_1=a_1$ the controller is a deadbeat controller,

$$u_1^{(3)}(i)=-a_1^T x_1(i)+ma_2^T x_2(i)+mu_2(i) \qquad (13)$$

which forces $\xi(i)$ to go to zero in 2n discrete sampling time.
This controller type can be also obtained by application of the optimal regulation problem with r=0 cost functional described below (Yamane, 1992).

Here a class of optimal regulation problem is considered to avoid magnitude of plant control inputs from increasing urgently (Yamane 1992).

3. Regular Optimal Controller

The closed loop plant applied by the controller results in

$$x_1(i+1)=(A_1-bk_1^T)x_1(i)+b\{k_2^T x_2(i)$$

$$+mu_2(i)\} \quad +bk_1^T\{x_1(i)-\hat{x}_1(i)\} \qquad (14)$$

The reference model can be rewritten by

$$x_2(i+1)=(A_2-bk_2^T/m)x_2(i)$$

$$+b\{k_2^T x_2(i)+mu_2(i)\}/m \qquad (15)$$

Now we define $\delta x(i)$ and $\delta u(i)$ by

$$\delta x(i)=x_1(i)-mx_2(i) \qquad (16)$$

$$\delta u(i)=u_1(i)-u_1^{(1)}(i) \qquad (17)$$

It is required to regulate both of $\delta x(i)$ and $\delta u(i)$ to zero.
When the gains of controller satisfy the model matching condition of Eq(8), the controller can be changed into

$$u_1(i)=-k_1^T\{x_1(i)+e(i)\}+mk_1^T x_2(i)$$

$$-mk_1^T x_2(i)+k_2^T x_2(i)+mu_2(i)$$

$$=m(a_2-a_1)^T x_2(i)+mu_2(i)$$

$$-k_1^T\{\delta x(i)+e(i)\} \qquad (18)$$

From Eqs (17) and (18) one obtains a next plant input deviation equation

$$\delta u(i)=-k_1^T\{\delta x(i)+e(i)\} \qquad (19)$$

Furthermore subtracting Eq(12) from Eq(11) and instituting Eq(19) we have a plant state deviation equation

$$\delta x(i+1)=(A_1-bk_1^T)\delta x(i)-bk_1^T e(i)$$

$$=A_1\delta x(i)+b\delta u(i) \qquad (20)$$

$$\xi(i)=-c_1^T\delta x(i)$$

The input and state equations of Eqs (19) and (20) are correspondent to ones when instituting $k_2=0$ and m=0 into Eqs (3) and (14).

Here we introduce the optimal regulation problem to determine the optimal gain k_1 of the plant deviation system of Eqs (19) and (20).
A cost functional with quadratic form ($0\le Q$ and $r>0$) is given by

$$J=\sum_{i=0}^{\infty}\{(y_2(i)-y_1(i))^2+r(u_1(i)-u_1^{(1)}(i))^2\}$$

In general, we have

$$J=\sum_{i=0}^{\infty}\{\delta x^T(i)Q\delta x(i)+r\delta u^2(i)\} \qquad (21)$$

The optimal regulation problem with r>0 in the cost function is called a regular optimal control problem.
The regular optimal gain k_1 is expressed as

$$k_1=(r+b^T Pb)^{-1}b^T PA_1 \qquad (22)$$

where P means a solution of Riccati equation

$$P=A_1^T PA_1+Q-A_1^T Pb(r+b^T Pb)^{-1}b^T PA_1 \qquad (23)$$

4. Irregular Optimal Controller

In a special case when in the weighting coefficients of cost function,r, is zero and Q is diagonal,the feedback gain of the regular optimal controller is reduced to the gain of the deadbeat controller. As an iterative algorithm of obtaining the solution of Riccati equation we have (Lee, 1964)

$$P_{j+1}=A_1^T P_j A_1+Q-A_1^T P_j b(r+b^T P_j b)^{-1}b^T P_j A_1 \qquad (24)$$

$$j=0,1,2,3,\cdots$$

For $P_0=0$, one obtains $P_1=Q$ where P_1 is symmetric. Recalling that $N^T P_1 b=0$ the first term of the right hand side of Eq(24) becomes

$$A_1^T P_1 A_1=(N+ba_1^T)^T P_1(N+ba_1^T)$$

$$=a_1 b^T P_1 ba_1^T+N^T P_1 N \qquad (25)$$

Substituting $A_1=N+ba_1^T$ the third term of the right hand side of Eq(24) turns out to be

$$A_1^T P_1 b(b^T P_1 b)^{-1}b^T P_1 A_1=a_1 b^T P_1 ba_1^T \qquad (26)$$

At last Eq(24) for j=1 becomes

$$P_2=N^T P_1 N+Q \qquad (27)$$

It is noted that P_2 is also symmetric.
By repeating this way as $i=2,3,\cdots$. Pj ($j=3,4,\cdots$) is concluded to be symmetric.

Therefore we have $N^T P_j b=0$ $j=1,2,3,$. Remembering this fact, Eq(22) results in

$$k_1 = a_1 \qquad (23)$$

In this way it is shown that the gain of the deadbeat controller is directly yielded by the regular optimal controller gain in a special case where Q is diagonal and r is zero.

5. Two Stepwise Optimal Controller

It is frequently pointed out that applications of the deadbeat controller to the plant offers a very rapid following of the plant output to the output of reference model with any kind of input signals but motivates to generate relatively huge plant input signals. From the reason to avoid magnitude of plant input signals from increasing urgently, the deadbeat controller is replaced by the regular optimal controller. However this replacement deteriorates the property with the fastest tracking to the desired output which is the best feature of the deadbeat controller. This trade-off invokes to switch complementarily in part to the deadbeat controller.

For this purpose two different cost functions are chosen to switch each gain of the controller at a time when the deadbeat observer finish the tasks.

In the first half part of operation it is natural to take the both weighting coefficients of cost function positive numbers. Eventually in the transient period the plant control input is forced to the pure feedforward controller signal as well as the plant output is regulated to the reference model output. In the last half part of operation a complete achievement of the control effects is preferable to assigning the zero weighting coefficient of cost function concerning the plant input difference term. This assignment causes the output error to approach to zero in the fastest time. Resultantly the deadbeat control input fulfils the control objective.

6. Numerical Examples

Consider a second order discrete system. For Σ_1 we choose $a_1=(-1,-2)^T$, $c_1=(2,1)^T$ $x_1(0)=(1,1)^T$ and for Σ_2 , $a_2=(-0.08,0.4)^T$, $m_1=1, x_2(0)=(0,0)^T, u_2(i)=\sin(i/5)$ for any i. For the deadbeat observer the initial state, $x_1(0)=(0,0)^T$, is assigned.

In time interval, $0<=i<=2$, we have

$$u1(i)=(0.828,1.414)\hat{x}_1(i)+$$

$$(0.0916,0.986)x_2(i)+u_2(i)$$

and in time interval $3<=i$

$$u_1(i)=(1,2)\hat{x}_1(i)+(-0.06,0.4)x_2(i)+u_2(i)$$

Four different types of controller are cited in the examples. The first type indicates the pure feedback controller of Eq(12) (white round marks). The 2nd one is correspondent to the deadbeat controller of Eq(13) (white square marks). The optimal feedback gain $k_1=(-0.828,-1.414)^T$ is obtained for the cost functional with Q=I and r=1. This gain implements the 3rd type controller (white triangular marks). The 4th type controller means the two step wise controller (white cross marks). The reference model input signal is drawn by balck round marks line.

Fig.1 shows that four different types of controller signals are going to converge to the same one and the 4-th type optimal controller constrains the plant input signal being higher than the deadbeat controller in the transient time interval. Fig.2 illustrates that the plant output of 4-th controller converges faster than other three results.

Fig.3 depicts the plant state norm of the 4-th type is following those of the reference model as the same as the plant output results are shown in Fig.3.

7. Conclusion

It was shown that the two stepwise optimal controller proposed here exhibits the both special features of regular and irregular optimal controller. By an application of regular optimal controller in the first transient term the state and output of the plant not only follow the state and output of the reference model but the plant input also approaches the pure feedforward control input.

References

Athans, M. and P. L. Falb (1966). Optimal Control, McGraw-Hill.

Lee, R. (1964). Optimal Estimation, Identification and Control, MIT Press.

Wolovich, W.A.(1974). Linear Multivariable Control Systems, Springer Valag.

Yamane, Y. and P.N.Nikiforuk (1990). Direct Design of Model Matching Controller, The 4-th International Symposium on Application of Multivariable System Techniques, 398-403.

Yamane, Y. and P.N.Nikiforuk (1991). Feedforward Compensation for Feedback Control System Based on Model Matching Technique,The 3rd International Congress Condition Monitoring and Diagnostic Engineering Management, 336-340.

Yamane, Y. and P.N.Nikiforuk (1992). A Design of Optimal Controller for Nonminimum Phase Plant Following the Reference Model Output, The 7th IFAC Symposium on Information ControlProblems in Manufacturing Technology,658-660.

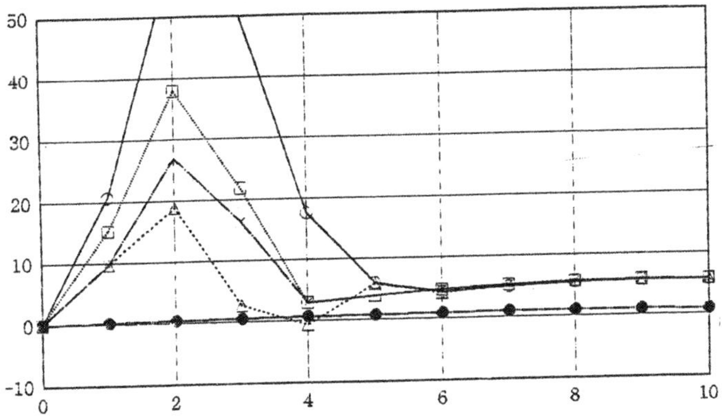

Fig.1 Input signals of plant and reference model

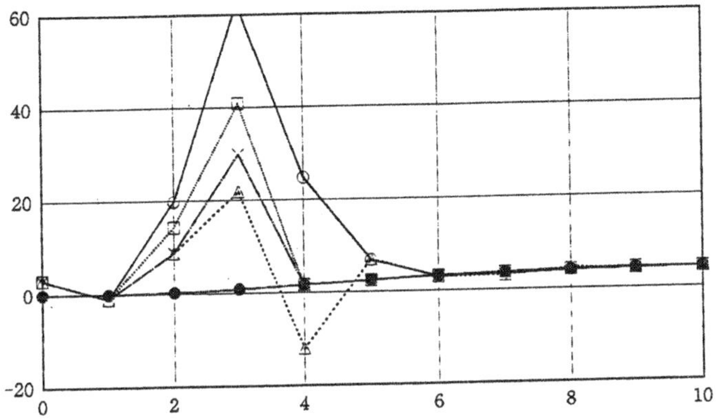

Fig.2 Output signals of plant and reference model

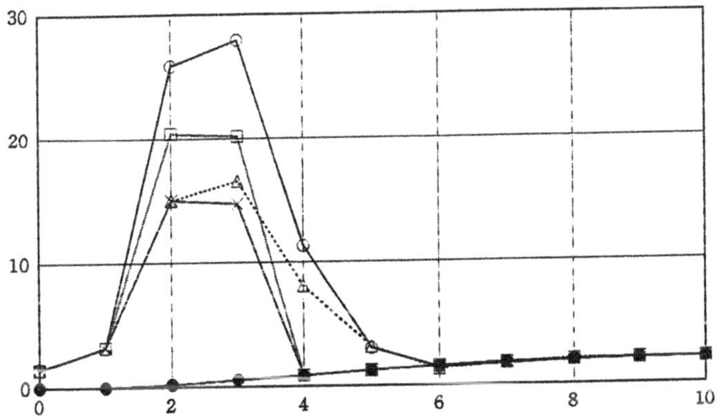

Fig.3 State norms of plant and reference model

REGULATING CHARACTERISTICS OF EXACT MODEL MATCHING

K. Ichikawa

Sophia University, 31-8, Shoan 3-chome, Suginami-ku, Tokyo 167, Japan

Abstract. Exact model matching (EMM) control system is such a control system that the controller is synthesized so that the resulting closed loop system transfer function matches exactly with the transfer function of a given reference model, where the reference model is settled so as to clear the given design specifications. Thus, EMM control system is essentially a tracking system. The regulating characteristics is evaluated by the disturbance effect. The disturbance effect to the original EMM control system is often intolerably large. By introducing a disturbance prediction technique, the regulating characteristics is exceedingly improved. Thus, a new design method for EMM with both perfect tracking and satisfactory regulating characteristics has been established.

Keywords. Exact model matching; disturbance effect; prediction; regulating characteristics.

INTRODUCTION

The idea and method of exact model matching (EMM) was introduced in the literature (Ichikawa, 1985, 1987, 1989). EMM control system is such a control system that the controller is synthesized so that the resulting closed loop system transfer function matches exactly with a given reference model. The reference model transfer function is settled so as to clear the given design specifications subject to a relative degree condition (Melsa and Schultz, 1967). EMM has several excellent properties. For example, it can be employed irrespective of the stability of the plant itself in contrast with PID control. Also, it can be applied to multivariable control. Furthermore, decoupling control is achieved in an extremely easy way.

The definition of EMM can be restated as follows: EMM control system is such a control system that the controller is synthesized so that the plant output tracks exactly the reference model output for any outer reference input signal. That is, EMM control system is essentially a tracking system. After some transient period, tracking is quite perfect. However, not only the satisfactory tracking characteristics but also the satisfactory regulating characteristics are required to any control systems. The regulating characteristics is evaluated by the effect of disturbances. In the first part of this paper, the disturbance effect to the original EMM control system is analyzed. The disturbance effect is not so small as is desired in some time. A new technique for predicting disturbances, by which the disturbance effect is practically suppressed, is then demonstrated. Thus, a new design method for EMM with both perfect tracking and satisfactory regulating characteristics has been established.

In this paper, we restrict our attention to linear, discrete time, single-input single-output systems. The reasons why discrete time systems are considered are that the computation required to generate the control signal is carried out within a digital computer in almost all cases and the prediction algorithm is quite definite in the discrete time systems. The arguments in this paper, however, can be transcribed to continuous time systems, and even to multivariable systems.

A BRIEF SURVEY OF EMM

The following assumptions are in effect.

(A1) Plant assumptions
The plant is described by

$$y(k)=t(z)u(k); \quad t(z)=r(z)p^{-1}(z) \qquad (1)$$

where $p(z)$ is an n-th degree monic polynomial, while $r(z)$ is an m-th degree polynomial with $0 \leq m \leq n-1$. The polynomials $r(z)$ and $p(z)$ are relatively prime, and $r(z)$ is a stable polynomial. The inverse of the highest term coefficient of $r(z)$ is denoted by k_I.

(A2) Reference model assumptions
The reference model is described by

$$y_m(k)=t_m(z)v(k) \tag{2}$$

where $v(k)$ is an outer reference input signal and $y_m(k)$ is a reference model output. The relative degree of $t_m(z)$ is not less than n-m.

At least the above two assumptions are also required in adaptive control, because adaptive control is an extension of EMM.

Let us consider the situation immediately after the k-th sampling instant. At this time, $y(k)$ and $v(k)$ are surely available, and we are going to calculate $u(k)$ according to the control law which will be mentioned later. Let $q(z)$ and $\gamma(z)$ be any monic stable polynomials of degree n-1 and n-m, respectively. The physical meaning of $q(z)$ is the characteristic polynomial of the reduced order observer for the plant state, and that of $\gamma(z)$ is á scalar version of the interacter for the plant. The convenient choice is that $q(z)=z^{n-1}$ and $\gamma(z)=z^{n-m}$.

The Diophantine equation

$$k(z)p(z)+h(z)r(z)=q(z)[p(z)-k_I\gamma(z)r(z)] \tag{3}$$

yields unique solutions for $k(z)$ and $h(z)$ if the degree of these solutions is restricted to not more than n-1. In particular, the degree of $k(z)$ is not more than n-2 because $m \leq n-1$. The formal method for solving Diophantine equation depends on the theory of eliminant matrix (Wolocich, 1974), but an extremely convenient method based on division algorithm was developed by the author (1987, 1989). The control law which achieves EMM is

$$u(k)=q^{-1}(z)k(z)u(k)+q^{-1}(z)h(z)y(k)+k_I\gamma(z)t_m(z)v(k) \tag{4}$$

(Ichikawa, 1985, 1987, 1989). In the following, somewhat different explanation than before is presented. Since $\gamma(z)t_m(z)$ is a proper transfer function, $\gamma(z)t_m(z)v(k)$, or $\gamma(z)y_m(k)$, is available. The control law (4) can be rewritten as

$$u(k)=\frac{q(z)}{q(z)-k(z)}\left\{\frac{h(z)}{q(z)}y(k)+k_I\gamma(z)y_m(k)\right\} \tag{5}$$

Thus, the EMM control system has the structure as shown in Fig.1. The closed loop system dynamics is described as follows.

$$y(z)=\frac{\dfrac{r(z)}{p(z)}\cdot\dfrac{q(z)}{q(z)-k(z)}}{1-\dfrac{r(z)}{p(z)}\cdot\dfrac{q(z)}{q(z)-k(z)}\cdot\dfrac{h(z)}{q(z)}}\cdot k_I\gamma(z)y_m(z)$$

$$=\frac{r(z)q(z)}{p(z)q(z)-p(z)k(z)-r(z)h(z)}\cdot k_I\gamma(z)y_m(z)$$

$$=\frac{q(z)r(z)}{q(z)k_I\gamma(z)r(z)}\cdot k_I\gamma(z)y_m(z) \tag{6}$$

It is seen that the characteristic polynomial of the closed loop portion of Fig.1 is $q(z)k_I\gamma(z)r(z)$, which is of degree 2n-1. That is, the realization for the closed loop portion of order 2n-1 can be obtained. In the literature, the state space representation with 3n-2 state vector is presented for the matched system (Narendra and Valavani, 1978; Sastry and Bodson, 1989), but it is redundant. Anyhow, the characteristic polynomial is stable and thus the EMM control system is internally stale. Furthermore, (6) reduces to

$$y(z)=y_m(z) \tag{7}$$

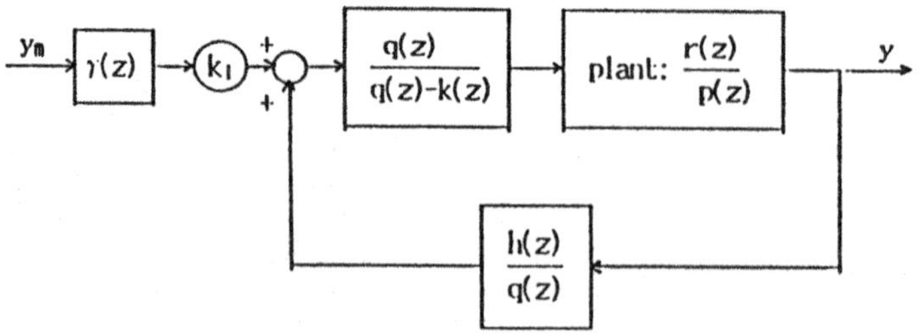

Fig.1 Structure of EMM control system

It is interesting to notice that $u(k)$ specified by (4) can be synthesized from $u(k)$ and $y(k)$ through n-1 (not 2n-2) dimensional system as shown below. Let

$$k(z)=k_{n-2}z^{n-1}+\cdots+k_0 \qquad k=[k_{n-2} \ \cdots \ k_0]^T \tag{8}$$

$$h(z)=h_{n-1}q(z)+h_{n-2}z^{n-2}+\cdots+h_0 \qquad h=[h_{n-2} \ \cdots \ h_0]^T \tag{9}$$

$$q(z)=z^{n-1}+q_{n-2}z^{n-2}+\cdots+q_0 \tag{10}$$

Then, (4) is rewritten as the following state space representation.

$$\xi(k+1)=\begin{bmatrix} 0 & 0 & \cdots & 0 & -q_0 \\ 1 & 0 & \cdots & 0 & -q_1 \\ \cdot & 1 & \cdots & & \cdot \\ \cdots & & & & \cdot \\ 0 & 0 & & 1 & -q_{n-2} \end{bmatrix}\xi(k)+ku(k) +hy(k)$$

$$u(k)=[0 \ \cdots \ 0 \ 1]\xi(k)+h_{n-1}y(k) +k_1\gamma(z)y_m(k) \tag{11}$$

Eq.(7) means that $y(k)\to y_m(k)$ as $k\to\infty$. As a matter, if $q(z)=z^{n-1}$ and $\gamma(z)=z^{n-m}$, $y(k)=y_m(k)$, $k\geqq 2n-1-m$ (Ichikawa, 1989). Anyway, tracking of $y_m(k)$ by $y(k)$ is perfect after passing some transient period.

REGULATING CHARACTERISTICS

Suppose that an unknown disturbance $d(k)$ is applied to the above-mentioned EMM control system. We assume

(A3) Plant is described by

$$y(k)=t(z)\{u(k)+d(k)\} \tag{12}$$

The following relation is easily obtained.

$$y(z)=y_m(z)+\frac{q(z)-k(z)}{q(z)k_1\gamma(z)}d(k) \tag{13}$$

That is, the disturbance effect is $[q(z)-k(z)][q(z)k_1\gamma(z)]^{-1}d(k)$. We notice that the disturbance never causes instability, but the disturbance effect may not be so small as is desired. If $d(k)=0$, $k\geqq k_0$ for some k_0, the disturbance effect decays to zero with speed specified by the zeros of $q(z)\gamma(z)$.

SUPPRESSION OF DISTURBANCE EFFECT

Suppose for the moment that the disturbance $d(k)$ is detected and available. Clearly, the control law specified by

$$u(k)+d(k)=q^{-1}(z)k(z)\{u(k)+d(k)\} +q^{-1}(z)h(z)y(k)+k_1\gamma(z)y_m(k) \tag{14}$$

achieves exact model matching without any deviation due to the disturbance. The control law (14) can be rewritten as

$$u(k)=q^{-1}(z)k(z)u(k)+q^{-1}(z)h(z)y(k) +k_1\gamma(z)y_m(k)-d(k)+q^{-1}(z)k(z)d(k) \tag{15}$$

Now, since $d(k)$ is unknown in reality, we cannot employ the control law (15). Then, we replace $d(k)$ by its prediction $d_p(k)$. That is, the following control law is employed.

$$u(k)=q^{-1}(z)k(z)u(k)+q^{-1}(z)h(z)y(k) +k_1\gamma(z)y_m(k)-d_p(k)+q^{-1}(z)k(z)d_p(k) \tag{16}$$

Let us investigate the method of predicting $d(k)$ with satisfactory accuracy. Since $q^{-1}(z)k(z)$ is strictly proper, the signal $q^{-1}(z)k(z)d_p(k)$ can be synthesized from the past predictions $d_p(k-1)$, $d_p(k-2)$, $\cdots$, $d_p(k-n+1)$.

The system described by

$$y_1(k)=r(z)p^{-1}(z)u(k) \tag{17}$$

is called an inner model. Since $r(z)p^{-1}(z)$ is strictly proper, $y_1(k)$ can be synthesized from the past data $u(k-1)$, $u(k-2)$, $\cdots$, $u(k-n)$, and hence $y_1(k)$ is available. Therefore, $e(k):=y(k)-y_1(k)$ is also available. Clearly

$$e(k)=r(z)p^{-1}(z)d(k) \tag{18}$$

By using a division algorithm, we obtain the form

$$p(z)r^{-1}(z)=f_{n-m}z^{n-m}+\cdots+f_1z+s(z)r^{-1}(z) \tag{19}$$

where $s(z)$ is some m-th degree polynomial. Then, from (18) we obtain

$$d(k)=f_{n-m}e(k+n-m)+\cdots+f_1e(k+1) +s(z)r^{-1}(z)e(k) \tag{20}$$

where the last term of (20) is available, because $s(z)r^{-1}(z)$ is proper and $e(k)$ is available. However, $e(k+1)$, $\cdots$, $e(k+n-m)$ are unknown, because these are future values.

Let us predict these future values by the extrapolation principle. We provide a proposition.

Proposition 1 An i-th degree polynomial, which passes i+1 point $\{k-i,e(k-i)\}$, $\cdots$, $\{k,e(k)\}$ in the k-e(k) plane is described by the equation

$$e(k+j)=[1 \ j \cdots j^i] \times[a_0e(k)+a_1e(k-1)+\cdots+a_i(e(k-i)] \tag{21}$$

where $a_s=P^{-1}e^{s+1}$, $s=0,1,\cdots,i$, with e being a unit vector and P being a matrix

$$P=\begin{bmatrix} 1 & 0 & \cdots & 0 \\ 1 & -1 & \cdots & (-1)^i \\ 1 & -2 & \cdots & (-2)^i \\ \cdots & \cdots & \cdots & \cdots \\ 1 & -i & \cdots & (-i)^i \end{bmatrix} \tag{22}$$

Proof: We obtain

$$\begin{aligned} &[\,1 \;\; 0 \;\; \cdots \;\; 0\,]a_0=1 \\ &[\,1 \;\; 0 \;\; \cdots \;\; 0\,]a_1=0 \\ &\cdots\cdots\cdots\cdots\cdots \\ &[\,1 \;\; 0 \;\; \cdots \;\; 0\,]a_i=0 \end{aligned}$$

Therefore, (21) holds for $j=0$. Likewise, we can prove that (21) holds for $j=-1, -2, \cdots, -i$. This completes the proof. □

For any positive integer j, (21) provides the prediction $e_p(k+j)$. For convenience of practical use, $e_p(k+j)$ are listed in Table 1 for small numbers of i and j. From (20), we can obtain the prediction $d_p(k)$ as follows.

$$\begin{aligned} d_p(k)=&f_{n-m}e_p(k+n-m)+\cdots+f_1e_p(k+1) \\ &+s(z)r^{-1}(z)e(k) \end{aligned} \tag{23}$$

Thus, the control law (16) has been completed.

A NUMERICAL EXAMPLE

Consider a plant described by

$$\begin{aligned} y(k)=&(z-0.3)(z^2+0.3z+0.02)^{-1} \\ &\times\{u(k)+d(k)\} \end{aligned}$$

where $d(k)$ is an unknown disturbance, but it is assumed for simulation studies as

$$\begin{aligned} d(k)=&0.2+\{1-2\exp(-0.1k)+\exp(-0.2k)\} \\ &\times\sin[2\pi k/\{10(1+(k-10)/50)\}] \end{aligned}$$

That is, the disturbance consists of a step function with magnitude 0.2 and an almost sinusoidal function with both amplitude and frequency being slowly varying functions.

The reference model is assumed as

$$t_m(z)=0.75(z^2-0.25)^{-1}$$

The outer reference input signal is quite arbitrary, but it is assumed for simulation studies as

$$v(k)=1+0.2\sin(2\pi k/50)$$

Table 1 Predicted Values, $e_p(k+j)$

j / i	1	2	3
0	e(k)	e(k)	e(k)
1	2e(k)-e(k-1)	3e(k)-2e(k-1)	4e(k)-3e(k-1)
2	3e(k)-3e(k-1)+e(k-2)	6e(k)-8e(k-1)+3e(k-2)	10e(k)-15e(k-1)+6e(k-2)

For convenience, $q(z)$ and $\gamma(z)$ are taken as z and z, respectively. The Diophantine equation yields the solutions

$$k(z)=0.3,\quad h(z)=0.sz+0.2$$

The control law becomes

$$\begin{aligned} u(k)=&0.3u(k-1)+0.3y(k)-0.02y(k-1) \\ &+zy_m(k)-d_p(k)+0.3d_p(k-1) \end{aligned}$$

Since

$$p(z)r^{-1}(z)=z+(0.6z+0.02)(z-0.3)^{-1}$$

we obtain

$$d_p(k)=e_p(k+1)+(0.6z+0.02)(z-0.3)^{-1}e(k)$$

where

$$\begin{aligned} e(k)=&-0.3e(k-1)-0.02e(k-2) \\ &+y(k)+0.3y(k-1)+0.02y(k-2) \\ &-u(k-1)+0.3u(k-2) \end{aligned}$$

We only need $e_p(k+1)$. The number i can be taken arbitrarily, but we take $i=2$ here. Then,

$$e_p(k+1)=3e(k)-3e(k-1)+e(k-2)$$

The simulation results are shown in Fig.2. The result of control for the original EMM is shown in (a), while the result of control using prediction is shown in (b), where the disturbance effect is fully suppressed except the effect of step function component contained in the disturbance.

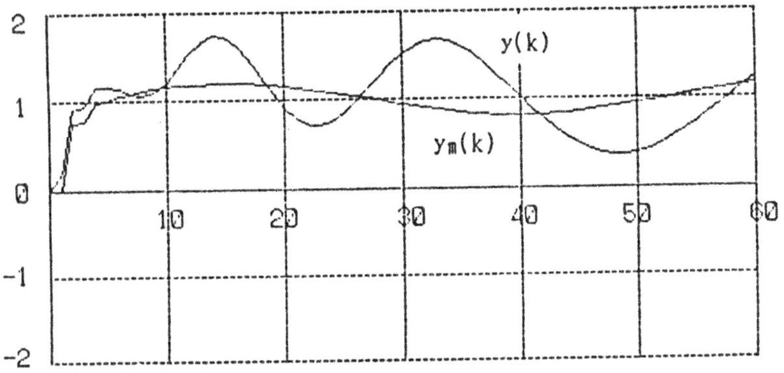

(a) without consideration of disturbance

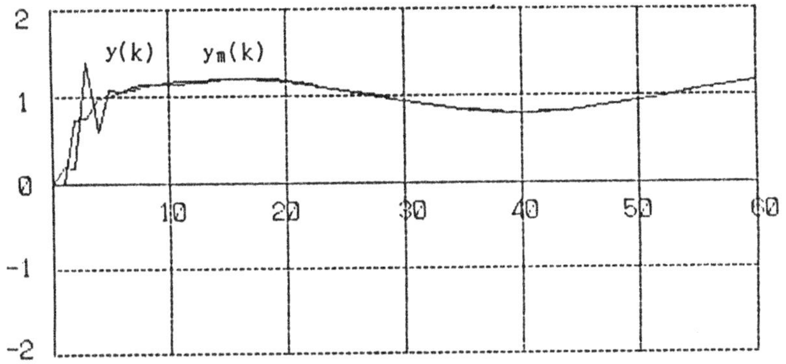

(b) suppression of disturbance effect

Fig. 2 Control of a plant with unknown disturbance

CONCLUSIONS

The regulating characteristics of EMM control systems is examined, and it was shown that the regulating characteristics can be improved by applying a prediction technique to disturbances. The simulation studies show that the effect of smooth components contained in the disturbance can be fully suppressed. Since the tracking characteristics of EMM is perfect from the beginning, it can be concluded that a new design method for EMM with both perfect tracking and satisfactory regulating characteristics has been established.

REFERENCES

Ichikawa, K. (1985). Control System Design Based on Exact Model Matching Techniques. Springer-Verlag, Berlin.

Ichikawa, K. (1987). The idea and method of exact model matching. J. SICE, 26, 974-984. (in Japanese)

Ichikawa, K. (1989). Theory of Control System Design. Gijutsushoin, Tokyo. (in Japanese)

Melsa, J.L., and D.G. Schultz (1967). Linear Control Systems. McGraw-Hill, New York.

Narendra, K.S., and L.S. Valavani (1978). Stable adaptive controller design —— Direct control. IEEE Trans. on Autom. Control, AC-23, 570-583.

Sastry, S., and M. Bodson (1989). Adaptive Control. Prentice Hall, Englewood Cliffs.

Wolovich, W.A. (1974). Linear multivariable systems. Springer-Verlag, Berlin.

VARIABLE STRUCTURE REGULATORS FOR INDUSTRIAL PROCESS CONTROL

C. Scali*, A. Landi**, G. Nardi* and A. Balestrino**

*Department of Chemical Engineering, University of Pisa, Via Diotisalvi 2, Pisa, 56126, Italy
**DSEA - Department of Electrical Systems and Automation, University of Pisa, Via Diotisalvi 2, Pisa, 56126, Italy

Abstract
VS-PI controllers, with only two parameters variable with the error, are analyzed for low and high order systems; limitations of performance are clearly pointed out. For high order systems the VS-PI controller is augmented with a suitable filter with constant parameters; this allows to maintain high performance also in the presence of process uncertainty and input saturation. This is shown by a comparison of results with advanced controllers designed according to the IMC procedure.

Keywords: Variable Structure Control, Non Linear Plants, Robustness, Process Control

INTRODUCTION

The majority of process regulators are of the PID type, in force of their simple and reliable structure. Beside standard regulators a variety of configuration are now implemented by including special features such as adaptation, self-tuning and prediction. The main goal is to achieve desirable response also when process parameters are poorly known. A large class of processes can be controlled by using variable structure controllers (Utkin, 1974), (Harashima and others, 1985). In this way system dynamics become very fast and low sensitivity to external disturbances and parametric variations is assured. The most important drawback of this class of regulator is due to chattering in the input and the output signal. Therefore a lot of suitable modifications to the variable structure approach were proposed in order to overcome this difficulty. Among them a non linear standard regulator with a PID algorithm (NLSR), based on a variable structure logic, has been proposed in Balestrino and others (1989, 1990a). Its application in electrical drives was successful with an improvement of the transient responses, an overshoot reduction and an effective high insensitivity to external disturbances and parametric variations (Balestrino and Landi, 1990b). Moreover an on-line parameter estimation as in self-tuning controllers is not necessary. Some questions arise: which class of process is suitable to be controlled with these regulators? What happens in presence of input saturations and in which sense is it allowed to compare NLSR with conventional or with advanced controllers? Is it possible to provide for the NLSR simple and general tuning rules? In this paper we will try to answer previous questions in a systematic way.

VS-PI CONTROL

The structure of the controller is the classical one with proportional and integral actions. The overall logic of the NLSR is as follows:

a) for very large errors the I control action vanishes and the P action is the global one; the proportional gain must be chosen with upper bounds within the stability region. The input saturation gives a further bound on the maximum proportional action allowed.

b) for small errors the P action vanishes and the I action grows and becomes predominant.

The general form of this modified standard PI controller is as follows:

$$C(s)^* = K_c(e) + K_i(e)/s \qquad (1)$$

where e is the error variable, i.e. the difference between the desired output r and the actual response y: $e = r-y$. Coefficients in (1) are non linearly dependent on the error e; a particularly simple choice is:

$$K_c(e) = K_{co} / (1+c_p|e|^m) \quad (2a)$$

$$K_i(e) = K_{io} / (1+c_i|e|^n) \quad (2b)$$

Different coefficient variations as function of the error can be chosen; Dote and Saitoh (1991), propose an exponential law. The smoothed transition from high gain to steady-state proportional action due to (2a) eliminates the typical drawback of relay control due to chattering. In most plants the constant integral gain produces the undesirable phenomenon of overshoot, in the time responses. An automatic variable integral gain, with an inverse dependence on e, ensures that integral actions is very small whenever the error is large. In this way wind-up and overshoot problems are effectively removed.

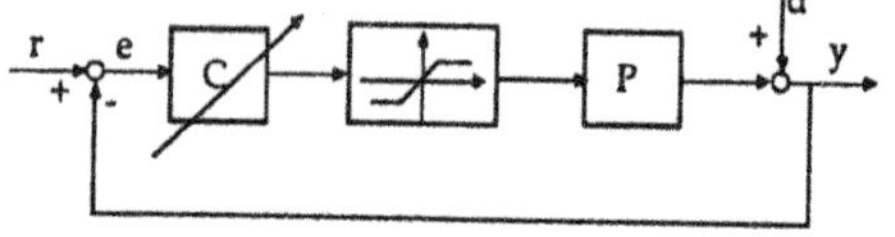

Fig. 1. The reference scheme for VS-PI control

LOW ORDER SYSTEMS

The issues presented in the introduction are illustrated by means of a simple example. For a first order system with a model:

$$\tilde{P}(s) = \tilde{k} / (\tilde{\tau}s+1) \quad (3)$$

a PI controller:

$$C = K_c (1+\tau_i s)/\tau_i s \quad (4)$$

is able to give error free response at steady state for step like (type-one) inputs. In the case of perfect matching between the process and the model ($P=\tilde{P}$), by choosing the controller integral time constant equal to the process time constant ($\tau_i=\tilde{\tau}$), the system pole is canceled by the zero added by the integral component. By increasing the proportional gain K_c, set-point tracking resposes are obtained, which become faster and maintain a desirable overdamped response. In the case of a mismatch between model and process ($P\neq\tilde{P}$), time responses may present an overshoot. This inconvenience is exalted by the presence of saturation on the process input: in this case the corresponding responses deteriorate both in the nominal and in the uncertain case. The extent of this phenomenon depends on specific values of gains, time constants and saturation levels. The situation is depicted in Fig. 2 for typical values of parameters

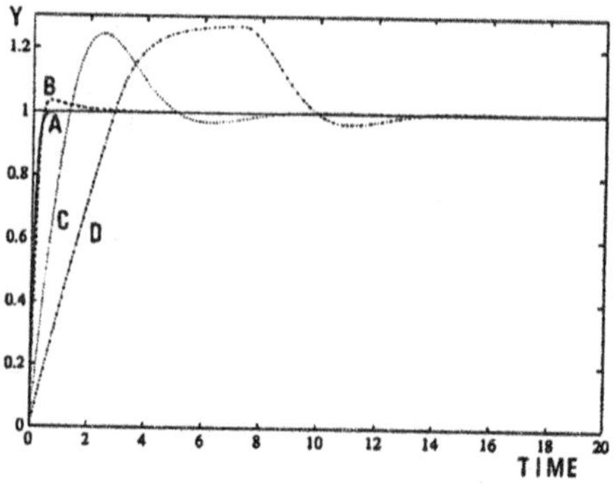

Fig. 2. Time responses for first order system with PI control:
$\tilde{P}=1/(s+1)$; $P=1/(10s+1)$; $K_c=10$; $\tau_i=\tilde{\tau}=1$.
A: nominal, (no sat); B: nominal, (sat=4)
C: uncertain, (no sat); D: uncertain, (sat=4).

In the case that controller parameters can vary as a function of the error (e) according to the logic described in the previous section, increasing the proportional part (u_p) and reducing the integral component (u_i) for large errors, time responses can be positively affected. At the expense of a slightly slower time response in the nominal case, noticeable improvements can be obtained when there is a large variation in the process time constant. This happens in both cases of saturated and not saturated control action: time responses present almost negligible overshoot (Fig. 3). Equal final values for K_c (=10) and τ_i (=1) are used; tuning of VS-PI algorithm is not critical, both for coefficients and exponents of (2). Values of controller parameters are:
$K_c=10/(1-0.9|e|^{0.5})$, $K_i=10/(1+250|e|^4)$.

Results can be easily explained by considering that in the case of a PI regulator with constant parameters, the control action generated by the integral component (u_i) is larger than the *optimal* value (for $\tau_i=\tilde{\tau}$). Therefore it leads the system beyond desidered values, up to the point that the sign of the control action change (overshoot conditions). Analogously, when saturation is present, the control system works at maximum allowed values, thus generating an offset in the output response until the effect of a change of sign in the control action becomes active.

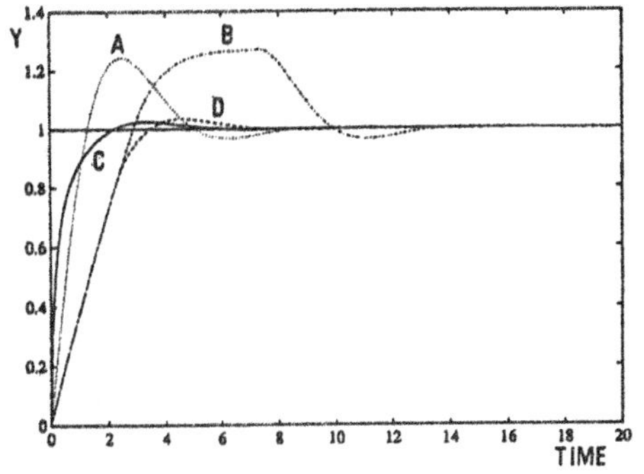

Fig. 3. Comparison of time responses between PI (A,B) and VS-PI (C,D) for a first order system in the uncertain case. A,C: no sat; B,D: sat = 4.

Instead, in the VS-PI structure the integral action is very small when the error is large and so the global control action is generated mainly by the proportional component ($u \approx u_p$). The resulting response, with or without saturation constraints, reaches steady state with small overshoot. The situation is illustrated in Fig. 4, where control actions generated by the two controllers are reported for the case of uncertain system and presence of saturation. To be noted the smoothed transition from P to I control effected by the adopted algorithm.

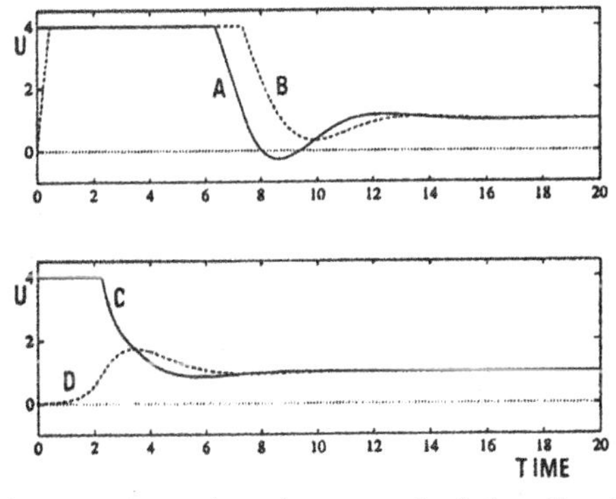

Fig. 4. Control actions required by the two controllers for first order system.
PI: total (A), integral (B)
VS-PI: total (C), integral (D)

Some observations about previous results are in order at this point:
- It is evident the positive influence on the time response of a decrease of the integral component, especially when saturation is present. This inconvenience can be only partly reduced by adding an anti-windup device in the PI structure, while it is almost completely avoided by the VS-PI.
- In the examined case (first order systems) there are two favourable characteristics which permit to take full advantage of the VS-PI potentiality. With respect to stability: a gain margin equal to infinity allows to increase the proportional gain in the initial part of the response (and consequently u_p up to saturation values). With respect to performance: the closed loop response under proportional control has a monotonic behaviour towards steady state conditions. This fact allows to switch to integral action (to eliminate offset) with a law which is based only on the output error (e).

HIGHER ORDER SYSTEMS

The situation generally changes in the case of high order systems. Here the maximum proportional gain is limited by the stability condition. The closed loop response under proportional control may be not monotonic, that is the output response does not contain all informations about system dynamics. Consequently an algorithm which is based only on the output error may not be enough to allow an efficient switch from P to I control. Results similar to those previously shown are still obtained, by using the same control structure, for the case of systems described by one dominating time constant, whose dynamic characteristics do not differ very much from a first order process.

Previous considerations allow to realize inherent limitations of a simple control algorithm as the one given by equations (2) and of a simple PI structure. Qualitatively speaking, two ways can be followed to overcome these difficulties:
- adopting a more complex algorithm in a PI structure to account not only for the output error but also for its derivatives, in order to have *informations* about the states of the systems,
- adopting a more complex controller structure, for example adding a filter F to the C* controller, in order to *reduce* the high order system to a first order-like system.

First method introduces new parameters which must be chosen to weight different contribution of error derivatives to the value of the controller constants, so tuning becomes much more difficult. The second possibility has been explored here and will be described in the sequel.

Actually the basic idea of increasing the controller order as a consequence of the system order is not new: it is a result of analytical feedback design (Newton, Gould and Kaiser, 1957). More recently (Morari and Zafiriou, 1989), introduced the Internal Model Control (IMC) design method in which:

- the nominal process ($\tilde{P}$) is canceled by the nominal controller ($\tilde{Q}$), guaranteed to be stable, causal and optimal according to the ISE criterion,
- a filter F is added to give physical realizability to the global controller $Q=\tilde{Q}F$ and tuned to give desired performance in the presence of uncertainty.

The global controller in a IMC scheme (Q) leads to an equivalent controller in a feedback structure: $C=Q/(1-\tilde{P}Q)$. This controller can be decomposed in terms of a standard PI structure augmented by a filter of suitable order. In our case the filter (F') is added to the VS-PI controller with the scope of reducing the high (n) order process (P) to a *new* process P'=PF', which is characterized by one dominant time constant and n-1 smaller time constants, which is the favourable situation for the VS-PI controller. Obvsiously this reduction of order is not perfect in the presence of uncertainty in model parameters, but anyway a positive effect of the filter acts also in this case. In addition saturation on the plant input must be properly taken into account . The methodology of design and a comparison of results with IMC is now presented in the case of a third order system.

CASE STUDY

Let the model of the process be:

$$\tilde{P}=1/(\tilde{\tau}s+1)^3 \qquad (5)$$

with $\tilde{\tau}=10$. For a quantitative analysis of results assume that the process time constant may vary by a factor 2 with respect to the nominal case, to originate a faster process ($\tau=5$) and a slower process ($\tau=20$); in addition the input saturation value is taken =2.

<u>IMC Design</u>

The IMC controller is:

$$Q=(\tilde{\tau}s+1)^3/(\lambda s+1)^3 \qquad (6)$$

The filter parameter λ allows to account for model uncertainty and its value determines closed loop dynamics: smaller values of λ corresponds to faster response and to larger sensitivity to process uncertainty. The corresponding time responses for the case $\lambda=10$ are reported in Fig. 5 for the three situations of nominal, faster and slower process.

In this case output response do not change in the presence of saturation because the required control action does not reach saturation level (Fig. 6). It can be noted that the smoothed response exhibited in the nominal case is substituted by underdamped responses, with high overshoot (more than 30%) in the case of slower process.

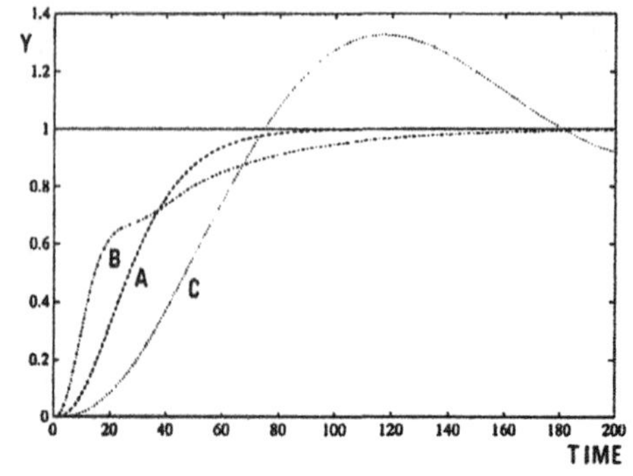

Fig. 5. Responses for a third order system; IMC controller ($\lambda=10$); A: nominal, B: fast, C: slow.

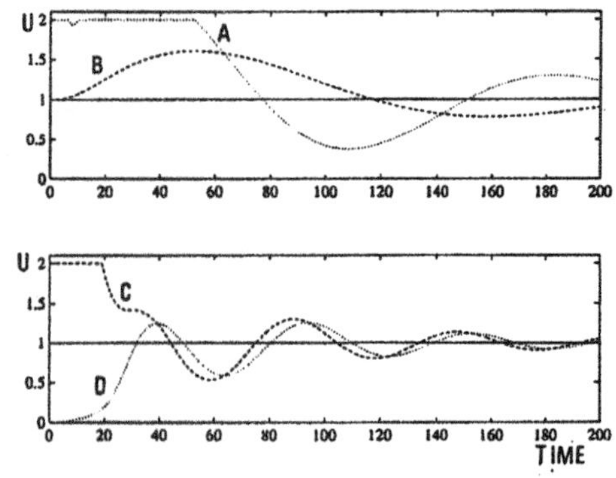

Fig. 6. Control actions required by the two controllers for third order system (sat = 2): IMC, total action: $\lambda=5$ (A), $\lambda=10$ (B); VS-PI: total (C), integral (D).

If the closed loop speed is increased, by adopting the value of $\lambda=5$, corresponding time responses are reported in Fig. 7. The faster response obtained in the nominal case deteriorates to give higher overshoot (more than 40%) for the slow process. In this case, the required control action reaches saturation levels (Fig. 7) and the corresponding time responses become worse; to be noted that also in the nominal case some overshoot is presented (Fig. 8).

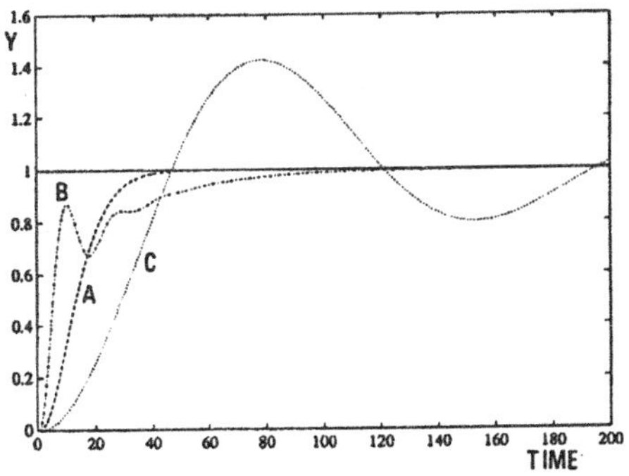

Fig. 7. Responses for a third order system; IMC controller ($\lambda=5$), no sat;
A: nominal, B: fast, C: slow.

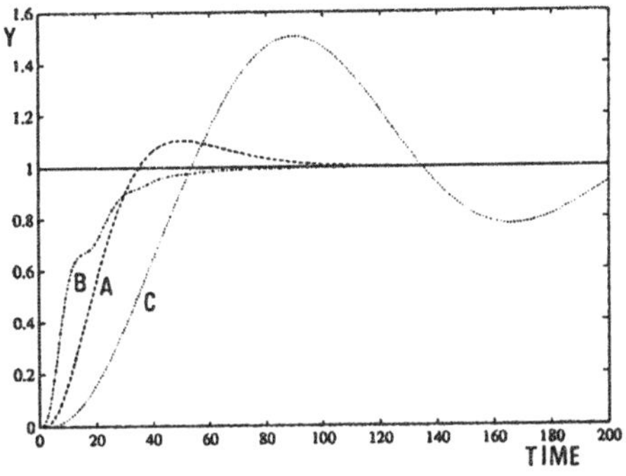

Fig. 8. Responses for a third order system; IMC controller ($\lambda=5$), sat $=2$;
A: nominal, B: fast, C: slow.

VSC Design and Results

The feedback controller equivalent to the IMC one is:

$$C=\frac{K_c(\tilde{\tau}s+1)^3}{\tau_i s(\frac{\lambda^2}{3}s^2+\lambda\ s+1)}$$

and becomes equivalent to a a PI controller plus a filter (F_{22}) having degree equal to 2 in numerator and in denominator ($K_c=\tilde{\tau}/3\lambda$; $\tau_i=\tilde{\tau}$). In our approach we want to *transform* the original into a third order system with a dominating time constant. This is possible by adding to the VS-PI controller a filter:

$$F_{33}=(\tilde{\tau}s+1)^3\ /\ (\alpha s+1)\ (\beta s+1)^2$$

If the ratio $\alpha/\beta > 10$ the resulting new process P' has desired characteristics:
$P' = PF_{33} = 1\ /\ (\alpha s+1)\ (\beta s+1)^2$

In particular, when the dominant time constant is chosen of the same values as one of the open loop system ($\alpha=\tilde{\tau}$), the scope is reached by the same order filter as in the IMC case (n=2). This is the situation which has been studied for a more immediate comparison.

For the VS-PI controller, parameters vary inside a stability region which is determined as a preliminary part of the design (Fig. 9); in the K_c-K_i plane the region is obtained as intersection of stability regions corresponding to nominal and uncertain systems. The control algorithm is:
$K_c=2.9/(1-0.5|e|^{0.5})$; $K_i=0.6/(1+250|e|^4)$.

Results for the three cases of nominal, fast and slow process, without input saturation are reported in Fig. 10; for the same cases, in the presence of saturation, results are given in Fig. 11. It can be noted that the speed of response in the nominal case is comparable with previous examples, very small variations are shown in the presence of uncertainty and, also when input saturation is active, the overshoot is negligible.

The corresponding control action is reported in Fig. 6, separated into a global and a P component, to be compared with the global control action required by IMC. Results presented here have been obtained adopting the control configuration of Fig. 12; the filter F has been placed in the feedback path: this allows some improvement in time responses and reduce oscillation in control action (Nardi, 1992). All previous considerations about filter effectiveness remain unchanged.

Applications to minimum phase systems (RHP zeros and delays) are possible by augmenting the VS-PI controller with a compensator, thus generating a Smith predictor-like structure and are the object of actual studies.

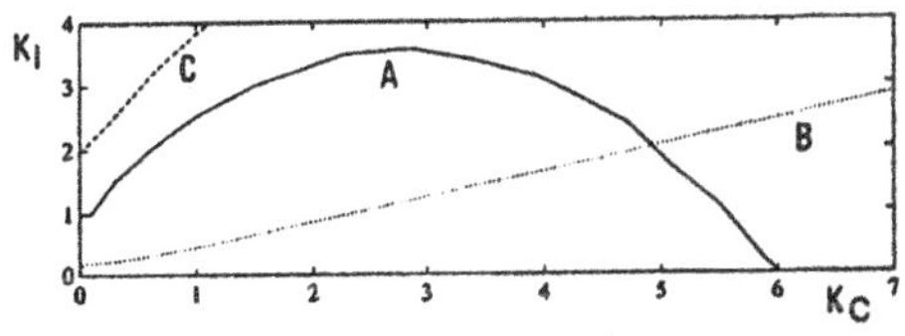

Fig. 9. Stability regions in the K_c, K_i plane.
A: fast, B: slow, C: nominal.

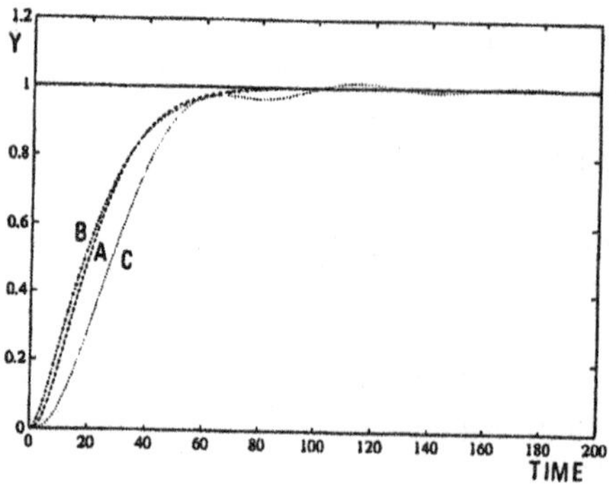

Fig. 10. Responses for a third order system; VS-PI control no sat, A: nominal, B: fast, C: slow.

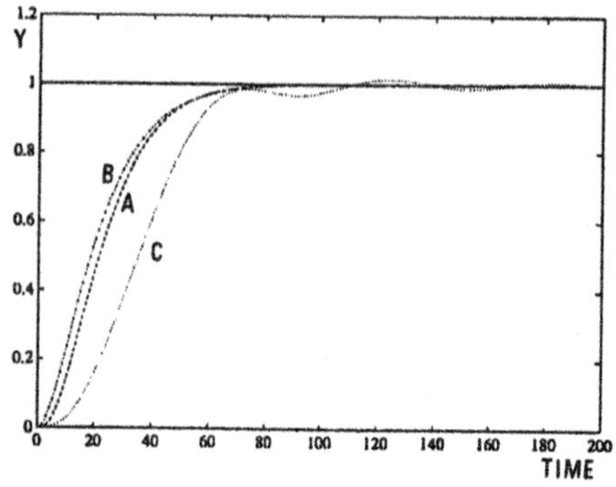

Fig. 11. Responses for a thir' 'tem; VS-PI control with saturatio
A: nominal, B: fas

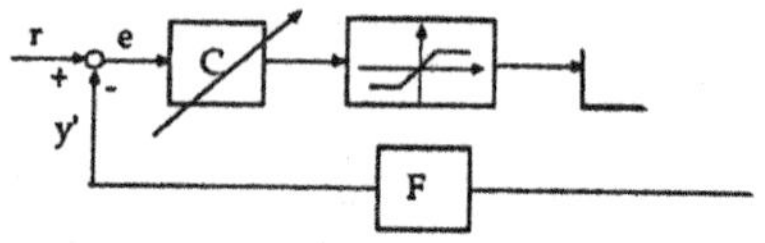

Fig. 12. The reference scheme for VS-PI control with filter in the feedback path

CONCLUSIONS

Limitations of performance of the simple VS-PI structure, which has only two parameters variable to keep integral action to zero for large errors have been clearly pointed out. The methodology presented in the paper allows to extend positive features of VS-PI control structure to high order systems, which are reduced to first order-like systems by adding a suitable filter with constant parameters. The comparison with IMC, which uses an equivalent structure, is appropriate and shows advantages of the proposed method in the case of large parameter variations and presence of input saturation.

Ease of tuning is maintained, as parameters variation lies in a stability regions determined by robust stability considerations, and values of coefficient are not critical.

ACKNOWLEDGMENTS

This work has been developed with the financial support of MURST 40% and CNR.

REFERENCES

Balestrino, A., M. Innocenti and A. Landi (1989). Variable structure conventional controllers. IFAC Proc. Low Cost Automation, W 187-191.

Balestrino, A., A. Brambilla, A. Landi, C. Scali (1990a). Non linear standard regulators. IFAC Proc.World Congress Vol. 8, 20-25.

Balestrino, A. and A. Landi (1990b). Intelligent variable structure control in electrical drives. IEEE Proc. Int. Workshop on Intelligent Motion Control, 719-722.

Dote, Y. and T. Saitoh (1981). Stability analysis of variable structured PI controller by fuzzy logic for servo system, IECON '91, Vol. 1, 363-365.

Harashima, F, H. Hashimoto and S. Kondo (1985). Mosfet converter-fed position servo system with sliding mode control. IEEE Trans. on IE, 32, 238-244.

Morari, M. and E. Zafirou. Robust process control. Prentice Hall, Englewood Cliffs.

Nardi, G. (1992). Tesi di Laurea. Chem. Eng. Dept., University of Pisa.

Newton, G. C., L.A. Gould and J.F. Kaiser (1957). Analytical design of linear feedback controls. J. Wiley & Sons, New York.

, V.I. (1974). Sliding modes and their 'ication in variable structure systyems. Ed. 'IR, Moscow.

LOW COST DISTRIBUTED CONTROL SYSTEM BASED ON SERIAL FIELD BUS

F.J. Suárez, D.F. García, J.R. Arias and J. García

University of Oviedo, Department of Electrical, Electronic, Computers and Systems Engineering, Campus de Viesques s/n, 33204-Gijón, Spain

Abstract. This paper presents the strategies adopted in the development and implementation of a small scale low-cost distributed control system and its more important characteristics. The communications way adopted was the serial field bus, which confers great economy to the system. The structure used was the star type, where a master unit handles communication with slave units. The system was developed under personal computer's environment and includes several utilities, in local mode from the process control stations and in remote mode from the central control room of the plant. As a result, we have achieved a reliable and modular control system, capable of powerful additional possibilities with only small efforts.

Keywords. distributed computer control; serial communications; acquisition systems; software communication protocols.

INTRODUCTION

Instead of the big distributed control systems that are been introduced in the great industrial plants -all with several complex processes- including expert systems, powerful communications ways and local area networks, we have developed a simple and economic distributed control system that permits us to grapple the problems of little industrial plants, with a small number of control loops. As these plants can not afford too expensive systems, the economic factor is one of our most important design objectives.

The distributed character of the system, based on a very low-cost bus confers a great reliability to the system without an important additional cost.

Another important characteristic of the system that we will detail later, is its modular structure, in hardware and software, which permits a great flexibility of the work and many more possibilities for future developments of the system.

HARDWARE STRUCTURE OF THE SYSTEM

The system presents a typical star structure, with a central processor station or "master unit" from which communication lines go out towards the local processor stations or "slave units". These are commissioned to control the different loops of the plant as can be seen in Fig. 1.

The master processor station is located in the central control room of the plant. It receives information from all local processor stations and permits coordination of all the system control activities: set points, sample time variation, modification of the different loop controllers, etc.

To obtain a central station with possibilities for tracking and coordination, we used a processor equipment powerful enough, having been acceptable a personal computer (PC AT) with a VGA graphics adapter. This was the processor equipment chosen for the central station.

For the local processor stations, although we could have used any processor equipment that incorporated an A/D-D/A acquisition system and communication possibilities, we used PC AT equipments, even for the central control processor. There are several reasons: these equipments possess serial communication lines and it results very simple to equip them with high quality acquisition systems and moreover, they have a powerful microprocessor, memory, hard disk, etc. To all this advantages, we must add the great understanding of this equipments in industry with their moderate and decreasing prices.

Using personal computers as local processor stations makes possible an efficient control of the loops of the plant and also provides a comfortable interface with the user.

In the implemented system, every local station (PC AT) was equipped with a simple, low-cost acquisition system sufficient to govern one control loop. It results very simple to adapt the stations to any loop number or to a more complex acquisition system -for example for multivariable control-, because the processing power of the PC-AT equipment is more than enough.

A negative aspect is related with the practical impossibility for the PC's to work on environments where high temperature, dust, high moisture and vibrations prevail. For such cases specially robustized IBM compatible PC's are available, although those devices are quite expensive.

The general structure of the distributed control system is shown in Fig. 1.

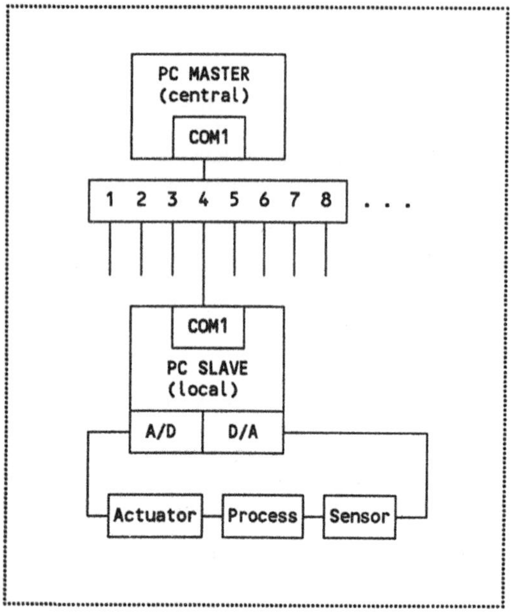

Fig. 1. General structure of the distributed control system

The acquisition system used was also developed in our Department especially for this project and its main features are the following:

- A/D-D/A conversions with 12 bit of precision
- Unipolar and bipolar modes with several possible scales
- 1 microsecond D/A conversion speed
- A/D conversion rate up to 30 conversions per second
- 8054 Programmable timer
- Interrupt management
- Programmable I/O port addresses
- Very low cost

The general structure of the acquisition system is shown in Fig. 2.

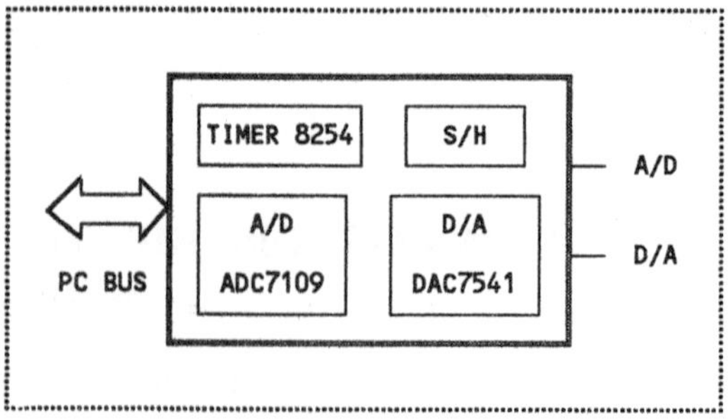

Fig. 2. General structure of the developed acquisition system

As controller algorithms we have used the typical PIDs, for several reasons:

- Dedicated one or multi-loop controllers are applicable to a wide range of industrial processes resulting in lower investment and maintenance costs.
- They do not require an extensive knowledge of the process parameters.
- It is not difficult for plant operators to understand the operation of PID controllers or correct the effect of eventual parameters changes.

More complex control structures as several feed-forward compensations, cascade regulators, two or more variables ratio control, multivariable control, etc, can be configured without any additional cost or difficulties.

SOFTWARE STRUCTURE OF THE SYSTEM

All the software of the system was developed in C programming language. This medium level language is perfectly valued for our needs due to that it presents several interesting advantages of the low level languages (interrupts handling, access to the input/output ports of the machine, etc.) and of the structured form of the high level ones.

Besides, the C language generates an executable code of less length than that of the high level languages, with less execution time.

In the communications area -a high priority in the system developed- this langauge is the most frequently used by professionals.

All the programs that we designed, not only the corresponding to the central station but the corresponding to the field units too, were structured in separated modules. The development of a maintenance file containing all the necessary orders to obtain a single executable file, was a available possibility in C language. So, it results very easy to modify the software in order to adapt the system to any specific application.

The modules are the follows:

- Modules for the main programs
- Modules for basic communication routines and functions
- Modules for messages transmission
- Modules for messages reception
- Modules for user interface
- Module for graphic representation
- Modules for control routines and functions (for local stations only)
- Modules of additional procedures and functions used

Among the commercial C compilers, we selected the MicroSoft C V6.0, which is one of the compilers that offers more possibilities. Some of its advantages are: the easy interrupt handling; the direct access to the I/O ports and to the absolute memory addresses; on-line assembler code to make critical-time routines, powerful libraries, etc.

The programs that we developed were compiled under the MS DOS operative system in personal computers.

SYSTEM COMMUNICATIONS

The communications structure adopted was that of typical star topology with a Master-Slave organization. The central station takes the master function and the local stations take the slave function, attending any possible orders sent from the master unit.

The communications way used was the serial line one, which confers great economy to the system.

A very important characteristic of the implemented system is that in order to avoid any dependence on the Operative System employed and optimize the communication speed, the basic management of the serial communications port was made programming the PC's Asynchronous Communication Chip (8250) directly. This was made through interruptions, and these have place when information from the port inputs or outputs.

The PC AT equipment that constitutes the master unit must have at least one serial port (COM1) and must be linked to an expansion connector to provide eight free serial connectors in parallel with the input one.

The expansion connector can not be a simple junction of wires, because all the messages of the slave units must be attended by the single serial port of the Master unit and that connector would give place to shortcircuits in the bus wires. To solve this problem, two solutions were proposed.

The first one was to employ a special board for the PC Master that would supply to the computer several communications Chips, one per each slave unit. This solution needed additional Hardware and Software and so, the system resulted more complex and expensive.

The other one was to employ a simple circuit based on diodes to implement the connector. The circuit uses only one diode per serial port and an additional resistor. This circuit can be seen in Fig. 3, where we represent all the bus connections.

In the slave units, it will be enough to have one serial port (COM1), besides the corresponding acquisition system.

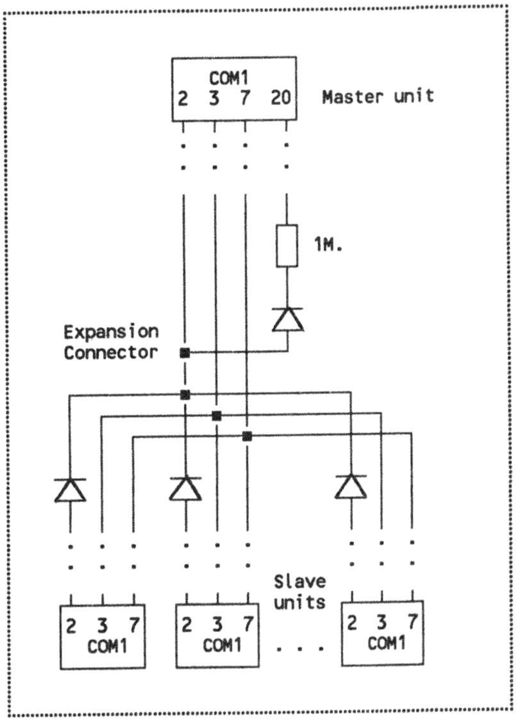

Fig. 3. Bus connections

The length of the serial Bus is limited by the capacity of the wire (2500 pF maximum), and so the maximum length varies from near 1 Km. for 110 baud rate, to 80 m. for 19200 baud rate.

Communications protocol

In data transmission, the communications protocol refers to the structure and procedures adopted in the different messages that link the master station with the remote stations.

For communications in the distributed control system, we designed a simple protocol that has the following characteristics:

- Messages are bidirectional, from master to slave and from slave to master.

- Dialogue is always initiated from the master and the slaves can only to send a message, if they are previously requested by the master.

So the communication is of the Half-Duplex type.

- When the system work in supervisory mode, the master station requests alternatively the slave stations to obtain information from the processes by them (polling algorithm).

- The messages are composed of several bytes transmitted in asynchronous mode according to the EIA RS-232C electric norm, at a 19200 maximum baud rate.

- For each byte of information, as shown in Fig. 4, the bits transmitted are the following:

  · 1 start bit
  · 8 information bit
  · 1 parity bit
  · 1 stop bit

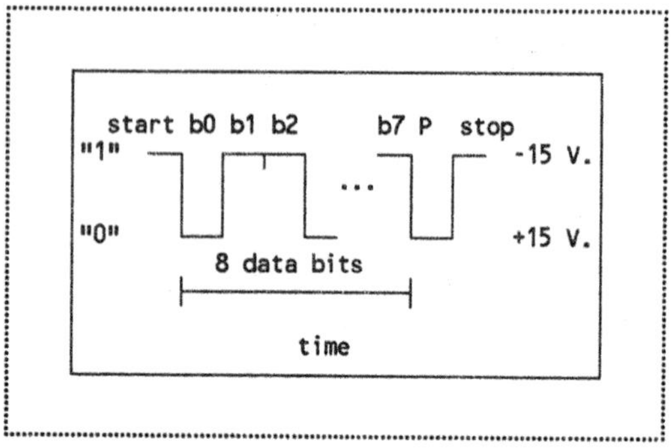

Fig. 4. Frame Format

- The protocol is software totally and so control bytes are used in the communication. Because of this, it is enough to use a communication line with only three wires - transmission, reception and common-, which is better for the economy and reliability of the system.

- For the verification of the messages we used two strategies: the horizontal parity check-up in each transmitted byte and the vertical parity check-up (Parity Byte) in each transmitted message.

Messages structure

The set of bytes to transmit is named message or telegram and has two parts well differentiated, the head and the data field. The head appears in all messages but the data field only appears in the messages that send a numeric value or an increment/decrement information to any variable of the local stations.

The head contain three bytes:

1. Message length.
2. Station code, which identified the station that send the message.
3. Function code of the message.

The data field follows the head and consists of an ordered list of Ascii characters (bytes) that represents a numeric value (including a sign, a decimal point and an exponent if they are necessary) or an increment/decrement symbol (+/-). In the first case, the receptor station converts this character list in a numeric value as to assign it to the corresponding variable. In the second case, it only increments or decrements the corresponding variable in a fix value. The data field do not exist in some messages.

Examples of messages types are showed in Fig. 5.

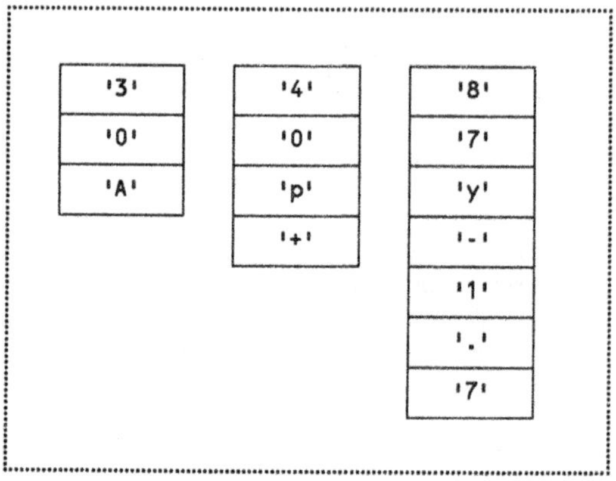

Fig. 5. Different messages types

With the just message structure commented, a very precise value can be sent to the local station with only about ten bytes in the message.

The messages from the central/master station can be of several types:

- Enquire to a local station for variable's value sending in order to be monitorized in the central station.
- Sending of the set point and sample time variable values.
- Sending of the increment/decrement order for the regulator parameters.
- Sending orders to change the control structure to opened loop or closed loop.
- Sending orders to switch on and switch off the local control mode.

From the slave stations it is possible to send the value of any variable of the local control to the master unit, but these must have a previous enquire from it.

Transmission phases

In the sending of the message, besides the proper bytes of the message, other bytes like the control one and the vertical parity are transmitted. The phases of transmission are the following:

- Transference of the control byte that indicates the beginning of the message or STX (Start of text).

- Transfer of the message bytes (head plus data field, if it exists).

- Transference of the control byte that indicates the end of the message or ETX (End of text).

- Finally, transference of the vertical parity byte takes place. This byte is calculated making the XOR logical function with each byte of the message, including the ETX byte. The receptor station also calculates its own vertical parity with the received bytes and at the end it compares it with the vertical parity byte that finally it receives. Only if both parity bytes are equal the message is confirmed to the emitting station.

If the message is received without errors, the receptor station confirms the message to the emitting one sending back the ACK (Acknowledge) control byte. On the other hand, if there were errors, the message is not confirmed and it sends back the NAK (Negative Acknowledge) control byte.

The Ascii and hexadecimal codes of the control bytes are shown in Fig 6. In Fig. 7, the communications protocol is schematized.

Symbol	Ascii	Hex
ACK	6	6
NAK	21	15
STX	3	3
ETX	4	4

Fig. 6. Control Bytes in the Protocol

SYSTEM POSSIBILITIES

Following, we mention the possibilities of control offered by the developed distributed system, both from the local station and from the central station.

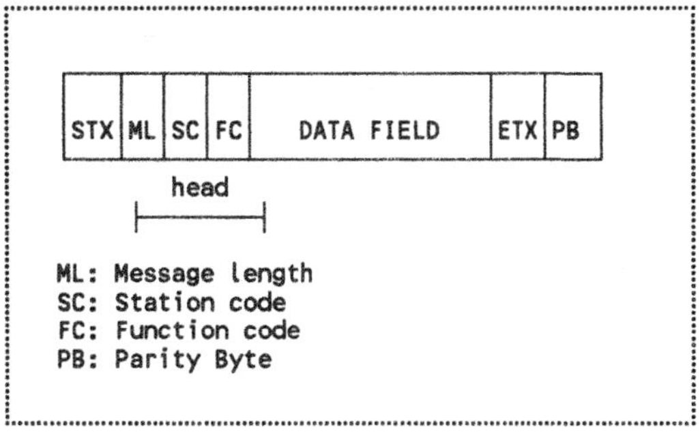

Fig. 7. Communications Protocol

Control from local stations

From the local station, all the control characteristics can be modified:

- Work in opened loop
- Work in closed loop with PID regulator
- Increment/Decrement of the regulator parameters
- Change of the set point
- Change of the sample time
- Graphic and alpha-numeric representation of the process information.

Supervision and control from central station

Besides all the local control possibilities, the following ones are included from the central station:

- Supervision of the current state of the loops.
- Alarm management.
- Graphic and alpha-numeric presentation of each process information.
- Simple and easy-handled interface with the process.
- Switch on and switch off the local control mode.

In this moment, we are developing new possibilities for the system, all for the central station, in order to take maximum advantages from the MASTER PC. Some examples are:

- Identification procedures
- Controller design
- Statistic process analysis

The use of RS-485 network as communications way is also being studied.

EXAMPLES

In Figs. 8,9,10,11,12, we present some examples of the actual representations in the central station. The processes used were linear/first order plants and several types of set point were applied.

CONCLUSIONS

The distributed system presented in this paper is considered as an alternative to conventional equipments within a small-scale, low-cost applications. The combination of versatile and powerful personal computers as work stations and a very simple communications way seems to provide a very reasonable and reliable solution to many application problems of this type.

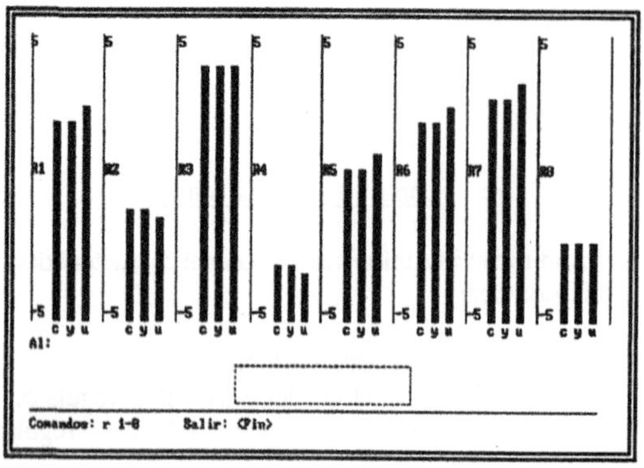

Fig. 8. General supervision of the system.

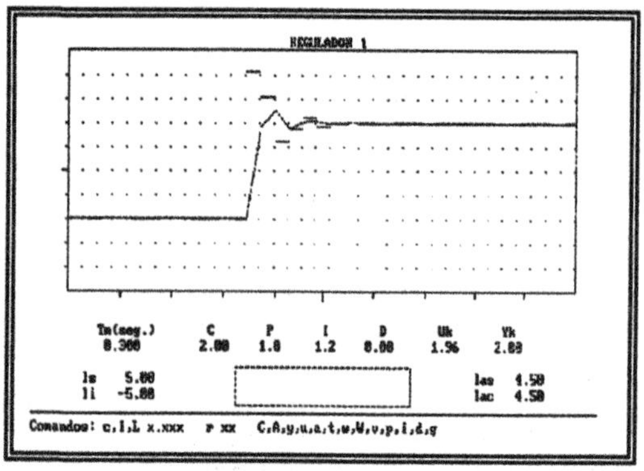

Fig. 9. Control and supervision of a specific loop (set point change)

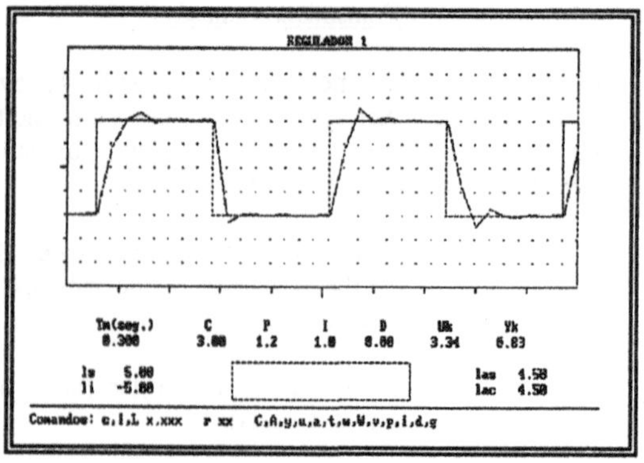

Fig. 10. Control and supervision of a specific loop (square wave reference)

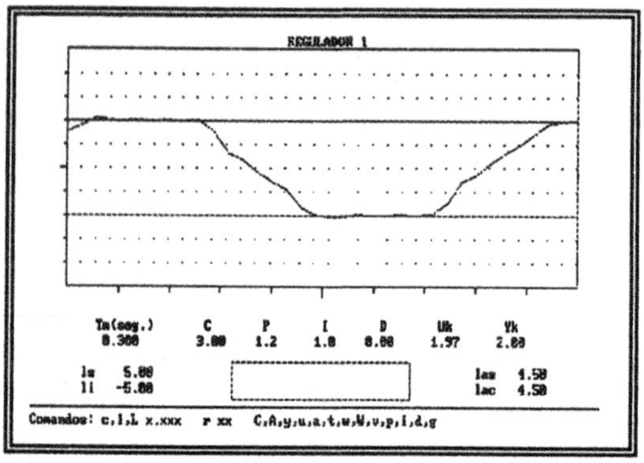

Fig. 11. Control and supervision of a specific loop (ramp reference)

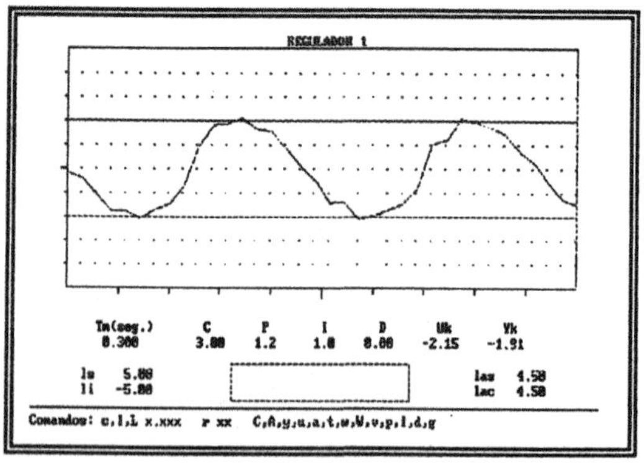

Fig. 12. Control and supervision of a specific loop (sine wave reference)

REFERENCES

SUAREZ ALONSO,F.J.(1990). "Sistema de control distribuido basado en microprocesadores independientes comunicados mediante un bus de campo serie". Univ. Oviedo, Proyecto Fin de Carrera del DIEECS.

PALMGREN,B.,and AASMA,F.(1988). "Microprocessor based development tool, design and realization of SISO controllers". IFAC Workshop on Real Time Programming,Benicasim,Spain.

CREUS SOLE,A.(1987). "Control Distribuido de Procesos industriales". Mundo Electrónico,174,p.65-70.

JOURDAIN,R.(1986). "Solucionario del Programador para IBM PC/XT/AT y compatibles". Editorial Anaya.

AGUADO,A.(1989). "Personal Computers as a tool for the analysis, design and implementation of control systems". IFAC Symposium on Low Cost Automation, Milano,Italy.

APPLICATION LAYER FOR BITBUS BASED LOW COST TECHNOLOGICAL NETWORK

D. Hrubý, K. Fabian and J. Kalina

Institute of Automation and Communication, Slovak Academy of Science, Severná 5, 974 00 Banská Bystrica, Czechoslovakia

Abstract Effective use of automation resources requires that they are interconnected by a communication system. To interconnected field devices such as sensors, actuators and intelligent process controllers; Fieldbus is used. The main idea of Fieldbus is to replace point to point links from each sensor or actuator to it's controller by a serial communication system. In this article we present the field bus from the application layer point of view.

Experimental testbed of system consists of some sensors, actuators and PC machines. As a communication network Intel's Bitbus was chosen. Distributed system is based on the architecture of standard host PC with Master Bitbus controller and many remote slaves Bitbus modules (max 250).

Each slave module represents microcontroller i8344 with on board memory RAM, EPROM, serial communication interface RS-485 and 8-bit PC-BUS extension which allows controlling of many standard I/O industrial PC cards.

PC operates as an extension of a master node. When compared to other commercial field buses, it is seen that Bitbus is at least near enough the IEC standards requirements. Practical experience performed on experimental system have shown that commercial Bitbus software library is cumbersome and is not well suited for hard real time applications. Therefore new software modules were implemented in order to fulfill distributed real-time constraints. The flexibility of the microcontroller i8344 allows to implement specialized Logical Link Control sub-layer and a Application layer.

Keywords Fieldbus, Industrial networks, Distributed control, Microprocesor control, Computer communication.

1. Introduction

In traditional process control system, sensors and actuators are connected by point to point links to the controlling devices. When the system becames very large - typically with several hundreds of sensors and actuators - the control by one central device presents problems and disadvantages:

- the software complexity increases rapidly, much faster then the number of attached devices,
- all functions as control, monitoring, protection are concentrated in one device which is a single point of failure.

An effective solution in this case is to decompose this level of automation into multiple controllers each one performing one of the above functions. The whole control system is now controlled by multiple devices which must have access in real time to the sensors and actuators. A communication network used for most applications must appear as an "ideal" information channel without loss of information, without senors and without information delay to the user. At the lowest level of communication network in factory automation hierarchy is placed the Fieldbus network. Fieldbus networks are special kind of networks intended to link sensors and actuators to controllers.

Field buses essentualy differ from other networks in some ways:

- they are submitted to severe time constraints
- they are installed in harsh industrial enviroments
- field buses transfer usually short messages
- very low attachment cost is required
- a predominant part of the traffic is cyclic.

2. System architecture

Solution described in this paper uses Intel's Bitbus, widely accepted as an efficient, cost-efective, and casily configured system for large-scale data acquisition tasks. Up to 250 multitasking modules, each with multichannel I/O capacity, can be linked to the PC via single plug-in card and a single twisted-wire cable which can be up to 13.2 km long. Intel's Bitbus protocol is an IEEE standard, supported by a number of manufactures of industrial control devices and data acquisition systems. Bitbus offers similar electrical integrity and performance to a token bus network but with a simpler master/slave protocol. Bitbus interconnect supports up to 250 nodes and three bit rates dependent on application performance requirements.

3. Operating environment

Operating environment is composed of IBM PC and the IPCX 344 board, that provides the Bitbus gateway to PC. Based on Intel's highly integrated 8044 (an 8051 microcontroller and an SDLC controller on chip) the iPCX 344 board extend's the real-time control capability of the PC via the Bitbus interconnect.

With this iterface the user can utilize the human interface and application software of the PC and extend the I/O range of the PC to include real-time distributed control. The 8044 microcontroller and on-chip firmware provides a simple interface to PC in one package, and high performance communications and control capabilities for a control system.

The ISO OSI reference model specifies that the communications functions are structured in 7 layers interacting via well defined interfaces. In Bitbus communication model 3 layer architecture is adopted; (Fig. 1) composed of physical, link and application layers. This choice can be justified as folows: communication on a field bus are real-time and a very local. These two facts imply that connection oriented services - as provided by the transport and session layers - are not used. Furthemore, faster response times can be achieved if certain layers of the full architecture are bypassed.

Physical layer:

The physical layer is described in terms of physical and electrical characteristics of the transmission medium and of the electrical signaling method used. For maximum reliability and to facilitate standardization International Electrotechnical Commission (IEC) mechanical board and connector specifications and the Electronic Industry Association (EIA) RS-485 Electrical Specification were chosen.

Link layer:

The link layer performs and controls transmission service functions required by higher layers. It's main functions are:

- acces to the transmission medium
- addressing
- protection of the messages against loss
- transmission control

The data link interface refers to device to device transfer of frames on the Bitbus. On Data link level of the Bitbus interface a subset of IBM's Serial Data Link Protocol (SDLC) with the REJECT option is implmented. The standard frame format transferred across the Bitbus is shown in Fig. 2. The information field carries the Bitbus message.

This communication level is fully implemented in firmware DCM 44 INTEL's BITBUS software [1].

4.Application layer

The Application Layer provides services which are directly applicable by the user. For added reliability, the Bitbus incorporates error checking at the message level in addition to error checking provided by SDLC at the data link level. The message control interface defines the format and functions of messages transmitted in frames across the Bitbus (Fig. 3). The protocol requires that for every message transmitted across the bus a reply message must be transmitted in return. This is a typical Master-Slave protocol and centrally controlled field bus.

The Bitbus networks can be configured in two ways, either as distribute I/O systems with centralized control, or as distributed control system.

From this point of view we propose and implement two application layers for user tasks on PC.

Distributed I/O application layer:

Distributed I/O systems are easy to design. Code that runs on the remote Bitbus board is not required because the network is controlled by the host system. Each Bitbus board comes with a built-in set of procedures known as Remote Acces and Control (RAC). The host sends out commands to the remote nodes and uses these RAC procedures to collect data from remote sensors ar to send data to remote actuators. Input/output Bitbus interface (IOBI) is a set of procedures called from application program to perform communication, remote acces and control functions over the Bitbus interconnect. Most of these procedures are standard RAC commands implemented on PC.

IOBI application layer consist of two sublayers. Lower sublayer is a set of four procedures that enables communication over the interface between PC and IPCX 344 board. These procedures are:

- **interface_init:** performs any necessary interface initialization
- **msg_recieve:** recieves one Bitbus message from another node on the interface
- **msg_transmit:** transmits one message to any node on the interface
- **interface_status:** checks status of interface

This sublayer can use two methods when transferring messages:

iteractive transfer via the processor or transfer via DMA.

Methods can be selected with procedure interface-init.

All these procedures are operating-system-independent.

In higher sublayer are following functions are implemented:

- read/write/logical functions to I/O port
- read/write to memory
- control and information functions

First group of procedures enable application tasks send or receive in one Bitbus message information to/from on 8-bit I/O port at remote node (procedures and_io, xor_io, read_io, write_io, update_io), one 16-bit I/O port (and_wio, or_wio, xor_wio, read_wio, write_wio, update_wio) or up to six (limited by maximum size of Bitbus messages) 8-bit I/O port (write_sio, read_sio). Using concatenated procedures reduces traffic on the Bitbus.

Second group of procedures enables read and write information direct to memory of remote nodes. These procedures are:

read_code_mem, write_code_mem, read_ext_mem, write_ext_mem, read_int_mem, write_int_mem.

Range of memory is limited by the maximum size of one Bitbus message.

Last group of procedures is concerned with the remote node identification and configuration (node-info), software reset at node (node-reset), acces protection at node (protect_RAC) set and check of a Bitbus interface parameters (set-interface, probe-interface).

Distributed control application layer.

Bitbus network can also be used to implement powerful distributed control system. With distributed control, each board functions as a controller performing a set of dedicated tasks. On a periodic basis, the master sends messages to remote boards to collect process control data from sensors. New updated values are then sended to actuators. The built-in multitasking executive on the 8044 Bitbus microcontroller allows up to 7 user tasks to run on the node at the same time. In this case a remote node not only monitors the status of multiple sensors, it can also locally execute user developed control algorithms.

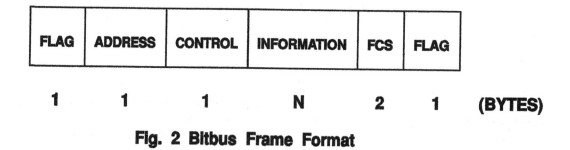

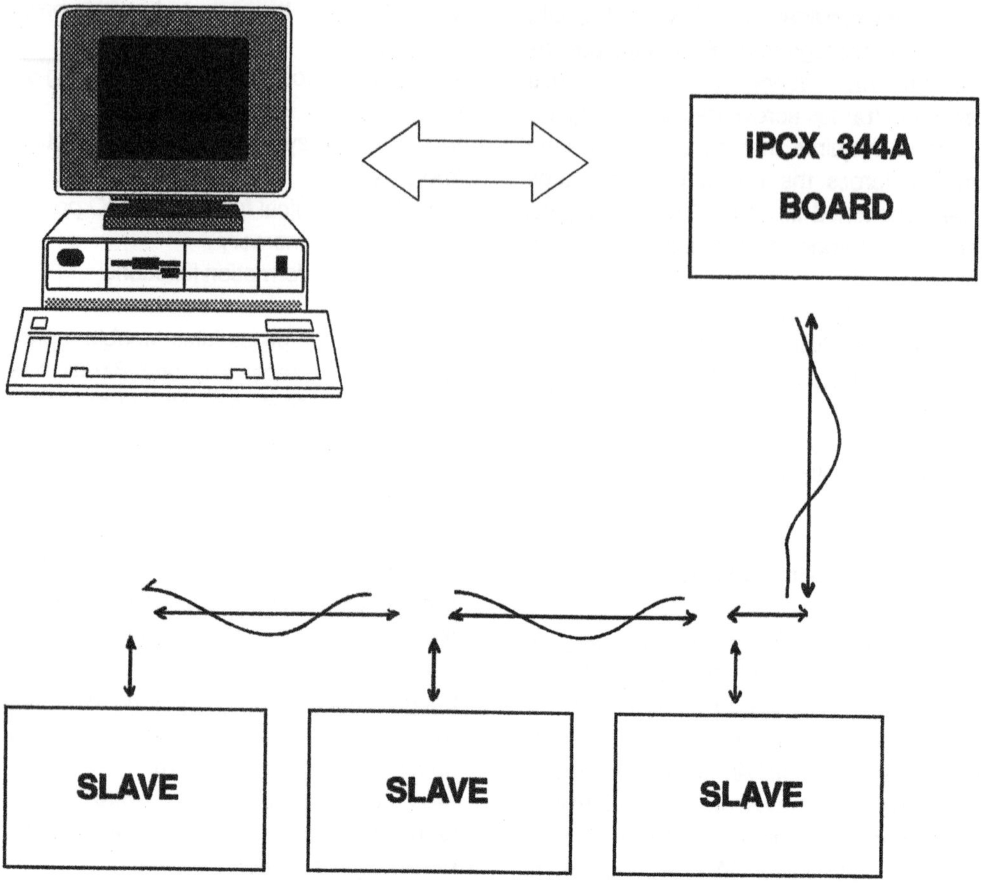

Fig. 1 Operating Environment

FLAG	ADDRESS	CONTROL	INFORMATION	FCS	FLAG	
1	1	1	N	2	1	(BYTES)

Fig. 2 Bitbus Frame Format

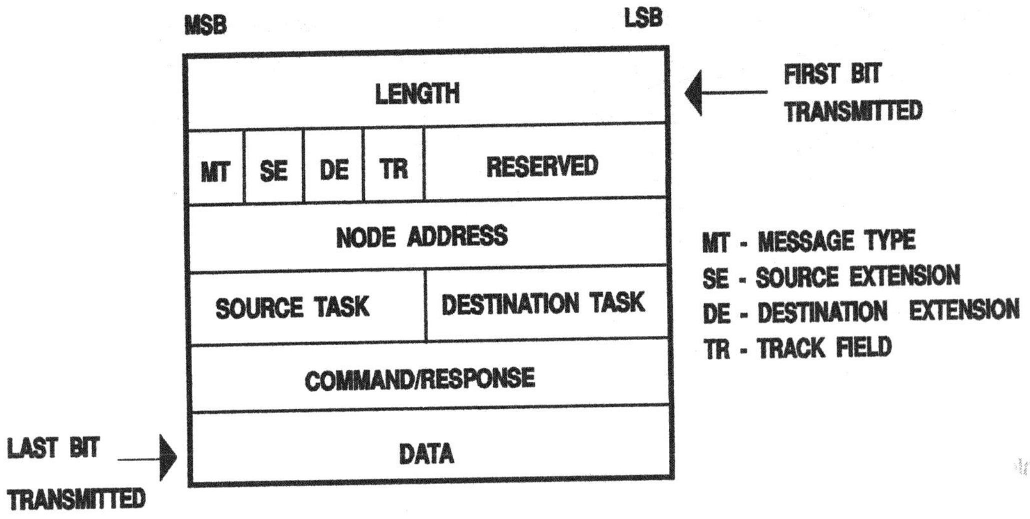

Fig. 3 Bitbus Message Format

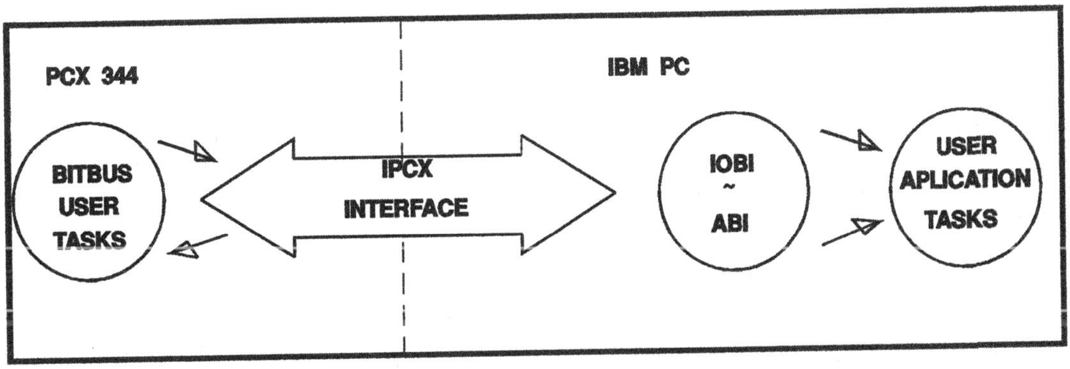

Fig. 4 Exchanging Information between a Bitbus Interconnect and IBM PC System

In this way traffic on the Bitbus is reduced to a minimum with a consequent increase in system operating speed. In this case the Bitbus messages between master and remote nodes carry a information of higher semantic level.
Application Bitbus Interface (ABI) is a set of procedures called from application program on PC to create user tasks on remote nodes, perform remote acces to sensors and actuators, and control functions over the Bitbus interconnect.
ABI layer is composed from two sublayers. Low sublayer is the same as in the case of IOBI layer. Higher sublayer provides services to download code to a remote node (task_load) create and delete tasks at a remote node (create_task, delete_task), receive the most recent state of sensors (read_sensor), update the state of actuators (write_actuators), configuration of the remote (config_sensor, config_actuator), identification sensor and configuration of equipment on the remote node (node_info), set and check of a Bitbus interface (set_interface, probe_interface).

Conclusion

The presented application layers were developed to be employed in multisensory systems. The chosen application is numerically controlled developmnet multisensory system [4], in which all sensors and actuators are linked using Bitbus interconnect. This testbed system is intended for practical and easy development of new control and communication algoritm's, and new configuration approaches. Application layers was implemented using C++ lanquage.

References:

[1] - Decotignie J.D., Pleinevaux P., Field buses in manufacturing automation: Study of the application layer requirements, (1988).

[2] - Ulloa G., Fieldbus Application Layer and Real-time Distributed Systems, IECON 91 Tokyo, Japan, 1991.

[3] - Distributed Control Modules Databook, INTEL 1988.

[4] - Havlík Š., Hrubý D., Kalina J., Pokorník P., An Integration of the Multisensory System, 1992.

LOW COST SIGNAL SAMPLING USING LOCAL INTEGRALS

M.D. van der Laan, W.A. Halang and P. Koblížek

University of Groningen, Department of Computing Science, P.O. Box 800, 9700 AV Groningen, The Netherlands

Abstract. An alternative method for signal sampling is presented, which is based on practical requirements. Samples are obtained by local integration of the signal. The method can easily be realised in standard low-cost hardware. The local integration is to be carried out by voltage-to-frequency converters. A special VFC design, presented in this paper, minimizes the non-linearity error. The resulting overall system has low complexity. It reduces the work load of the analysing computer, and relaxes time constraints.

Keywords. Analog-digital conversion; data acquisition; digital signal processing; approximation theory.

INTRODUCTION

Most sensors produce analogue output signals. To enable digital processing, the signals must be sampled. Usually, signal sampling is carried out through point measurements at equidistant times. The sampling frequency, which is the inverse of the time between two subsequent samples, must, at least, be twice as high as the so-called Nyquist frequency f_N, which is the highest frequency occurring in the signal. This is expressed by Shannon's sampling theorem (Shannon, 1949; Jerri, 1977), which states that, if:

1. a signal $x(t)$ is sampled by an ideal sampler,
2. the frequency spectrum of the signal is bandlimited, i.e., it has a maximum frequency f_N, and the signal is sampled at a frequency higher than $2f_N$, and
3. the signal is observed during infinitely long time,

then it is possible to reconstruct the signal from its samples without error, by applying the interpolation formula:

$$\hat{x}(t) = \sum_{n=-\infty}^{+\infty} x(n\Delta t)\, \frac{\sin(2\pi f_N t - n\pi)}{2\pi f_N t - n\pi} \qquad (1)$$

where $\hat{x}(t)$ is the reconstruction of the original signal $x(t)$, and Δt the time between two successive samples. This reconstruction is equivalent to filtering the sampled signal with a perfect low-pass filter.

However, *the requirements for this theorem cannot be met in practice*, because signals are not perfectly bandlimited and sampling can only be carried out during finite time. Therefore, the value of this theorem for real sampling applications is limited.

REQUIREMENTS FOR ALTERNATIVE SAMPLING METHODS

In practice, a signal is observed during a finite time interval $[0, T]$. The signal is to be represented by a finite sequence of numbers, say N. Sampling can now be generalised as any operation which extracts N real numbers (coefficients) from the signal in $[0, T]$.

It is possible to put the sampling problem into a geometrical perspective. Shannon (1949) already mentioned the geometrical representation of signals, but this point of view is not found in more recent literature. Let the signal be a vector in the infinite dimensional function space $\mathcal{C}[0, T]$, the set of continuous and bounded functions on $[0, T]$. The sampled signal, which is completely characterized by the N coefficients, is a vector in an N-dimensional subspace of $\mathcal{C}[0, T]$. Sampling reduces to an approximation problem: Given the vector $x(t)$ in $\mathcal{C}[0, T]$, find the approximating vector $\hat{x}(t)$ in the subspace, which is closest to the original vector. Shannon's sampling theorem is no more than a special case of this theory: the signals are approximated in the space of bandlimited functions (Moddemeijer, 1990).

A vector in the subspace can be represented by:

$$\hat{x}(t) = \sum_{n=0}^{N-1} c_n\, w_n, \quad c_n \in \mathbf{R} \qquad (2)$$

where $\{w_0, \ldots, w_{N-1}\}$ is a basis of the subspace.

Depending on the nature of the signals in a particular application, one should choose a subspace which provides good approximation quality for the signals under consideration. Given a basis, the c_n's have to be extracted from the signal. The most general linear operator for this purpose can be expressed as:

$$\mathbf{O}_n : \begin{cases} C[0,T] \to \mathbb{R} \\ x \to \int_0^T x(t)\,\lambda_n(t)\,dt \end{cases} \tag{3}$$

Here the quantities λ_n are sums of piecewise continuous functions and at most countably many Dirac distributions (Halang, 1984).
With this general theory, a lot of sampling operators can be deduced. However, the operator is to be implemented by a physical system, which will introduce additional requirements (Van der Laan, 1992):

- **Approximation properties**
  The optimal sampling operator minimizes the "distance" between the two vectors. In a normed vector space the distance equals the norm of the error signal. Different norms can be used, e.g., the commonly used mean square norm. However, in order to be sure that the error never exceeds a certain limit, the operator should be optimised with respect to the minimax norm:

$$\|x\|_\infty = \sup_{t\in[0,T]} |x(t)|. \tag{4}$$

- **Linearity**
  The operator should be linear (in the mathematical sense). Only in this case linear operations on an analogue signal can be approximated by linear operations on the sampled signal.

- **Time-invariance**
  As it is non-economical to have a different procedure for each of the N samples, it is recommended to implement the sampling with a single procedure, initiated at N different times. Linear time invariant operators can be realised using linear filters followed by point measurements.

- **Causality**
  For real-time applications, convolution can only be realized by a linear filter if the impulse response is causal, i.e., future values are not to be used. If the time conditions are weak, also semi-causal filters are allowed. These filters use only a finite part of the future input signal, which can be realised by delaying the input.

- **Equidistant evaluation**
  In order to avoid additional timing data to be stored with the samples, the procedure should be repeated at equidistant times. If both time invariance and equidistant evaluation are required, the sampling operator can only handle stationary signals.

- **Compact support**
  It is desirable, but not necessary, for the sampling operator to have compact support. Hence, the influence of erroneous samples remains local. If the support cannot be compact, the support must damp out rapidly.

- **Suppression of noise and other distortions**
  The operator should possibly be insensitive to other distortions commonly present in practical systems, such as noise or spikes. E.g., one should avoid point measurements because they are highly sensitive to both imperfections.

THE LOCAL INTEGRAL SAMPLING OPERATOR

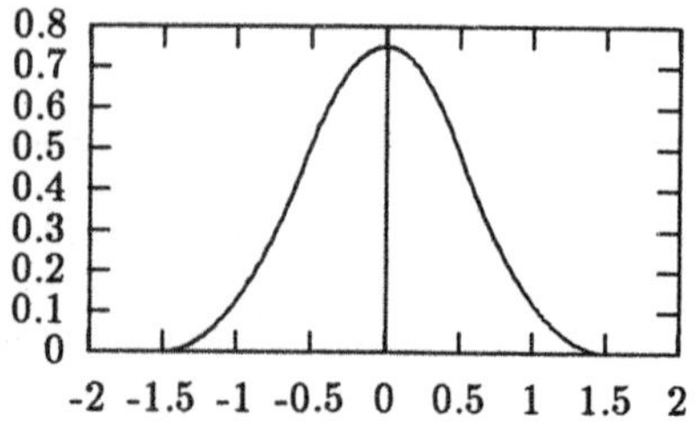

Fig. 1. The second-order B-spline

In many approximation problems, polynomial *splines* give good results. Therefore, we consider a space of polynomial splines as a subspace of the signal space. An appropriate set of basis functions are B-splines as defined by the following equation:

$$b_{(m)}(t) = \frac{1}{m!}\sum_{k=0}^{m+1}(-1)^k\binom{m+1}{k} \left(t+\frac{m+1}{2}-k\right)_+^m \tag{5}$$

in which:

$$(x)_+^m = \begin{cases} x^m & x \geq 0 \\ 0 & x < 0 \end{cases}$$

and m is the degree of the spline. The B-spline of degree 2 is shown in Fig. 1. Splines have some advantageous properties. First, they are easy to evaluate, because every spline consists of polynomial pieces. Moreover, splines have a compact support.
In Shannon's sampling theorem, as well as in traditional spline approximation theory, the coefficients are obtained by point measurements at equidistant times. However, it was pointed out that point measurements should be avoided, because of the high sensitivity to noise and spikes. It is better to use an integrating measurement, which is less sensitive to those imperfections. This is also in agreement with the practical situation, because a physical device cannot take point measurements, but will always locally integrate the signal. To extract the coefficients, the following operator will be used:

$$c_n = \frac{1}{\Delta t}\int_{n\Delta t}^{(n+1)\Delta t} x(t)dt \tag{6}$$

where again $\Delta t = T/N$ is the sampling time.

Combining the local integrals (6) with a B-spline representation, the total sampling operator becomes:

$$\hat{x}(t) = \sum_{n=-m}^{N-1} c_n \cdot b_{(m)} \left(\frac{t}{\Delta t} - n - \frac{m+1}{2} \right) \qquad (7)$$

The m basis functions occurring additionally in this representation are necessary to achieve a uniform approximation quality on $[0, T]$. The corresponding coefficients must be calculated from the signal outside $[0, T]$ or by other methods.

This linear and positive spline-approximation operator has been investigated in great detail by Halang (1980, 1984). It shows good approximation quality with respect to the minimax norm, and it possesses a number of advantageous properties. Important for sampled data processing is its "variation diminishing" property which provides inherent suppression of spikes and high-frequency noise, and its compact support which bounds the effect of erroneous data. It is easily verified that the operator is time-invariant and that the coefficients are calculated at equidistant points in time. The operator is semi-causal. It may be concluded that this local integral operator meets the requirements of the previous section.

Traditionally, sampled data are filtered in order to reduce noise. Calculations (Halang, 1984) showed that the inherent noise suppression by the above operator is better than that of a digital FIR filter applied to (pointwise) sampled data. In contrast to digital filtering no additional arithmetic operations are necessary, reducing the redundancy required in the sampled data. Consequently, the sampling frequency can be lower.

Moreover, the operator has decreased sensibility to power-frequency disturbance, provided that the integration period is a multiple of the power frequency. The local integration can be seen as convolution with a rectangular kernel, followed by point measurements. Since the Fourier-transform of a rectangular kernel is a sinc-function, the sensitivity to power frequency noise can be reduced by choosing the integration time Δt in such a way that the zeros of the sinc-function are exactly at multiples of the power frequency.

Since the samples obtained by the integral sampling operator differ from "normal" (point) samples, the existing theory for digital processing of analogue signals is no longer valid. Therefore, new theory has to be developed, which is subject for ongoing research. In practical applications, this will not cause problems, because the point samples are always approximated by small integrals.

IMPLEMENTATION IN HARDWARE

Analogue integration can be implemented using Voltage-to-Frequency-Converters (VFC's). The input signal is transferred into a pulse train with a linear relationship between input voltage and output frequency. Integration then reduces to pulse counting. The accuracy of integration depends on the number of pulses within an interval and, therefore, on the dynamic range of the VFC. This range should be high enough to retain the information content of the original signal in the outgoing pulse train.

The operation principle of a Voltage-to-Frequency Converter is to charge an integrating capacitor with the applied input voltage until the voltage at the capacitor has reached a given level. Then, an output pulse is generated and the capacitor is discharged, to be loaded again by the input voltage. Following the same principle, in some designs the integration is inverted, i.e., with the input voltage the capacitor is discharged to zero, before being charged again for the next cycle. Only if the discharge time is negligibly small, a linear relationship between the input voltage and the output frequency is obtained. However, in practice the discharge time cannot be neglected, which introduces a nonlinearity error.

A solution to this problem is based on a dual slope concept, in which the capacitor is charged *and* discharged with the input voltage. A dual slope VFC circuit is displayed in Fig. 2. The central part of this circuit consists of the integrating amplifier (opamp 5). The integration capacitor is not charged by the full input voltage U_x, but by the voltage $U_{mid} + U_x$, where $U_{mid} = U_{cc}/2$. On the other hand, the capacitor is discharged by the voltage $U_{mid} - U_x$. Because of the unipolar power supply, it is obvious that $|U_x| < U_{mid}$. The charging and discharging of the capacitor is switched at the time when the output voltage reaches the reference levels U_l or U_h:

$$0 < U_l < U_{mid} < U_h < U_{cc}$$

The input of the integrator is controlled by two analogue switches $S1$ and $S2$. Assume the capacitor is being charged, i.e., switch $S2$ is closed and the input is $U_{mid} + U_x$. The output of the integrator is fed into a comparator (opamp 6). If the output voltage reaches U_h, the comparator output goes down. This transition opens the charging switch $S2$, closes the discharging switch $S1$ and changes the reference level of the comparator to U_l by closing $S4$. The capacitor is being discharged with $U_{mid} - U_x$ until the new reference level is reached. Then, the comparator output goes up, thereby starting a new charging phase.

The circuit was realised using discrete components. It showed good linearity, compared to the traditional VFC concepts. Of course, the results could not compete with commercial VFC's, because of the technology used. However, we think that the circuit, realised in a single integrated circuit, has better linearity than existing VFC's.

The total sampling system is drawn in Fig. 3. Every sensor is connected to a VFC, which should be located as close as possible to the sensor. The VFC-s' digital output pulse trains are transported over a single pair of wires to the computer. A special count/storage module counts the pulses and stores the counter-values at the end of every integration period into a register. Each channel has one of these modules associated with it. The computer simply accesses the registers to obtain the samples. The only time constraint is that the values must be read before the next

integration period is finished.
This sampling system has the following advantages:

- Distributed processing of analogue data and (noiseless) digital transport to the processor via low cost cables.
- Reduced communication complexity, since the sampling units do not require a central clock.
- Decreased processor load; a datum is transferred only once per integration interval.
- Easy calibration of VFC by supplying DC voltage.

CONCLUSIONS

1. The here presented local integral sampling system provides an alternative to traditional point measurements. It was shown that the method has a number of advantageous properties with respect to the suppression of noise, spikes and power frequency disturbance. Because additional digital filtering is no longer required, the sampling frequency can be lower than usual, thus relaxing the capacity requirements of the transport channels and the analysing computer.
2. The dual slope VFC shows good linearity. This circuit can be realised as a low cost, single chip device.
3. The sampling operator can be implemented using VFCs and count/storage modules. This implementation requires only standard (low cost) hardware. Due to the fact that the communication complexity between the count/storage modules and the computer is low, the latter can be used more efficiently for analysing the signal.

REFERENCES

Halang, W.A. (1980). Approximation durch positive lineare Spline-Operatoren konstruiert mit Hilfe lokaler Integrale. *Journal of Approximation Theory* 28, 161–183.

Halang, W.A. (1984). Acquisition and representation of empirical time functions on the basis of directly measured local integrals. *Interfaces in Computing* 2, 345–364.

Jerri, A.J. (1977). The shannon sampling theorem – its various extensions and applications: A tutorial review. *Proc. of the IEEE* 65(11), 1565–1598.

Laan, M.D. van der (1992). Towards alternative strategies for signal-sampling. In *Proc. of the 13th. Symposium on Information Theory in the Benelux*, 81–88.

Moddemeijer, R. (1990). Sampling and linear algebra. In *Proc. of the 11th. Symposium on Information Theory in the Benelux*, 118–125.

Shannon, C.E. (1949). Communication in the presence of noise. *Proc. of the I.R.E.*, 10–21.

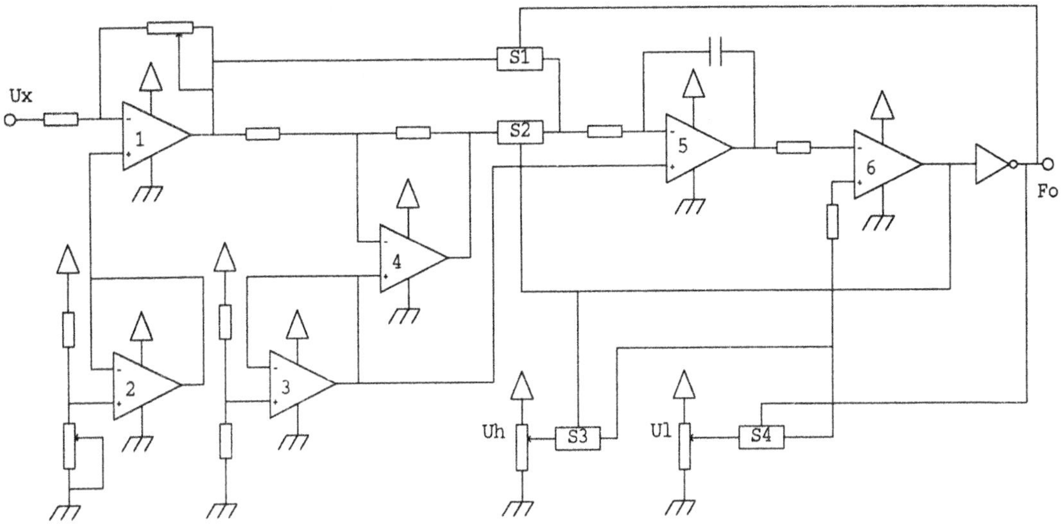

Fig. 2. The dual slope VFC concept

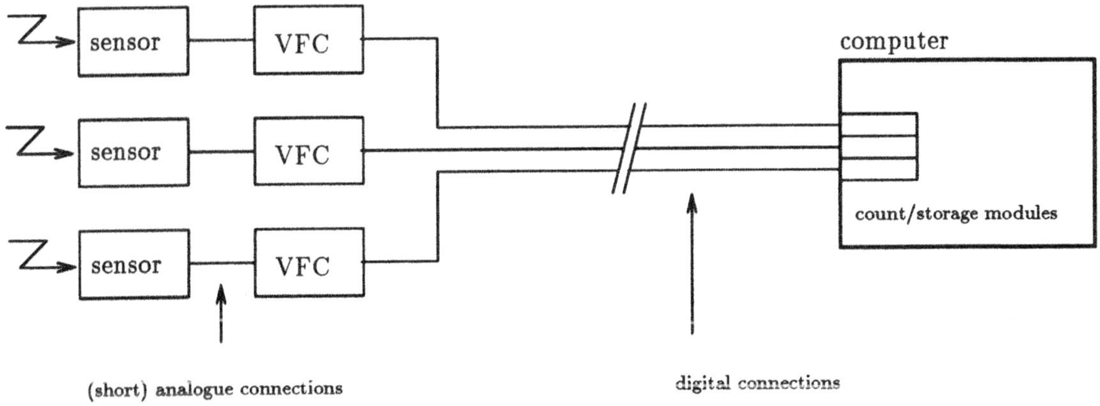

Fig. 3. Block diagram of sampling unit

DIRECT CONTROLLING OF SCALE MODELS FROM PC-MATLAB

J. Houška and B. Sládek

Department of Control, Faculty of Electrical Engineering, Czech Technical University, Karlovo nám 13, 121 35 Prague, Czechoslovakia

Abstract. For direct access to real plants and their scale models from MATLAB the RT–TOOLBOX was created. It is a high-level software tool for data acquisition, rapid prototyping and testing of control algorithms in MATLAB enviroment using real systems and their scale models. This paper presents basic features of RT–TOOLBOX and experience with its use. Tasks, supported by RT–TOOLBOX involve both simple access to A/D and D/A convertors and more complex tasks like periodical I/O operations, signal filtering, discrete controllers and data recording which run in background with different sampling rates.

Keywords. Computer control, control system design, real-time computer systems, software tools, education

INTRODUCTION

Implementing and testing modern control algorithms is expensive, time-consuming activity. This is due to a wide range of tasks which have to be done simultaneously.

Low-lewel input/output operations must be performed in regular time intervals often with multiple sampling frequencies.

High-level self-tuning algorithms on the other side, are so computation intensive that they can be hardly be performed synchronously with the sampling frequency.

Identical problems appear in education of control engineering. Experiments with real plants and their scale models are valuable source of students experience. There are many ways how to perform them:

Simulation. It gives fast results in comfortable enviroment. If these results are not verified immediately with a real plant, this approach can lead to results which can be hardly implemented in practice. This approach is often responsible for the lack of confidence in modern control algorithms.

Traditional programming languages. Languages like PASCAL, C or OCCAM provide high flexibility together with the possibility to use special hardware like transputers or signal processors. Very low support is available for design and implementation of the control algorithm itself. Consequently students concentrate on debugging the programs rather than on controlling the plant.

Cross-compilers. They provide the missing link between high-level design tools like MATLAB [1] and special hardware like transputers and signal processors (Peretti, 1992). This very hopeful way has two major drawbacks, namely problems with portability between hardware platforms and necessity to work in different environments.

Special data acquisition software. More comfort is available here but the obtained data must be transferred between at least two different enviroments. This brings communication difficulties and slows down the whole process which must run off-line consequently.

Data-acquisition tools. If they are integrated directly into a traditional high-level design tool they combine high support for design with possibility to process data on-line. This is the basic need for implementation of adaptive controllers.

Despite of some implementation problems, the last mentioned approach has been used in this work. As a high-level enviroment, the PC–MATLAB has been chosen. The MATLAB itself sets de-facto standard in CADCS and is available across wide range of hardware platforms. Its version, PC–MATLAB , has been chosen for its availibility, simple extensibility and also for its cost.

[1] MATLAB is a registered trademark of MathWorks, Inc.

CONCEPTS

There are two classes of tasks characteristic for modern control algorithms:

Data processing. Tasks like data pre-processing, control loops and data recording are time critical but usually with low computational demand.

High-lewel procedures. Identification, controller design, filter design, etc. are computational intensive but can be performed on-line with respect to I/O operations.

In the ideal case, each task would run concurrently with other tasks in real-time enviroment allowing their mutual synchronisation. Parallel execution is necessary condition of modular design and testing of the whole control system. Debugging and modifying programs which are written this way saves significant amount of time and effort.

Ideal controller module

It would act as follows:

- Get data from A/D converters
- Process it with no delay
- Send data to D/A converters
- Store it for later use
- wait till the end of the sampling period
- repeat the above action

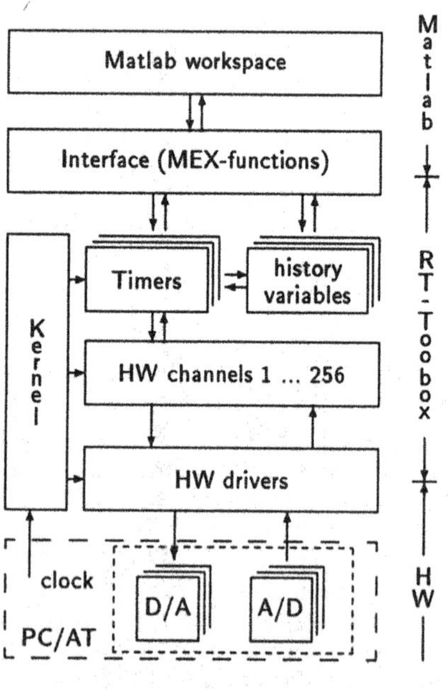

Figure 1: Structure of RT–TOOLBOX

This can not be implemented under MATLAB directly because:

- MATLAB is a typical single user single task enviroment.
- Although there exists a built-in function for time measuring, it is not sufficient to provide the real-time enviroment.
- The interpretation language of MATLAB is too slow for real time applications. Although it does some precompilations, it cannot manage more than one SISO polynomial controller with sampling frequency 10–20 Hz.

To get to these demands as close as possible following structure has been chosen: up to one MATLAB program running on foreground with a set of low-lewel preprogrammed tasks running on background (see Figure 1).

IMPLEMENTATION

The RT–TOOLBOX (Fig. 1) consists of the following function blocks: user interface, kernel, internal variables and hardware interface.

Timing

Because of the sequential nature of MS–DOS, parallel run of all tasks must be provided by means of RT–TOOLBOX .

Each *timer process* is specified by the period of activation, by the priority and by the time remaining to its next execution. Execution of a process is requested if the time to next activation is less or equal zero. Execution of a task is started if it is requested and and no task with a higher priority is currently being executed. Any task with lower priority is then stopped until all task requests with higher priority are finished.

Using this strategy, a task can be lost completely only if total time demand is higher than the 1, i.e. if

$$\sum_{i=1}^{p} t_i \frac{T_i}{T_0} > 1, \qquad (1)$$

where
p is the number of processes to start,
t_i is time demand of the i-th process,
T_i is the period of the i-th process,
T_b is the basic period of the kernel.

This situation is detected and reported as an error. In addition to this the user can check the actual overhead of RT–TOOLBOX to prevent these problems.

Overload detection fails in the only case (rare in practice), if the task with highest priority can itself cause overload situation. This is the case if very high order transfer functions or a very short sampling periodes are used. This situation can be recognized only by very high amount of time given to the background.

The delay between the task request and its execution is varying and it can not be predicted exactly. As the toolbox has been mentioned for educational purposes, this delay is not considered to be harmful.

Algorithms

The current set of preprogrammed algorithms includes:

Data transfer. Data are passed from input hardware channels to a shadow variable or from a shadow variable to output hardware channels.

$$Y_j(z) = X_j(z), \qquad (2)$$

where
Y is the shadow variable in input process or hardware port in output process,
X is the hardware port in input process or shadow variable in output process,
n is the number of A/D or D/A channels used.

Discrete transfer function. For data filtering as well as for the SISO discrete controller this discrete transfer function can be used

$$Y_j(z) = \frac{B(z)}{A(z)} X_j(z), \qquad (3)$$

where
A, B are polynomials describing the discrete transfer function.

Note that the discrete transfer function in z operator

$$\frac{Y(z)}{U(z)} = \frac{B(z)}{A(z)} = \frac{b_{n-m}z^m + \ldots + b_n z^0}{a_0 z^n + \ldots + a_n z^0} \qquad (4)$$

is implemented as

$$y_k = \sum_{i=0}^{n} \frac{b_i}{a_0} u_{k-i} + \sum_{i=1}^{n} -\frac{a_i}{a_0} y_{k-i}. \qquad (5)$$

This realization is nonminimal, as it uses $2n - 1$ states to represent a system of n-th,order. This results in having $n - 1$ non-observable and non-controllable states in the system realisation. Because these states are stable they cause no problems. For a more detailed discussion see (Kučera, 1979).

Multivariable transfer functions. They provide a realization of MISO polynomial controllers. A MIMO polynomial controller has to be implemented as a set of MISO controllers.

$$Y(z) = \sum_{j=1}^{n} \frac{B_j(z)}{A(z)} X_j(z), \qquad (6)$$

where A, B_j are the polynomials describing the multivariable discrete transfer function. The realization is similar to the SISO transfer function.

Data structures

Timer variables. To describe process state, timer variables are used. Parameters available to the user involve:

- **timer number** which is used to identify the process and to set its priority.
- **period** of execution
- **time** remaining to next execution
- **channel numbers**
- **type of task** e.g. input/output, polynomial transfer function etc.
- **parameters of task** e.g. numerators/denomitors of a transfer function.

Links. By default, a *shadow variable* is linked to a hardware channel. This assignment can be changed by *linking* the channels explicitly. If a timer is linked to another timer then it reads input data from the output of this timer or writes the output data to its input in the same way as if it used hardware channels.

History variables. To provide data recording, history variables are introduced. The history variable is a matrix of one dimension equal to the dimension of the timer shadow wariable and the other dimension determining the number of samples to hold.

If the history variable is linked to an output timer, each time the timer process is executed the result is stored into the history variable and the history variable counter is incremented. Input timers use history variables in a similar way to fetch predefined input values.

Interfacing

User interface. It is provided by a set of external MATLAB functions (MEX–files). The data access is asynchronous from the user's point of wiew with internal synchronization. The workspace of RT–TOOLBOX and of the MATLAB are separated. Main advantage of this approach is higher stability of the RT–TOOLBOX kernel, no memory allocation conflicts and guaranted integrity of read/written data.

Hardware interface. A set of hardware-dependent *hardware drivers* for data transfer between physical devices and logical hardware channels is used. When running a program with different I/O hardware only the hardware driver must be changed.

EXAMPLES

Examples in this section are typical students excercises. The main criterion for their choice is the need to demonstrate capabilities of RT–TOOLBOX at the limited given space.

Example 1

Simple PID Controller. PID running in foreground is the simplest but most illustrative controller which can be implemented.

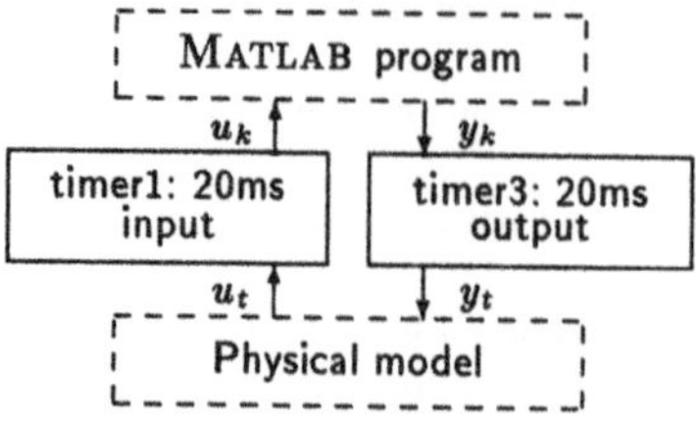

Figure 2: Simple PID controller

```
% Loading of the kernel & driver
% for the PCL-812 card
rtload('pcl812');

% Controller constants
Ts = 0.1; kp = 0.5; kd = 0.5;

% Set point, initialization
w = 0.3; e1 = 0;

%--------- T i m e r s -------------
%       Nr  type   period  channel
rtdef( 1 , 'in' ,   Ts    ,   11  );
rtdef( 3 , 'out',   Ts    ,    1  );

% Control loop; 500 steps
rtstart;

for i = 1:100;
  % wait for next sampling time
  while rtrd(1,'t')<i; end;

  % read the system output
  y = rtrd(3,'y');

  % compute the controller output
  e = w - y;
  u = kp*e + kd*(e-e1)/Ts;
  e1 = e;

  % write the controller output
  rtwr(1,'y',u);
end;

rtunload;
```

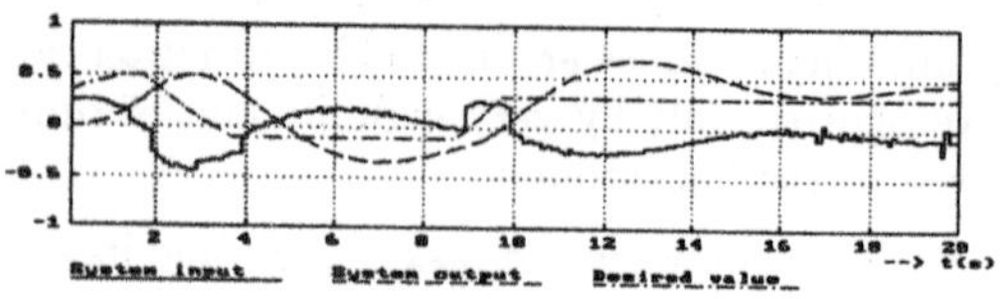

Figure 3: Data from PID

Example 2

Polynomial controller. PID controller from Example 1 is implemented as discrete polynomial controller running on background with digital filters on controller input and output.

This structure provides a platform for implementation of modern control algorithms.

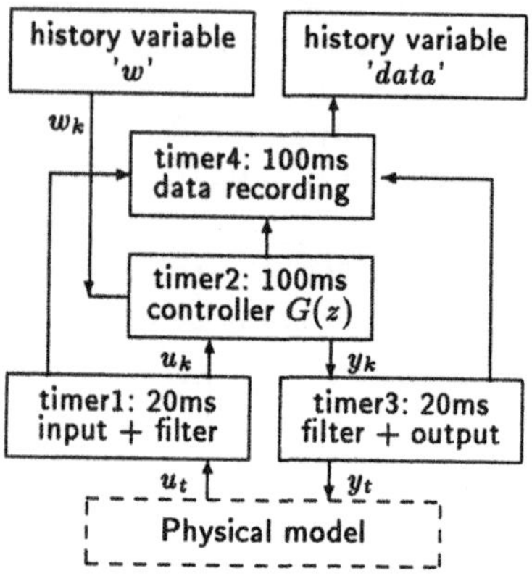

Figure 4: Polynomial controller on background

```
% Controller constants
bc = [kd kp]; ac = [0.001 1];
[bc,ac] = c2dm(bc,ac,Ts,'matched');
bc = [bc;-bc];

% Filter constants
Tf = Ts/5;
[bf,af] = butter(3,Ts/Tf);

%--------- T i m e r s -------------
%     Nr  type    period  chan. param.
rtdef( 1 ,'tfin'    ,Tf , 11 , bf,af)
rtdef( 2 ,'mvtfout',Ts , 255, bc,ac)
rtdef( 3 ,'tfout'  ,Tf , 1  , bf,af)
rtdef( 4 ,'in'     ,Ts , [255 254] )

%-------- History variables --------
%         name     size
rtdef( 'data' , 200,2 )
rtdef(  'w'   , 200,1 )
```

```
%--------- L i n k s ---------------
%           Source target channel
% controller
rtlink(      2    ,  'w' ,  1     )
rtlink(      2    ,  1   ,  2     )
rtlink(      3    ,  2            )
% recorder
rtlink(      4    ,  2   , 254    )
rtlink(      4    ,  1   , 255    )
rtlink(    'data',  4             )

% setpoint
load w
rtwr('w',w)

% Control loop; 200 steps
rtstart;
while rtrd(4,'t')<200; end;
data = rtrd('data');
rtunload;
```

Example 3

Real-time simulation. Simulation is used to compare system parameters obtained by LQ identification with behaviour of the real system from Example 2.

To switch between real-time experiment and simulation, only the definition of timer 1 has to be changed.

The is example demonstrates that this approach is even more convenient than the use of MATLAB language, especially if the influence of different sampling periods has to be simulated.

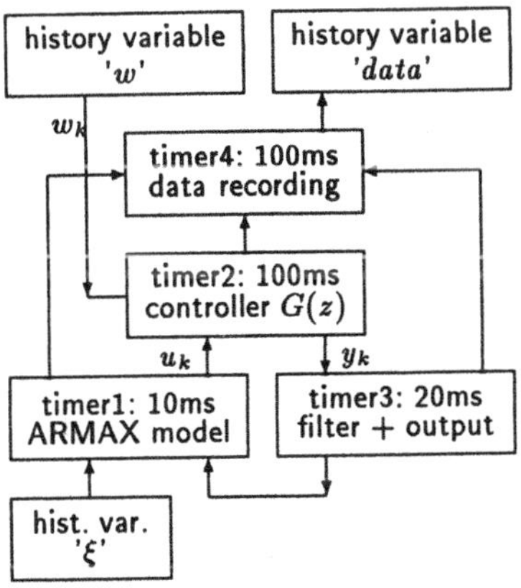

Figure 5: Simulation

```
% ARMAX model parameters
Tm=Ts/10; bm = [0 0 1]; am = [1 0 0];
[bm,am]=c2dm(bm,am,Ts,'zoh');
bm=[bm;zeros(bm)]; bm(2,1)=1;
```

```
%--------- T i m e r s -------------
%      Nr  type  period  chan. param.
rtdef( 1 ,'mvtfout',Tm , 255, bm,am)
rtdef( 2 ,'mvtfout',Ts , 255, bc,ac)
rtdef( 3 ,'tfout'  ,Tf , 1  , bf,af)
rtdef( 4 ,'in'     ,Ts ,[255 254]  )

%-------- History variables --------
%         name     size
rtdef( 'data' , 200,2    )
rtdef(  'w'   , 200,1    )
rtdef( 'ksi'  , 200,1    )

%--------- L i n k s ---------------
%           Source target channel
%  controller
rtlink(      2    ,  'w' ,  1     )
rtlink(      2    ,  1   ,  2     )
rtlink(      3    ,  2            )
% recorder
rtlink(      4    ,  2   , 254    )
rtlink(      4    ,  1   , 255    )
rtlink(    'data',  4             )
%model
rtlink(      1    ,  3    , 1     )
rtlink(      1    , 'ksi' , 2     )

% external disturbance = 0
ksi=zeros(200,1);
rtwr('ksi',ksi)
```

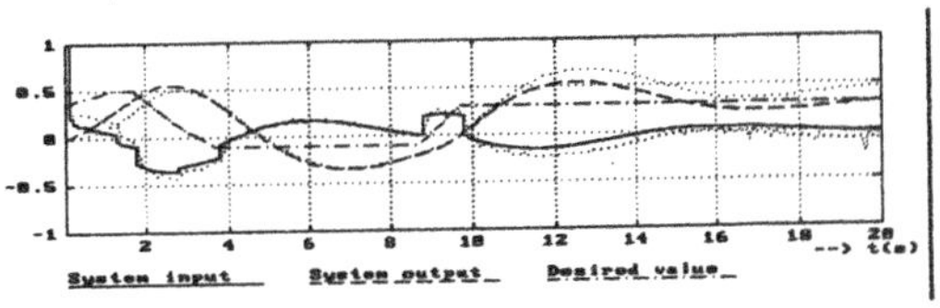

Figure 6: Simulated vs. real data

CONCLUSIONS

The first release of the RT–TOOLBOX had been used in MSc. courses in the Dept. of Control, Faculty of Electrical Engineering of Czech Technical University in Prague for half a year.

Although this version did not contain all the features available in the later version created recently, which is subject to extensive tests now, following conclusions can already be made:

- Time, spent with data acquisition has been reduced to minimum, more tasks can be solved in given time.
- extensive use of scale models of real systems, described in (Horáček, 1991) instead of analog computers and digital simulation makes laboratory excercises more interesting and close to practice.

- the possibility of testing new algorithms in the early stage of its development decreases the risk of using inadequate simplifications.
- developing and testing of new control algorithms (fuzzy control) becomes more comfortable.

For students work, the RT–TOOLBOX proved to be an excellent enviroment. Fore more serious research work, memory limits of PC–MATLAB could be rather narrow. To solve this source of possible problems, possibility of transfer to higher versions of MATLAB , such as AT–MATLAB and 386–MATLAB are now being solved.

Acknowledgment. This work was supported in part by the TEMPUS project JEP–0886-92/2.

REFERENCES

Åström, K.J. and B. Wittenmark (1984). *Computer Controlled Systems, Theory and Design*. Prentice Hall, Englewood Cliffs.

Horáček, P. and J. Houška (1991). *Control Systems Laboratory in Control Engineering Education*. in: Proceedings of 20th IGIP international Symposium on Engineering Education, Dresden.

Houška, J. and B. Sládek (1992). *Real Time Toolbox for PC-Matlab*. in: Proceedings of 18th IFAC/IFIP Workshop on Real Time Programming WRTP'92, Bruges.

Kučera, V. (1979). *Discrete Linear Control*. Academia, Praha.

MathWorks, The, (1991). *PC–Matlab v. 3.5i, User's guide*. The MathWorks, South Natick.

Peretti, G., Milek J. and Guzella, L. (1992) *CTRL-Lab — a Tool for DSP-Based Implementation and Testing of Control Algorithms.* in: Proceedings of SICICI'92, Singapore.

A GUIDE TO THE DESIGN OF A DIGITAL CONTROLLER

R. Caccia and A. De Carli

Department of Computers and Systems Sciences, Via Eudossiana 18, I-00184 Rome, Italy

Abstract: It is possible to design a control strategy by different procedures with a different complexity. Better results are obtained by dedicated controllers, rather than general purpose controllers like PID. The design of a dedicated control strategy requires in general the knowledge of the mathematical model of the system to be controlled and a mathematical formulation of its specifications. In this paper a simple design procedure that does not require a heavy mathematical support is proposed.

INTRODUCTION

A *Low Cost Automation* approach gives the opportunity to join not only end users to specialists of controlled systems, but also designers of controlled systems to researchers in Control Theory Area. This second aspect is particularly important since it allows to transfer results of researches into the applications and to improve the procedures that refer to ideal systems and ideal operating conditions.

The design of a control strategy makes up only the first step of the realization of the controlled system. The second step is the implementation of the device that makes the selected control strategy operate. The closer the connection between these two steps, the more efficient is the attained realization.

Control strategies can be divided into two classes. The first one includes procedures that do not require the knowledge of the dynamic behavior of the system to be controlled. These procedures are based on empirical criteria (see PID regulators, for example) and although can be very simply applied, they allow a limited influence on the behavior of the controlled system. The second class includes procedures based on the knowledge of the dynamic behavior of the system to be controlled. According to the systematic tools these procedures allow to design dedicated strategies that can have a strong influence on the behavior of the controlled system.

The dedicated control strategies are the most suitable ones for innovative problems or for series products because they allow to attain an optimum cost/performances ratio. It's impossible to attain this goal by applying a general control strategy.

The design of a dedicated control strategy needs the knowledge of systematic procedures proposed sometimes in an unjustifiably complicated way. The application of this kind of control strategies can provide obvious advantages in the behavior of the controlled system; nevertheless they are not diffused because they are proposed in an abstract and unusual mathematical framework. This aspect discourages industrial designers from acquiring innovative methods. It is possible to obtain better results without employing the proposed procedures as well. It is fundamental to master the meaning of the new control strategy and to work out easier and more efficient design procedures.

The aim of this paper is to propose a procedure for the design of a dedicated digital control strategy that can be handled by designers without the knowledge of abstract mathematical approaches. These procedures can bring end users closer to experts in controlled systems too.

For the correct application of the proposed procedure it is suitable that the designer is up to date of the advantages of various control strategies on the performances of controlled systems. Moreover, the designer familiar with a programming language like MATLAB™ should be able to solve the computations required by the procedure easier.

The importance of the proposed approach is due to the fact that digital techniques make the implementation easier. The difficulties in the design of digital control strategies remain although some improvements in the behavior of a controlled system can be obtained by applying empirical techniques also.

FRAME OF THE PROBLEM

The choice of the control strategy depends on the knowledge of the dynamic behavior of the system to be controlled, on the foreseen operating conditions, and on the performances required by the controlled system, which are called specifications. The definition of the specifications is a really delicate problem. End users give in fact qualitative information often not referring to the operative conditions of the controlled system. The designer of the control system tends to link the specifications to the procedures to follow for the design of the control strategy sometimes without understanding their meaning in the realization of the controlled system. This status quo makes it difficult to define specifications that are good for them both.

When fixing specifications, it is suitable to assume that the performances of the controlled system should be better than the ones of the uncontrolled one, and to find out the main aspects of the behavior of the controlled system that must be obtained by the chosen control strategy.

In a controlled system the most relevant aspects

concern the steady-state and the transient behavior, the possibility to have the controlled variable depending on the command variable only, since other factors, e.g. disturbances, don't influence the controlled variable. Specifications should take all these aspects into account.

As regards the steady-state behavior, it should be fixed if the controlled variable must reach the required value without error. It should then be determined if the controlled system is to have a dynamic behavior faster than the non-controlled one. The requirements on the steady-state and the transient behavior are more important than the ones regarding disturbances. They both can be worked out by defining the required step response of the controlled system.

The shape of the step response of the controlled system is a meaningful specification. When defining this shape it is suitable to know the one of the step response of the non-controlled system, to avoid requiring an impossible dynamics.

The simplest way to define the specifications for the controlled system is to fix the location of some relevant points of its step response. It is then necessary to verify that the selected points can be interpolated by a smooth curve, typical of low-pass systems. A spline procedure is so used to trace out the shape of the step response: the interpolating curve is in fact continuous up to the second order derivative. Fig. 1 shows the step and the impulse response of the system to be controlled, the points used to characterize the required step response and the spline curve.

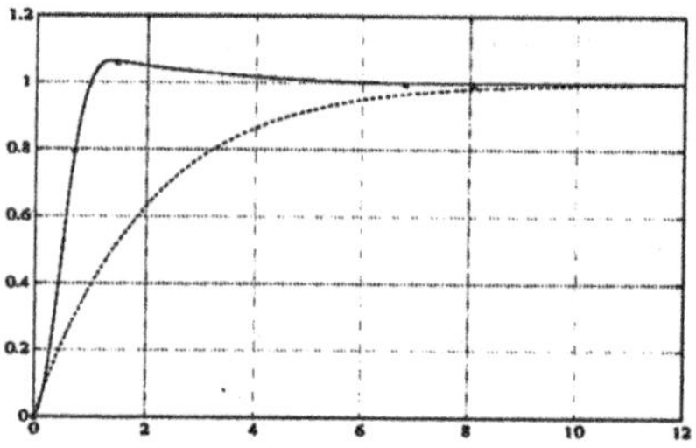

Fig. 1 - Open loop (continuous) and required response (asterisks), spline curve (dashed).

Once it is verified that the specifications are fixed in a suitable way, it is possible to use the spline coefficients for the analytical computation of the impulse response, i.e. the derivative of the step response. In such way uncertainties due to the computation of the impulse response from the step response are reduced. A similar procedure gives the shape of its impulse response if the only step response of the plant is known or measurable.

In the control system shown in fig. 2, if the transfer functions of the controlled system and the plant are

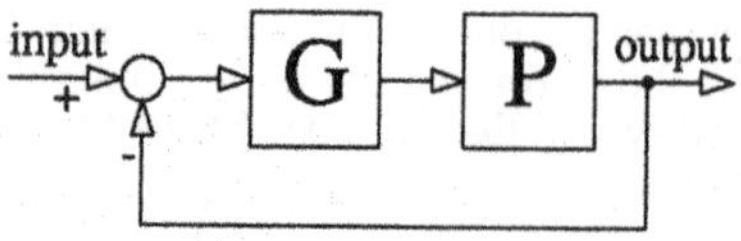

Fig. 2 - Control scheme with one compensator G.

known, being

$$W = \frac{G \cdot P}{1 + G \cdot P} \tag{1}$$

transfer function G, whatever variable (z or s) should be, is given by:

$$G = \frac{1}{P} \cdot \frac{W}{1 - W} \tag{2}$$

As the analytical representations of W and P are not given, it is suitable to follow a numerical procedure for computing the mathematical model of the controller. The numerical procedure needs the discretization of the impulse response of the controlled system and of the plant and applies a deconvolution procedure. It can be computed by a macro instruction that is generally available in high level languages (for example Matlab).

To use the deconvolution result for implementing the controller, further computations should be carried out to reduce uncertainties due to the discretization and to the quantization. The mathematical model obtained in such way can be easier implemented by analog or digital circuits.

Digital controllers are the most interesting technical solution if their realization is carried out by a low cost hardware.

Therefore a procedure will be presented for deducing the mathematical model. This latter should be the simplest one which does not reduce the quality of the controlled system.

To attain this goal, it is necessary to take into account that in a digital controller not only the sampler, but also the zero holder and the anti aliasing filter A(s) are present. Figure 3 shows the block diagram of a closed loop system with a digital controller G(z).

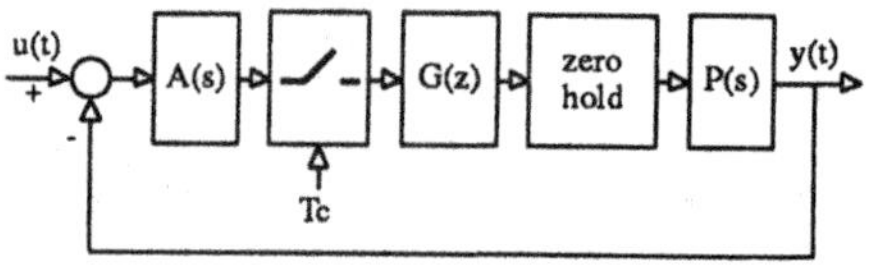

Fig. 3 - Control scheme with sampling, hold device and anti-aliasing filter A(s).

Once the system to be controlled and the filter are both characterized by the same model the corresponding shape of the step response can be easily computed by convolution and so the block diagram of figure 4 is equivalent to that of figure 3.

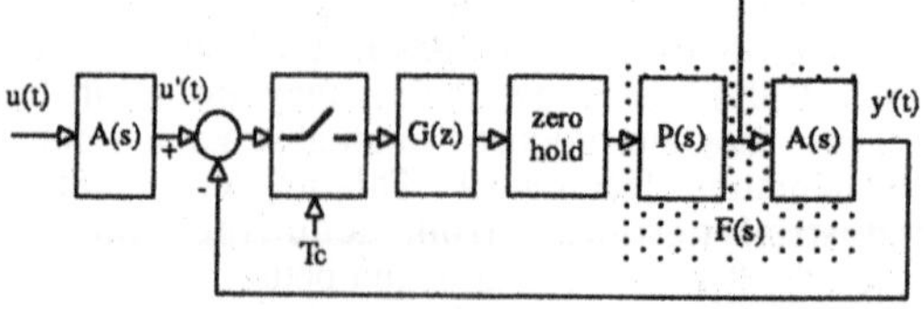

Fig. 4 - Control scheme with anti-aliasing filter included in the plant.

Since the relation

$$\frac{y'(s)}{u'(s)}=\frac{y(s)\cdot A(s)}{u(s)\cdot A(s)}=\frac{y(s)}{u(s)} \tag{3}$$

is valid, the controller G(z) to be synthesized can be obtained imposing the same specifications to the problem of fig. 5:

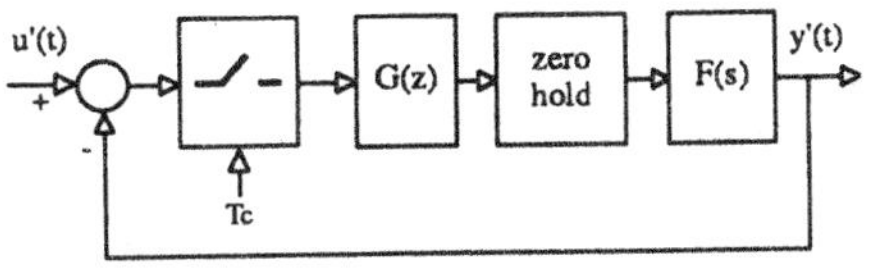

Fig. 5 - Equivalent control scheme.

Since the system in Figure 5 is equivalent to that in Figure 6, the digital controller to be synthesized is the

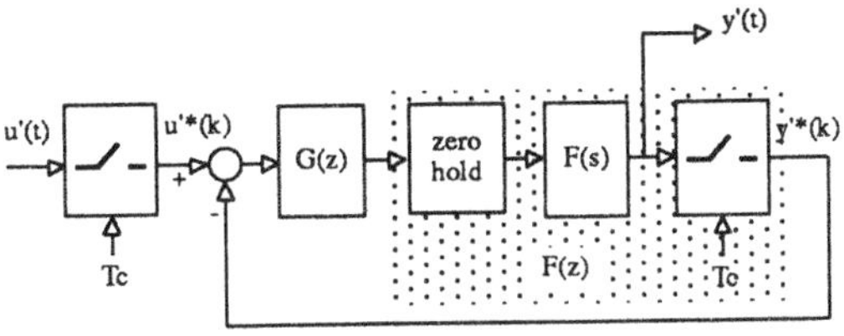

Fig. 6 - Equivalent control scheme.

one that allows the system of Figure 7 to have a step response matching the required one in the sampling points.

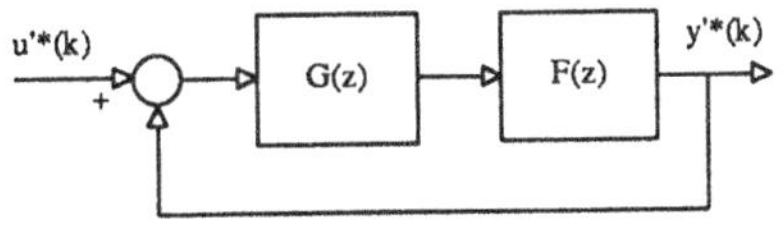

Fig. 7 - Discrete time control scheme.

Considering the hold device, the discrete model can be easily obtained from the sampled values of the step response by considering the transfer function of the hold device, that is

$$\frac{1-e^{-sT_c}}{s} \tag{4}$$

F(z) is given by:

$$F(z)=\frac{z-1}{z}\cdot Z\left\{F(s)\cdot\frac{1}{s}\right\}= \\ =s(0)+\frac{s(1)-s(0)}{z}+\frac{s(2)-s(1)}{z^2}+\ldots \tag{5}$$

The samples of impulse response F(z) result that

$$F(z)=f(0)+\frac{1}{z}\cdot f(1)+\frac{1}{z^2}\cdot f(2)+\ldots \tag{6}$$

In terms of MATLAB instructions, being the samples of step response of F(s) stored in a row vector Fstep it results

Fstep = [Fstep(0) Fstep(1) Fstep(2) .....]

The impulse response F(z) stored in the variable Fz is given by the difference between Fstep vector and the same vector delayed by one step:

Fz = [Fstep(0) Fstep(1)-Fstep(0) .....]

that is

Fz = [Fstep 0] - [0 Fstep]

The same procedure gives the impulsive response W(z)

Wz = [Wstep 0] - [0 Wstep]

The numerical controller is obtained by a direct synthesis procedure processed by a numerical procedure. Being

$$W(z)=\frac{G(z)\cdot F(z)}{1+G(z)\cdot F(z)} \tag{7}$$

if the number of samples of the impulse response is large enough to represent both W and P, the impulse response G(z) is given by

$$G(z)=\frac{1}{F(z)}\cdot\frac{W(z)}{1-W(z)}= \\ =g(0)+\frac{1}{z}\cdot g(1)+\frac{1}{z^2}\cdot g(2)+\ldots \tag{8}$$

In terms of MATLAB instructions:

```
numG = Wz;
temp= [1-Wz(1)  Wz(2: n) ];  % 1-W(z)
denG= conv( Fz , temp );     % F(z) (1-W(z))
[ fut , G_z ] = deconv(numG,denG,n);
```

Deconv(num,den,N) function yields the division between two polynomials of this shape. Number N of the significant samples (the ones characterized by a high signal/noise ratio) is a finite number, because it is necessary to realize a strictly stable numerical controller. This function yields two values. The first one is the number of the non-zero samples for negative time. When this number is greater than zero the controller cannot be physically carried out and the synthesis must be repeated with different parameters and/or specifications. The second one is the quotient vector containing the samples of the impulse response G(z). By computing them it is necessary to identify a mathematical model for G(z) in the form

$$x(k)=\mathbf{\Phi}\cdot x(k-1)+\mathbf{\Psi}\cdot u(k-1) \\ y(k)=\mathbf{C}\cdot x(k)+\mathbf{D}\cdot u(k) \tag{9}$$

or

$$G(z)=\frac{num[G]}{den[G]}= \\ =\frac{b_{n-1}z^{n-1}+\ldots+b_1z+b_0}{z^n+a_{n-1}z^{n-1}+\ldots+a_1z+a_0} \tag{10}$$

During the identification procedure the denominator coefficients a_i must be first computed; then the numerator coefficients b_i can be computed.

The main problem is to impose the degree of the system to compute the denominator: if the degree is too low, its impulse response does not match very rapid varying shapes; if the degree is too high there are problems due to the presence of very close pole-zero couples. The designer's experience can work out a good trade-off solution with few attempts.
An automatic procedure for determining the degree of the system is illustrated in Figure 8.

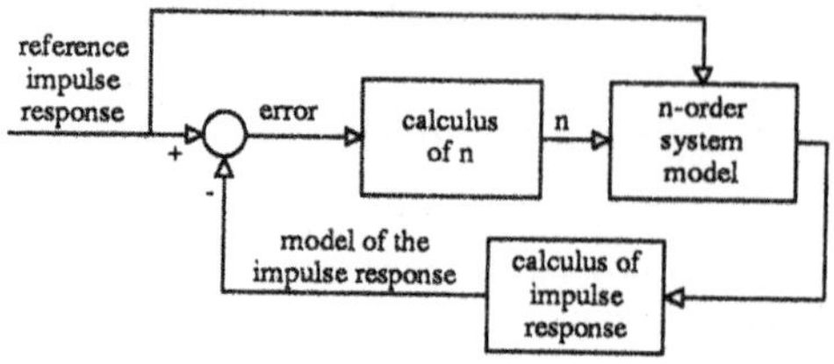

Fig. 8 - Algorithm for the identification of the system.

The algorithm starts with a first order attempt and a step by step increase of the degree is executed until the difference between the impulse response and the modelled one minimized.

To compute the denominator and the numerator coefficients it must be remarked that

$$G(z)=\frac{b_{n-1}z^{n-1}+\ldots+b_1 z+b_0}{z^n+a_{n-1}z^{n-1}+\ldots+a_1 z+a_0}= \\ =g(0)+\frac{1}{z}\cdot g(1)+\ldots+\frac{1}{z^k}\cdot g(k) \qquad (11)$$

where g(k) is the last non-negligible sample of the impulse response. The denominator coefficients ai and the values of the impulse response are related as follows

$$\begin{bmatrix} g(0) & g(1) & . & g(n-1) \\ g(1) & g(2) & . & g(n) \\ . & . & . & . \\ g(m-1) & g(m) & . & g(n+m-2) \end{bmatrix} \cdot \begin{bmatrix} a_0 \\ a_1 \\ . \\ a_{n-1} \end{bmatrix} = \begin{bmatrix} -g(n) \\ -g(n+1) \\ . \\ -g(n+m-1) \end{bmatrix} \qquad (12)$$

By the least squares method it is possible to solve this set of equation and compute the denominator of G(z).

The b_i coefficients of the numerator polynomial it must be noticed that (11) yields

$$\begin{cases} b_n = g(0) \\ b_{n-1} = g(0)\cdot a_{n-1} + g(1) \\ b_{n-2} = g(0)\cdot a_{n-2} + g(1)\cdot a_{n-1} + g(2) \\ \quad .. \\ b_0 = g(0)\cdot a_0 + g(1)\cdot a_1 + .. + g(n) \end{cases} \qquad (13)$$

The designer must claim the value of the degree N of the transfer function (that is the degree of the denominator polynomial) that must identify the response and the number M of the linear equations. Such method also shows the frequency response of a system whose step response is only given.
It is important to know the mathematical model of G(z) for the realization of a controller with a closed loop structure that can be easily implemented by a computing device as a microprocessor [3].

EXAMPLE

The proposed procedure has been applied to design the controller of a plant chracterized by the following transfer function

$$P(s)=\frac{1}{2s+1} \qquad (14)$$

Figure 9 shows with a dashed line the open loop step response of the plant and with a solid line the desired closed loop step response of the controlled plant.

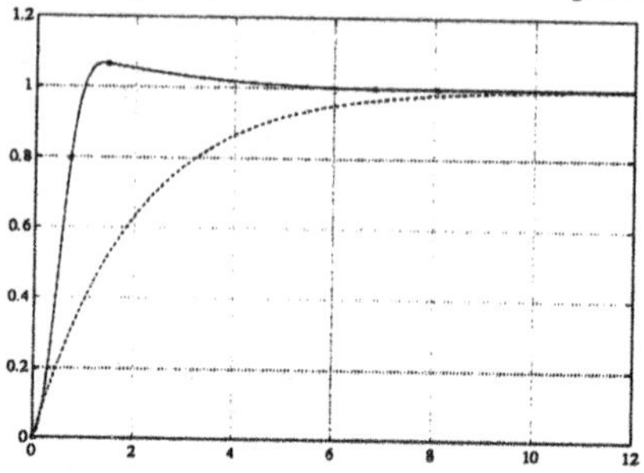

Fig. 9 - Plant and required step response.

The Interpolation has been effected by means of a spline function which allows to work out very easily the continuous shape of the impulsive response. From this latter, the sampled step response is deduced as illustrated in Figure 10.

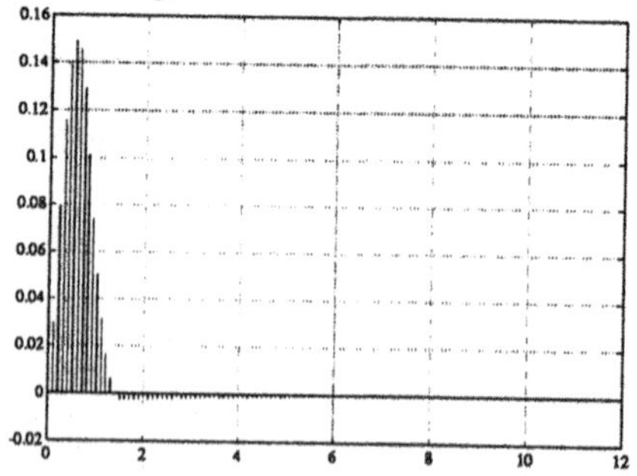

Fig.10 - Impulse response samples of closed loop system.

The sampled impulsive response of the controlled is worked out by applying a dedicated program which implements the proposed procedure and processes the samples of the closed loop and the open loop impulsive response. Figure 11 illustrates the sampled impulsive response of the controller.

The implementation of the controller on a microprocessor device could be effected by using this samples directly according to an open loop filter

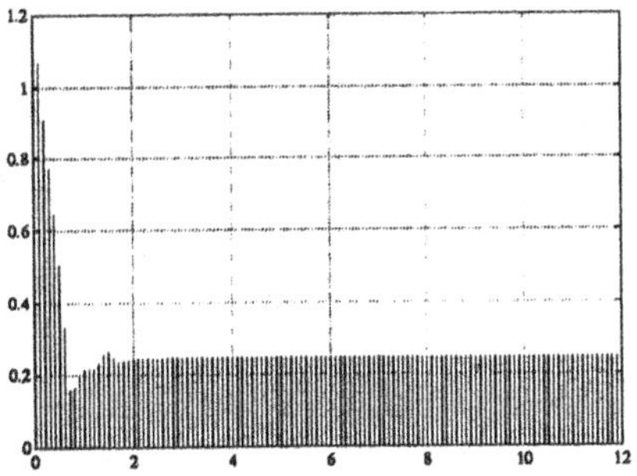

Fig.11 - Impulse response samples of controller.

realization. In order to optimize the computing time and the accuracy, a closed loop filter realization is better indicated. In order to attain this target the mathematical model of the sampled impulsive response should be deduced. By processing these samples according to the procedure of identification previously indicated, the coefficients of the discrete time transfer function are computed. It results:

NUMERATOR	DENOMINATOR
b(0) = -0.0068	a(0) = -0.2401
b(1) = -0.1536	a(1) = 0.5149
b(2) = -0.2197	a(2) = -0.4152
b(3) = -0.0809	a(3) = 0.2261
b(4) = -0.0324	a(4) = -0.3121
b(5) = -0.1614	a(5) = 0.5191
b(6) = 0.2910	a(6) = -1.2923
b(7) = 0.6017	a(7) = 1

A further optimization of the implementation is obtained eliminating the natural modes characterized by a lower gain In this way, the order of the transfer function is reduced and consequently, also the computing time of the digital controller. This approach allows to work out the transfer function of the controller in which only the dominant natural modes appear. The new set of the transfer function coefficients results

NUMERATOR	DENOMINATOR
b(0) = 0.0284	a(0) = 1
b(1) = 0.4548	a(1) = -3.5103
b(2) = 0.3442	a(2) = 4.4834
b(3) = -0.9428	a(3) = -1.9732

Figure 12 compares the impulse response of the original controller with the simplified one.

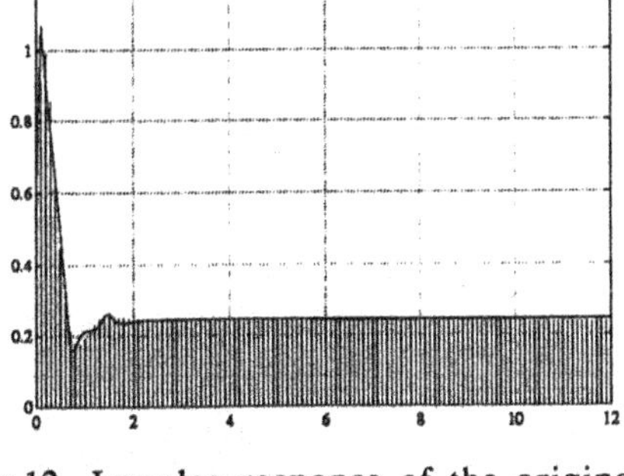

Fig.12 - Impulse response of the original and simplified controller.

The discrepancy is really very low and does not influence the overall behavior of the controlled plant, as shown in Figure 13 in which the desired step response is also illustrated with a bold line.

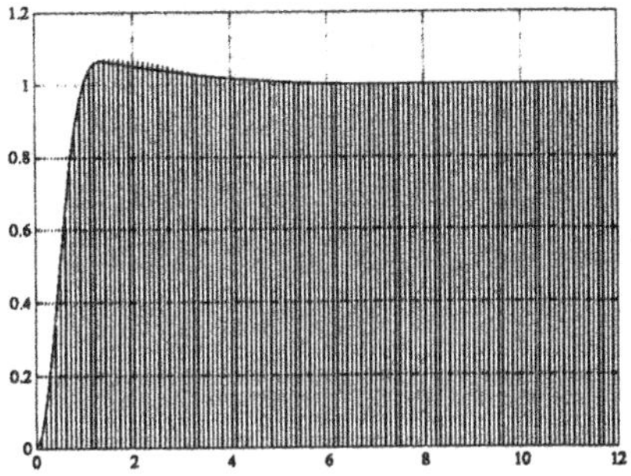

Fig.13 - Desired and obtained step response.

CONCLUSIONS

Good results are obtained by the controlled system if specifications are defined so that they can be really obtained by the controlled system. This feature is independent from the procedure applied in the design of the control strategy. It is in fact possible to design the control strategy by various procedures characterized by a different complexity.

The most meaningful specification regarding the steady state and the transient behavior is given by the step response of the closed loop system. Besides this, the specification regarding the input-forcing and the disturbance-output behaviors is important.

The shape of the step response of the controlled system can be fixed without the analytical support. It is fundamental to define specifications that can be really satisfied.

As shown in this paper the application of direct procedures for the design of a control strategy does not necessarily require the formulation of analytical models. The iterative application of a direct synthesis procedure allows to improve the results to attain other specifications.

The development of alternative procedures that simplify the design of the controller is very important, because the use of only PID controllers limits the quality of the performances achievable by the controlled system.

This paper has been developed with the sponsorship obtained by the Ministry of University and Scientific and Technological Research for 1992.

BIBLIOGRAPHY

[1] ISERMANN R.: *Digital Control*, Springer-Verlag, Berlin,1990.

[2] MATLAB *Control Toolbox, Identification Toolbox, Signal Toolbox*, The Math Works.

[3] CACCIA R., DE CARLI A.: *A Criterion For Qualifying The Implementation of Digital Controllers*, Proceedings of the IFAC Symposium on "Intelligent Components and Instruments for Control Applications", Malaga 1992.

[4] DE CARLI A., KOZLOVIC A.: *A New Approach for the Direct Synthesis of a Digital Controller*, Proceedings of the Proceedings of the 2nd IFAC Symposium on LCA "Techniques Components and Instruments Applications", Milano 1989.

GRAPHICAL METHOD FOR THE ROBUST ANALYSIS OF P.I.D. CONTROLLERS

J. Quevedo

ETSEIT, Univ. Politècnica de Catalunya, c/Colom, 11. 08222 Terrassa, Spain

Abstract. The difficulty involved in characterizing industrial processes with reliable parametrized models, and their generally variable, non-linear behavior, makes it necessary to consider the robustness of controlled processes as a key aspect in the design of control systems.

In this paper a useful method is presented for analyzing the robust stability of processes with structured perturbations in their model's parameters. The process controllers considered must have a linear structure and fixed parameters.

This method has been implemented with specific software, called ANAGRO, which makes it possible to graphically observe whether there exists any risk of instability in the controlled process, enables graphical comparison of various PID controllers, and aids in selecting the most robust one.

ANAGRO also provides a index of the robustness of various controllers for the same process, and allows the analysis of the relative stability (minimum damping value) of the system.

Keywords. Robust analysis, PID controllers, robust design, CACSD.

INTRODUCTION

Due to perturbations in process behavior, the robustness of a controller is fundamental for limiting possible degradations of the operating performance of the controlled process.

Modifications of perturbations in the process, as asserted by Boyd and co-workers (1990), include both physical changes in process behavior and the imprecision or simplification of the process model used for the controller's design or adjustment, and constitute entirely the degree of uncertainty of the model representing the process.

Therefore, it is essential to take into account the degree of uncertainty in the process when analyzing the performance of a particular control system and in designing new robust control algorithms. At present, this aspect is being intensely studied by various automatic control research groups (see the survey by Dorato (1987) and Fu (1989) among others).

Our paper introduces a software package for graphically analyzing the robust stability of control systems with linear structure and fixed parameters (generally PID) acting upon processes with a particular degree of uncertainty. This software is based on the method described by Gerry (1988) and Kompass (1989) and includes several extensions to it.

The proposed method of analysis considers the process to be represented by a know, structured model and that the process uncertainty is inserted in the model's parameters (structured perturbations) for three important reasons :

Firstly, because it is familiar to the control specialist to limit the changes in process behavior to the model's basic parameters (one may often hear ,"... this process is critical because its gain varies by a factor of 1 to 10, or because its time delay may vary from between half and double its time constant... ").

Furthermore, the application of parametric process identification and estimation methods based in their input and output, as in the case of IES (T. Escobet, 1988), provides a fixed model structure and parameter values defined by their mean value and standard deviation, which may be used to define the structured perturbations of the process .

Thirdly , because the results of robustness analysis with structured perturbations are usually exact (necessary and sufficient conditions) while those due to non-structured perturbations (test of maximum singular value of structured singular value) are generally conservative (sufficient condition only) and not very useful for determining their real robustness limits.

In Section 2, below , the method for graphical robustness analysis is described , along with its advantages, and its possible field of application. In Section 3 the basic characteristics of the ANAGRO software, which implements the analysis method, are presented, and the results obtained are shown in Section 4. Lastly, Section 5 presents the main conclusions of this work.

GRAPHICAL METHOD FOR ROBUSTNESS ANALYSIS

The graphical method for robustness analysis consists of drawing on a plane the curve which delimits the stable and unstable regions of the controlled process when there is an uncertainty in the two process parameters (their gain and delay time, for example). The axes on the robustness graph (Figure 1) correspond to the variations in the uncertainty parameters relative to their nominal values, and the dotted polygon in the figure corresponds to a particular safety area for the uncertain parameters.

The delimiting curve of the stable and unstable regions has been calculated by Routh-Hurwitz stability classical method, which gives the **n** stable interval for one of the parameters (p_1) which corresponds to **i=1,..,n** discrete values for the other parameters (p_i).

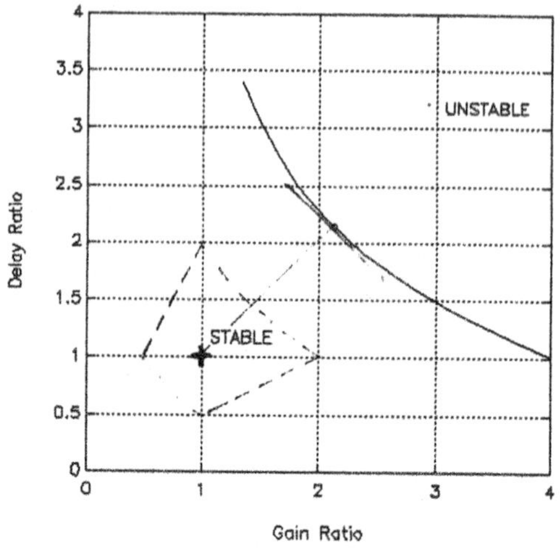

Fig. 1. Graph of robustness which a safety factor of 2 in a controlled process.

This graphical method, previously introduced by Gerry (1988) and Kompass (1989), has been completed with the incorporation of a index of robustness which enables the measurement and comparison of different control algorithms for the same process , and includes the possibility of drawing a relative stability frontier (damping constant) for the controlled process.

The index of robustness consists in the radius of the maximum circle stable with center in the nominal values (1,1). This circle is tangent to the delimiting curve and the tangential point is drawing in the robustness graph (figure 1).

The projection of this tangential point gives information about the maximum variation allowed of two models parameters with full independence of uncertainty together.

To include the possibility of drawing a relative stability

frontier, a simple change of Laplace variable: s'= s-x allows to study, in the graph of robustness, the parameters region assuring an acceptable temporal response (minimum damping limited) for the controlled process.

Using this graphical method in a SCADA may be very useful on the industrial level, since it enables:

-studying graphical whether or not stability is ensured in the controlled process, and obtaining graphical information regards the existing margin of stability.

-analyzing and supervising the robustness of the process with a particular controller, which may be returned if it does not ensure stability with the set degree of safety.

-graphically and quantitatively comparing the robustness of different process control algorithms, thus making it a valuable element for aiding in the decisions of the process controller.

-tuning the controller's parameters, as Ackermann (1980), if the procedure is inverted, and instead of studying the robustness of a fixed controller with a perturbed process, the robustness of a controller with adjustable parameters acting upon a know, fixed process is studied.

ANAGRO

ANAGRO is a program, written in Matlab language, for the graphical robustness of PID controllers acting on single-input single-output systems with low order models.

This program produces the robustness plot described above, based on the following information:

- choice of the reduced-order model structure which best represents the system.

- choice of the two parameters which characterize the uncertainty of the system.

- definition of nominal values for the model and controller parameters.

- selection of the desired type of analysis: absolute or relative stability. For the latter, the minimum allowable damp for the process response must be fixed.

ANAGRO allows the user to select the reduced order model among the most usual structures: first-order system with or without time delay; second or third order system with a three-value time constant. The graphical analysis of robustness may be performed with all the possible combinations of two parameters for each of these models.

ANAGRO provides, optionally, tuning the parameters of the PID controller which minimize the following performance indexes: a 1/4 damping ratio, the integral of the square of the error (ISE), the integral of the absolute magnitude of the error (IAE), or the integral of time multiplied by absolute error (ITAE) proposed by Ziegler-Nichols (1942), López et al. (1967) and Rovira(1981).

Furthermore, this program allows the user to perform real-time control of industrial processes with standard inputs and outputs (0-10 v), using the PID controller chosen through this analysis.

RESULTS

In order to describe the performance of ANAGRO, two examples are presented, which demonstrate the main features of the program.

First Order System with Time Delay

Let a dynamic system be characterized by a first order model with time delay. Two of the parameters of this system are supposed to be either time varying or partially unknown (e.g. the gain and the time delay) and one parameter is known and fixed (e.g. the time constant)

$$G=\frac{Ke^{-\tau s}}{1+Ts}$$

The nominal values of the parameters were considered: K=3; T=10s ; t=2s.

The robust stability of this system has been analyzed with three possible PID controllers, tuned with the empiric techniques of Ziegler-Nichols [ZN], López el al. (minimizing the ISE) [LO] and Rovira (minimizing the IAE) [RO] yielding the following parameters (table 1):

Table 1 Three PID parameters for a first order system with time delay

parameters	ZN	LO	RO
Kc	2.0	2.28	1.47
Ti	4.0	2.63	14.00
Td	1.0	1.11	0.80

The results obtained with these three techniques may be compared by means of the robustness plot shown in figure 2. It is apparent that the PID controller tuned by the technique of Rovira (IAE), gives the greatest robustness to the controlled system.

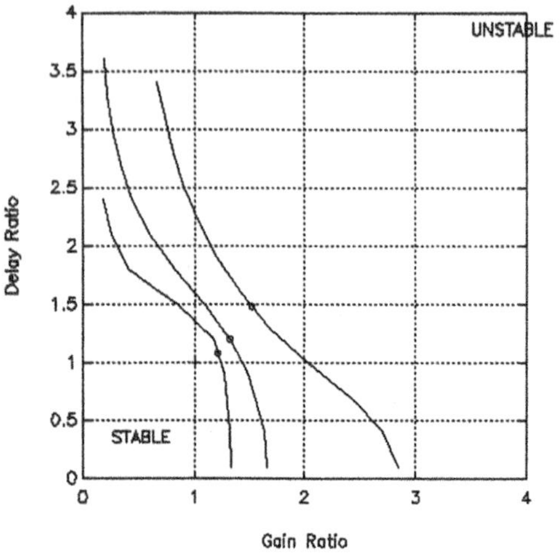

Fig. 2. Robustness plot of three PID controllers for a first order system with time delay

The program produces a numeric output of the robustness index and the maximum allowable range of variation of the uncertain parameters, as shown in table 2.

As mentioned above, a "good" robust controller form the point of view of absolute stability does not necessarily produce an "acceptable" time response for the controlled system. In order to ensure this aspect, it is necessary to analyze the relative stability of the system.

Figure 3 shows the robustness plot of the above described system, with a PID controlled by the Rovira (IAE) technique and with a minimum damping of T=10 s (x= T-1= 0.1 s-1). The results obtained is, in course, more restrictive than in the previous analysis, with a robustness index of only 0.44, with an allowable variation of 32% in the gain and 30% in the time delay.

It is also important to remark that, in this case, there exists a second frontier which limits the minimum values of the model parameters in order to obtain the set damping.

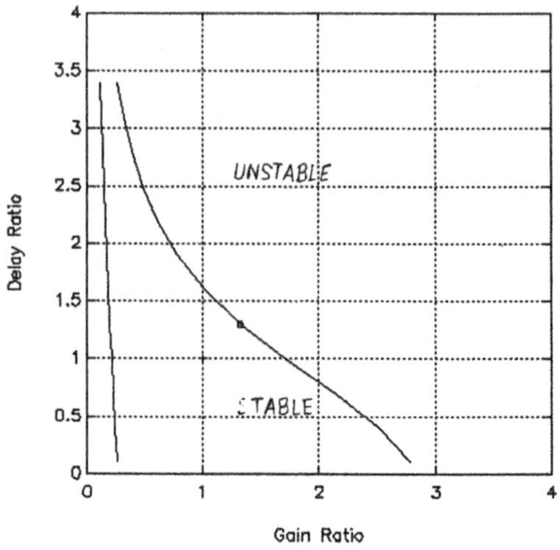

Fig. 3. Robustness plot of the controlled system with relative stability (x=0.1)

Third Order System

This second case analyzes the robustness of a system characterized by a third order transfer function with a multiple time constant of uncertain value and uncertain static gain. Thus,

$$G=\frac{K}{(1+Ts)^3}$$

If a unit nominal gain and a nominal time constant of 10 seconds are considered, and a PI controller is applied with a proportional gain of Kc=1 and an integral time of Ti=10s, the robustness plot of figure 4 is obtained. As may be observed, if the multiple time constant (T) is kept invariable and equal to nominal value (10s), the gain of the system may increase up to 100%, thus maintaining the controlled system stable.

Table 2 Results of the three PID Controllers

	robustness index	max K	max t
Ziegler-Nichols	0.371	31%	20%
López (ISE)	0.218	20%	8%
Rovira (IAE)	0.705	52%	48%

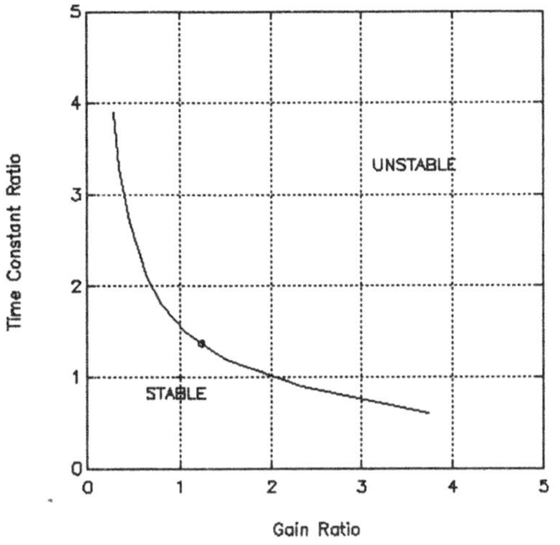

Fig. 4. Robustness plot of a third order system with a PI controller

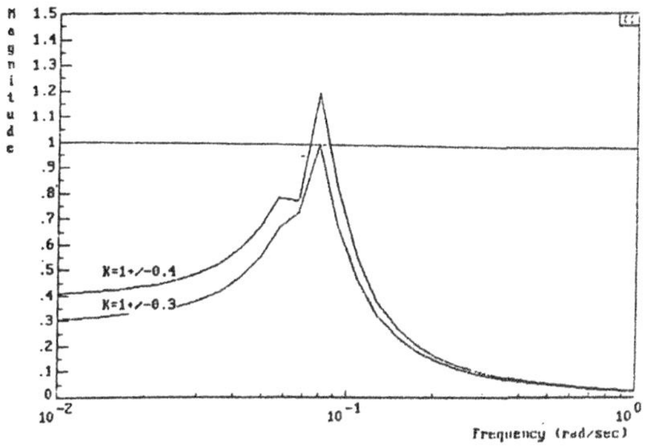

Fig. 5. Test of the structured singular value for real parameter variations of a third order system

If these results are compared to those obtained (figure 5) for the same case with the test of the structured singular value for real parameter variations obtained using the CC software (P. Thompson,1988), it is observed that a variation of 30% in the gain of the system, keeping the multiple time constant invariable (T=10s) fulfills the condition of this robustness test (magnitude <1). Conversely, a higher variation (e.g. of 40%) violates this condition and therefore, the stability of the whole system cannot be guaranteed with this test. The real limit is far from this value. Indeed, as observed in the robustness plot, it is of 100%.

CONCLUSIONS

The method presented in this paper constitutes a useful tool for the analysis of robustness in controlling processes with structured uncertainty in parameters of the model.

The most main features of this method are:

- the accuracy of the output information, as compared to the generally poor information of the robustness tests based on the small gain theorem for system with structured or unstructured perturbations.

- the graphical display of results, which allows the user to define the uncertainty zone (or the safe zone) with no condition at all. This represents an advantage over other methods which impose restrictive conditions in order to analyze the robustness of system with structured perturbations (e.g. Kharitonov, 1978; Barlett and co-workers.,1987)

- the relatively low computation time of the robustness plot which is useful for future application in real time process supervision.

The main limitation of this method is the graphical representation of more than three uncertain parameter of high order models.

One of the future extensions of this work will be to broaden the scope of ANAGRO by adding new models to the library (second and third order systems with time delay), as well as studying others cases of relative stability or performance requisites of the controlled system.

Acknowledgement. This work has been supported by the Spanish Board for Research and Technology (CICYT) under contract ROBO736/91, and the paper has been partially supported by the Group for Study and Research in Automatic Control (CERCA).

REFERENCES

Ackermann, J. (1980). Parameter space design of robust control systems. IEEE Trans. on A. C., vol. AC-25, nº 6, 1058-1072.

Barlett, A.C. and co-workers (1987). Root locations of an entire polytope of polynomials: it suffices to check the edges. Proceedings of ACC (Ed.), Minneapolis, 1611-1616.

Boyd S. and co-workers (1990). Linear controller design: limits of performance via convex optimization. Proceedings of IEEE (Ed.), Vol. 78, nº 3, 529-574.

Dorato, P.(1987). A historical review of robust control. IEEE Control Systems Magazine (Ed.), Vol. 7. 44-47.

Fu, M. and co-workers (1989). Introduction to the parametric approach to robust stability. IEEE Control Systems Magazine, pp. 7-11.

Gerry, J.P. (1988). Find out how good that PID tuning really is. Control Engineering.

Kharitonov, V.L. (1978). Asymptotic stability of an equilibrium position of a family of systems of linear differential equations. Differentsyalnye Uravnenya (Ed.), 1483- 1485.

Kompass,E.J. (1989). PID control software: features and comparisons. Control Engineering, pp. 155-158.

Thompson, P.M. (1988). Program CCV. 4.01, Reference Manual, Vol. II, Systems Technology Inc.

PC-BASED EQUIPMENT FOR SYSTEM IDENTIFICATION AND CONTROL

E. Bertran and G. Montoro

Department of Signal Theory and Communications, Universitat Politecnica de Catalunya, P.O. Box 30.002, 08080-Barcelona, Spain

Abstract:

Personal computer-based equipments have interest as much in measurement and control as in educational field.

The behavior and applications of a low-cost, versatile and expandable equipment are shown in this paper. The equipment is useful for identification and control applications and it has a great potential/cost ratio, compared with other market available equipments. Mainly it has two particular characteristics: at first, the acquisition board is autonomous and capable to adapt unknown input signals (automatic level adjustments), and at second, an adaptive filter for interference cancelling is available. The equipment, in general, is intended to facilitate the analysis of systems and to design closed loop control laws.

Keywords: *Data acquisition, signal processing, control equipment, identification, adaptive systems.*

1.- General System:

The overall system consists of an autonomous microprocessor-based data acquisition board (ADS) connected to a conventional personal computer, via RS-232 connection. The control strategy debugged with PC could be translated later into an internal EPROM, becoming an autonomous digital controller. So, there are three performance alternatives:

a/ The control algorithm is held in the EPROM.

b/ The control algorithm is computed by the PC.

c/ The control algorithm is resident in EPROM and scheduled by PC.

Therefore, this equipment is versatile and expandable, as well in hardware as in software.

There are available many algorithms to generate the set-points to be sent to the plant (digital control applications). This control signal can be obtained directly from the computation of the input signals (time domain) or from the frequency characteristics (frequency domain).

When the equipment is working plugged to the PC, its operations are selected following the options offered by a menu tree.

The equipment offers the following possibilities:

-Scope: For signal visualization (time domain) the equipment works as a digital oscilloscope.

-Spectrum analyzer (via FFT): The Discrete Fourier Transform of the samples obtained from input signals are carried out in order to visualize the spectrum.

-System identification: Via many methods as FFT's output/input ratio, Samulon step response method, RLS (recursive least squares), RELS (recursive extended least squares), RGLS (recursive generalized least squares), RML (recursive maximum likelihood), RIV (recursive variable instrumental), LMS (least mean square).

-Digital filtering: Designed in S-domain or in Z-domain.

-Direct digital control: Is the closed loop implementation of previous possibility.

-Noise cancelling: With an adaptive algorithm, LMS, usual in signal processing applications.

2.- Hardware characteristics:

Previous to sampling input signals, an analog preprocessing in order to obtain the best measurement window of input signals, known or not, is carried out. This aspect is important due to input signals are normally unknown. Usual market available equipments have not this performance, so, if input signal has unexpected amplitude and offset levels, data acquisition could be false.

Programmable amplifier parameters are sent from PC to microcomputer that controls the acquisition board (ADS). The following codes are sent: start communication code, number of samples, sampling period, output analog step amplitude and offset, amplitude scale, and d.c. level. When the start communication code is received the sampling is activated, and the samples are held in a dedicated RAM.

The user can choose (Bertran,1988), by using the PC:

1/ The sampling rate (until 5 or 10 Mhz).

2/ The scale amplitudes (manual or automatic adjustment).

3/ The offset level (manual or automatic adjustment).

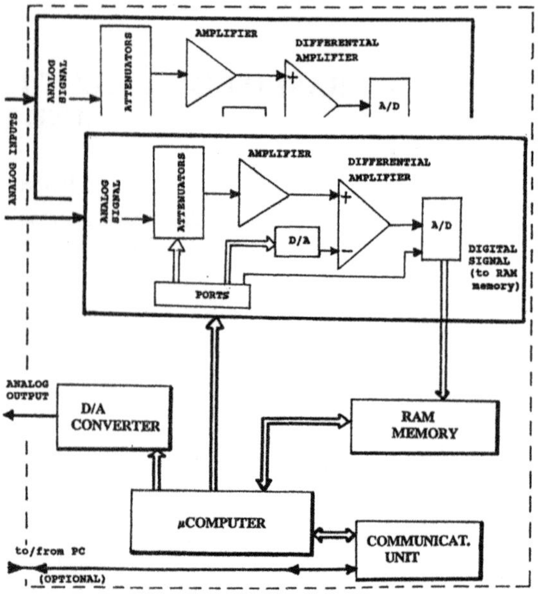

figure 1: Hardware configuration.

3.- Software characteristics:

3.1.- Basic operation:

The equipment supports as conventional digital control algorithms as others more advanced algorithms for signal processing and control applications.

Data acquisition runs according to the following steps:

-PC sends to ADS the chosen operation options.

-ADS samples the analog signals present in channels 1, 2 or both. It works according to operation codes sent by the PC. If automatic adjustments are required the ADS obtains iteratively the best parameters.

-ADS sends to PC the required number of samples, the used amplitude scale and the offset level.

All the operations are selected by the user following a menu tree. The main menu request the parameters to establish and to check the communications between PC and ADS.

3.2.- Signal representation:

PC represents the received signal, and a menu is displayed. Options are: time signal representation (figure 2) or frequency representation (figure 3). In this case usual FFT problems at the higher part of frequency window have not been hidden.

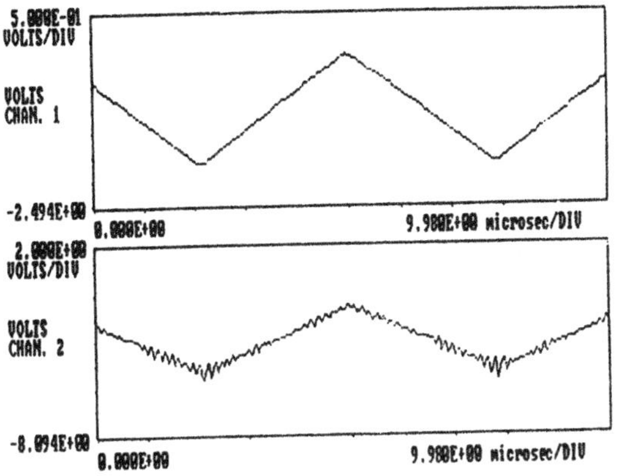

figure 2: Time domain representation.

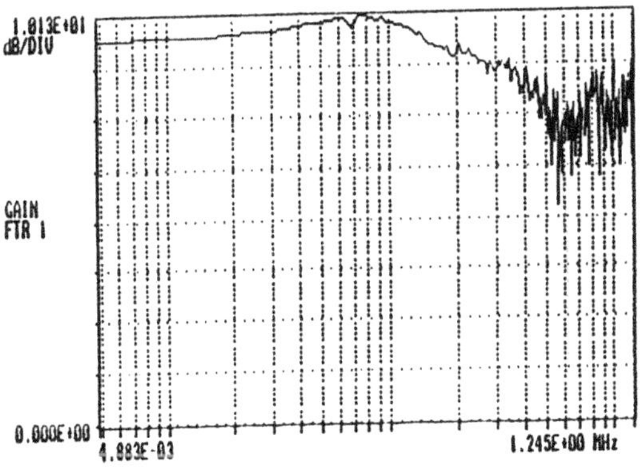

figure 3: Spectrum representation.

3.3.- Digital filtering and control:

In digital filtering there are two options: first one is based on an IIR discrete time filter designed in continuous time domain, and computed by using a bilinear transform (Oppenheim, 1989) of the transfer function H(s):

$$H(s) = \frac{Ks + C}{s^2 + As + B}$$

being

$$s \leftrightarrow \frac{2}{T_s}\frac{1 - z^{-1}}{1 + z^{-1}}$$

The other way to implement the digital filters, is introducing the coefficients of a Z-domain designed filter, with transfer function H(z):

$$H(z) = \frac{A_0 + A_1 z^{-1} + A_2 z^{-2} + A_3 z^{-3}}{1 + B_1 z^{-1} + B_2 z^{-2} + B_3 z^{-3}}$$

In figure 4-a, an interfered signal present at channel-1 is shown. By programming the band-pass filter,

$$H(s) = \frac{3960s}{s^2 + 3960s + 45,53e6}$$

the interference has been separated (figure 4-b).

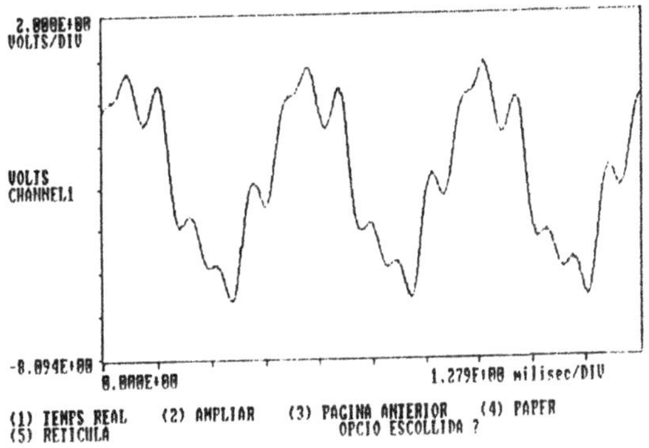

figure 4-a: Filter input signal.

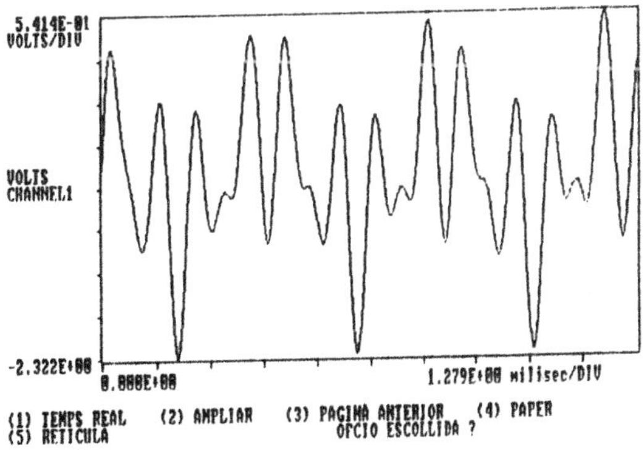

figure 4-b: Filter output signal.

3.4.- Parameter identification:

There are a lot of techniques by which the parameters of a system model can be obtained, so in this software the more usual algorithms are available: frequency, gradient and least squares methods.

1/ Frequency identification methods are normal FFT's output/input ratio and a variation of this ratio computation based on Samulon methodology. Results are frequency plots (see figure 3).

2/ Gradient methods implemented in the equipment are based on the LMS algorithm, and the results can be chosen between an IIR or an FIR filter. In both cases the results are transfer function parameters.

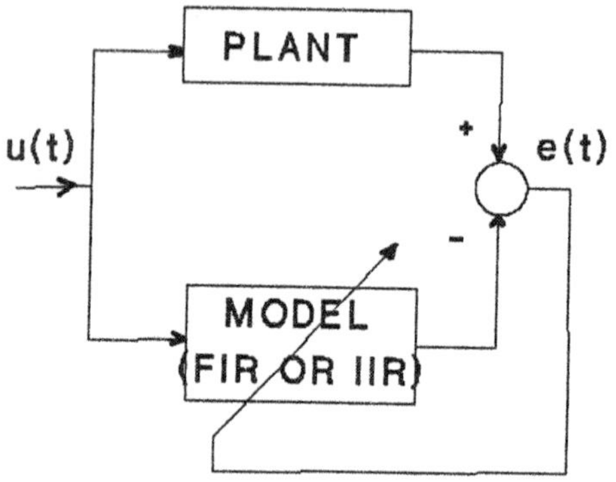

figure 5: Gradient identification structure.

3/ Least squares methods (LS) implemented are RLS (recursive), RELS (extended), GLS (generalized), RIV (instrumental variables) and RML (maximum likelihood). (Strejc, 1980).

Except for the first, all the methods can identify a model for noise.

Figure 6 shows the results of a LMS-based identification, with a 20 coefficients FIR structure and $\mu=0.013$

Figure 7-a shows the acquired samples of a step response to be identified by using the RML method.

Identified transfer function H(z) (which MATLAB simulation is shown in figure 7-b) has been:

$$H(z) = \frac{(-6,99e-2)z^2 + (0,2)z - (9e-2)}{z^2 - (1,84)z + (0.88)}$$

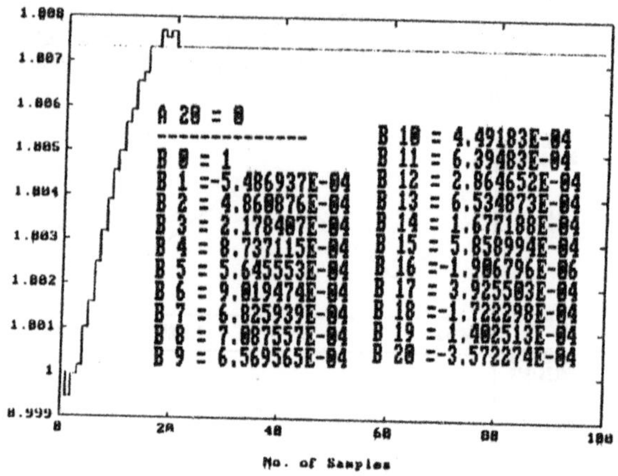

figure 6: LMS identification.

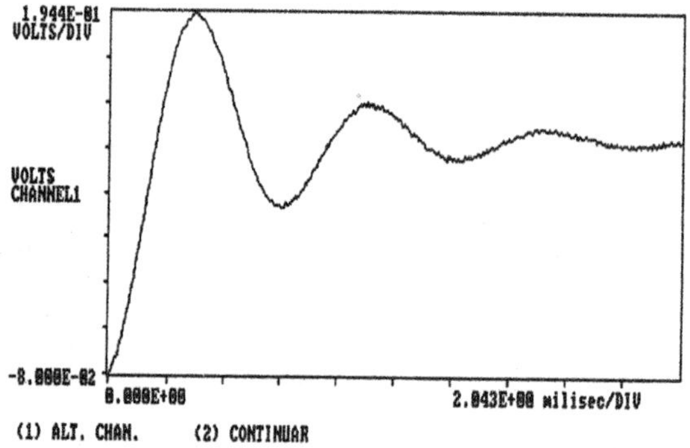

figure 7-a: Second order identification via RML.

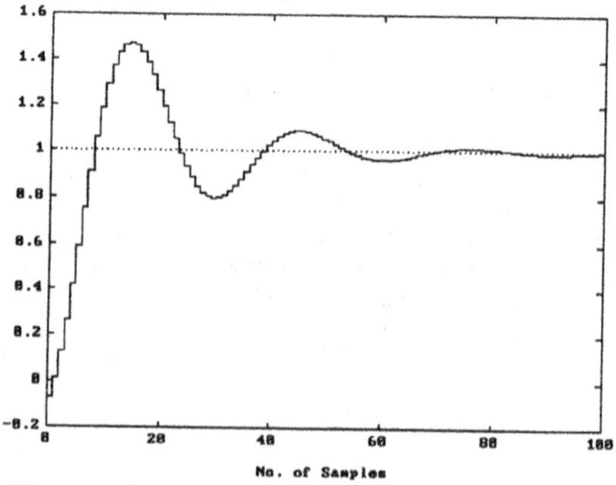

figure 7-b: Step response simulated from obtained model.

3.5.- Noise cancelling:

Sometimes the elimination of additive noise when the input signal is corrupted by an interference (as is the case of supply interferences) is an important problem. There is a conventional alternative for cancelling a single frequency interference, a notch filter, but is difficult to design a fixed notch filter tuned to the frequency of the interference if this is close the signal frequency. Adaptive canceller filters offer easy control of bandwidth and tracking of noise characteristics (Widrow,1985) (Haykin,1986). It operates as an adaptive notch filter which bandwidth, convergence speed and precision is determined by a parameter μ.

Noise adaptive cancelling option included in the software eliminates interferences via adaptive LMS algorithm. Consist of an adaptive filtering of channel-1 according to the reference noise present in channel-2 (see figure 8). The original signal and the additive noise are uncorrelated, and the reference noise supplies a correlated version of the interference.

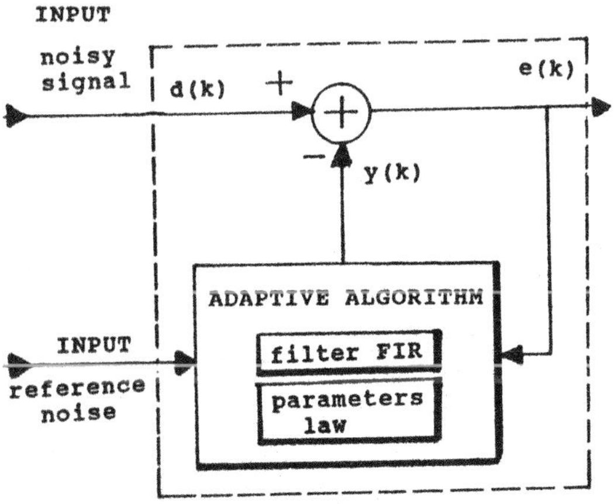

figure 8: Adaptive noise canceller.

So, channel-1 input is a signal contaminated by an interference:

$n(t) = A.\cos(w.t+\theta) =$
$= Ac.\cos(w.t) + As.\sin(w.t)$

and channel-2 input must acquire samples from reference interference:

$nr(t) = B.\cos(w.t)$.

The filter has a linear transfer function with a FIR structure, and ten coefficients are estimated with a recursive law. Algorithm for LMS filter is:

-Noisy input signal (samples from channel-1): $d(k) = s(k) + n(k)$

-Interference model reference (samples from channel-2): $nr(k)$

-Estimated interference:

$$y(k) = W_0(k).nr(k)+W_1(k).nr(k-1)+\dots$$
$$\dots+ W_9(k).nr(k-9)$$

-Parameter computation law:
$W_i(k+1) = W_i(k) + \mu.e(k).nr(k-i)$

-Filter output signal:
$e(k) = d(k) - y(k)$

Results when input signal is a squared one, contaminate with a 50 Hz sinusoidal noise, and adaptive filter parameter μ=0.0001 are shown (figures 9-a and 9-b).

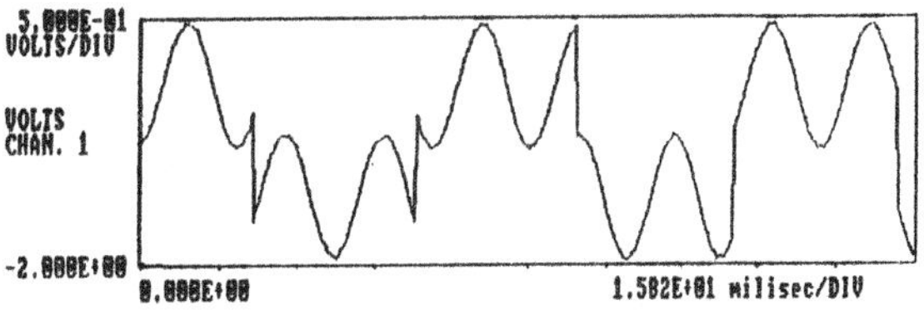

figure 9-a: Noisy signal.

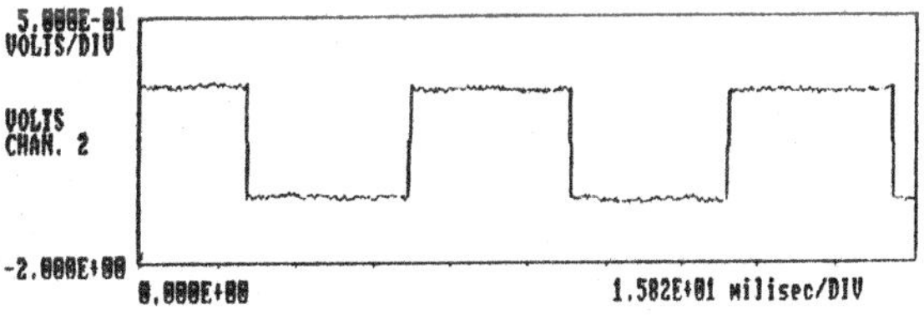

figure 9-b: Adaptive filter output.

4.- References:

Bertran, E. and A. Guimerà (1988). *Computer-based System for Control and Circuit Analysis*. Proc. of 36 I.S.M.M. International Conference, June.

Coteau, J.; Grant, D. and S. Wurcer (1986). *Feed Analog Signals to IBM PC-Compatible Personal Computers*. EDN, April.

Foley, C. (1984). *Use a PC and DFT to extract Data from Noisy Signals*. EDN, April.

Haykin, S. (1986). *Adaptive Filter Theory*. Prentice-Hall.

Oppenheim, A.V. and R.W. Schafer (1989). *Discrete Time Signal Processing*. Prentice-Hall.

Samulon, H.A. (1951). *Spectrum Analysis of Transient Response Curves*. Proc. of I.R.E., February.

Strejc, V.(1980). *Least Squares Parameter Estimation*. Automatica, 16, 535-550.

Widrow, B. and S.D. Stearns. (1985). *Adaptive Signal Processing*. Prentice-Hall, 1985.

N-OPTIMUM AND SUBOPTIMUM FEEDBACK CONTROL OF SYSTEMS WITH VARIABLE PARAMETERS

P. Pivoňka and P. Vavřín

Department of Control Engineering, Faculty of Electrical Engineering, Technical University of Brno, 612 66 Brno, Božetěchova 2, Czechoslovakia

ABSTRACT

At present we can witness increasing interest in the design and implementation of adaptive control algorithms using methods that make it possible to control dynamic systems while only a minimum of a priori information on the properties of such systems is available. An n-optimum and suboptimum control algorithm was proposed to be used as an algorithm of the self-tuning regulator, with minimum demand on the hardware and the software.

KEYWORDS

Adaptive control;Self-tuning regulators;Direct digital control;PID control

INTRODUCTION

Implementation of adaptive control requires deep knowledge and much experience. No particular increase in the quality of control can be expected in dynamic systems that are linear and deterministic, with known and almost invariable parameters. It is not certain whether the increased costs will be returned by implementation of a more complicated system. Application of adaptive control algorithms to a real technical object, however, may bring about problems due to highly sophisticated mathematical tools, extremely laborious exact methods, and to assumptions that may not be proved. Mathematical models often differ from real objects and therefore they must be indifferent ways modified and specified. It is not easy to find a sufficiently general and reliable method of approximation that would lead to the required agreement between a real process and its mathematical model. Acceptable mathematical - physical derivation requires restrictions which may be the cause of differences between the mathematical model and real systems. Verifications of real systems show that better results are often obtained in relatively simple models of technological processes than in sophisticated models where a large number of parameters must be estimated. In the focus of interest is the design and implementation adaptive control algorithms using methods that make possible control of dynamic systems with a minimum of a priori information on the properties of these systems. However, such methods are prevailingly based on the algorithm of PID regulator that can be set in many different ways. The task is to design an algorithm that would be comparable in simplicity and robustness with the PID regulator, at the same time improving the quality of transfer, and that would be accomplished after a finite number of sample periods. Duration of transfer would be as short as possible, with the smallest possible oscillation. The algorithm should be applicable to systems with transfer lag and the computation should be fast enough. A control algorithm is presented that can easily be applied to specific dynamic systems and control of specific technological processes. The algorithm makes it possible to simplify adaptive control and its application, while implementation is not very demanding.

MATHEMATICAL TOOLS

In the design of the regulator the method of discrete state space was applied, extended to cover the solution of continuous dynamic systems with transfer lag. By solving discrete state equations the course was determined of the state trajectory of a continuous dynamic system with (or without) transfer lag during the transfer from an arbitrary initial state to a required final state.

A linear, stationary and continuous n-th order dynamic system can be described by equations

$$\dot{\underline{x}}(t) = \underline{Ax}(t) + \underline{Bu}(t)$$
$$y(t) = \underline{Cx}(t) \qquad (1)$$

By transformation of (1) into a discrete form we get as the discrete model in the form

$$\underline{x}(k+1) = \underline{f}(\underline{x}(k),u(k)) \qquad (2)$$

Dead time is assumed, $\tau \in \langle 0,T)$; where T is the sample period. The system output in interval $\langle k,k+\tau)$ is further influenced by the variable u(k-1); in the following interval $\langle (k+\tau),(k+1))$ by the variable u(k). In next step, the state of the system is therefore the function of both these variables:

$$\underline{x}(k+1) = \underline{f}(\underline{x}(k),u(k),u(k-1)) \qquad (3)$$

The values of the state variables at sample times T are given by equations:

$$\begin{aligned} \underline{x}(k+1) &= \underline{Kx}_{r+1}(k) \\ \underline{x}(k+2) &= \underline{Kx}_{r+1}(k+1) \\ &\vdots \\ \underline{x}(k+n) &= \underline{Kx}_{r+1}(k+n-1) \end{aligned} \qquad (4)$$

where

$$\underline{x}_{r+1}(k) = \left[x_1(k),\ x_2(k),\ldots,\ ,x_n(k),u(k),u(k-1)\right]^T \qquad (5)$$

and $\underline{K}$ is matrix $\underline{K}(n,n+2)$

If a transfer lag occurs in the system, stabilization cannot be achieved at the (k+n)-th step, but only after the expiration of the time of the transfer lag τ. The system of equations describing behavior of the system, taking into account the transfer lag effect, is as follows:

$$\begin{aligned} \underline{x}(k+1) &= \underline{Kx}_{r+1}(k) \\ \underline{x}(k+2) &= \underline{Kx}_{r+1}(k+1) \\ &\vdots \\ \underline{x}(k+n) &= \underline{Kx}_{r+1}(k+n-1) \\ \underline{x}(k+n+1) &= \underline{Kx}_{r+1}(k+n) \end{aligned} \qquad (6)$$

From the course of the trajectory the required sequence of n manipulated variables can be calculated for n-optimum control, where n is the order of the dynamic system. This sequence of manipulated variables can be modified in case the manipulated variables must be limited for physical or technological reasons.

Solving equations (6), we get the control vector

$$\underline{u} = \underline{Px}_{r-1}(k) + \underline{Qx}(k+n+1) \qquad (7)$$

where

$$\underline{u} = \left[u(k),\ u(k+1),\ldots\ ,u(k+n)\right]^T$$

$$\underline{x}_{r-1}(k) = \left[x_1(k),x_2(k),\ldots,\ x_n(k),u(k-1)\right]^T$$

The control algorithm given by equation (7) is an open control loop from which the n -optimum control algorithm can be derived in closed control loop

$$u(k+i) = \underline{g}_i\underline{x}_{r-1}(k+i) + \underline{h}_i\underline{x}(k+n+1) \qquad (8)$$

If limitation is assumed during control and regulation

$$|u(k+i)| \le u_{max} \qquad (9)$$

conditions can be derived for transfer from an arbitrary initial state to a given final state. To compensate errors occurring during control and regulation (errors due to disturbance variables, inaccurate estimate of the system controlled) a so called differential manipulated variable was derived. The difference between the state variables of the system model not affected by disturbances on the one hand and the real system on the other hand is the differential vector

$$\underline{e}_{r-1} = \underline{x}_{m(r-1)} - \underline{x}_{r-1} \qquad (10)$$

The differential manipulated variable Δu

can be derived from (10). Using standard norm, we get

$$\min_{\Delta u} \| \underline{Ke}_{r-1} \| \qquad (11)$$

Then the integration element is introduced.

$$u_d(k+i) = u_d(k+i-1) + \Delta u(k+i) \qquad (12)$$

ADAPTIVE VERSION OF ALGORITHM

For identification purposes the analytic method of identifying dynamic systems or recursive least squares identification was used.

For the purpose of practical verification the n-optimum and suboptimum control was limited to the model of a dynamic system of second order with transfer lag.

We get equations in the form

$$\begin{aligned} u(k+i) &= g_{i1}x_1(k+i) + g_{i2}x_2(k+i) + g_{i3}u(k+i-1) + \\ &\quad + h_{i1}x_1(k+3) + h_{i2}x_2(k+3) \\ u_d(k+i) &= c_1\Delta x_1(k+i) + c_2\Delta x_2(k+i) + \\ &\quad + c_3\Delta u(k+i-1) + u_d(k+i-1) \end{aligned} \qquad (13)$$

The differential vector for a second-order reduced model with transfer lag can be regarded as a PID regulator. Block diagram of adaptive control system is in Fig.1.

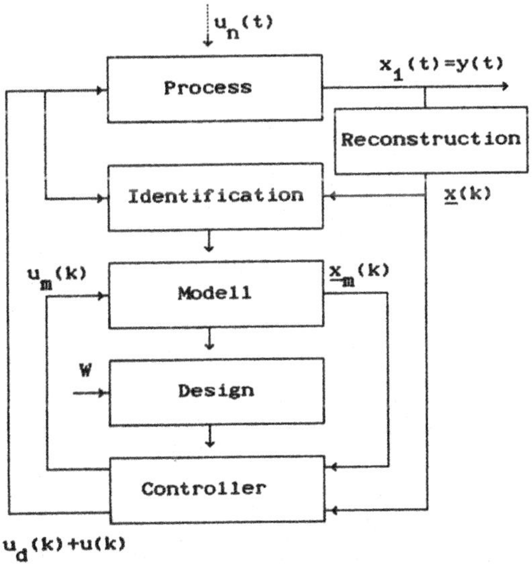

Fig.1. Block diagram of adaptive control system.

EXAMPLES OF APPLICATION

Fig.2 illustrates behavior of a typical

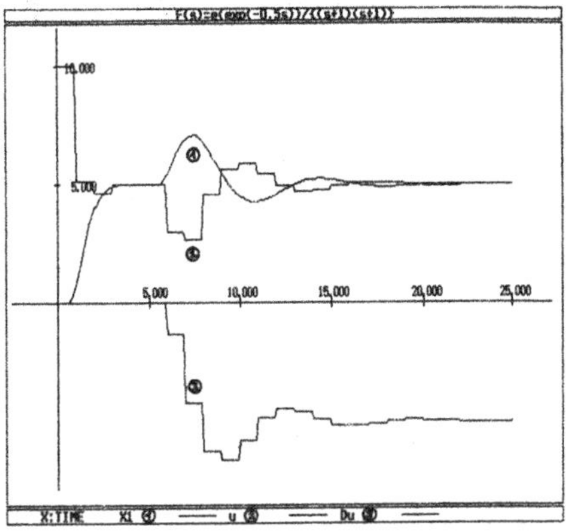

Fig.2.

n-optimum control for a given second-order transfer function with transfer lag. Output value, differential vector and manipulated variable are shown. The system is defined by the transfer function:

$$F(s) = \frac{e^{-0.5s}}{(s+1)(s+1)} \qquad (14)$$

Adaptive control with a third-order transfer function is illustrated in Fig.3.

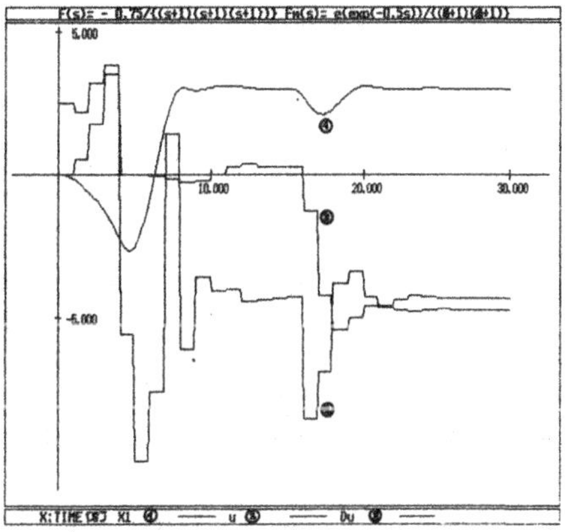

Fig.3. $F(s) = \dfrac{-0.75}{(s+1)^3}$; $F_m(s) = \dfrac{e^{-0.5s}}{(s+1)^2}$

During the first four sample periods matrix K is identified and transfer to another required value takes place. At time t = 15s the disturbance starts being effective.

The algorithm was found to be fast enough. It was tested e.g. by being used to control the movement of a robot arm in real time.

The progress of the movement is shown in Fig.4. It should be noted, however, that

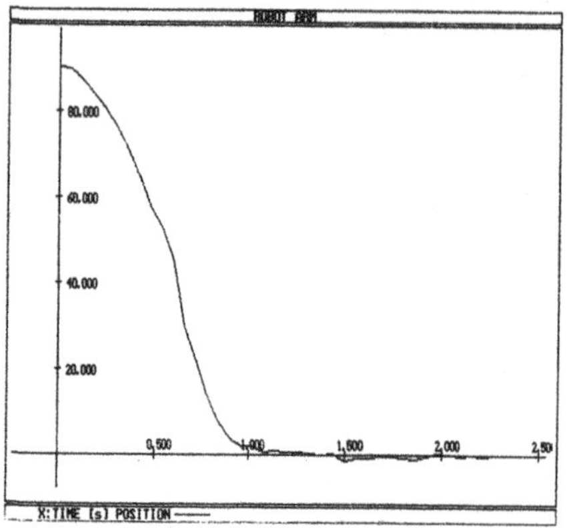

Fig.4.

scanning of the movement of the arm was strongly disturbed by noise because an unsuitable sensor was used.

CONCLUSION

Simulations calculations and verification on real systems showed that the proposed control algorithm displayed good properties not only in n-optimum and suboptimum control of second-order systems with (or without) transfer lag (when an accurate apriori estimate of the system controlled was assumed), but also in the control of systems of the third or fourth order when the estimate of system parameters was only approximate or none. In addition to the control of overdamped systems the control was also successfully tested on systems with the pole in the right-hand half of complex plane p, and on systems whose response to unit entrance signal was of oscilating nature. The control algorithm proposed can easily be modified to fit a specific continuous dynamic system. It provides for a considerable simplification of problems of self-adaptive control, and its implementation is not very demanding.

References:

[1] Pivoňka,P. : N-optimum and suboptimum Control.PhD Thesis. Dept. of Automatic Control,FE VUT, Brno, 1988.

[2] Isermann,R.: Stand und Entwiklungs-tendenzen bei adaptiven Regelungen. Automatisierungstechnik,4,1987.

[3] Pang, G.H.K.,MacFarlane, A.G.J.: An expert systems approach to computer-aided design of multivariable systems. Springer Verlag,Berlin, 1987.

[4] Warwick,K. : Implementation of self-tuning controllers. Peter Peregrinus LTD, London, 1988.

[5] Aström,K.J., Wittenmark,B. : Computer controlled systems. Pretince - Hall, Englewood Cliffs, 1984.

EXPERIMENTAL RESULTS WITH SIMPLIFIED ALGORITHM OF ESTIMATION THE PARAMETERS OF A DYNAMICAL SYSTEM

E. Jezierski

Politechnika Lodzka, Instytut Automatyki, Stefanowskiego 18/22, 90-924 Lodz, Poland

Abstract: The paper deals with the three estimation methods of parameters, i.e. non-recursive least-squares method, recursive least-square method, and Kaczmarz's projection algorithm. These methods were experimentaly tested on estimation the parameters of a fast DC motor. The tests showed that the simple projection algorithm gives a little less precise estimates in comparison to the least-squares methods, but is significantly faster. This algorithm can be successfully applied in those situations where occur rapid changes in the description of a system dynamics, what is typical for example in robotics. It can be useful in some applications of indirect adaptive control schemes.

Keywords: adaptive control, d.c. motors, electric drives, least-squares estimation, process parameter estimation, real time computer system.

INTRODUCTION

On-line estimation of process parameters is a key element in adaptive control (Giri, Dion, Dugard, 1991). A basic technique for parameter estimation is the least-squares method. This method can be applied if the model of the system has the property of being linear in the parameters that should be estimated. If the number of parameters is equal to n then the recursive updating of the model needs the elementary operations (additions, multiplications and divisions), the number of which is proportional to the second power of n (e.g. Isermann, 1988). When the number n grows then the method is rather time consuming. This can restrict some applications of adaptive control systems or force the designers to apply the powerfull and expensive control units, especially if dynamics of control object is very fast. Åström and Wittenmark (1989) suggested to use in such cases more simple Kaczmarz's projection algorithm.

An example of a fast and complicated system is an industrial robot equipped with electric drive. The motors are working in rather difficult conditions characterized by big and rapid changes of load torques and resultant moments of inertia acting on their shafts. There is a growing need for investigation in the field of adaptive control of electric drive and some works both theoretical and experimental were done previously (e.g. Canudas, Åström, Braun, 1987; Kelly,1987; Bassi, Benzi, 1988; Bassi, Benzi, Scattolini, 1990). However in these works the only use of the classical least-squares method was discussed.

The main aim of the paper is to show the experimental results of estimation the parameters of a fast DC motor that is applied in a typical drive system of an industrial robot.

PROJECTION ALGORITHM

The projection algorithm, similarly to the least squares method, can be used for estimation the parameter vector $\mathbf{p}$ in a following mathematical model of a process

$$y(i)= \mathbf{p}^T\mathbf{w}(i) = \sum_{j=1}^{n} p_j w_j \tag{1}$$

where $\mathbf{w}(i)$ is the i-th sample of an input vector, while $y(i)$ is a value of a scalar output at the same time. The method is based on the following idea. One measurement $y(i)=\mathbf{p}^T\mathbf{w}(i)$ determines the projection of the parameter vector $\mathbf{p}$ on the vector $\mathbf{w}(i)$. From this interpretation it follows, that n measurements are required to determine the vector $\mathbf{p}$ uniquely, under condition that $\mathbf{w}(1)$, $\mathbf{w}(2)$, $\ldots,\mathbf{w}(n)$ span R^n.

In order to obtain the recursive form of the algorithm let us assume that the estimate $\hat{\mathbf{p}}(i-1)$ is available and the new measurements $\mathbf{w}(i)$ and $y(i)$ are obtained. Since $y(i)$ contains information only in the direction $\mathbf{w}(i)$ in a parameter space, it is natural to update the estimate as

$$\hat{\mathbf{p}}(i) = \hat{\mathbf{p}}(i-1) + \beta\mathbf{w}(i), \tag{2}$$

where the scalar β is chosen so that

$$y(i)= \hat{\mathbf{p}}^T(i)\mathbf{w}(i) = \hat{\mathbf{p}}^T(i-1)\mathbf{w}(i) + \beta\mathbf{w}^T(i)\mathbf{w}(i) \tag{3}$$

From this relationship the value β is equal to

$$\beta = \frac{1}{\mathbf{w}^T(i)\mathbf{w}(i)}[y(i) - \hat{\mathbf{p}}^T(i-1)\mathbf{w}(i)] \tag{4}$$

Substitution β from (4) into updating formula (2) gives the recursive form of the projection algorithm

$$\hat{\mathbf{p}}(i) = \hat{\mathbf{p}}(i-1) + \frac{\mathbf{w}(i)}{\mathbf{w}^T(i)\mathbf{w}(i)}[y(i) - \hat{\mathbf{p}}^T(i-1)\mathbf{w}(i)] \tag{5}$$

In practical applications the above formula is slightly modified in a following manner

$$\hat{\mathbf{p}}(i) = \hat{\mathbf{p}}(i-1)+ \frac{\mathbf{w}(i)}{\alpha+\mathbf{w}^T(i)\mathbf{w}(i)}[y(i)-\hat{\mathbf{p}}^T(i-1)\mathbf{w}(i)] \tag{6}$$

where the scalar value $\alpha>0$ is a constant that should be taken to be comparable with the average value of the scalar product $w^T(i)w(i)$, it makes the algorithm more robust. Updating the parameter vector $\hat{p}(i)$ requires in every step only $3n$ multiplications, $3n$ additions and n divisions.

EXPERIMENTAL RESULTS

An experiment of comparison some methods of parameter estimation was performed on a fast DC motor that is used in an industrial robot. The dynamics of a drive system with such a motor is described by the set of following equations

$$\frac{d\Theta(t)}{dt} = \Omega(t) \tag{7}$$

$$\frac{d\Omega}{dt} = \frac{c_t\phi}{J}I_a(t) - \frac{T_l}{J} \tag{8}$$

$$\frac{dI_a}{dt} = -\frac{c_v\phi}{L_a}\Omega(t) - \frac{R_a}{L_a}I_a(t) + \frac{1}{L_a}u(t) \tag{9}$$

where:

- Θ - angular position of the rotor
- Ω - angular velocity of the rotor
- I_a - armature current
- $c_t\phi$ - torque constant of the motor
- $c_v\phi$ - voltage constant of the motor
- J - resultant moment of inertia brought to the shaft of the motor
- T_l - external load torque acting on the shaft
- R_a - armature resistance
- L_a - armature inductance
- u - supply voltage.

Usually in control system of robots it is acceptable that the parameter of the motor i.e. R_a, L_a, $c_v\phi$, and $c_t\phi$ are constant. However, the mechanical values J and T_l can vary in a wide range as 1:10 or more. If that is the case than the relationship (8) gives possibility to estimate both the resultant moment of inertia and the external load torque. The equation (8) after discretization can be written approximately in the form

$$\frac{\Omega(i+1)-\Omega(i)}{\Delta t} = \frac{c_t\phi}{J}I_a(i) - \frac{T_l}{J} \tag{10}$$

where $\Omega(i)$ and $I_a(i)$ denote velocity and armature current in the i-th instant respectively, while Δt is a sampling period. Introducing the following denotations $p_1=1/J$, $p_2=T_l/J$, $w_1=c_t\phi I_a(i)$, $w_2=-1$, and $y(i)=(\Omega(i+1)-\Omega(i))/\Delta t$, the last equation takes the standard form

$$y(i) = p_1w_1(i) + p_2w_2(i) = \mathbf{p}^T\mathbf{w}(i) \tag{11}$$

Laboratory Stand

The laboratory stand that was used in physical experiments is schematically shown in Fig. 1. The tested motor (M) is controlled by a velocity regulator. The motor is mechanically coupled with the DC load generator (G) working in a current mode which extorts a desired external torque. The tachogenerator (TG) and the 10-bits encoder are connected to the shaft of the motor. The system is controlled by a typical PC/AT 286/287 10MHz computer equipped with digital 16-bits I/O cards and a fast multi-channel A/D converter cards working in DMA mode. All executable programs were written in C language.

The laboratory stand contains also the two analog channels equipped with transducers (Hall-effect converter LEM LF50-P) and active low-pass filters. It gives possibility to measure currents of both machines, or only one current and the shaft velocity. The last element of the stand is a digital storage oscilloscope Tektronix 2230 with double-channel isolation unit Tektronix A6902B. The oscilloscope is connected to the central unit by a serial interface, and is normally used to collect the values of these additional analog signals which are not need in a real-time data processing and control.

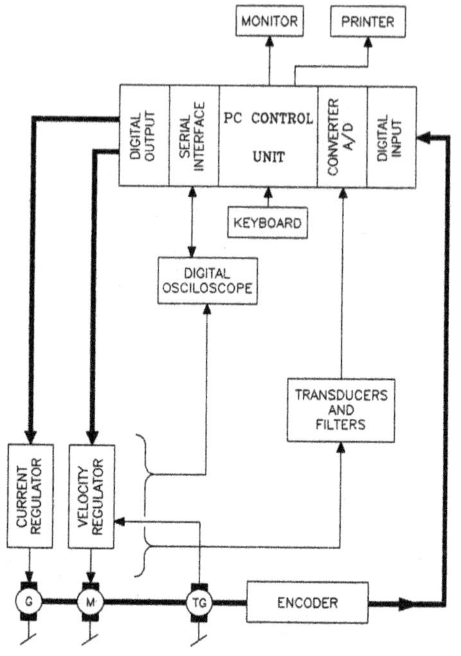

Fig. 1. Diagram of laboratory stand

The DC motor of the type PZTK 88-35 TRR, that is used in IRp-6 industrial robot, was under the tests. The characteristic parameters of this motor are

nominal armature voltage	35 V
nominal armature current	5.5 A
nominal rotational speed	3,200 r.p.m.
torque constant	0.105 Nm/A
inertia of the rotor	$1.45*10^{-4}$ kgm^2
armature resistance	0.56 Ω
electrical time constant	2.4 ms
mechanical time constant	7.6 ms

The motor works in its standard electrical configuration - as a complete drive system SUN-05-KT with velocity regulator and a PWM transistor power amplifier. The nominal modulation frequency in the four quadrant supply unit is 4 kHz. The role of the laod generator (G) performed the second DC machines of the same type (PZTK 88-35 TRR) controlled by a current regulator. The torque generated is proportional to its armature current according to the relationship $T(t)=c_t\phi I_a(t)$.

Compared Methods of Parameter Estimation

There were three methods of estimation the parameters of the motor tested during physical experiments, i.e. non-recursive least-squares method (NRLS), recursive least-squares method (RLS), and projection algorithm (PA).

In the first case the estimate of the parameter vector is calculated on the basis of the last m measurements using the following relationship

$$\hat{\mathbf{p}}(i) = \left[\mathbf{W}(i/m)\mathbf{W}^T(i/m)\right]^{-1}\mathbf{W}(i/m)\mathbf{y}(i/m) \quad (12)$$

where the vector $\mathbf{y}(i/m)$ and the matrix $\mathbf{W}(i/m)$ are of the forms

$$\mathbf{y}(i/m) = \begin{bmatrix} y(i-m+1) \\ y(i-m+2) \\ \dots \\ y(i-1) \\ y(i) \end{bmatrix} \quad (13)$$

$$\mathbf{W}(i/m)=\left[\mathbf{w}(i-m+1)\ \mathbf{w}(i-m+2)\ \dots\ \mathbf{w}(i-1)\ \mathbf{w}(i)\right] \quad (14)$$

The least-squares method is the most often applied in its recursive version. Then the updating formulae for the correction vector $\mathbf{k}(i)$, the new estimate vector $\hat{\mathbf{p}}(i)$, and the covariance matrix $\mathbf{P}(i)$ are as follows

$$\mathbf{k}(i) = \frac{\mathbf{P}(i-1)\mathbf{w}(i)}{\lambda+\mathbf{w}^T(i)\mathbf{P}(i-1)\mathbf{w}(i)} \quad (15)$$

$$\hat{\mathbf{p}}(i) = \hat{\mathbf{p}}(i-1) + \mathbf{k}(i)[y(i)-\hat{\mathbf{p}}^T(i-1)\mathbf{w}(i)] \quad (16)$$

$$\mathbf{P}(i) = \frac{1}{\lambda}[\mathbf{1} - \mathbf{k}(i)\mathbf{w}^T(i)]\mathbf{P}(i-1) \quad (17)$$

where the forgetting factor $\lambda \in (0,1>$.

Results of Experiments

Each experimental test contained an execution of the program prepared for a separate method of estimation. In an elementary loop of the main part of each program the following operation were performed:

-sending the input data for the velocity and current regulators, calculated on the basis of desired profiles of the rotational velocity and the laod torque;

-measurements the real values of the rotor velocity and armature current of the motor;

-filtering the digital data;

-estimation the parameter vector $\hat{\mathbf{p}}$, and calculation the estimates of the torque load and the moment of inertia.

There were six series of experiments performed for all combinations of the two desired velocity profiles (cases *v1*, and *v2* presented in Fig. 2.) and the three desired armature current profile of the load generator (*I1*, *I2*, and *I3*). Fig. 3. shows the desired profiles of $I_{ad}(t)$ in the cases *I2* and *I3*, while in the case *I1* the load current was chosen to be *0*.

Figs. 4 and 5 show the typical real shapes of rotational velocity (measured by a tachogenerator) and armature current of the motor in the variant *v2-I3*. Especially in the profile of armature current one can see a large number of short-time disturbances that are caused by choppering work of a PWM power amplifier. These disturbances were not sufficiently rejected by an analog filter placed before the A/D converter. Unfortunately the least-square method of estimation is very sensitive to high-amplitude disturbances. To obtain better results of estimation a simple method of filtering was applied, namely the average values of every pair of samples were taken for the further calculation.

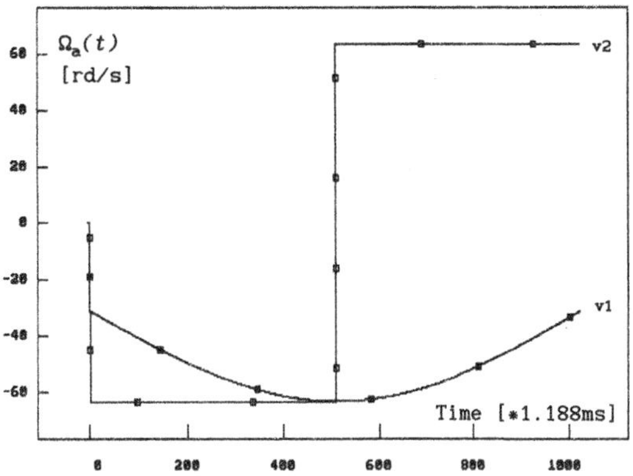

Fig. 2. Desired velocity profiles

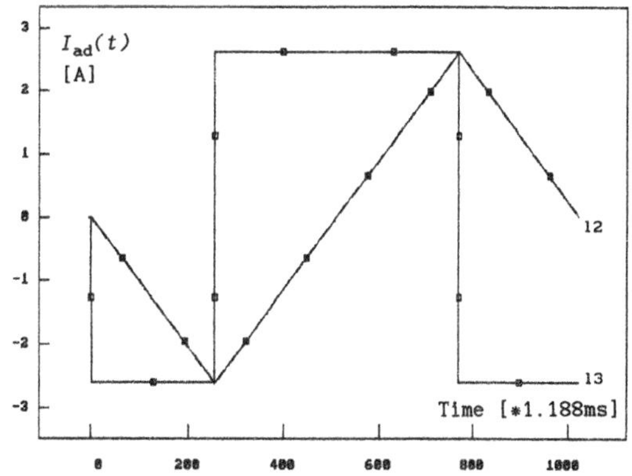

Fig. 3. Desired armature current of load generator

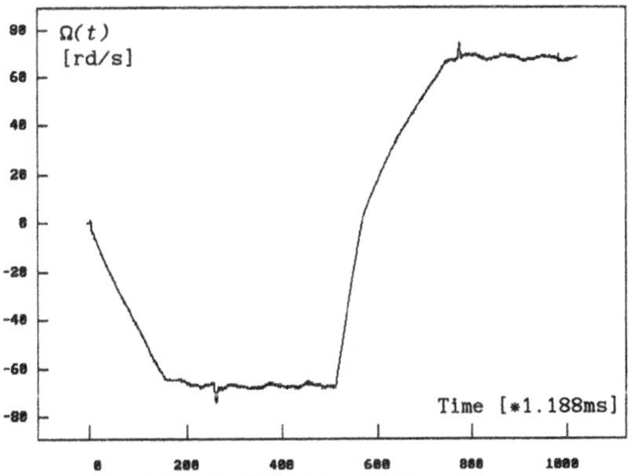

Fig. 4. Velocity of the rotor

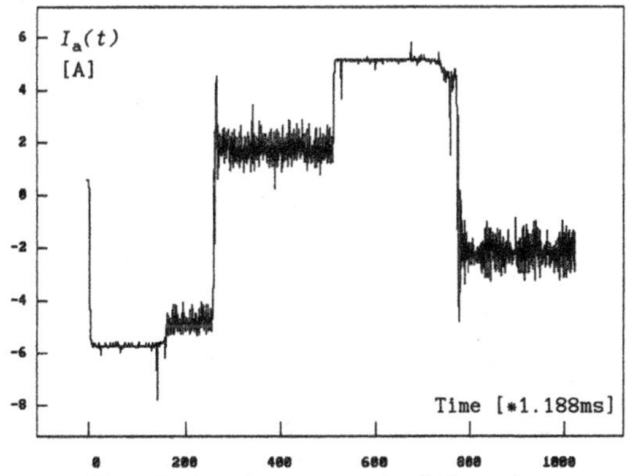

Fig. 5. Armature current of the motor

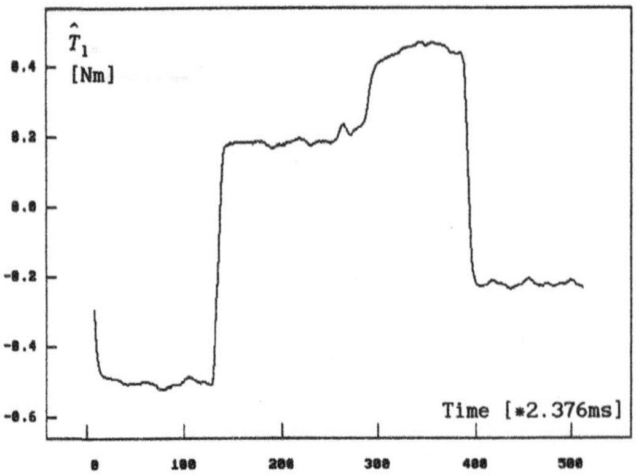

Fig. 6. Non-recursive LS estimation, m=8

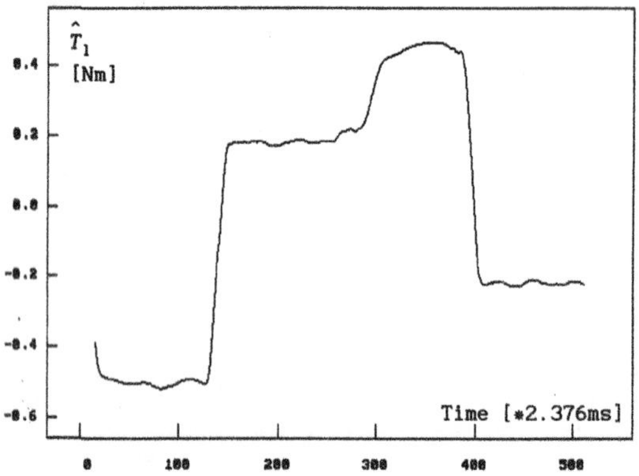

Fig. 7. Non-recursive LS estimation, m=16

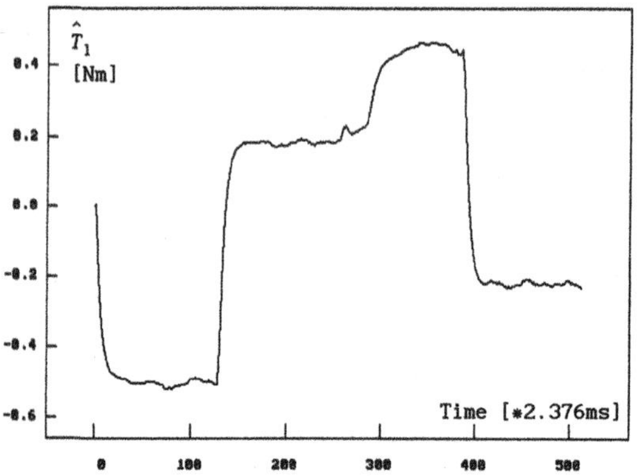

Fig. 8. Recursive LS estimation, λ=0.80

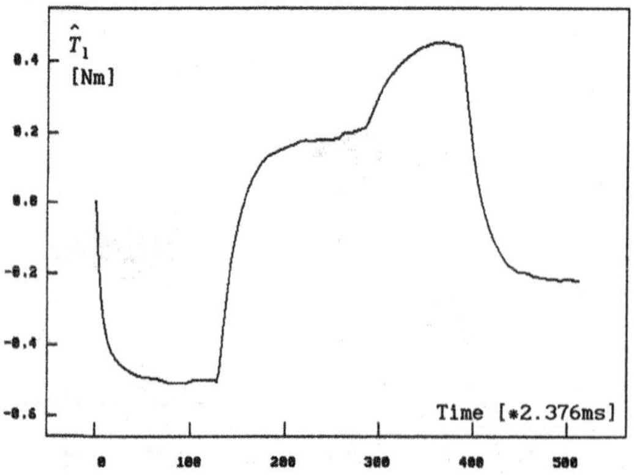

Fig. 9. Recursive LS estimation, λ=0.95

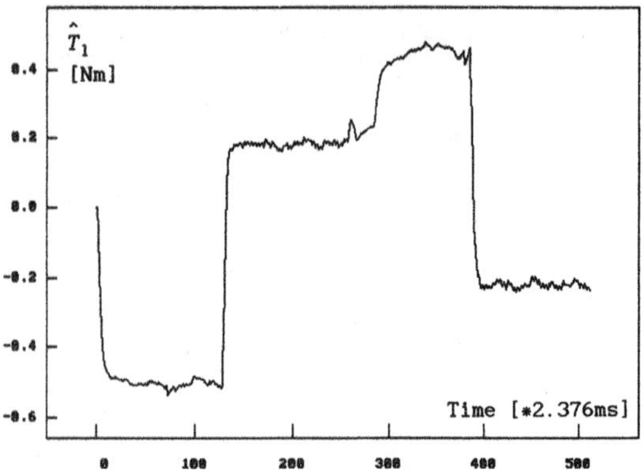

Fig.10. Estimation by projection algorithm, α=0.5

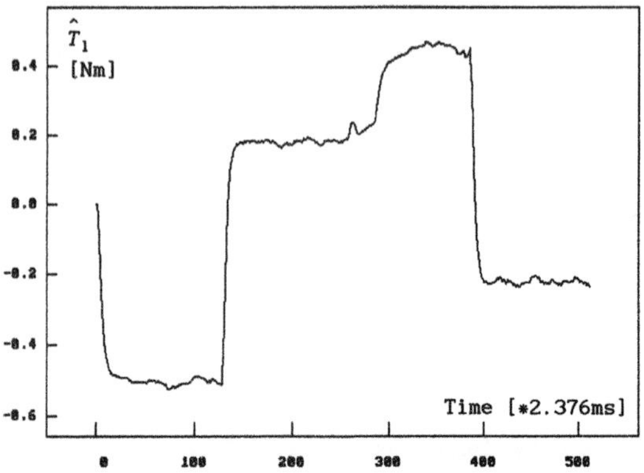

Fig.11. Estimation by projection algorithm, α=2.0

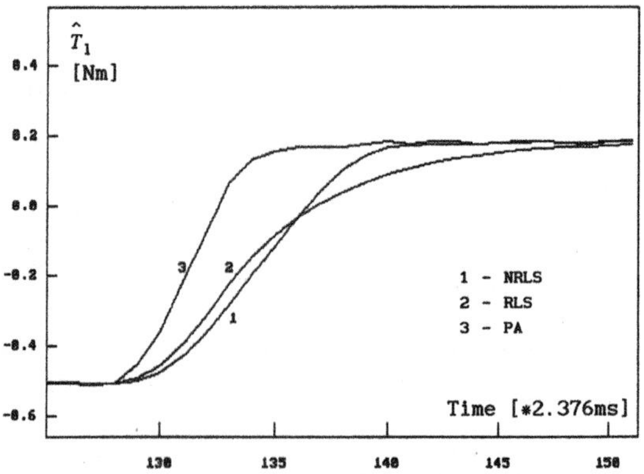

Fig.12. Comparison of estimation methods

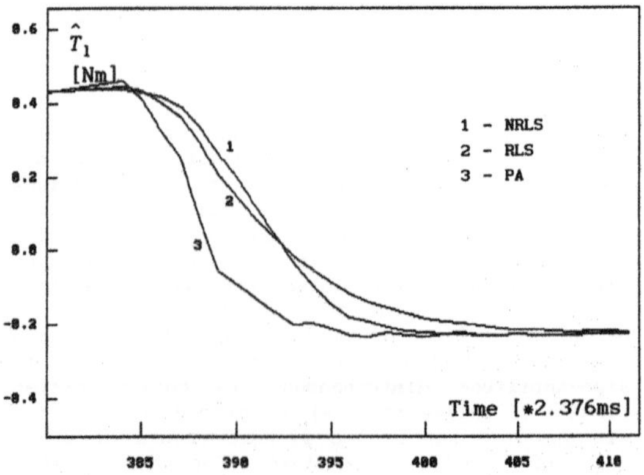

Fig.13. Comparison of estimation methods

The exemplary results of estimation are presented in Figs. 6.-11. These results were obtained for the variant *v2-I3* and for the three different methods mentioned previously. The only estimates of a resultant torque load are shown there, because the estimates of inertia moment were very similar (the value of $\hat{J}(t)$ was nearly constant). For the first look there is an important difference between the profiles of $\hat{T}_1(t)$ presented in these pictures and the profiles of $I_{ad}(t)$ that was shown in Fig. 3. (the case *I3*). However the explanation is quite simple: when the rotational velocity is positive then the torque extorted by the load generator is increased by a friction torque, while in the opposite situation the above values subtract.

Fig. 12 and 13 show some fragments of $\hat{T}_1(t)$ taken from Figs. 6, 8, and 10. Both situations presented here correspond with the positive and negative edges of the desired load torque respectively. These last two figures prove that the projection algorithm is faster in comparison with non-recursive or recursive least-squares method.

The projection algorithm can be successfully applied in such situations where the other two methods are too slow to give good estimates. An example is shown in Fig. 14. where the estimation of the torque friction is presented. The profile of rotational velocity was shown in Fig. 2. - the case *v1*. The initial condition for the parameter vector $\hat{\mathbf{p}}$ was chosen to be *0*. One can see that after short initial period in the profile of the estimate $\hat{T}_1(t)$ occur some regular oscillations. These oscillations coincide with the angular position of the rotor, it means that there was a misalignment of the axes of both machines.

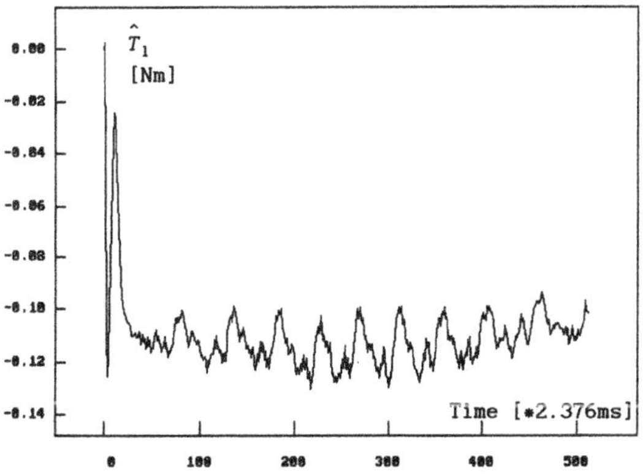

Fig.14.Estimation of friction torque, variant *v1I1*

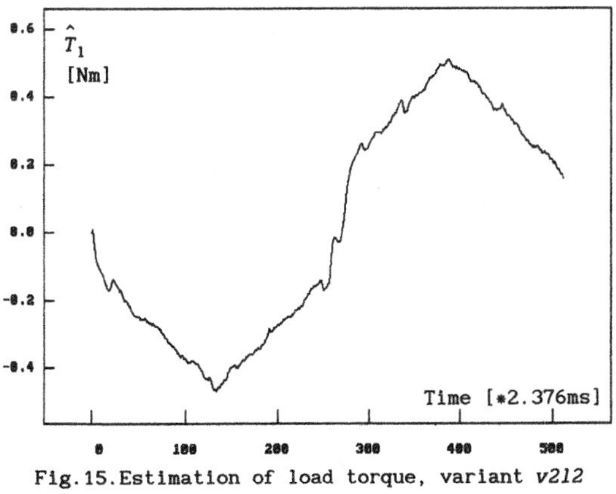

Fig.15.Estimation of load torque, variant *v2I2*

The last picture shows the estimation of a load torque in the case when the command signals for velocity and current regulators were rectangular and triangular versus time respectively. The slope of the presented estimate is distinctly rippled, what was caused by the character of the friction torque described above.

CONCLUSIONS

The experimental tests on estimation methods of DC motor parameters showed that the simple projection algorithm gives a little less precise estimates in comparison to the least-squares methods, but is significantly faster. The Kaczmarz's projection algorithm can be successfully applied in some situations where occur rapid changes in the description of system dynamics. This is typical in robotics. Indirect adaptive control of a robot needs to estimate a relatively big number of parameters, up to 10. Fortunately these parameters occur in linear dependencis in the description of a manipulator dynamics (Craig, 1988). The simulation tests (Bartoszewicz,1991) showed that the projection algorithm can give good results in the case of self-tuning control of a robot. The experimental results presented in this work confirm a usefulness of the algorithm.

ACKNOWLEDGMENT

The author wants to express his gratitude to D. Zarychta who assisted in preparation the laboratory stand and took part in some experiments.

REFERENCES

Åström,K.J., and B.Wittenmark (1989). *Adaptive Control*. Addison-Wesely Pub. Comp., Reading, MA.

Bartoszewicz,A (1991). Self-tuning control of a manipulator with electric DC drives (in Polish). *Proc. XI Nat. Conf. of Automatics, Bialystok-Bialowieza, 17-20 Sept. 1991*, 455-462.

Bassi,E., and F.Benzi (1988). Digital signal processor-based adaptive algorithms for DC position control drives. *Proc. Int. Conf. on Drives, Motors, and Control, Birmingham, 30 Nov.- 1 Dec. 1988*, 5.11-5.16.

Bassi,E., F.Benzi, and R.Scattolini (1990). On-line identification and pole-placement based controller for electrical drives in industrial applications. *Proc. IEEE Int. Workshop on Advanced Motion Control, Yokohama, 29-31 March 1990*, 18-24.

Canudas.C., K.J.Astrom, and K. Braun (1987). Adaptive friction compensation in DC-motor drives. *IEEE J. of Robotics and Automation, vol. RA-3*, 681-685.

Craig,J.J. (1988). *Adaptive control of mechanical manipulators*. Addison-Wesley Publ. Comp., Reading, MA.

Giri,F., J.M.Dion, L.Dugard, and M.M'Saad (1991). Parameter estimation aspects in adaptive control. *Automatica, vol. 27*, 399-402.

Isermann,R. (1988). *Identifikation dynamischer Systeme*. Springer-Verlag, Berlin.

Kelly,R. (1987). A linear-state feedback plus adaptive feed-forward control for DC servomotors. *IEEE Trans. on Ind. Electron., vol. IE-34*, 153-157.

Kumar,P.R. (1990). Convergence of adaptive control schemes using least-squares parameter estimates. *IEEE Trans. Autom. Control, vol. AC-35*, 416-424.

SOFTWARE PACKAGE FOR MODEL IDENTIFICATION AND PARAMETER ESTIMATION OF SYSTEMS

T. Escobet and J. Quevedo

EUPM. Universitat Politècnica de Catalunya, c/Av. Basse de Manresa, 61-73, 08240 Manresa, Barcelona, Spain

Abstract. In this paper is presented a new algorithmic process of automatic selection of the model best structure. This software package, called IES, permits to identify automatically the order and delay of a model using a combination of t-Student test and a modified Akaike's criterion. The exploration of the model structure requires two phases, one for approximation and the second for adjustment. IES software uses some MATLAB functions and performs in DOS Operating System with a Personal Computer.This software package has been used in some dynamic processes with good results.

Keywords. Computer-aider system design; identification; parameter estimation.

INTRODUCTION

Throughout the last decade, in the field of automatics we have been witnessing important and constant efforts in the implementation of algorithms in the computer, with the main object of placing everything that concerns advanced automatics within the reach of a high number of users. For this reason, the field of computer-aided control system design (CACSD) is undergoing a phase of growth and is currently an important research area.

There is an eminently numeric group of CACSDs dedicated to study of models of control systems: MATLAB, MATRIX, CTRL-C, PIM, amongst others. They have introduced a high number of estimation methods and are availed of certain analysis and identification criteria for providing the control engineer with the modeling of the processes. This works interactively with the user.

In recent time, in order to improve modeling and to facilitate analysis, expert systems are being incorporated to the previous programs together with powerful graphic interfaces. In this area, we could mention programs such as SEXI (Gentil, 1990); ESPION (Haest, 1990) and SEISM (Monsion, 1991). These consist of a numeric module which contains the different estimation algorithms, and a symbolic module consisting of a knowledge base which is set up on the whole through a declarative program. This fact means that such programs require a considerable capacity of memory.

In this paper, we shall be describing the program that has been developed under the name of IES - Identification and Estimation of Systems- whose purpose is to automatize the techniques of identification, estimation and analysis of the processes of the first group of CACSD which has been mentioned, specifically in the MATLAB program.

First of all, we shall discuss some of the identification criteria studied, and describe some of the tests carried out for choosing a solid identification criterion.

Next, we shall describe the program that has been developed, indicating above all the steps that have been taken in determining the order of the model.

Finally, we shall consider how the IES program is capable of correctly identifying a simulated system and of providing a model which adapts satisfactorily to experimental results.

STUDY OF THE IDENTIFICATION CRITERIA

Identification criteria

In using the different techniques for the estimation of parameters, it is necessary to know the order of the system; therefore, the number of parameters that may be identified has to be determined in advance. When working with a real process, this type of information is, in most cases, unknown. It is for xs reason that, throughout the last decades, the problem of determining the order has been the object of considerable attention. Upon referring to bibliography, we may appreciate the wide variety of methods or criteria that have been proposed in this sense.

We have divided these methods into five groups:

1. Tests based on the direct study of the co-variance matrix
- Singularity of an information matrix, (Woodside, 1971)
- Instrumental Determinant Ratio test (IDR), (Wellstead, 1978)
- Estimation error Variance Norm test (EVN), (Young, 1980)

2. Tests based on the variance of the residue
- F-test, (Söderstrom, 1977)
- Akaike's Final Prediction-Error criterion (FPE), (Akaike, 1969)
- Akaike's Information theoretic Criterion (AIC), (Akaike, 1974)
- Modified version of Akaike's test, (Edmunds, 1985)

- Wilks criterion, (Söderstrom, 1977)
- Criterion Autoregresive Transfer function (CAT), (Parzen, 1974)

3. Test based on the analysis of poles and zeros
- Test of pole-zero cancellation, (Söderstrom, 1975)

4. Tests based on the study of correlation
- Correlation test, (Sinha, 1983; Ljung, 1988)
- Cross correlation test, (Ljung, 1988; Söderstrom, 1989)

5. Test of standard deviation of the parameters
- t-Student test, (Iserman, 1973)

Identifiers studied

First of all, a prior selection of the different criteria and methods of identification located in bibliography was carried out. In the course of this selection, the following aspects were taken into account:

- Analysis of the comparative studies carried out by different authors (Van der Boon, 1974; Unberhauen, 1974);

- Facility and time of calculation;

- The possibility of implementing them in a specific software, such as MATLAB or others.

From all of these, six criteria have been chosen, with their being subjected to study on a comparative basis.

The criteria analyzed are those minimizing the functions:

1. FPE test

$$FPE(M(\theta,Z))=\frac{1+(\hat{n}+\hat{m})/N}{1-(\hat{n}+\hat{m})/N}V(\theta,Z)$$

2. AIC test

$$AIC2=N\log V(\theta,Z)+2(\hat{n}+\hat{m})$$

3. Modified AIC

$$AIC16=N\log V(\theta,Z)+16(\hat{n}+\hat{m})$$

4. F-test

$$f=\frac{V_1(\theta_1,Z)-V_2(\theta_2,Z)}{V_2(\theta_2,Z)}*\frac{N-(\hat{n}_2+\hat{m}_2)}{(\hat{n}_2+\hat{m}_2)-(\hat{n}_1+\hat{m}_1)}$$

5. EVN test

$$EVN(\theta)=\frac{V(\theta,Z)}{Z(\hat{n}+\hat{m})+1}\sum_{i=1}^{2(\hat{n}+\hat{m})+1}\hat{P}_{ii}$$

and those evaluating the obtaining of significant parameters, according to the t-Student test:
6. t-Student test

$$ta_i=\hat{a}_i/\sqrt{\hat{P}_{ii}}$$
$$tb_i=\hat{b}_i/\sqrt{\hat{P}_{ii}}$$

where V(Θ,Z) represents variance of the error; Θ, estimate of the parameter vector; Z, information vector to contain the measured values of the input and output signals; M(Θ), model corresponding to the parameter vector Θ; N, number of data points; and P an element of the diagonal of co-variance matrix.

Results obtained

In order to select the identification criteria, we have simulated a series of models, M, of different orders and delay. The models studied have been of the ARX and ARMAX types, which can be represented in general by the equation in discrete time:

$$A(q^{-1})\ y(k)=q^{-r}\ B(q^{-1})\ u(k)+C(q^{-1})\ e(k)$$

With the polynomials $A(q^{-1})$, $b(q^{-1})$ and $C(q^{-1})$ of the order n, m and p respectively, and considering r as the pure delay of the system. For the simulation, as u(k) we used a PRBS (+1) sequence; as e(k) we used a white noise sequence uncorrelated with the input and with variance σ; and N=500.

The representative models studied have been:

M1:

$$(1-1.5q^{-1}+0.7q^{-2})y(k)=(q^{-1}+0.5q^{-2})u(k)+0.5(1-1.5q^{-1}+0.7q^{-2})e(k)$$

M2:

$$(1-1.5q^{-1}+0.7q^{-2})y(k)=(q^{-5}+0.5q^{-6})u(k)+0.5e(k)$$

In figures 1,2 and 3, it is shown how the first five of the identification criteria under study vary upon considering different structures and using IV4 as an estimation method.

The structures have parallels in the values of n and m with the relation:
- n=i, m=0, 1, ..., i-1 and i= 1,2,3,... in the figures 1 and 3.
- n= i, m= i-1 and i= 1,2,3,... in the figure 2.
and a pure delay of r=1 in the case of figures 1 and 2 and r=5 in figure 3.

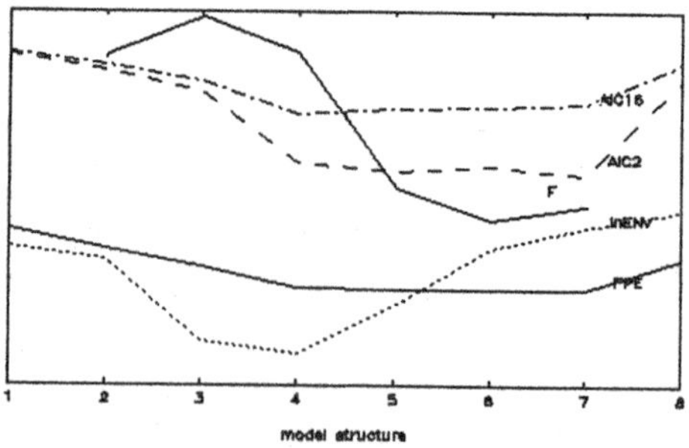

Fig. 1 M1, Identification criteria values

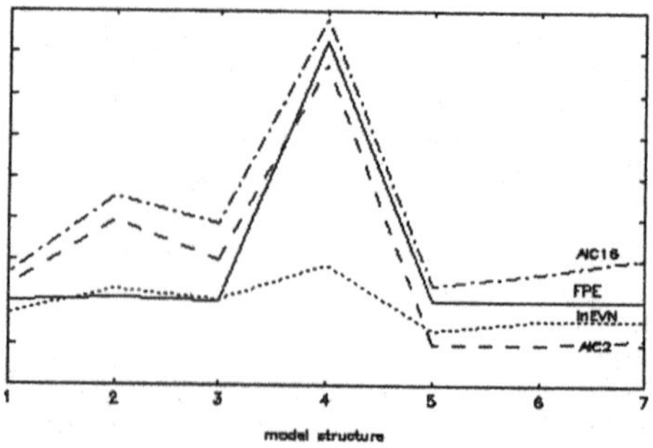

Fig. 2 M2, Identification criteria values

TABLE 1. The values of t-Student

Structure	1	2	3	4	5	6	7
ta	-2.49	-1.57 1.47	-2,53 2.49 -1.49	-3.45 9.95 -14.47 15.81	-51,44 18.35 -8.17 5.13 -3.28	-4.81 0.16 1.10 -1.02 1.12 -1.07	-2.54 -1.80 2.85 -0.81 1.25 0.48 -0.83
tb	1.83	0.18 -0.98	0.37 -2.08 1.56	0.50 -0.84 2.01 -0.73	2.21 -0.47 0.86 -0.27 24.06	2.32 0.95 0.68 0.28 26.01 3.72	2.07 1.71 0.55 0.55 21.26 6.93 0.92

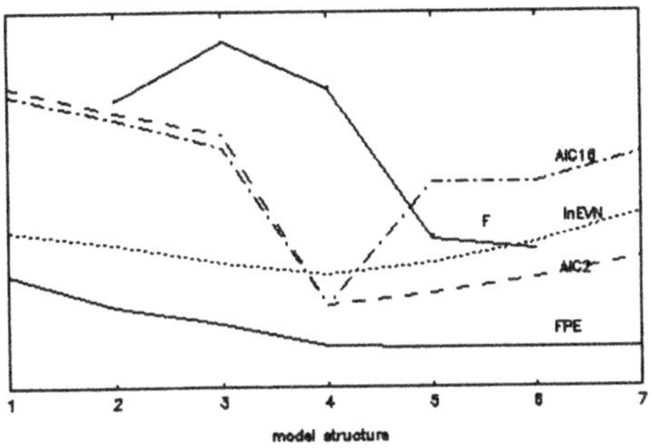

Fig. 3 M2, Identification criteria values

It is observed that the criterion in the modified version of Akaike is the only one which correctly identifies the structure of the simulated model, n=2 and m=1; the rest select models of higher order.

In those cases where the order of delay is not identified correctly (Figure 2), all the criteria studied present oscillations with local minima, a fact that makes convergence in the repetitive algorithms difficult. Under these circumstances, the analysis of the sixth criterion is interesting (table 1). It is observed that the values of ta and tb start to become significant as from the structure 5, equivalent to n=5 and m=4, identifying a value of r=5 which coincides with the simulated model.

Similarly, a study has been made as to how to determine the order of the stochastic part, figure 4, concluding that the most solid criterion is also AIC16.

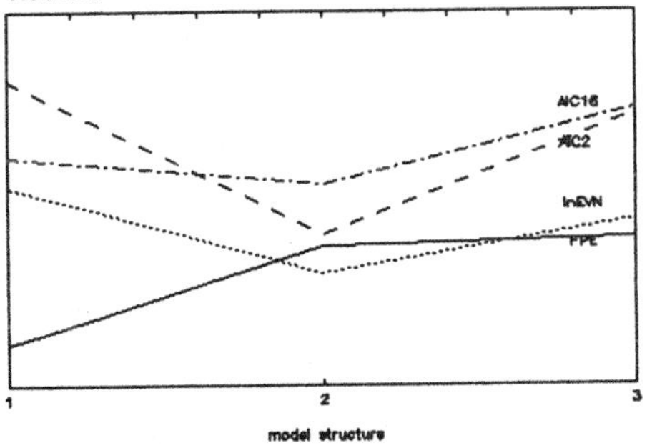

Fig. 4 Identification criteria values

DESCRIPTION OF THE IES PROGRAM

General Features

The software developed has been programmed in the Matlab language and runs on personal or compatible computers with MS-Dos.

In this framework, new modules have been generated which permit:

- Identifying automatically the order and the delay of a model which performs in a way similar to that of real process, merely from the knowledge of the input and output data of latter.

- Facilitating the communication with the users as from a manu-driven graphic interface.

The IES uses the different algorithms for the estimation of parameters contained in MATLAB program.

At the present time, like most CACSDs currently on the market, IES allows identifying systems whose characteristics are: linear, SISO and in open loop. Work is now underway to extend its field of action with a view to incorporating the study of closed loop and systems including static non-linearities.

Identification of the structure of the model

This is the most important part of the program, and it is carried out schematically as follows:

- consideration of a certain order for the model which, in our specific case, starts with a reduced value of n=1, m=0 and r=1;

- estimation of the value of the parameters of the model, using the IV4 method;

- calculation of the value of the identifier, AIC16 or t-student;

- taking a decision in accordance with the value determined, either increasing or decreasing that order, or accepting the model as valid.

The delay and the order are identified separately:

- for calculating the value of delay, r, the value of r is increased successively until attaining, in accordance with the criterion of t-Student, permanent and significant values of the parameters estimated in the polynomials $A(q^{-1})$ and $B(q^{-1})$;

- for determinating the value of n and m, a model with the same number of poles and zeros minimized by the AIC16 criterion is sought in a first approximation; once this has been determinated, the number of poles and zeros is then varied, taking the model that minimizes the AIC16 identifier as valid.

IES also allows identifying the stochastic part of the process. For the determination of order, p, the same criterion is used as in the deterministic part, AISC16; and PEM is used for estimating the value of the parameters.

Validation and analysis of the model

The software developed allows the validation of the identified model using the following methods:

- Simulation of the model and comparison of the real data against the simulated data;

- Study of the placement of the poles and zeros of the model;

- and analysis of the correlation.

APPLICATIONS OF THE IES PROGRAM

Results of the IES program with a simulated system

The following model has been simulated:

$$(1-1.5q^{-1}+0.9q^{-2})y(k) = (q^{-2}+0.4q^{-3})u(k) + 0.5(1-1.0q^{-1}+0.2q^{-2})e(k)$$

- by considering u(k) as a PRBS (+1) sequence, and e(k) as a white noise sequence.

The results obtained from the study on the simulated data are shown on Table 2 and Table 3. Here one may observe that:

- The value of the delay identified, which provides a significant value of all the parameters (table 2), is 2.

- The order of model which gives a minimum AIC (table 3) is when n=2 and m=1.

The estimated values of the parameters are:

$$A(q^{-1}) = 1-1.5047q^{-1}+0.8984q^{-2}$$
$$B(q^{-1}) = 0.9887q^{-1}+0.4136q^{-1}$$

- The output signals obtained from the simulated data are highly similar and close to the real ones. The real output and the simulated output may be visualized on Figure 5.

TABLE 2. Delay Determination

n m r	ta_1	ta_2	tb_1	tb_2
1 0 0	0.6501*	---	2.2493	----
1 0 1	-6.1536	---	-0.2161*	----
2 1 1	-170.8757	102.4429	-0.6302*	27.9146
1 0 2	10.0552	---	-2.4044	----

* not significative parameters.

TABLE 3. Order Determination

n m	1 0	2 1	3 2	2 1	3 1	2 2	2 0	1 1
AIC	615.4	-83.49	-60.62	-83.49*	-67.85	-68.35	-39.92	129.8

*minimum value of Akaike criterion

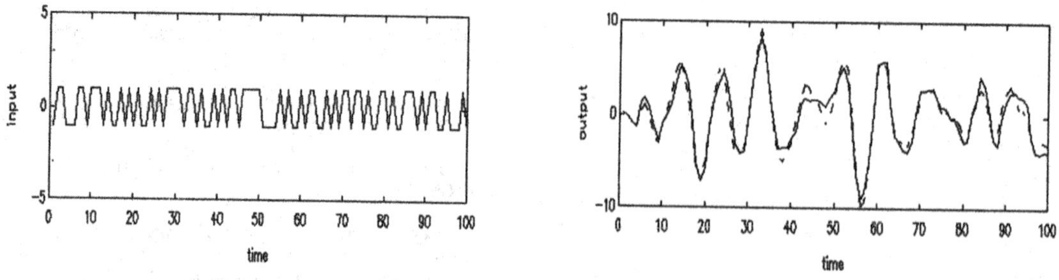

Fig. 5 Input and output signals with a simulated model

Results of the IES program with a real process

As a test, we set up a benchmark formed by a power amplifier, a DC electric motor activated by permanent magnets, and a tacho-dynamo to generate a current proportional to the rotation speed of the axle. For capturing the signals, a PC was used incorporating a "PC-Labcard" data acquisition card.

A continuous signal was sent to the motor amplifier, on which a noise with a Gaussian distribution of 15 Hz band width had been superimposed. As the output signal, the reading of the rotation speed of the axle was taken. The period for acquisition of the signals was of 50 Hz.

TABLE 4. Delay and order determination with a real process

Delay Determination

n	m	r	ta1	tb1
1	0	1	-50.04	17.99

Order Determination

n	m	AIC
1	0	-881.388
2	1	-897.224
3	2	-605.643
2	1	-897.224
3	1	-885.667
2	2	-871.457
2	0	-914.086*

*minimum value of Akaike criterion

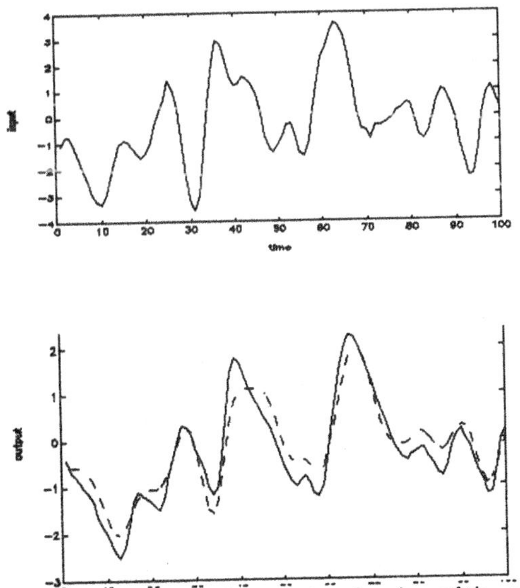

Fig. 6 Input and output signals with a simulated system

The results obtained with the IES program are given on table 4. From these it will be noted that:

- there is no delay in the response of the system

- the order of the model is n=2 and m=0

- the estimated value of the parameters is:

$$A(q^{-1})=1-1.207q^{-1}+0.3625q^{-2}$$
$$B(q^{-1})=0.1173q^{-1}$$

Comparison between the real data from the process and the simulated output from the estimated model may be visualized on Figure 6. From here it may be deduced that the model identified performs in a similar way to the real model under the working conditions studied.

CONCLUSIONS

In this article, a new tool for the identification and estimation of dynamic systems has been presented. The program developed (IES) delivers automatically, from the data obtained with the instrumentation of the process, a mathematical model which performs similar to the real one.

With the purpose of achieving solid identification, a comparative study was carried out on different identification criteria selected for our program: the t-Student test for the calculation of the pure delay of the process, and the modified version of the Akaike criteria for determining the order.

IES is able to identify linear or linearizable systems from real data, with or without inherent noises on the signals, as has been used for identifying other dynamic processes such as hovering systems (Quevedo, 1991) and a cardio-vascular system (Escobet, 1989), with good results.

Acknowledgement. This work has been supported by the Spanish Board for Research and Technology (CICYT) under contract ROB0736/91, and the paper has been partially supported by the Group for Study and Research in Automatic Control (CERCA).

REFERENCES

Akaike, H. (1969) Statistical prediction identification. Ann. Inst. Statist. Math, 22, 203-217.

Akaike, H. (1974) A new look at the statistical model identification. IEEE Transactions on automatic control, 12.

Edmunds, J.M. (1985). Model order determination form state-space control design methods. Int. J. Control, 41, 941-946.

Escobet, T. (1989). IES Disseny d'un programa per a la identificació i estimació de sistemes dinàmics. Engineer's Final Project (ETSEIB).

Gentil, S. and co-workers (1990). SEXI: An Expert Identification Package. Automatica, 26, 803-80².

Haest, M. and co-workers (1990). ESPION: An Expert System for System Identification. Automatica, 26, 85-96.

Iserman, R. (1973). Identification and system parameters estimation proceeding. 3rd IFAC Symposium.

Ljung, L. (1988). System Identification, theory for the user. Prentice Hall.

Paizen, E. (1974). Some recent advances in time

series modeling. IEEE, Transactions on automatic control, 12.
Quevedo, J. and co-workers (1991). Heigth Control of a Hovering System with a Degree of Freedon. 1st ECC, Grenoble.
Sinha, N.K. and Kuszta, B. (1983). Modeling and identification of dynamic systems. Ed. Van Nostrand Reinhold Company.
Söderstrom, T. (1975). Test of pol-zero cancellation in estimate models. Automatica, 11, 537-541.
Söderstrom, T. (1977). On model structure testing in system identification. Int. J. Control, 26.
Söderstrom, T. and Stoica, P. (1989). System Identification. Prentice Hall International.
Wellstead, P.E. (1978). An Instrumental Product Moment Test for Model Order Estimation.
Woodside, C.M. (1971). Estimation of the order of Linear Systems. Automatica, 7, 727-733. Automatica, 14, 89-91.
Young, P., Jakeman, A. and McMurtrie, R. (1980). An Instrumental variable method for model order identification. Automatica, 16, 281-294.

A KNOWLEDGE-BASED CLOSED-LOOP PROCESS IDENTIFIER VIA CORRELATION PATTERN RECOGNITION

V.P. Deskov*, G.M. Dimirovski*, N.E. Gough** and I.H. Ting**

*Laboratory of ASE, Electrotechnical Faculty, P.O. Box 574, St Cyril & Methodius University, Y-91000 Skopje, Republic of Macedonia

**Control Systems Group, Faculty of Science & Technology, Wolverhampton Polytechnic, Wolverhampton WV1 1SB, UK

Abstract: An objact-oriented, knowledge-based, correlation-matching expert system for pattern recognition and typical identification of linear dynamic models of industrial object processes has been designed. It represents an advanced stage of research and development results in AI techniques for automation and systems engineering applications. It is aimed at recognition and typical identifiaction of approximate process models within a PC CADCS environment for hierarcical controls of interconnected multivariable objects.

Keywrds: Artificial intelligence, pattern classification and recognition, identification, expert system, typical models, rule-based inference.

INTRODUCTION

Process identification in plant environments under normal operating conditions involves, practically, all the science and art of systems modelling,signal processing and model identification (Isermann,1980; Raybman & coworkers,1983). A number of profound research studies from various points ofview have been devoted to this subject.Correlation matching is one of matured theoretical methodologies that is widely used in industrial applications.

It is well known, however, that human expertise, knowledge and experience are indispensible, in addition to theoretical knowledge on methods and models in identification. The process of arriving at an appropriate model, acceptable with respect to the purpose and accuracy defined, and yet simple enough to enhence practical control algorithm design, also requires iterative heuristic selection and evaluation.

Artificial inteligence (AI) techniques for certain purposes in process identification have appeared in the literature occasionally for quite some time (Al -Tiga & Gough, 1971).However, following the developments in computing sciences and information tehnologies, the use of AI techniques in process identification(and control) has grown considerably during the last decade and, in particular, following Astrom's paper on *Expert Control* (1984).Apart from his coworkers, to authors' knowledge, several important contributions have emerged recently.Monison and coworkers (1988) have presented their work aimed at expert systems for industrial process identification; Gentil and coworkers (1985,1989,1990,1991) have completed a well-structured, knowledge-based expert identification technique; and Haest and coworkers (1990) have also rereported an expert system for process identification. Recent surveys on the use of expert systems in control have been provided in Gough and others (1990), Dimirovski (1991) and Stanley (1991).

Similar research on using AI techniques for process model identification started in 1987 at the Automation & Systems Engeneering Laboratory of Skopje University (Crvenkovski & Dimirovski,1987; Dimirovski, 1988).It continued thereafter (Dimirovski & coworkers,1989,1991) in close collaboration with the Control Systems Group at Wolverhampton Polytechnic in conjuction with the joint complementary work on computer-aided analysis and design in Control Systems (Gough & coworkers, 1989,1991; Dimirovski & coworkers, 1990, 1991). An early result of this research was a package, which used a simple pattern recognition method aimed at linear dynamic models, with typical auto- and cross-correlations, and a rather restricted group of several types of monotonic non-linear static input-output characteristics most often met in the process industries. The dynamic process identification was converted into an expert system procedure, called ESTIO, by using a simple rule-based inference mechanism combining forward and backward chaining search algorithms.

THE IDEA, BACKGROUND AND CONCEPTUALIZATION

In the current course of research (Deskov & Dimirovski, 1991), the same task was formulated as an objective to build an expert analyst, known as ESTIOB (Expert System for Typical Identification of Objects) as part of a CAD environment aimed at analysis and design of a practical class of hierarchical con trol systems (Dimirovski & others, 1991). The background expertise in typical identification of linear, lower-order, dynamic models in an industrial environment was found in the research monograph of Raybman & coauthors (1983), based on correlation matching methodology. Our study has resulted in an established group of well-defined classes of models distinguished by input auto-correlations and corresponding cross-correlations, in addition to our previous findings on the possibility of using simple pattern recognition techniques. This result, in tourn, has led us to an appropriate composition of a set of qualitative recognition and quantitative com putation algorithms, combined with a rule-based inference mechanism.

The qualitative recognition algorithm performs the classification task and the overal supervision task, wich directs and coordinates the computational algo rithms. Intermediate results comming out of these algorithms,after being saved in a data base for the object to be identified, are converted into generalized strings. Then they are processed by a string-matching to lead to the type of knowledge representation of the corresponding classified correlations Only then appropriate inpulse weighting functions (sequences) and their corresponding differential (difference) equations and transfer functions are identified in the form of mathematical expresions. The time-discretised models could then be passed over to a procedure for the analysis of the characteristic input-output modes in terms of the characteristic patterns and vectors (Gough & Al-Thiga, 1985) using CAD techniques known as CBSL (Ting & co workers, 1990) and CHIOMOD (Dimirovski & co. 1991).

To complete the consistency requirements from the theoretical viewpoint (Raybman & coworkers), it is presumed that the controlled object has been excited by an additive pseudo-white noise signal through the command input. An appropriate signal-to-noise ratio must be established beforehand.Then the input and output signals of the controlled object are carrying the information needed for correlation mat ching identification. This way, apart from a fiew examples with unusualy odd cross-correlations, the ESTIOB analyst preserves its performance within the qualitative and quantitative accuracy of Raybman's expertise.

In order to achieve considerable processing speed and portability to follow the principles of object-oriented programming, the software of this AI identification technique has been written in ANSI C/C++. Currently, it is in its experimental stage of development, and research is in further progress.An additional requirement that has to be met due to limited resources in Skopje, is that of the adequacy to standard PC hardware which is widely available.

In addition to identification applications, ESTIOB is envisaged to be used in conjunction with steady-state multivariable decoupling control and for multimodel control methodology at the local, dynamic control level within the two-level control of interconnected objects (Dimirovski& coworkers, 1991).

SYSTEM THEORETICAL BACKGROUND

If a close-loop linear stationry dynamic system is asumed which is excited by additive pseudo-white noise on its command input (Fig.1), then the input auto-correlation function (r_{xx}) and the input-to-output cross-correlation function (r_{xy}) of so defined an object can be connected via Fredholm's integral

$$r_{xy}(t) = \int_0^{\infty} g(\tau)\, r_{xx}(t-\tau)\, d\tau, \quad -\infty < t < +\infty \qquad (1)$$

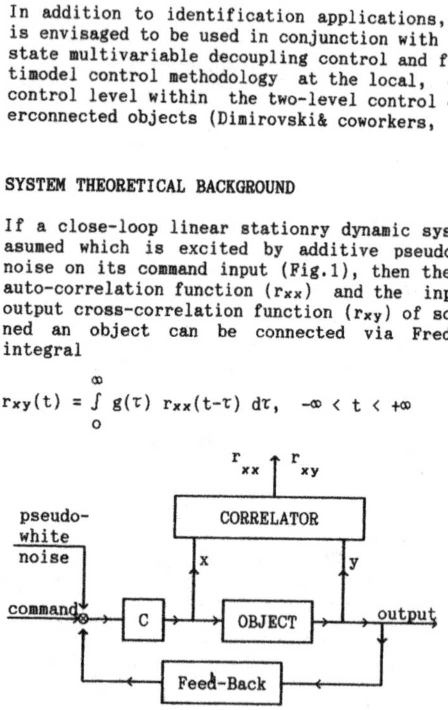

Figure 1 - *Block schematic diagram of the object in its working environment*

where g(t) denotes the weighting function of the object. Further on, the correlation functions can be expressed as

$$r_{xx}(t) = \begin{cases} r^{+}_{xx}(t), & t \geq 0 \\ r^{-}_{xx}(t), & t \leq 0 \end{cases}$$

and (2)

$$r_{xy}(t) = \begin{cases} r^{+}_{xy}(t), & t \geq 0 \\ r^{-}_{xy}(t), & t \leq 0 \end{cases}$$

It is recalled that, bearing in mind the auto-correlation function is symetrical by definition, the following two results are valid

$$r^{+}_{xx}(t) \equiv r^{-}_{xx}(-t), \quad t \geq 0 \qquad (3)$$

and, however,

$$r^{+}_{xy}(t) \neq r^{-}_{xy}(-t), \quad t \geq 0. \qquad (4)$$

By combining of Eqns.(1), (2), (3) and (4)

$$\Delta r_{xy}(t) = \int_0^{t} g(\tau)\, \Delta r_{xx}(t-\tau)\, d\tau, \quad t \geq 0, \qquad (5)$$

where

$$\Delta r_{xy}(t) = r^{+}_{xy}(t) - r^{-}_{xy}(t), \quad \text{and}$$

$$\Delta r_{xx}(t-\tau) = r^{+}_{xx}(t-\tau) - r^{-}_{xx}(t-\tau).$$

If now Laplace tranformation is applied on Eqn.(5), it follows that

$$G(s) = [R_{xy}(s)-R_{xy}(s)] \;/\; [R_{xx}(s)-R_{\wedge\wedge}(s)] \qquad (6)$$

Thus, it is obvious from Eqn.(6) that for known correlation functions $r_{xx}(t)$ and $r_{xy}(t)$, G(s) and, by appropriate inverse Laplace transformations, g(t) as well as the differential equation of the object of interest can be identified.The problem is how to to derive the analytical expressions of the correlation functions provided the measured correlation sequences are available.

THE ESTIOB DESIGN AND IMPLEMENTATION

The leading idea of the ESTIOB project was to produce an object identifier able to perform identification of objects while operating in closed loop configuration (Fig. 2). That is an process object identifier which would not demand laboratory environment but rather it will complete the identification task during normal operation of the object. In order to accomplish this objective, an insertion of pseudo-white noise with appropriate signal/noise ratio at the command input of the system is needed. The sensored signals at the input and the output (Fig.2-x,y) of the considered object are fed in a correlator for appropriate set of correlation time-sequences to be produced. And, afterwords, these sequences are passed over to the ESTIOB system which automaticaly produces representational symbolic models of the object according to the knowledge built in generalized Symbolic Model Database (SMD).

Taking into account that for real-world objects of interest quite a constrained set of correlation function timplates can be defined (Raybman 1983) a method of pattern matching is feasible. Namely, a database of symbolic expression timplates with general coefficients in the form of strings can be established. The string timplates are later expanded in sequence timplates via expression parsing. Finally, the generated sequence timplates are matched with the mesured correlation sequences and by use of an appropriate criterion the decission on matching is made.

Some of the actual coefficients in relationships' expression, needed for the expansion process, are available directly from the measured sequence (for example the value $r_{xx}(0)$ is almost always one of the expression coefficients) and some are obratined through quite a trivial computation (division, substraction or, at most, a function call). However, in some cases, the determination of a number of coefficients involves an algorithm for solving a set of non-linear equations (Percinkova & Dimirovski, 1990). Recently, a number alternative methods, like modified genetic algorithms, are also observed for this purpose, yet no final choise is made so far.

Meanwhile, a qualitive test of the mesured correlation sequence is made for the purpose of reducing the set of candidate symbolic expressions. For example, if the measured correlation sequence has a zero-cross, it is non-realistic to match it with sequence timplate produced from the expression symbolic timplate of the form A*exp(-a*t) as it does not have zero crosses. Also, a number of quantitive characteristics are derived during the qualitive analysis. For example, where does the zero cross appear? If it appears 'too far' related to other time-scale representants of the correlation, it should be treated as a noise and neglected.

The matching process gives as result the best matching set of symbolic correlation expressions available in the database, as well as their actual coefficients for the observed object.On the other hand, the symbolic expression timplates of the correlation functions (CF) are related to the symbolic expression timplates of impulse respose (IR),transfere function (TF) and differential equation(DE) as well as to the algorithms for the computation of their coefficients (ALG), which is given in the algorithm base also in the form of generalized character strings. After the actual coefficients are computed using ALG, they are inserted in the corresponding places of the CF, IR, TF and DE string timplates instead of the general coefficients. Thus both the symbolic and actual models of the object process are identified.

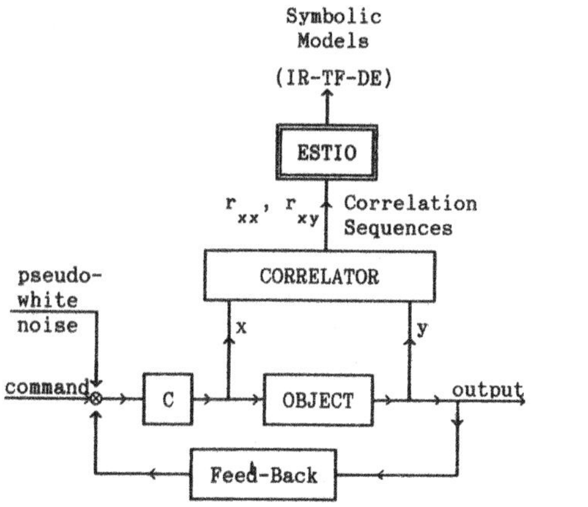

Figure 2 - *Principal scheme of ESTIO instalation*

Conceptual block diagram of ESTIOB mechanism is depicted on Fig.3. First of all, the measured correlation sequences are inspected for the qualitive characteristics such as presence of extrema, zero crosses, periodic components, etc. These data are coded into a character string (KEY) used later as a key for the database of symbolic models (SMD). The key implicitly determines the list of candidate model timplates considered to match the measured sequences. Also, this procedure generates a list of positions of significant characteristic values (SIGNI) in the mesured correlation sequences, which are used later for the purpose of computation of the actual expression coefficients according to the ALG. Following this step, the recognition process, closely related to the knowledge given in SMD, takes place.

The SMD consists of one datafile named INDEX that contains overall information on supplied model timplates and several model datafiles (MD) containing the timplates and corresponding algorithms.The timplates and algorithms are given in the form of sequence of character string expressions. First six expressions in MD are the model timplates (for: $r_{xx}(t\geq0)$, $r^{+}_{xy}(t\geq0)$, $r^{-}_{xy}(t\leq0)$, G(s), g(t) and the differential equation) with general coefficients. From seventh expression on, the algorithm for the computation of the actual coefficients is given. The INDEX file records contain the KEY field, carrying coded information on qualitive characteristics of the model, and a field with t he file-name of the corresponding MD.

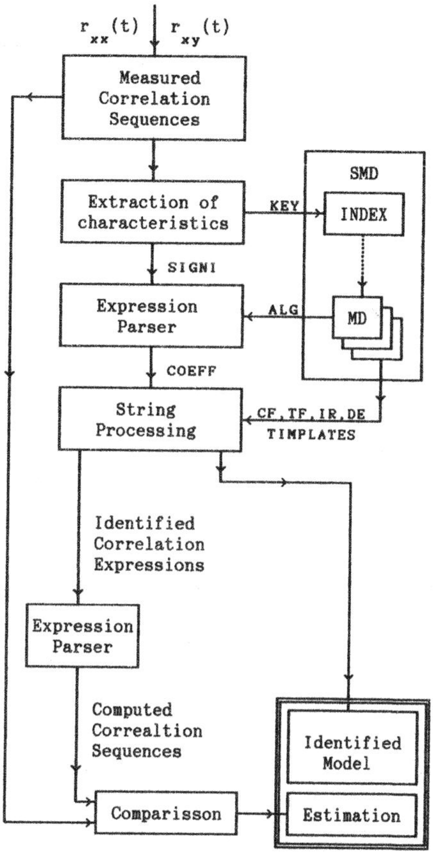

Figure 3 - *Conceptual block diagram of ESTIOB mechanism*

The proces of recognition begins with KEY matching. If the generated KEY matches one of those supplied in the INDEX, the corresponding MD is opend and the algorithm in it executed. The algorithm makes use of values supplied in the SIGNI list and, via the expression parser, computs the actual coefficients. The computed coefficients are now converted into character strings and inserted in the corresponding places instead of general coefficients in the model timplates (first six expressions in MD). The first three expressions (corresponding to: $r_{xx}(t\geq0)$, $r^{+}_{xy}(t\geq0)$ and $r^{-}_{xy}(t\leq0)$) are now expanded into time-sequences and comparred with the measured ones. An estimation on simmilarity is generated, and the process described is repeated for all candidate model timplates available in SMB. Finally, the model which fits best the measured sequences is choosen.

The identified model is practicaly a set of six expressions given in a form of character strings. The expressions obtained are the actual auto-correlation, function (for $t\geq0$), cross-corelation function (two expressions; for $t\geq0$ and $t\leq0$), transfere function, the impulse response and the differential (diference) equations.

CONCLUSIONS

A knowledge-based, correlation-matching expert system for pattern recognition and typical identification of linear dynamic models of industrial object processes has been designed. It is implemented in a

PC computing environment in C/C++ language under MS-DOS operating system.

In this paper, it is shown that object process dynamic behaviour recognition on the grounds of recognising the class of its auto- and cross- correlations is a well-defined problem for a rule-based inference machanism. Then the expertise knowledge on dynamic behaviour and related auto-/cross- correlations is easily imbedded into a knowledge-base.

A particular feature of this AI technique is extensive use of conversion from quantitative into qualitative system representants (extraction of the sequence characteristics) and vice versa (string processing and expression parsing), as well as quantitative estimation of the infered results. Considering this, no standard expert system shell is convenient for development of so specific an expert system.

ESTIOB is aimed at recognition and typical identification of approximate process models within a PC CADCS environment for hierarchical controls of interconnected multivariable objects. It can produce both symbolic and concrete approximate system models of all types.

Acknowlwdgements: Authors would like to express their gratitude to British Council and Ministry of Sceince of Republic of Macedonia, which have supported in part this resaech. Thanks are also due to EMO Electrical Co., Ohird, for their partial sponsorship.

REFERENCES

Al-Thiga, R.S.A. & N.E.Gough (1971), "Artificial intelligence applied to the discrimination of the order of multivariable linear systems". Int.J.Control, 20, 6, 961-969.

Astrom, K.J. (1984),"Expert control". Prepr. 9th IFAC World Congress, Budapest, H, , - .

Ashby, W.R. (1964), "The next ten years" Keynote Address. In J.T.Tou & R.H.Wilcox (Eds.)Computer and Information Sciences - Learning, Adaptation and Control in Information Systems, Spartan Books, Wasington D.C., 2-11.

Crvenkovski B.L. & G.M.Dimirovski (1987), An Expert System Technique for Identification of Linear Dynamic Objects" (in Macedonian). Res.Rep. MSUES-03 /ASE-ETF/87, St.Cyril & Methodius University, Skopje, Rep. of Macedonia.

Deskov V.P. & G.M.Dimirovski (1991),"ESTIOB Analyst: a generalised AI technique for identification of linear dynamic models ofcontrolled objects" (in Macedonian). Res.Rep.CASEIS-05/ASE-ETF/91, St.Cyril & Methodius University,Skopje, Rep. of Macedonia.

Deskov, V.P., N.E.Gough, G.M.Dimirovski & I.H. Ting (1991), "Interactive package for analysis of dynamic multivariable systems via characteristics input-output modes" (in Macedonian). Proc. XXXV Yugoslav ETAN Conference, Ohrid, Rep. of Macedonia, The ETAN Association, Beograd, YU, VIII, 11-18.

Dimirovski, G.M. (1985), "On man-machine aspects of interactive computer-aided methodologies for research and education".Automatika (Zg), 26, 217-222.

Dimirovski, G.M. (1988), AI Techniques in Engineering Applications: An Experience in Typical Identification of System Models. Res. Rep. PSCE/R388 - University of Bradford, Bradford and SCET/R102 - Wolverhampton Polytechnic, Wolverhampton, U.K.

Dimirovski G.M., B.L.Crvenkowski, N.E.Gough,R.M. Henry, B.R.Percinkova & D.M.Joskovski (1989), "AI techniques in engineering: Results in expert systems for typical identification of system models with personal computers". Proc.3rd ETAI Theme Symp. (G.M.Dimirovski, Ed.), Ohrid, Rep. of Macedonia, The ETAI Society, Skopje, 2, 157-166.

Dimirovski, G.M., B.L.Crvenkovski & D.M.Joskovski (1989), "Expert system for recogniton and typical identification of dynamic process models".Proc. 2nd IFAC/IMACS/IFORS Symp. on Advanced Information Processing in Automatic Control (M.G.Singh, Ed.), Nancy, France, I, 222-227.

Dimirovski, G.M. (1991),"Towards intelligent control and expert computer-aided control engineering" Invited Plenary Paper. Proc. XXXV Yugoslav ETAN Conference, Ohrid, Rep. of Macedonia, The ETAN Association, Beograd, YU, I, 27-42.

Dimirovski, G.M., N.E.Gough, R.M.Henry, B.R.Percinkova, V.P.Deskov, I.H.Ting, O.L.Iliev, D.M.Joskovski & N.Sadaoui (1991), "Personal computer environment for CADCS of hierarchical controls of inter connected systems". Prepr. 5th IFAC/IMACS Symp. on Computer-Aided Design in Control Systems (H.A.Barker, Ed.), Swansea, U.K., 45-50.

Gentil, S. & F.Rechenmann (1985), "Identification des procedes et intelligence artificiel". Proc. Congres AFCET - Automatique des outils pour demain, Toulouse, France.

Gentil, S., A.Barraud & K.Szafnicki (1990), "SEXI: An expert identification package". Automatica, 26, 4, 803-809.

Gough, N.E. & R.S.A.Al-Thiga (1985), "Characteristic patterns and vectors of discrete multivariable control systems". Arab.J. Sci. & Eng., 10, 3, 253-264.

Gough N.E., I.H.Ting, N.Sadaoui & G.M.Dimirovski (1990), "A survey of knowledge-bassed systems in control". Proc. 2nd Symp. Warren Springs Laboratory on Use of Personal Computers in Industrial Control (A.W.Self, Ed.), Stevenage, U.K., The Microcomputer Unit, London, 53-60.

Gough N.E., I.H.Ting, N.Sadaoui & G.M.Dimirovski (1991), "CAD of discrete multivariable control systeme using singular characteristic patterns and vectors". Proc. 5th IFAC/IMACS Symp. on Computer-Aided Design in Control Systems (H.A.Barker, Ed.), Swansea, U.K., 502-507.

Haest, M., G.Bastin, M.Gevers & V.Werts (1990), "ESPION: An expert system for system identification". Automatica, 26, 1, 85-95.

Isermann, R. (1980), "Practical aspects of process identification". Automatica, 16, - .

Monison, M., B.Bergeon, A.Khadad & M.Bonsard (1988), "An expert system for industrial process identification". Proc. 1st IFAC Workshop on Artificial Intelligence in Real-Time Control, Swansea.

Percinkova, B. & G.M.Dimirovski (1990), "A new method for solving multivariable static systems: 1 - Theoretical Foundations; & 2 - Computer Implementation" (in Macedonian). Sceintific Meeting on Theoretical Research in Natural and Technical Sciences in Macedonia (Acad. B.Popov, Gen.Chairman) Macedonian Academy, Skopje, Paper 2.18.

Raybman, N.S. (ed.), S.A.Anisimov, I.S.Zayceva & A.A.Yarlov (1983), Typical Linear Models of Controlled Objects (in Russian). Energoatomizdat, Moscow.

Szafnicki, K. & S.Gentil (1991), "An object orioriented knowledge base for process identification". Proc. 5th IFAC/IMACS Symp. on Computer-Aided Design in Control Systems (H.A.Barker, Ed.), Swansea, U.K., 188-193.

Stanley, G.M. (1991), "Experiences using knowledge-based reasoning in online control systems" Invited Plenary Paper. Proc. 5th IFAC/IMACS Symp. on Computer-Aided Design in Control Systems (H.A.Barker, Ed.), Swansea, U.K., 11-19.

Ting I.H., N.Sadaoui, N.E.Gough & G.M.Dimirovski (1990),"A portable software language for simulation of discrete multivariable control systems based on convolution algebra". Proc. UKCS Conf. on Computer Simulation (K.G.Nock, Ed.), The UKCS & University of Sussex, Brighton, U.K., 175-180.

SENSOR ACTUATOR TO CONTROL THE TEMPERATURE EFFECTS INTO A SYSTEM BASED ON A QUARTZ CRYSTAL

J. Arguelles*, M.J. López, M. Martínez** and S. Bracho****

**Time Frequency & Electronic, Burgos, Spain*
***University of Cantabria, Electronics Department, Santander, Spain*

Abstract. In many applications voltage sources with high stability frequency and fast warm up at any ambient temperature, stable over the operating temperature range, low phase noise and low cost are usually required. Quartz crystal oscillators are quite stable but for many applications, where highest stability rates are required, it is necessary to improve its stability by means of two techniques available. The best results are achieved by controlling the temperature of the oscillator. In this paper we show the design of a new controller type specially designed to avoid many of the long and short term stability problems that standard proportional controllers induce in the oscillator circuit. We have developed a variation of the controller to improve its warm up performances when required. Some applications are shown to prove the performances of the controller.

Keywords.: Temperature control, oven controlled crystal oscillator

INTRODUCTION

Current and future communication, navigation and instrumentation systems require small, high precision oscillators for timing and frequency reference. System specifications often require the oscillator to maintain high stability in various environments. Fast warm up at any ambient temperature, stability over the operating temperature range, low phase noise under vibration, high reliability and low cost are some of the typical requirements.

Quartz crystal oscillators are a good approach to achieve high stability frequency signals sources due to the high Q factor of quartz crystal resonators.

The quartz crystal is a piezo electric material which exhibits different behavior depending on the cut angle orientation with its crystallographic axes. There are many different types of cuts usually employed, those cuts can oscillate in different modes and overtones at the same time. Each kind of cut has a well defined behavior with its advantages and disadvantages for different applications (Salt, 1987).

Although quartz crystal oscillators are quite stable, in some cases where high stability sources are required, such kind of oscillators are not enough. In the next section we shall discuss some of the most common instabilities caused in quartz crystal oscillators (Vig, 1985).

One of the main frequency instabilities came from the frequency versus temperature behavior of the quartz crystal resonators. To reduce the effects of temperature changes on the crystal oscillator two different approaches have been used, the first is to compensate the frequency shifts due to the temperature by means of some circuitry that pulls the crystal frequency by the same amount but in the opposite direction. And the second is, to control the internal temperature in order to avoid the effect of the ambient temperature changes on the crystal (Frerking, 1978).

Highest stabilities are achieved by means of temperature control techniques. Usually the electrical configuration of those devices consists of four basic blocks: voltage regulator, oscillator, output buffer and thermal control. In this paper we shall show the design of a controller circuit, able to keep the crystal temperature constant and to avoid to deteriorate other circuit performances (Pincu, 1989).

This will lead us to analyze the controller design, and the other design and selection blocks both together in order to minimize the short term and long term stability problems, keeping a good warm up performance, with the less cost possible. To solve this problems, the controller design and the thermal design of the oven cavity and all the system should be done together.

First of all, we shall make a brief list of the main

instability causes in quartz crystal oscillators, remarking those that depend on the controller design. Then we shall describe the two traditional solutions to improve the stability in quartz crystal oscillators. This will help us to define the main points in controller design to be considered in oscillator design. Then, we shall describe our controller type and we shall show a variation of it useful to improve some performances of the system, remarking its main performances. This will let us make a comparison with other published controllers. Finally, we shall show the first applications of this research with some test results, to establish some conclusions.

QUARTZ CRYSTAL OSCILLATORS INSTABILITIES

In spite of the stability shown by quartz crystal oscillators, there are many causes for oscillator circuit instabilities, the most common came from 'load reactance changes'. For example in a typical high stability crystal with C_o=5pF, C_1= 4fF and C_L=20pF, a ΔC_L=10fF causes a $\Delta f = 1x10^{-7}$, C_o and C_1 are internal parameters of the crystal, C_L is the load capacitance and f is the frequency of the signal output. A ΔC_L aging of 10ppm per day causes $\Delta f = 2x10^{-9}$ per day oscillator aging.

Other usual instabilities factor is the drive level changes, for example typically $\Delta f = 1x10^{-9}$ per microwatt at a frequency operation of 5MHz in the 5th overtone in a AT-crystal. Also the increase in DC bias on a crystal increases the aging rate.

In spite of that the main frequency drifts came from the temperature versus frequency behavior of the quartz crystal resonators. There are many factors that determine this resonator frequency vs. temperature behavior. The primary factor are the cut angles, and among the secondary we have: overtone and mode, material impurities and strains, mounting stresses (magnitude and direction), bonding stresses (magnitude and direction), electrode geometry, blank geometry, drive level, interfering modes, load reactance (nominal value and temperature coefficient), temperature range of change, thermal history and ionizing radiation

Applied to quartz crystal units the term aging usually means a change in the resonant frequency of the unit, although other parameters, especially the motional resistance, may also change with time.

There are many causes of aging in quartz crystal units and the effects are complicated and varied. The following are some of the causes:

a) aging due to surface deterioration

b) aging due to surface contamination

c) aging associated with the electrodes

d) aging associated with physical effects: mechanical stresses are imposed upon the quartz plate by mounting structure and by electrodes, which in general have temperature coefficients of expansion different from those of quartz. All solids, except perfect crystals, undergo stress relaxation with time, especially at high temperature. Stress relaxation in the metal electrodes is a serious source of aging.

e) aging associated with the quartz

f) aging associated with holder defects: in some units vacuum encapsulating is used and in certain units such as the AT and SC-cuts, it is essential (Bottom).

With a good electrical design, and a right cut crystal selection can minimize most of usual oscillator instabilities, except for the frequency shift due to temperature changes. With a good crystal oscillator can be achieved stabilities versus ambient temperature about 5 to 10 parts in 10^6 can be obtained. Many applications require high frequency stability rates, in such cases two methods are available for eliminating or reducing the effects of temperature changes on the crystal oscillator, namely, temperature control, and temperature compensation.

The degree of temperature control required on a particular oscillator is determined primarily by the specification of the system in which it is to be used. Stabilities of approximately 5 parts in 10^7 can be obtained using plug-in crystal ovens with the oscillator circuitry external to the oven. Stabilities to several parts in 10^9 can be obtained with proportionally controlled ovens containing the crystal and oscillator circuitry, a proportionally controlled oven uses a temperature controlling system in which the power supplied to the oven is proportional to the heat loss. For stabilities better than 5 parts in 10^9 it is generally necessary to use a two stage oven. This may be a combination of two ovens with a single control circuit or two independent proportionally controlled ovens. In precision applications, both the crystal and the oscillator are usually packaged in the oven, and the oven is controlled by a thermistor sensor used in a bridge circuit with an operational amplifier. The amplifier drives a power transistor which controls the DC power to the oven heater. A crystal oven of this type is capable to control the temperature of the crystal to within 0.1°C over an ambient temperature range of -55°C to +75°C, and usually results in a stability of the frequency with the temperature in the order of $1x10^{-8}$. Various combinations of double ovens and hybrid arrangements using two heaters with a single control circuit are also available. The stability obtained from a double oven is often in the $1x10^{-10}$ region.

Temperature compensation of crystal oscillators is very practical to achieve frequency stabilities in the range of .5 to 10 ppm. With considerable care, compensation to 0.5 ppm is possible, but hysteresis effects tending to limit their frequency stability to parts in 10^7 over typical temperatures ranges. Temperature compensation is generally achieved by placing a voltage variable capacitor in series with the crystal. A voltage is then applied to the capacitor, which pulls the crystal frequency by precisely the amount that it drifted in temperature but in the opposite direction. The voltage is generated either by a thermistor analog network or by a digital system followed by a digital to analog converter. In general, this kind of oscillators offer moderate frequency stability, quick warm up, low power consumption and small size and weight.

It is clear from the above that if any crystal oscillator is to meet high stability requirements, then it must be an oven controlled oscillator, in which the crystal, and any other temperature sensitive components, are electronically maintained at a near constant temperature. From the list of instabilities causes in quartz crystal oscillators, we can understand the need to keep at constant temperature all those components which shift due to temperature since could affect the drive level across the crystal or they changes its load capacitance.

Due to different thermal coefficients of expansion of the materials involved, the controller design must regard any overstress on the crystal holder and any other thermal hysteresis that could happen during warm up time. The controller design also affects the aging of the different components in the system. Because of the high temperature kept during all the life of the system, and because of the overstress, a relaxation problems due to a wrong controller design may arise.

In this paper we have designed a temperature controller system able to keep the temperature of a quartz resonator constant when the ambient temperature is in the operating temperature range, (typ. -10ºC to 60ºC or -25ºC to 85ºC). In temperature controller and heater design the secondary effects which may produce other instabilities different from the thermal one have been considered.

THE CONTROLLER CIRCUIT. DESIGN REMARKS

In ovenized controlled crystal oscillators, the resonator and other circuitry sensible to temperature variations are enclosed in an oven cavity which is kept at almost constant temperature by means of the controller. As the ambient temperature increases, the current drove to the heater block, typically it is one or two power transistors, decreases proportionally so that the oven operating temperature remains constant.

Once the upper operating ambient temperature has been reached the heater turns off and the electronics continues to dissipate standing power, the power being supplied to the heater added to the standing power, proportional to the heat loss. Any further increase in ambient temperature causes loss of thermal control of the oven and subsequent large frequency shifts. The difference in upper operating ambient temperature and the oven set temperature is controlled by the standing power of the electronics and the thermal resistance between the oven and ambient.

In order to maintain high reliability of the oscillator, the oven set temperature should be set as low as possible consistent with thermally controlled operation at the desired upper ambient operating temperature. An offset of 10ºC to 20ºC above the maximum ambient temperature is typical, ensuring that the oven is not overheated by the standing power for high ambient temperatures.

During warm up time, and due to the different thermal coefficients of expansion of the quartz plate and the leads of its holder, some mechanical stress is imposed upon the crystal, those stresses are an important aging source, and also produce some frequency overshooting, being a cause of instabilities or simply increasing the warm up time to the final frequency stability. Those problems have produced, a few years ago, the development of a new doubly rotated cut of crystal, namely SC–cut (stress compensate).

The SC–cut was developed to obtain a resonator whose frequency would be independent of stress in the plane of the vibrating plate resulting from thermal stress and electrode changes. However, it has proved to have a number of other advantages, among these are: Superior warm up characteristics due to reduction of the effects of stress associated with thermal gradients in the plate. Reduced 'amplitude–frequency' effect for the same reason. Almost complete freedom from activity dips. A flatter frequency vs. temperature curve, especially at the turning points. Reduced tip over effect Better short term stability due to reduced response to thermal transients Reduced aging associated with relaxation of electrode stresses Relative immunity to the effects of radiation Less aging due to changes in mounting stresses

The added complexities of orientation and cutting are disadvantages of the SC–cut. Another disadvantage is that the SC–cut is very sensitive to air damping and therefore must be operated in vacuum to find optimum Q values.

The SC–cut has three thickness modes of vibration which are designated the a, b and c modes, these crystals also have two low–frequency face–shear modes. The a-mode is a quasi–longitudinal mode. It has a temperature coefficient of about –52 ppm/ºC. The b-mode is a quasi–shear mode having a temperature coefficient (T_f) of about –25 ppm/ºC. Within the range 0ºC to 100ºC, the frequency vs. temperature curve of the b-mode is almost linear. The c-mode is also a quasi–shear mode. A SC–cut plate is about 10% thicker than an AT–cut plate of the same frequency (Bottom).

The relative activities of the b-mode and c-mode can be varied a bit by the design of the plate but usually they are about the same. The SC–cut plate can be excited on either mode or on both modes simultaneously. Both modes are accompanied by the usual family of inharmonic overtone modes.

The flatter frequency vs. temperature curve is especially important for our controller design, because it lets us a larger variation of the oven temperature for the same frequency stability.

The time required for the system to achieve the final frequency stability is called warm up time. And it is one of the advantages of the compensation techniques over the control techniques. The warm up time is mainly determined for the thermal mass of the system, and the controller and heater design (Ho, 1985). The maximum current is consumed during the warm up, this amount of current is severely restricted in many projects, thus our controller should be able to fix it quickly.

As we explain before, the oven temperature is set 10ºC to 20ºC above the upper ambient operating temperature. And the frequency versus temperature behavior of a quartz crystal around the operating temperature is given by a 3rd order equation, which shows a flatter frequency versus temperature curve around the turning points. In Fig. 1 is shown this curve for a typical 10MHz 3rd overtone c-mode SC–cut crystal.

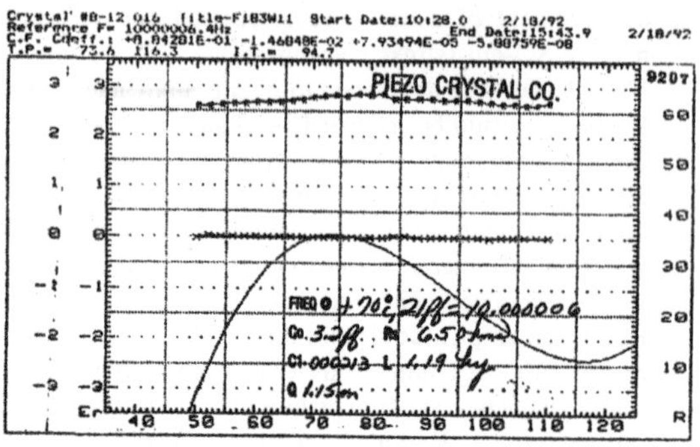

Fig. 1 Frequency vs. temperature curve for a 3rd o.t. Data from Piezo Crystal Co.

The coefficients of this curve are depending on some cut angles and some other different parameters of the crystal processing. This make that our controller design should be able to set the temperature quickly for each different unit.

Controller description

Our design is a proportional controller which senses the temperature, (V_t), at the crystal and measures the difference between this temperature and the turn over point of the crystal temperature, (V_s). And gives an output voltage (V_0) proportional to this difference. This voltage goes to a voltage to current converter which drives the current across the heater. The heater has also a second driver, which has a constant output to set the maximum current during warm up at any ambient temperature. In Fig.2 is shown a block diagram of the whole system.

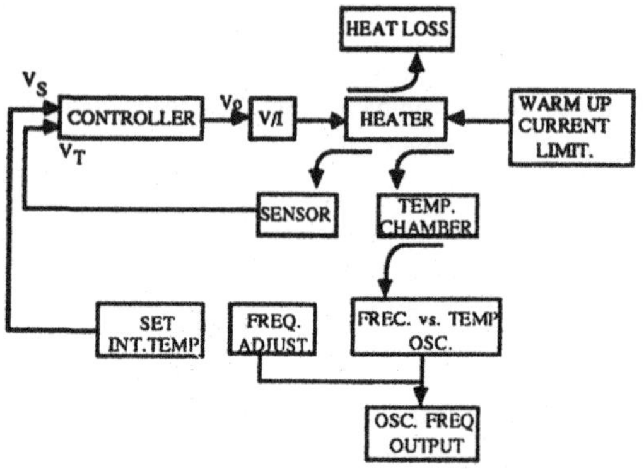

Fig 2. Block diagram of single slope temperature controller

The relationship between the voltage (V_t–V_s) and the output voltage of the controller, V_0 is linear, and it follows a single slope during all the warm up time, this means that the current supplied to the heater decreases in a linear way as the signal V_t reach the signal V_s, the internal temperature meet the turn over point temperature. This constant reduction in the speed of approximation to the target temperature avoids many of the problems typical of other controller configurations, such as the frequency overshooting or the aging problems related with the mechanical over stresses on the crystal holder. Those advantages are necessary when an AT–cut crystal is used in the oscillator circuit, and very useful in the case of SC–cut crystals when the unit is going to be switch on and off frequently, because of the aging rate.

Another relevant advantage of this system is the capability to use small thermal mass in the system, and to apply the heat source straight to the crystal, specially in small size oscillators, without inducing frequency overshooting.

All the points referred above have been measured in different physical designs that we shall show in the next section. An obvious disadvantage of this controller type comes from the low value of the slope which is required to avoid the problems above referred. In some oven designs this increases slightly the warm up time.

When the warm up time is a problem hard to solve due to the compromise with the frequency overshooting and mechanical overstresses, we have developed a variation of the controller which let us to increase the slope of the ratio between (V_t–V_s) and V_0 only during the beginning of the warm up, and once the system is about 20ºC to 30ºC from the target temperature, the controller changes the slope for a smooth approach to the final temperature. This variation of the previous explained circuit is called dual slope controller. In Fig.3 a block diagram of this controller type is shown.

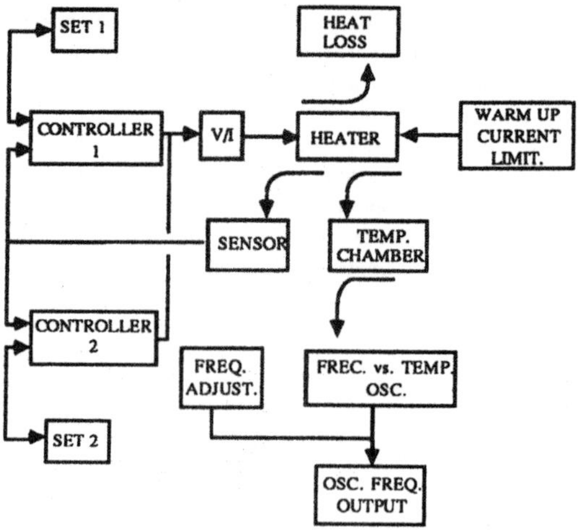

Fig. 3. Block diagram of dual slope temp. controller

Fig.4 shows a typical plot of the current consumed during warm up for three types of controller, single slope, dual slope and a standard proportional controller.

Other controllers comparison

From Fig4 the behavior of other controllers type can be understood. It is typical to provide the maximum current to the heater until the internal temperature is about 7ºC to 10ºC from the target temperature. This reduce the warm up time, but also induce a frequency overshoot which can be neglected if there is not a hard warm up time restriction, however its effect on the aging still remains. And its application with AT-cut crystals give worse results than the single slope approach, for the same complexity of the system.

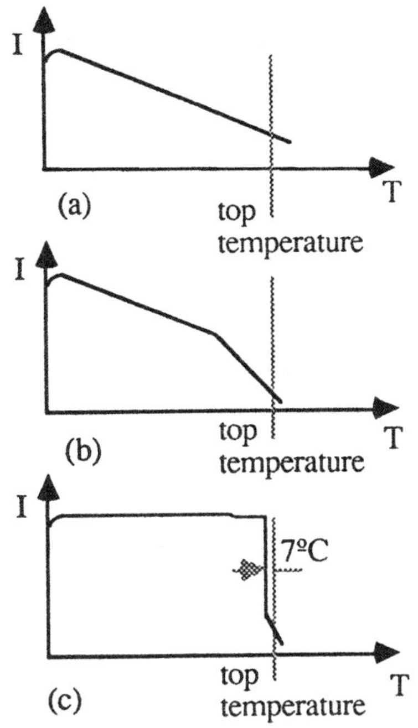

Fig.4.Current consumed by: (a) single slope controller (b) dual slope controller, (c) standard prop. controller

In most of the controller designs that have been reviewed, a certain self–oscillation risk exist in the controller circuitry, which can induce, by means of the power supply, side bands in the output of the crystal oscillator and destroy its short term stability. In our controller design any oscillation is not possible, because there is not any feedback from the heater to the inputs of the controller, that is the main cause for those self–oscillations.

APPLICATIONS

The research program, that is partially shown in this paper, has been successfully applied in three different ovenized controlled crystal oscillators, (OCXO). Those models yield different stability rates, since a few parts in 10^7 to a few parts in 10^9.

The first example is the OCXO model n tfe–MAA/1, in Fig.5.a shows its mechanical description. The oscillator circuit use a 3rd or 5th overtone AT-cut crystal to minimize the frequency shifts due to any changes in the load capacitance, and has been provided with an electronic frequency control to correct the aging frequency deviation along the years. A single slope controller is used, except when a short warm up time is required and then a dual slope design is preferred. Two power transistors are used as the heater system, and both are glued to the crystal enclosure in order to supply the thermal energy straightly to the crystal. In table 1 shows the electrical features of this system.

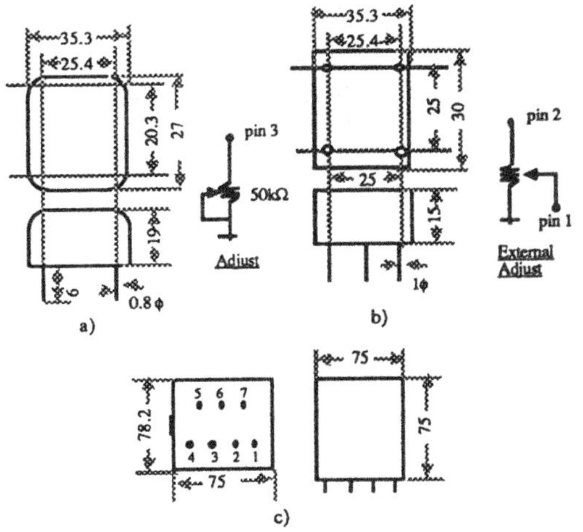

*All dimensions are in mm

Fig. 5. Mechanical description: a) MAA/1, b) MBA/1 c) Custom OCXO

The second example is the OCXO model n tfe–MBA/1, Fig.5.b shows its mechanical description. The oscillator use a 3rd overtone SC-cut crystal that is kept between 7ºC to 12ºC above the upper ambient temperature by means of a tape resistor wrapped around the crystal enclosure as the heater system. The controller is a single slope design fixed to yield a warm up time better than 180 s. with a stability of one part in 10^7. The aging rate is kept better than 5 parts in 10^{10} per day, for a frequency of the signal output between 5MHZ to 10MHz. An electronic frequency control circuitry is added to correct the aging along the years. In Fig. 6 is shown a typical warm up behaviour of this system. Table 1 shows the electrical characteristics of this model.

The third example is a custom OCXO, Fig.5.c shows its mechanical description, and table 1 shows the electrical features of this model.

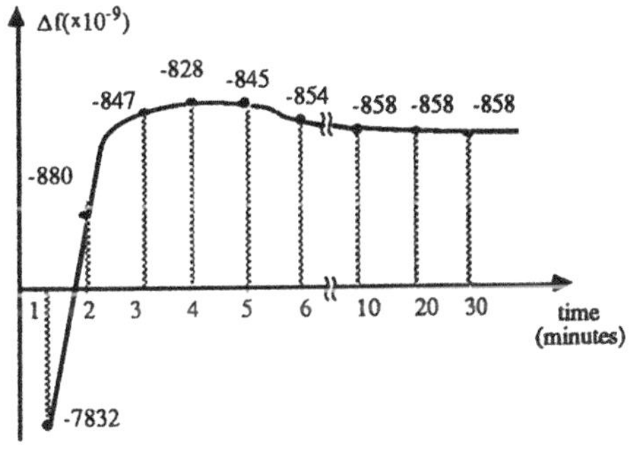

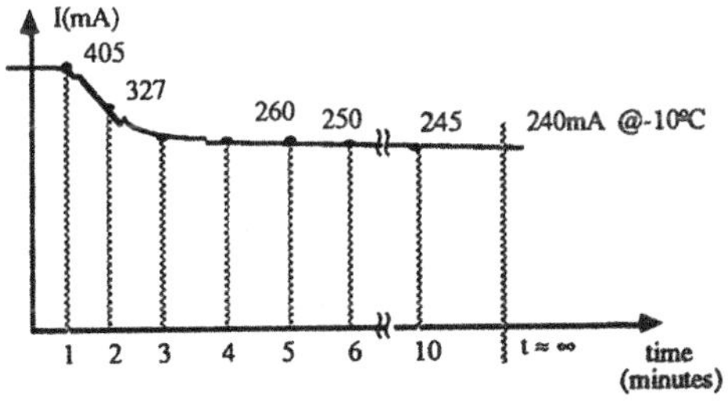

Fig. 6. Current consumed and frequency during warm up for model MBA/1

The stability versus ambient temperature, achieved in this design, has been better than 5 parts in 10^9. Some units have yielded stabilities of 6 parts in 10^{10}. The current consumed during warm up has been fixed to less than 400 mA to ensure the warm up time specification. An aging rate better than 2 parts in 10^{10} per day is assured in all the units.

MODEL	MAA/1	MBA/1	CUSTOM OCXO
Range Temperature	-10 to 60°C	-10 to 60°C	0 to 60°C
Stability vs Temperature	$\pm 1\ 10^{-7}$	$\pm 3\ 10^{-8}$	$\pm 5\ 10^{-9}$
vs Power Supply	5% $\pm 2\ 10^{-8}$	10% $\pm 1\ 10^{-8}$	10% $\pm 1\ 10^{9}$
vs Aging	$\pm 1\ 10^{-6}$/year	$\pm 5\ 10^{10}$/day	$\pm 2\ 10^{-10}$/day
Current Stedy state	160 mA	160 mA	150 mA
Warm up	500 mA	400 mA	400 mA
Warm up time to $1\ 10^{-7}$	5 min	3 min	10 min
Frequency	5 to 40 Mhz	5 to 10 Mhz	16.384 Mhz

Table 1

CONCLUSIONS

In this paper we introduce a new thermal controller for voltage sources with high stability in frequency. We have made a revision of the main contributions in this area, looking for the main instabilities causes in quartz crystal oscillators, and we have designed a new thermal controller that avoids most of those problems, this point being one of its advantages over other controller designs traditionally used for this kind of applications. The same controller can be modified to improve the time required to achieve the final frequency stability. This produces two controllers with a similar behavior, which are called single slope and dual slope controller. With both controllers the internal temperature is kept, almost constant, with a total variation better than 1°C around the turn over point temperature of the crystal. This allows us to make oscillators with stabilities between parts in 10^7 to parts in 10^9 for a ambient temperature range typical of –10°C to +60°C. The total cost of the controller components is about the 2% of the final price of each unit. This research program has been successfully applied in some products, three examples are shown in this paper to describe three different stability rates.

This research program has been partially supported by the Spanish Government by means of C.D.T.I. agency.

REFERENCES

Bottom, V.E. Introduction to quartz crystal unit design, Van Nostrand Reinhold Co.

Frerking, M.E., (1978). Crystal oscillator design and temperature compensation, Van Nostrand Reinhold Co.

Ho, J., (1985). Hybrid miniature oven quartz crystal oscillator, Annual Symposium of Frequency Control, 193-196

Pincu, Edry, (1989). Thermoelectric cooler/heater controlled crystal oscillator, 43rd Annual Symposium of Frequency Control, 44-46.

Salt, D., (1987). Quartz crystal devices, Van Nostrand Reinhold Co.

Vig, J., (1985). Quartz crystal resonators & oscillators, US Army Electronics Technology & Devices Lab., Fort Honmouth.

ON-LINE MEASUREMENT OF PAPER SHRINKAGE USING IMAGE ANALYSIS

A. Guesalaga*, A. Foessel*, M. Guarini* A. Cipriano* and H.W. Kropholler**

**Department of Electrical Engineering, Universidad Católica de Chile, Casilla 306, Santiago, Chile*
***Department of Paper Science, University of Manchester Institute of Science and Technology, Manchester M60 1QD, UK*

Abstract. The paper describes advances made in a computer method for the detection of paper shrinkage during the drying stages. The technique is based on two-dimensional Fourier transformations used to detect those periodic marks left on the paper by the wire during the formation stage. These marks are obtained from images of paper taken by low-cost CCD cameras for a later matching with those of the wire. Differences in separation of lines for the two cases are due to paper shrinkage. Finally some comments on hardware considerations of the sensor design and potentials of the method for the paper and graphic industry are given.

Keywords. Paper industry, sensors, paper shrinkage, image processing, Fourier analysis.

INTRODUCTION

During the drying stage in the manufacturing process, paper suffers serious dimensional variations. This is normally caused by two facts: fibre diameter reduction due to water removal and secondly, external forces acting on the sheet. In the formation process, fibres contain water external and internally. As the web passes through presses and dryers, water is gradually removed from the fibre. In the mean time fibres have formed bonds with their neighbours, which, as the water is taken out drag the contiguous fibres at the point of contact causing a general shrinkage of the sheet.

The adding of individual fibre shrinkages creates a reduction of the web in the cross direction. The opposite effect is normally observed in the machine direction, i.e. paper elongates. This is caused by the constant strain force of machine drives.

Dimensional changes of paper, and in particular irregular ones, are among the variables with greater effects on the physical characteristics of paper. This influences negatively quality parameters such as smoothness, uniformity and resistance (Galley, 1973). Because of this, a complementary sensor dedicated to monitor dimensional changes would be a very useful device during paper manufacture.

Further problems associated to irregular shrinkage are those related to dimensional stability: moisture changes in paper after production or during printing processes, can cause deformation of paper to a point where it becomes totally useless. This is specially valid in multiple stage printing processes where distorted images can occur due to displacements caused by dimensional deformation. Furthermore, irregular changes in dimensions can be caused by a wrong operation of the plant, particularly in dryers and drivers, so the monitoring of dimensional variables would be very desirable for this purpose too.

DESCRIPTION OF THE TECHNIQUE

The development of a novel method to measure these changes on-line has recently started (I'Anson and Kropholler, 1990, Guesalaga *et al*, 1992). The technique uses digital image processing, detecting wire marks left on the paper during the forming stage.

Among the elements in contact with paper in the manufacturing process, the formation wire is the the one that leaves the most regular and periodic mark of all. Changes in size of the wire pattern on the paper, can therefore be used to estimate the shrinkage (or elongation) that has occurred when compared to the original mould.

However, it is practically imposible to detect these marks by simple inspection, not only because of the minute size of them, but also because they are hidden under a mixture of other marks such as those left by felts, cylinders, or those with a "low-frequency" characteristics such as flocs or changes in grammage. This makes very difficult for the human eye to detect them and in the remainder of the paper, a computerized method to estimate shrinkage is described.

The limited resolution of the eye creates the need for a measuring technique with better sensitivity and greater processing capabilities than those of the human sensors in order to use them for analysis and control. The basic idea behind the method is to enhance the visibility of the wire mark through direct and inverse two-dimensional Fourier transformations.

Images containing these marks are generated by the transmitted light through the sheet, which is captured by an image grabber and later processed by a "number cruncher" computer, as figure 1 shows.

The first stage in the mathematical process is a Fourier tranformation applied to the original image, which is contained in frames of 512 by 512 pixels in 256 grey levels, that correspond to a paper area of aproximately 10x10 cms. Figure 2 shows the image created by transmitted light through a paper of 75[grms/mtr^2] with a coating base (starch). As the figure shows no marking can be seen.

Periodic marks on these images will generate strong spikes when the Fourier transform is applied, differentiating very clearly from those of random nature or non-periodic pattern. Figure 3 shows the Fourier transform of the previous image where some spikes can easily be noticed.

The method consists in selecting a reduced number of peaks in this spectra for a later transformation to the spatial domain, where a substantial improvement is obtained in the visibility of the wire mark, allowing for an estimation of the shrinkage when compared to the original wire pattern. Figure 4 shows the result of the inverse Fourier transform when only the stronger spikes have been selected for the backward transformation. Here, the pattern of a commercial formation wire is clearly recovered.

By measuring distance between lines or points in the processed image and comparing them with those in the real mesh, a value for the shrinkage at that point is obtained.

However, inverse Fourier transform is not really necessary nor convenient, i.e., it is possible to obtain spatial distances of periodic marks by looking at the position of the spectral spikes and using the following relation:

$$h = M/f \qquad (1)$$

where h is the period, in pixels, of a certain periodic mark, f is the distance (in pixels) between a given pair of frequency components and the origin, and M is the size in pixels of the image. In our case, we use images of 512 by 512 (M = 512).

Care must be taken during digital processing because of the well known "leakage-effect" (Lim, 1990), that causes a "spreading" of a given frequency over the whole spectrum, making the transform very hard to process and spikes difficult to be identified.

In order to solve this problem, two different mathematical approaches have been used. Firstly, leakage is reduced by using FIR filters, particularly a two-dimensional Kaiser window (Harris 1978, Huang 1972) that has shown to behave satisfactorily. Secondly and in order to improve the accuracy when measuring the frequency value of a spike, a weighted average of its position and those of its surrounding frequency components is used:

$$p_{eq} = \frac{M \cdot \sum_{i}^{n} \sum_{j}^{n} f(k+i,l+j) \cdot ((k+i)^2+(l+j)^2)^{1/2}}{\sum_{i}^{n} \sum_{j}^{n} f(k+i,l+j)} \qquad (2)$$

where p_{eq} is the equivalent pixel when the number of averaged pixels are given by the centre pixel $p(k,l)$ and the adjacent ones. The number of pixels considered are given by i and j ranging between $-n$ and n and forming a small square where typically n = 3. Function $f(x,y)$ is the grey value for pixel (x,y).

SENSOR DESIGN CONSIDERATIONS

In order to obtain a cross direction profile on line, an scanning apparatus has to be developed. Due to paper movement (up to 1500 mtrs/min), images would appear blurred if frames were acquired directly using fast commercial cameras. This problem can be solved via three alternatives:

- Cameras with very fast shutters (<1/100.000[secs])
- Rotatory cylinder containing camera and processor (fig 5)
- Image "freezing" using stroboscope light (fig 6)

The first option has been discarded because of the high cost of these type of cameras.

The second option seems to be a more feasible solution, however, reliability and maintenance costs could make it rather cumbersome. The camera unit rotates in synchrony with paper, so relative speed during acquisition is zero, obtaining sharp images.

The stroboscope light seems to be the simplest and least expensive of all. Nevertheless, possible problems related to flash energy, pulse width and camera integration are still to be checked.

Regarding pulse length, simple calculations give a necesary pulse width of aproximately 1[μsec] in order to avoid image distortion. This pulse width is possible to accomplish using commercial stroboscopes, however, the questions of light intensity and camera integration time are still to be answered.

MEASUREMENT TECHNIQUE

At present, shrinkage profiles are obtained carrying out tests off-line. Figure 7 shows these profiles for an experiment carried out in the UMIST pilot plant (Guesalaga *et al*, 1992) where images were taken every 2 cms (cross direction). The result are two curves, one for the cross direction shrinkage and the second for the machine direction shrinkage.

The application of this method to an industrial process depends largely on having a reliable method of estimating the dimensional variations of paper without the need of human intervention.

One of the initial approaches to achieve automatic shrinkage values was first to improve the visibility of the wire mark by carrying out two-dimensional autocorrelation of images. These method as figure 8 shows, turned out to be very effective in enhancing the wire mark, however, it didn't provide a simple mean of getting the desired deformation value.

A simple method has been developed by the authors which is based on a pattern recognition scheme. The two-dimensional spectra of the sample obtained from the process is correlated to the pattern of the wire spectra. By measuring the ratio between the distance of spikes to origin for both the paper impression and the wire, an estimate of the shrinkage is obtained. This procedure is repeated for every significant peak in the sample spectra that correlates positively with the wire spectra. Finally, an average ratio is computed in order to increase the accuracy and reliability of the method. Problems such as image rotation or false peaks are solved in the peak-correlation stage, where every peak in the sample Fourier transform must have its own counterpart in the wire spectra.

A further improvement of the method has been developed lately, consisting in a moving average scheme that was tried out with very good results.

One of the problem in identifying peaks related to wire marks is the background noise present in the image power spectra, which tends to hide these spikes. By averaging several power spectra images from consecutives frames, a substantial improvement is achieved as figure 9 and figure 10 show for single and 8 frames average respectively. The backgroung noise has been significantly reduced.

APPLICATION OF THE METHOD

As mentioned earlier, the method was tested for the first time in the pilot plant at UMIST using a dandy roll that leaves a particularly strong mark on the paper (Guesalaga *et al*, 1992). Since then, several grades of paper have been tested in our laboratories, obtained from different industrial plants.

Among the papers analized, a very demanding test was the one made for extensible grades. In these papers, shrinkage is forced using a special unit in the dryer section that reduces the speed of the sheet, compacting it and tending to erase any periodic mark. The method was tested with papers produced in the CMPC-Laja plant located in the south of Chile and the results were extremely good.

Other papers tested were light coated papers and different grades of boards. For papers with coating base (starch), the method has proven to perform satisfactorily for grammages up to 80 grs/m^2. The method didn't work for grammages higher than 200 grs/m^2, however, new improvements in the technique via image integration and auto correlation techniques makes it very likely to expand the aplication of the method to heavier grades too.

ACKNOWLEDGEMENTS

This research has been supported by grant from the Chilean Science and Technology Foundation (Grant FONDECYT 0692/91)

REFERENCES

Galley W., "Stability of Dimensions and Form of Paper", *Tappi Journal*, vol 56, n°11, (1973).

Guesalaga A., Kropholler H., Rodríguez F., "Medición de Encogimiento del Papel Usando Procesamiento de Imágenes", *VI Congreso Latino-americano de Celulosa y Papel*, Torremolinos, Spain, (1992).

Harris F.J., "On the Use of Windows for Harmonic Analysis with the Discrete Fourier Transform", *Proceedings of the IEEE*, vol 66, n°1, (1978).

Huang T.S., "Two-dimensional Windows", *IEEE Transactions on Audio and Electroacoustics*, vol 20, n°1, pp 88-89, (1972).

I'Anson S., Kropholler H., "Enhancing Visibility of Wire Mark by Image Analysis", *Journal of Pulp and Paper Science*, vol 35, n°6, (1990).

Lim J.S., "Two-Dimensional Signal and Image Processing", Prentice-Hall, New York, (1990).

Nuttall A.H., "Some Windows with Very Good Sidelobe Behaviour", IEEE Transactions on Acoustics, Speech and Signal Processing, Vol 29, n°1, (1981).

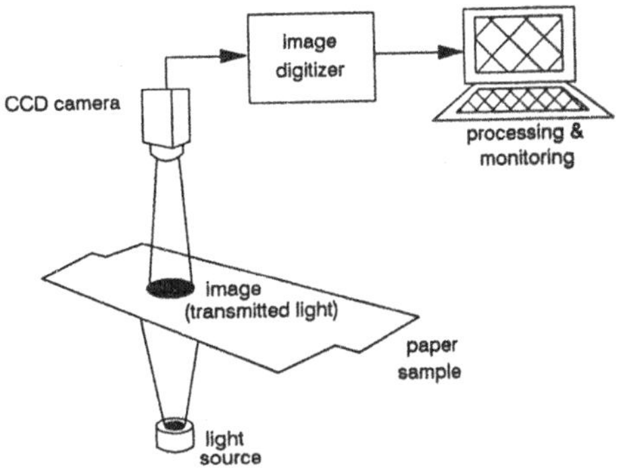

Figure 1 Laboratory sensor

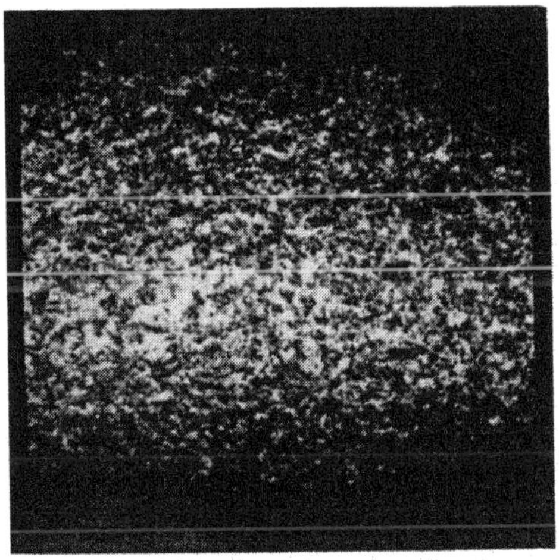

Figure 2 Original image (75[grs/m^2])

Figure 3 Two-dimensional Fourier transform

Figure 4 Filtered image

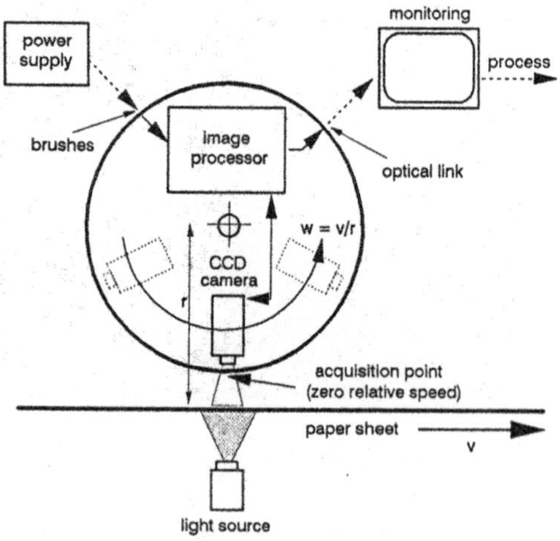

Figure 5 Rotating cylinder

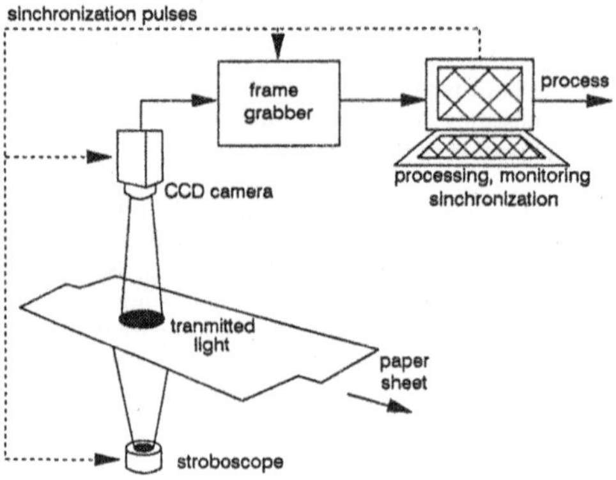

Figure 6 Stroboscope light

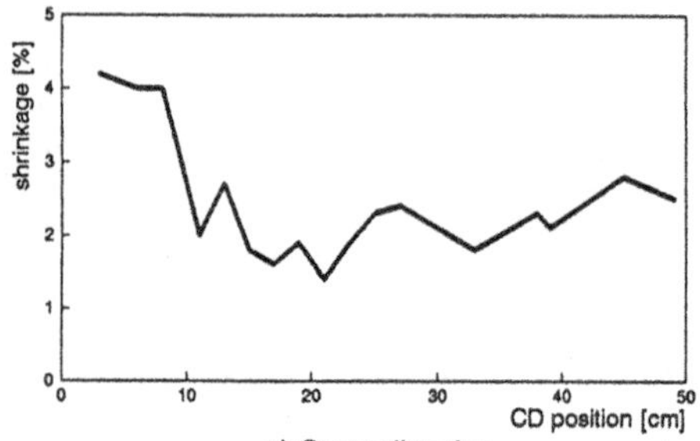

a) Cross direction

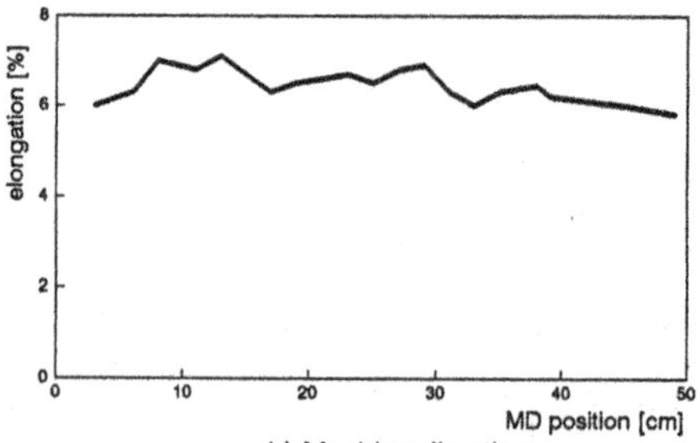

b) Machine direction

Figure 7 Dimensional variations profiles

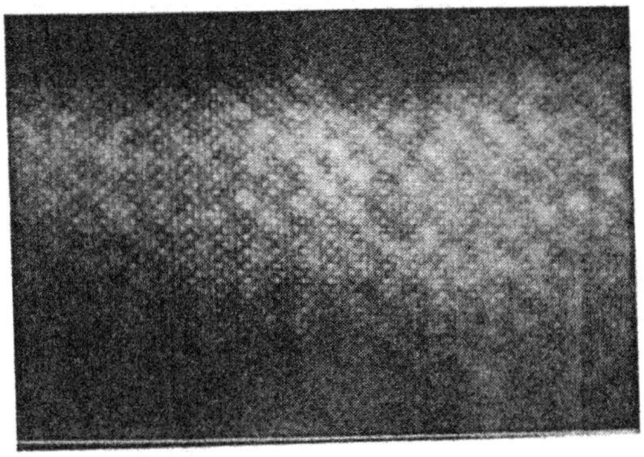

Figure 8 Autocorrelation

Figure 9 Single image

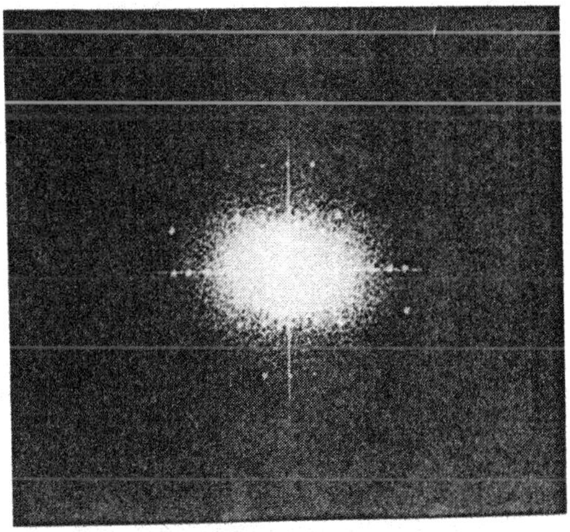

Figure 10 Averaged image (N=8)

AN INTEGRATION OF THE MULTISENSORY SYSTEM

Š. Havlík, D. Hrubý, J. Kalina and P. Pokorník

Institute of Automation and Communication, Slovak Academy of Sciences, Severná 5, 974 00 Banská Bystrica, Czechoslovakia

Abstract This papers presents a multisensory system for robotic application that uses communication network with specific requirements. The real system uncludes numerous sensors for measuring the mechanical quantities. There are:

- multicomponent force/torque, proximity/distance, noncontact position, tactile contact force.

The communication system should satisfy some given speed, time response and reliability requirements. Another set of problems is the architecture of the communication network between sensors/actuators and controller. Intel's Bitbus has been chosen. It is a master-slave type field bus, expecially suitable for transmission of short messages. The performance of this multisensory system is described.

Key words sensor network, distributed signal processing, field bus communication.

1. INTRODUCTION

In general, the high degree of factory automation based on the robotic application needs more and more sensors and actuators. These are physicaly distributed and must be integrated to the data acquisition and control system for monitoring, data processing and control.

When we will use classic low cost system - PC with special IO cards accomodated field of sensors and actuators for example A/D, D/A converters, digital and relay outputs, digital inputs, servo motor control etc., then this situation brings some major difficulties. Each sensor must be individually wired into PC, involving noise contaminated signals from sensors, costly cabling task. Essential problem is limited computing performance of PC machine for processing of input signals from sensors. These problems calls for other solution for intelligent distributed control system. Such system usualy consist of multiple control and data acquisition nodes which can provide control tasks and communication with host computer or with another nodes.

For our laboratory experiment was chosen Bitbus interconnect system. Bitbus is trade mark of Intel Corporation and it is an open standard for industry /IEEE 1118/. It can be easily integrated into many automation systems and Bitbus is a field bus system with worldwide suport by a large number of manufacturers of industrial control and data acquisition systems.

2. SENSORS AND LOW LEVEL DATA PROCESSING

General trends in sensing technology

Regarding general trends in advanord sensing technology we can see several most important features that directly result in the concept of an information-control system. There are:

- Minimization of the mass and dimension of the functional elements
- Processing the sensory data close to the place of measurement and bidirectional communication with the higher level data aquisition/control system

-Reducing the cost of fabrication, simple integration and maintenance.
- New quality of the sensing system oving to the availability of more powerfull computers and signal processing software (new performance characteristics, flexibility, self testing and error diagnostics, robustness, data nomitoring, fusion of several sensory information, etc.). These capabilities become even more essential in nowadays applications because of such an information system not only improves the quality and productivity but in end result it can simplify the system design and reduce its cost.

In general, the development in the sensing technology shifts the further the more functions into digital signal processing units that are built directly into a sensor body or in its near vicinity. Let us analyse now the requirements on the above system from the point of view of sensing the mechanical quantities by particular sensor and related signal processing according to the scheme in Fig. 1.

One component force/torque sensor

For the sensor built on the strain-gauge transducers we can write the characteristic in general form

$u(f) = a_0 + a_1.f + a_2.f^2 + .$ higher order terms (1)

where u - is the output voltage reading in analog/digital form, f - is the force/torque in a given point and direction, a_i - are coefficients. Neglecting second and higher order terms we can write for evaluation of the force

$$f = k.b_1.(u - u_0) \qquad (2)$$

where k - is the coefficient which represents force transformation of various physical scale factors, b_1 - is the sensitivity coefficient ($b_1 = 1/a_1$) of a particular sensor, and u_0 - is the bias signal/force.
By changing these coefficients we have the possibility to integrate various sensor very fast.

Multicomponent force-torque strain gauge sensor
By the similar way as above we can write for evaluation of the sensor readings

$$\mathbf{F} = \mathbf{K}.\mathbf{B}(\mathbf{U} - \mathbf{U_0}) \qquad (3)$$

where $\mathbf{F}$,$\mathbf{U}$ and $\mathbf{U_0}$ are force and signal vectors of components have to be measured. $\mathbf{K}$, $\mathbf{B}$ are matrices of corresponding dimensions.
Let us look at above relation in more details. **B** is the matrix of the calibration coefficients and directly relates to the particular sensor. By other words; each sensor has its calibration matrix estimated by experimental procedure. **K** matrix represents the transformation of the measured force and torque vector into other reference system. This is expecially useful in robotic applications where such a sensor is a part of the robot wrist. By giving the new parameters of transformation we can directly evaluate the contact forces when one changes the gripper or other tool.

Beside these principal function the processing of data from force and torque sensor can include:
- various algorithms for digital filtering
- elemination of the gravity and inertia effects. We save the forces/torques values sampled before contact to be substracted from subsequent ones under physical interaction in order to determine the net contact reactions
- comparing the actual forces/torques with the preset limit values given as thresholds. The movement should stop immediatly as soon as some threshold limit is exceeded.

Displacement sensor

The optical displacement sensor (PSD) gives two output readings u_1 and u_2. The actual distance, d, of the light reflected surface is calculated usint the relations

$$u = (u_1 - u_2)/(u_1 + u_2) \qquad (4)$$

$$d = b_0 + b_1.u + b_2.u^2 + b_3.u^3 + ...$$

where b_i coefficients of the polynom have been estimated by the experimental way.

Position sensor

The lateral position ,x, is comupted from the data of the incremental rotational optical counter multiplying by motion transformation ratio.

3. System Integration

This part of article explains the technical solution of two interconnect metod implementig Bitbus node for laboratory data acquisition and control system. In fig. 1 is block diagram of central system based on the PC. The IO cards are pluged into PC and their IO's are connected to the environment of sensors and actuators. PC services all functions for data acquisition from sensors and data processing.
The position sensor is connected to the card with frequency inputs and uses two inputs.
The displacement sensor is connected to the analog to digital converter through two A/D inputs.
The proximity sensor uses two A/D inputs.
The force/moment sensor uses two A/D inputs.
Servo motors are driven from D/A converters with power amplifiers. The central PC system measures and controls such physical phenomena as voltage, current, frequency. System was configured for real time position control with non visual sensors.
Own application developed SW provides multi tasking for high speed simultaneous acquisition, control, monitoring the robotic cell status and analysis capabilities of driven environment on PC 386, 486 platform.
First step to the distributed control was used in a two level architecture (fig. 3). Host PC is serially interconnected with remote slave module. Master Bitbus module is pluges into PC and slave Bitbus module controls all IO cards. Master and slave modules have standard architecture based on the microcontroler i8344 with opto-isolated serial interface. Modules were developed at UAKOM and thus are full, SW compatible with Intel cards iPCX344A and iRCB44.
The internal system bus of slave module is exteded with PC-bus interface.
This interface allows control of standard 8-bit IO cards. Slave module is piuged into single passive backblane with prevision cards.
The communication between master and slave modules is configured in synchonous mode, with data spedd 2,4 Mb/s and they are linked via a double twisted pair wire cable which can be up to 30 m long.
Slave module with IO cards performs only direct access to the I/O parts on site sensors and actuators and communicate via network.

Host PC performs high level data processing e.g. process control, real-time analysis of the robotic cell with grafic support. Communication SW has features as remote access of commands, respond for read, write of IO ports at a slave node. We can compare time needed for control a robotic cell via central architecture and with control via two level distributed architecture.
At Robotic application computing performance was degradated about 7 to 12 times. Limiting factor of computing performance was serial communication data rate between Master and Slave modules. This result can be improved by decision of control processing between slave node and host PC.
Best results can be achieved by using of distributed control architecture where each sensor and actuator carries its own compact slave Bitbus module (fig. 4). Each node contains local intelligence for control tasks and for communication. Up to eight tasks can be run for signal processing and communication with Master or with another slaves through master. The compact slave modules are build on the controler i8344 with specialized interface to the used sensor, actuators e.g. analogue inputs, analogue outputs, digital relay outputs, digital inputs.
The slave modules have opto-isolated RS 485 interface opto-isolated IO links of IO ports and isolated power supply for protection against electrical noise.
The slave modules allow to optimize SW protocol network performance for high speed system capability in the robotic application which requires reliable data transport layer and real time response.

4. Conclusion

We have described three methods of architecture for an integration of the multisensory system in the robotic cell.
In the practice, the architecture based on the intelligent distributed control was showed suitable. SW network allows the Bitbus slaves

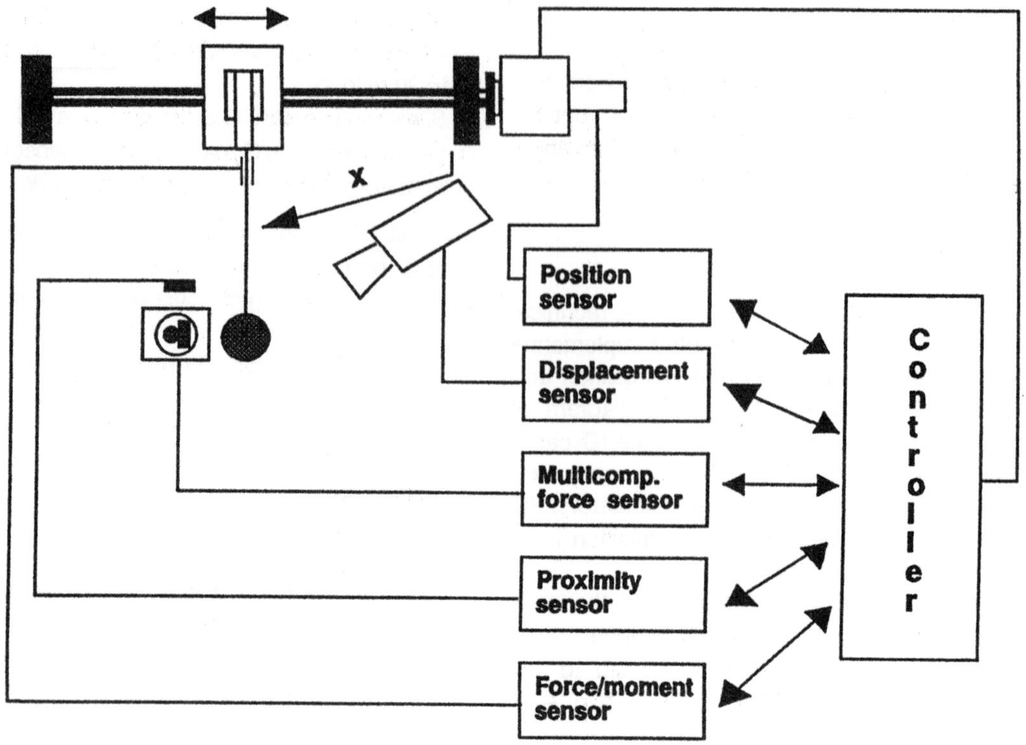

Fig. 1 Configuration of sensors/actuators in the robotic cell

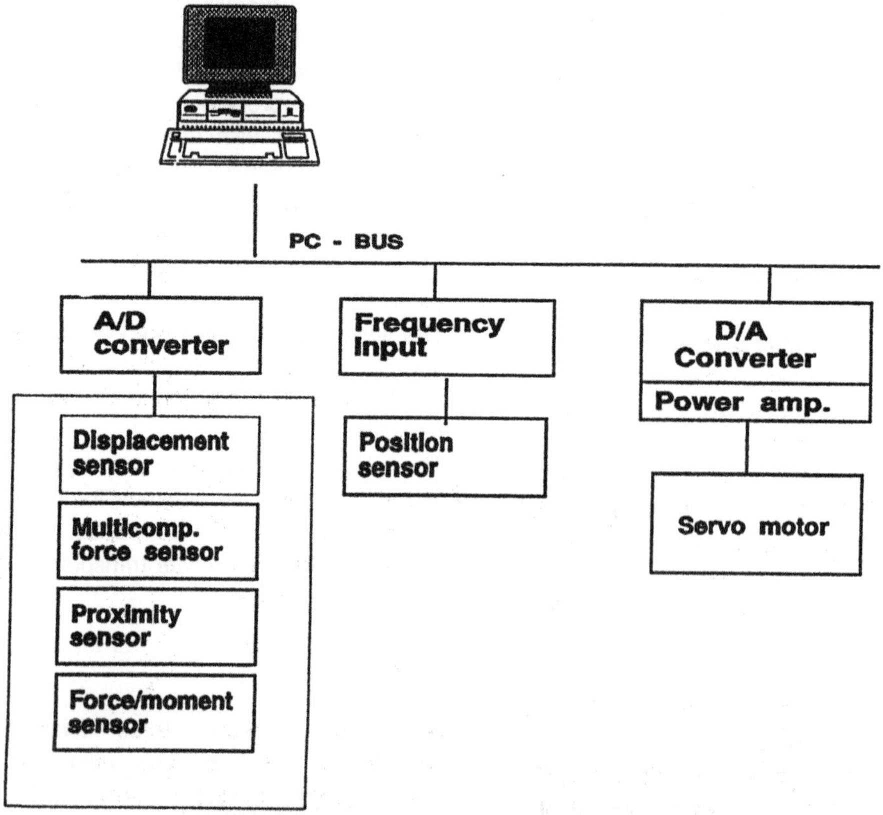

Fig. 2 The central architecture of data acquisition and control system for robotic cell

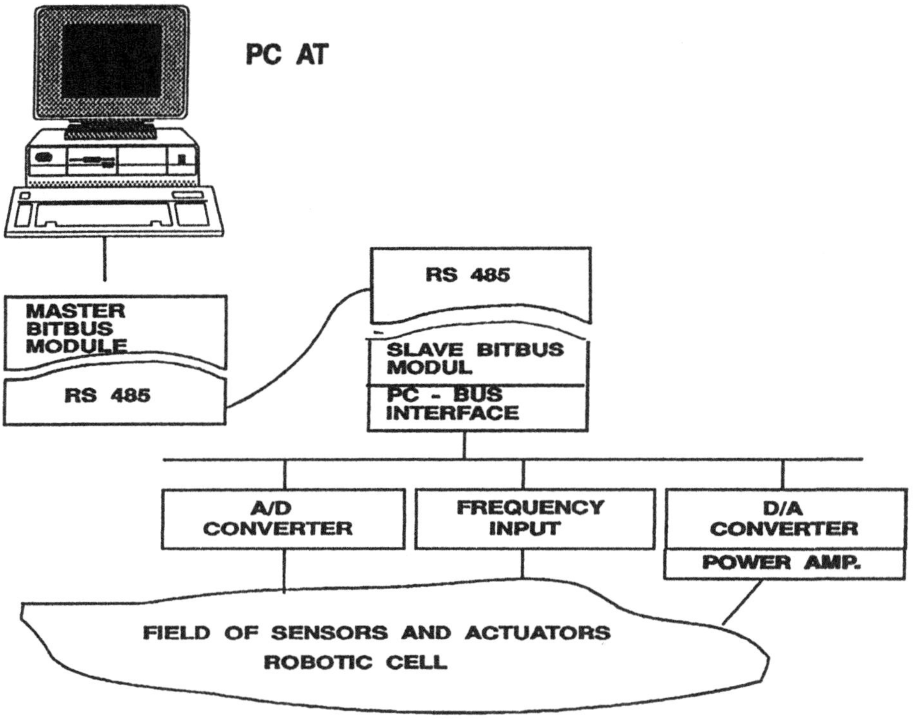

Fig. 3 The two-level architecture of data acquisition and system for robotic cell

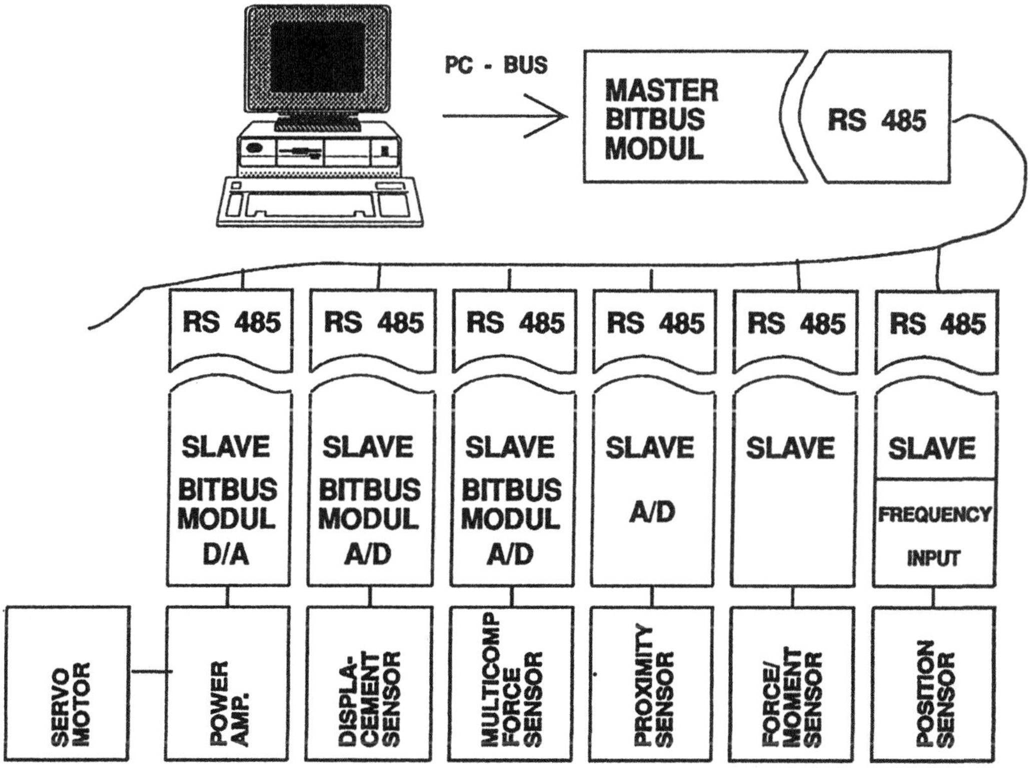

Fig. 4 The distributed architecture of data acquisition and control system for robotic cell

modules not only to respond commands of remote IO ports from PC. But control tasks can run independently in the data acquisition and control nodes for each sensor/actuator. All nodes must be scanned periodically /in our application about 50 ms/.

Bitbus slave modules allow easy system configuration. The user needs specific requirements accomodated to the remote sensor/actuator. Bitbus network modules are economically and technically superior with maximum flexibility in distributed control and data acquisition system.

REFERENCES:

Sintonen, L., Virvalo, T, (1988). The Hierarchy of Communication Networks in the Programmable Assembly Cell: An Expetimental Framework. IEEE Network Magazine, May 1988, Vol. 2, pp. 48 -54.

Oshman, M.K., (1991). Distributed control. Computer design, July 1991, pp. 23-25.

Virvalo, T., Puusaari, P., (1989). Developing intelligent actuator system: experimenting with Bitbus and the 8096. Microprocessors and Microsystems, May 1989, Vol. 13, pp. 263-270.

Pleinevaux, P., Decotignie, J.D., (1988). Time Critical Communication Networks: Field Buses, IEEE Network Magazine, May 1988, Vol. 2.

INTELLIGENT MULTIPROCESSING SYSTEM FOR DATA ACQUISITION

D. Ofrim*, F. Ceapoiu and M. Antonescu****

**O.F. Systems Ltd., Piata Romana nr 9, etaj 7, ap.31, Bucuresti, Romania*
***Department of Electrical Engineering, Master S.A., Bucuresti, Romania*

Abstract. The acquisition, processing, control and representation of experimental data of a process requiers intelligent and performant sistems. We present a multiprocessor structure with intelligent, multiprogrammed modules wich can process data and local algorithms. The processor modules provide the implementation of elements and techniques specifics for artificial intelligence, improving the possibilities of processing and analysing system's data.

Keywords. Multiprocessing systems; microprocessor; parallel processing.

INTRODUCTION

The acquisition, processing, control and representation of experimental data of a process requires the achievement and the connection with the experimental research equipment of intelligent, performing systems. These specific features increase the productivity.

The microprocessors and the microcomputers,with built-in elements of intelligence, can give a quickly and efficient answer for different requirements and are easily adaptable to the new work conditions, because they can be reprogramated without making any physical changes in the system of which they are a component, facilitating modifications and upgrades at a minimum cost.

The achievement and implementation of the

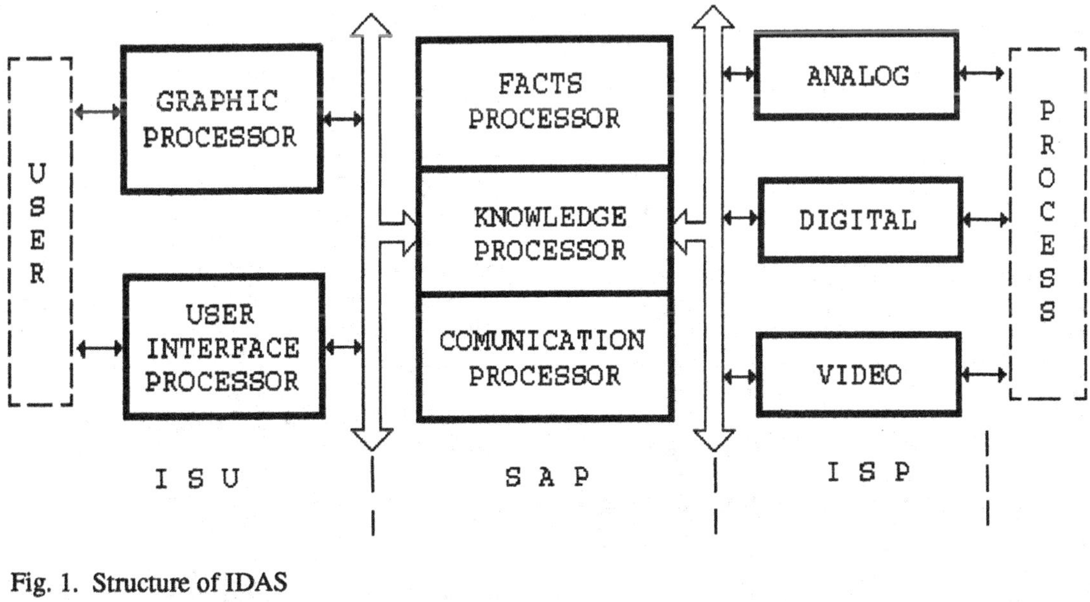

Fig. 1. Structure of IDAS

THE INTERFACE SUBSYSTEM WITH THE USER, ISU

Creating ISU into IDAS is necessary because a large quantity of informations offered to the user will be graphic processed so help him and improve his decisions.

The authors promoted specific tools for the developement of the system user interface, which ensure a fast and facil creation of complex interface for applications in industrial automation processes.

ISU includes a GRAPHIC processor necessary for a 2D/3D or alphanumeric representation of the data base (knowledge about the process) obtained from ISP through acquisition or drive system activities.

PM requires such architecture because the graphic processing needs too much time. In this case, the main processing unit processor will be hardly used and the global time response of IDAS will be reduced.

To answer the requirements of IDAS, ISU must provide different functions for representation and dynamic configuration of the data, integrate into an unitar way the graphic symbols with plottings and offer a total control over data for the user.

The autors used programming techniques based upon objects, hierarchical connections between objects to obtain a hierachical object-oriented subsystem with 4 levels, upper ... lower.

All the objects of this structure have attributes and by controlling them,ISU gets dynamic features managing.

THE SUBSYSTEM OF ANALISING AND PROCESSING, SAP

This element of IDAS, promoted by the authors, introduces ingenious methods for data analysing and processing using elements and specific methods in artificial intelligence, AI.

SAP contains: the processor for event base acquisition, the processor for knowledge and an optional processor for comunications.

The processor for event base acquisition, software implemented, performs the conection IDAS - process using a complex algorithm specific for each application of data acquisition or data processing.

The processor for knowledges and knowledge base maintenance represents a major qualitative jump we introduce into a conventional DAS to obtain a IDAS. This PM performs the qualitative processing for event base and introduces a new concept for experimental data processing: the quality of information. This concept reflects the quality of the studied process and supplies global and significant estimations for user.

Moreover, an IDAS, in this subsystem, can supply indications concerning the experimentations or the processing of the experimental context, the experiment researcher having an active partner in analysis and estimating the process.

In order to obtain a distributed multiprocessor system architecture using a IDAS, a communication processor has been introduced in the IDAS structure; it ensures the transfer of information to/from other systems or workstations. This processor which represents an opening of IDAS to other systems, will have to ensure, according to the type of application, the communication between more processing units; in this way it will be achieved a IDAS network or only a transfer of information between two informatic levels or between two systems distantly located.

APPLICATIONS

We will shortly present two applications for IDAS, that prove theirs features.

The first application was implemented into a test-stand of thermic engines.

The PC_FAST equipment is meant for the acquisition, simultaneously with 2 (Marker and Trigger) signals, and the processing of the fast variation parameters of thermic engines. SIP is represented by a processor module of original specialization which ensures the primary acquisition and processing of the data. A specialized software interface meant for the acquisition of experimental data from thermic engines has been built within USI.

The data gathered are transferred from SIP so as to make the fact basis which represents the process, being processed according to the knowledge base.

Another application is the system of analysis of mandibular cinematics. An ANALOGIC procesor is used for acquiring four analogic signals which represent the coordinates of a reference point(3) and its speed of mouvement. In this way the diagram of the mandibula mouvement can be drawn. The central processing subsystem processes and interprets facts, offering a diagnosis and treatment indications. USI ensures the representation and 2D/3D graphic processing, as well as the man-machine dialogue.

THE INTERFACE SUBSYSTEM WITH THE USER, ISU

Creating ISU into IDAS is necessary because a large quantity of informations offered to the user will be graphic processed so help him and improve his decisions.

The authors promoted specific tools for the development of the system user interface, which ensure a fast and facil creation of complex interface for applications in industrial automation processes.

ISU includes a GRAPHIC processor necessary for a 2D/3D or alphanumeric representation of the data base (knowledge about the process) obtained from ISP through acquisition or drive system activities.

PM requires such architecture because the graphic processing needs too much time. In this case, the main processing unit processor will be hardly used and the global time response of IDAS will be reduced.

To answer the requirements of IDAS, ISU must provide different functions for representation and dynamic configuration of the data, integrate into an unitar way the graphic symbols with plottings and offer a total control over data for the user.

The autors used programming techniques based upon objects, hierarchical connections between objects to obtain a hierachical object-oriented subsystem with 4 levels, upper ... lower.

All the objects of this structure have attributes and by controlling them,ISU gets dynamic features managing.

THE SUBSYSTEM OF ANALISING AND PROCESSING, SAP

This element of IDAS, promoted by the authors, introduces ingenious methods for data analysing and processing using elements and specific methods in artificial intelligence, AI.

SAP contains: the processor for event base acquisition, the processor for knowledge and an optional processor for comunications.

The processor for event base acquisition, software implemented, performs the conection IDAS - process using a complex algorithm specific for each application of data acquisition or data processing.

The processor for knowledges and knowledge base maintenance represents a major qualitative jump we introduce into a conventional DAS to obtain a IDAS. This PM performs the qualitative processing for event base and introduces a new concept for experimental data processing: the quality of information. This concept reflects the quality of the studied process and supplies global and significant estimations for user.

Moreover, an IDAS, in this subsystem, can supply indications concerning the experimentations or the processing of the experimental context, the experiment researcher having an active partner in analysis and estimating the process.

In order to obtain a distributed multiprocessor system architecture using a IDAS, a communication processor has been introduced in the IDAS structure; it ensures the transfer of information to/from other systems or workstations. This processor which represents an opening of IDAS to other systems, will have to ensure, according to the type of application, the communication between more processing units; in this way it will be achieved a IDAS network or only a transfer of information between two informatic levels or between two systems distantly located.

APPLICATIONS

We will shortly present two applications for IDAS, that prove theirs features.

The first application was implemented into a test-stand of thermic engines.

The PC_FAST equipment is meant for the acquisition, simultaneously with 2 (Marker and Trigger) signals, and the processing of the fast variation parameters of thermic engines. SIP is represented by a processor module of original specialization which ensures the primary acquisition and processing of the data. A specialized software interface meant for the acquisition of experimental data from thermic engines has been built within USI.

The data gathered are transferred from SIP so as to make the fact basis which represents the process, being processed according to the knowledge base.

Another application is the system of analysis of mandibular cinematics. An ANALOGIC procesor is used for acquiring four analogic signals which represent the coordinates of a reference point(3) and its speed of mouvement. In this way the diagram of the mandibula mouvement can be drawn. The central processing subsystem processes and interprets facts, offering a diagnosis and treatment indications. USI ensures the representation and 2D/3D graphic processing, as well as the man-machine dialogue.

CONCLUSIONS

The developement of the conventional DAS structure has been achieved using PM modules. This solution handled the increase of the system performances: answer speed and the possibilities to adapt itself to the process evolution.

The introduction of processor which include elements and methods specific to artificial intelligence, AI, represents the major qualitative jump we impose to a DAS to become an IDAS.

The usage of PM ensures a multiprocessor structure for IDAS.

REFERENCES

1. Brodlin, G - New Development in Measurement Science Relating to the Design of Intelli gent Equipment Proceedings of IMEKO, sept-oct 1980.

Waterman, D.A.- A guide toExpert Systems, Addison - Wesley Publishing Company, Inc - 1986

Winston, P.H. - Inteligenta Artificiala, Ed. Tehnica, 1982, Bucuresti

Hockney, R.W., Jesshope, C.R. - Parallel Computer Architecture, Programming and Algorithms

UNIFIED SELF-TUNERS FOR INDUSTRIAL APPLICATIONS

R. Prokop, Z. Turisová and A. Mészáros

Department of Process Control, CHTF-STU, Radlinského 9, 812 37 Bratislava, Czechoslovakia

Abstract: The use of self-tuning control in industrial applications has been a substantial effort for many years. As a consequence, there is much on-going research activity in the refining and tailoring of algorithms in order to accommodate the effective implementation on specific industrial applications. The contribution presents a method of single input-single output self-tuners applied in the metallurgical industry. The design of the self-tuners is based on the polynomial approach in discrete-time domain. The discrete-time model of the controlled plant is updated by recursive identification.

Key words: Discrete-time models, recursive identification, polynomial Diophantine equations, melting process

INTRODUCTION

The concept of self-tuning control (STC) with on-line identification has been enormously developed from the sixties. However, industrial applications have lagged behind theoretical developments. Owing to advances in computer technology, the use of STC is now becoming well established in various industrial applications. Many of the existing self-tuners are based on the pioneering works, e.g. Åström (1983), Isermann (1982), Clarke and Gawthrop (1979). Although there are no best self-tuners for all purposes, the reason to investigate more flexible, effective and robust algorithms is to be requisite.

There are several principles of adaptive control systems. Among them, self-tuners are probably the most frequently investigated. A common framework of self-tuners is simple. The adaptive control system consists of two loops. The inner loop is composed of an object and a linear regulator. Regulator parameters are adjusted in the outer loop containing a recursive parameter estimation and design calculation.

The aim of this paper is to demonstrate an application of single input-single output self-tuners based on the polynomial approach. The discrete-time synthesis of the linear regulator consists in solving of a pair of polynomial Diophantine equations. The right hand sides of the equations establish a performance criterion to be minimized. Thus, achieving an appropriate behaviour of the controlled plant is simple and flexible. The recursive identification of the discrete-time model is carried out in familiar way by sampling the input - output variables.

SYSTEM DESCRIPTION AND CONTROL DESIGN

Consider a single input - single output system represented by the discrete-time model (in the delay operator d):

$$a(d)\, y(k) = b(d)\, u(k) + c(d)\, n_{\sigma}(k) \qquad (1)$$

where:

$y(k)$, $u(k)$, $n_{\sigma}(k)$ is the output, input and white noise (zero-mean) sequences respectively

a, b, c are polynomials in d with conditions $a(0) \neq 0$; $b(0)=0$.

Equation (1) is the well-known ARMAX model, a special case for $c(d)=1$ is the ARX model.

We shall consider causal linear controllers which operate on the available data to generate the plant input. The scheme of the overall control system with the presence of noises is shown in Fig.1. It is convenient to view the plant as that part of the system which is completely unknown and to be found. All four random sources n_ρ, n_σ, n_φ, n_ψ in Fig.1 are pairwise independent zero-mean covariance-stationary white random processes with intensities ρ, σ, φ, ψ, respectively. The scheme in Fig.1 is too general and complex for the control of technological processes. We shall assume that the reference is a deterministic signal and it is not corrupted by the noise n_φ. Then, the reference generator is driven by the equation

$$f(d)\, w(k) = g(d) \tag{2}$$

with $f(0) \neq 0$. According to Fig.1 the control law is governed by

$$p(d)\, u(k) = -q(d)\, (y(k) + n_\rho (k)) + r(d)\, w(k) \tag{3}$$

As the plant, reference generator and controller are linear systems, each signal in the control system (Fig.1) can be divided into two components: the deterministic and the stochastic ones (in our case the stochastic part of the reference is absent). Thus, two different problems arise at the same time: first, we want the system to respond to the initial conditions in a suitable way (tracking of deterministic components); second, we want the system to behave optimally in the presence of noises. In the control parlance it means, that in the deterministic part of the system, the asymptotic tracking problem is to solve, i.e. find a controller such that

$$\lim_{k\to\infty} \{\lambda\, \Delta\, u_D(k) + \mu\, (w(k) - y_D(k))^2\} = 0 \tag{4}$$

If $\mu \neq 0$, the output is to follow asymptotically the reference. The stochastic part of the system is controlled in the sense of optimal tracking. The problem is to find a controller such that the ensemble average

$$J = \mathcal{E}\, \{\lambda\, \Delta\, u_S(k) + \mu\, y_S(k)\} \tag{5}$$

is finite and attains its minimum. In equations (4), (5) the subscripts D, S mean the deterministic and stochastic part of the signals; $\lambda.\mu > 0$ are weighting constants, Δ means the first difference ($\Delta\, u(k) = u(k) - u(k-1)$).
Moreover, the control system in Fig.1 must be asymptotically stable. The task of minimization of (4), (5) is called the LQG problem.

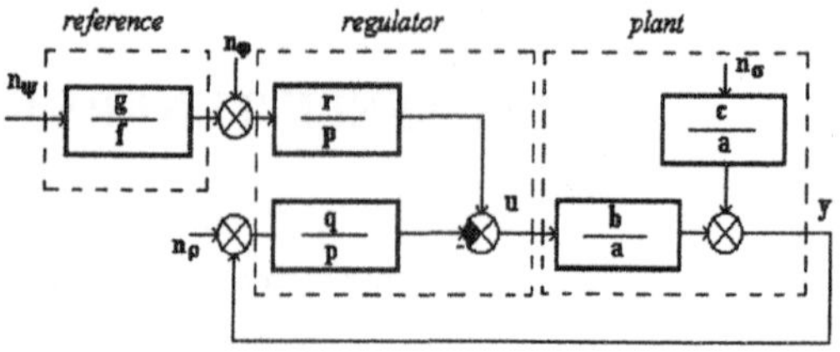

Fig.1: General scheme of control system

The solution of this problem is adopted from Kučera, Šebek (1984) where a deep analysis is performed. The most interesting result suggests, that both mentioned problems can be solved simultaneously. If $\varphi=\psi=0$, then the feedforward part of any controller which provides asymptotic tracking can be used to form the optimal controller. The stochastic control problem with minimization of (5) is given by the solution of the Diophantine equation:

$$a(d)\, p(d) + b(d)\, q(d) = h(d)\, m(d) \tag{6}$$

where the polynomials h(d), m(d) are given via "spectral factorization":

$$a_*(d)\, \lambda\, a(d) + b_*(d)\, \mu\, b(d) = h_*(d)\, h(d) \tag{7}$$

$$a_*(d)\, \rho\, a(d) + c_*(d)\, \sigma\, c(d) = m_*(d)\, m(d) \tag{8}$$

where $a_*(d) = a(d^{-1})$. Note that spectral factors h(d), m(d) are stable (or Hurwitz in general) polynomials and the left side of equation (6) is the feedback characteristic polynomial of the control system. The stability of h and m entails the asymptotic stability of the system. The feedforward part can be expressed also in the form of polynomial equation:

$$f(d)\, t(d) + b(d)\, r(d) = h(d)\, m(d) \tag{9}$$

Polynomial equations (6), (9) with spectral factors (7), (8) yield the regulator polynomials p, q, r under the following conditions:

a) polynomials a and b are relatively prime
b) polynomials f and b are relatively prime
c) the sequence $\frac{\lambda\mu(1-d)}{f}$ is stable

Note, that condition a) means that the plant is stabilizable and controllable. Moreover, the most frequent reference signal is a stepwise set point, where f=1-d, so that condition c) is fulfilled.

The synthesis equations (6), (9) cover also other criteria of discrete optimal control. These cases are more precisely analysed in Kučera, Šebek (1984), Kučera (1980), Šebek, Kučera (1982). If all noises in the control system (Fig.1) are omitted, the task of minimizing the goal function (LQ problem)

$$J = \sum_{i=1}^{\infty} \{\lambda\, \Delta\, u_i^2 + \mu\, (w - y)_i^2\}$$

can be solved as a spectral case of (6), (9) where m(d)=1. The well-known dead-beat (DB) servo problem can be solved through equations (6), (9) by the option h(d)=m(d)=1. Finally, if we choose h(d), m(d) stable polynomials with prescribed poles, the pole-placement (PP) is covered by (6), (9).

PARAMETER ESTIMATION AND ADAPTIVE CONTROL

The most commonly used approach to parameter estimation of controlled object is that of recursive least squares. By re-arranging (1), we can obtain

$$y(k) = z(k)\, \Theta(k) + \zeta(k) \qquad (10)$$

where:

$z^T(k) = [-y(k-1), \ldots ,-y(k-n_a), u(k-1),\ldots, u(k-n_b), \zeta(k-1), \ldots ,\zeta(k-n_c)]$ is the observation vector

$\Theta^T(k) = (a_1, \ldots , a_{na}, b_1, \ldots , b_{nb}, c_1, \ldots , c_{nc})$ is the parameter vector

and $\zeta(k)$ is the prediction error becoming equal to the noise sequence n_σ upon convergence.

In the case of an ARX model, data $\zeta(k-1), \ldots ,\zeta(k-n_c)$ and parameters $c_1, \ldots , c_{nc}$ in z(k) and Θ(k), respectively, are omitted. To avoid the burst effect and other drawbacks in recursive estimation, the algorithm LDDIF with directional forgetting was used (see Kulhavý, Kárný 1984).

The discrete self-tuner is then obtained by the combination of parameter estimation and control design. Owing to the unified approach of control design, we can easily change the criterion and the model of the controlled plant according to requisite circumstances. Thus, each derived algorithm (for a given structure of model (1)) has two main "knobs". The first one is for the option of the model (ARMAX or ARX); the second one is for the selection of the performance criterion (DB, PP, LQ, LQG).

Let us illustrate the control design for the commonly used SISO system of the second order. Consider model (1) in the form of the difference equation:

$$y(k)+a_1y(k-1)+a_2y(k-2) = b_1u(k-1)+b_2u(k-2)+n_\sigma(k)+ c_1n_\sigma(k-1)+c_2n_\sigma(k-2) \qquad (11)$$

First, we have to perform spectral factorizations (7), (8). There exist formulas for the second-order factorization (see e.g. Peterka) given by the following expression (for (7)).

$$a_*(d)\, \lambda\, a(d) + b_*(d)\, \mu\, b(d) = g_0 + g_1(d+d^{-1}) + g_2(d^2+d^{-2}) \qquad (12)$$

where

$$g_0 = \lambda\, (1+a_1^2+a_2^2) + \mu\, (b_1^2+b_2^2)$$
$$g_1 = \lambda\, a_1(1+a_2) + \mu\, b_1b_2 \qquad (13)$$
$$g_2 = \lambda\, a_2$$

For spectral factorization (8), expression (12) is similar with

$$g_0 = \rho\, (1+a_1^2+a_2^2) + \sigma\, (1+c_1^2+c_2^2)$$
$$g_1 = \rho\, a_1(1+a_2) + \sigma\, c_1(1+c_2) \qquad (14)$$
$$g_2 = \rho\, a_2$$

The spectral factors m(d), f(d) are then given:

$$\zeta_1 = \frac{g_0}{2} - g_2 + \sqrt{(\frac{g_0}{2} + g_2)^2 - g_1^2}$$

$$\zeta_2 = \frac{1}{2} - \zeta_1 + \sqrt{\zeta_1^2 - 4g_2^2} \qquad (15)$$

$$m_0 = 1 \quad m_1 = \frac{g_1}{\zeta_2+g_2} \qquad m_2 = \frac{g_2}{\zeta_2}$$

(h_0, h_1, h_2 in the same way for (14)).

The Diophantine equations of control synthesis (LQG) have the form (for the stepwise reference):

$$(1+a_1d+a_2d^2)(p_0+p_1d+p_2d^2) + (b_1d+b_2d^2)(q_0+q_1d) = (1+m_1d+m_2d^2)(1+h_1d+h_2d^2)$$
$$(1-d)(s_0+s_1d+s_2d^2+s_3d^3) + (b_1d+b_2d^2)r_0 = (1+m_1d+m_2d^2)(1+h_1d+h_2d^2) \qquad (16)$$

We can obtain the explicit formulas for p_i, q_i, r_i by simply equating the coefficients at like powers of d in the left and right side of eqs.(16). The resultant control law is given:

$$u(k) = r_0w(k)-p_1u(k-1)-p_2u(k-2)-q_0y(k)-q_1y(k-1) \quad (17)$$

It is clear that (17) covers the following possibilities:

a) LQG problem for m_i, h_i given by (13),(14),(15)

b) LQ problem for h_1=h_2=0 and m_i given by (13),(15)

c) DB problem for h_1=h_2=m_1=m_2=0

d) PP problem for a special option of h_i,m_i (for instance h_1=h_2=0 and m_i creating a stable m(d))

INDUSTRIAL APPLICATION

The self-tuner presented above has been applied to the control of a liquid alloy temperature in an aluminium melting plant. The melting gas-fired furnace is sized for an aluminium batch of about 9 tons. The metal batch is put into the furnace in a solid state of various shapes and figures (ingots, plates, sheets, scraps). The melting process consists of three stages. During the first stage, the metal is solid. The temperature of the furnace space and unmeasured temperature of the metal increases. In the second stage, the metal melts and the thermocouple plungs into the liquid alloy. The temperature of the alloy as well as of the furnace space remain approximately constant. During the third stage, the temperature of the liquid alloy increases further until the set point value is reached. The set-point value depends on the alloy grade in the interval $<720;\ 780^{o}C>$. At the end of the melting process, the content of the furnace is siphoned over to the next treatment (continuous lip pouring).

The questions of why, how and when to implement self-tuning control are arising. There is no reason to use a refined and ingenious algorithm during the first two stages. The feasible and optimal strategy in these stages is to heat as much as possible (i.e. the upper limit of the gas flow). The effective and flexible control algorithms are required for the last stage of the melt. Various circumstances and parameters of the process are changeable and varying. These are e.g. the amount of the metal batch, the kind of the alloy, the varying burning parameters (the temperature of the recuperator air, fouling of furnace lining, etc.). Moreover, various disturbances operate during the melting process. These are mainly the stirring of the alloy and the alloy refining. The aperiodic response of the controlled temperature is essentially required.
As a consequence, the proposed unified self-tuner was used as a tool of effective control.

IMPLEMENTATION AND RESULTS

The described adaptive control strategy has been implemented to computerized control of a melting process under the CPM operation system. Besides control, the process computer performs monitoring, supervisory functions and other necessary operations.
The industrial application of the control system has been preceded by testing the control algorithms in laboratory circumstances. The controlled object was an electrically heated bench-scale tin solder. A control event when controlling the temperature of the liquid-phase metal is illustrated in Fig.2. The LQG algorithm for λ=1 and μ=0.4 and the sampling period of T_v=40s has been used. The process was identified in the form of an ARMAX model with polynomial degrees n_a=n_b=n_c=2. In Fig.3, there is a result for the same process with an adaptive PID control performance which has been proposed by Maršík (1983). This adaptive PID strategy does not use any recursive identification procedure but an implicit parameter estimation. Fig.2 and Fig.3 show that the self-tuning strategy is superior to the other one. One of the reason for this is that thermal processes exhibit significantly different dynamic behaviour between heating and cooling modes. In such a case, regulators which are not supported by recursive identification may fail.
In Fig.4 and Fig.5 temperature control cases in an industrial aluminium melting furnace are demonstrated. The dead-beat algorithm and an ARX model have been used with sampling period T_v=100s. Being this an industrial-scale process with various disturbances, the control system behaviour is not ideal. Time instants, designated by O and by C, refer to opening and closing the furnace gate, respectively. At time t=45T_v, refining of the liquid-phase aluminium took place. There is a significant disturbance effect inside the furnace caused by pressure dropping. Nevertheless, the controller succeeded to maintain the temperature within the desired range.

CONCLUSION

The paper has offered and described a unified single input-single output discrete control algorithm based on the polynomial equation approach. This gives a possibility to choose both, various performance criterion to be fulfilled and various model structures. A combination of the control strategy with an appropriate recursive identification procedure has resulted in a robust self-tuning control system. Both, real-time laboratory and industrial applications were successful. According to a long-term evaluation, by implementing the proposed self-tuner for temperature control to the aluminium melting furnace, 15% of the fuel gas has been saved.

REFERENCES

Åström K.J. (1983): Theory and Application of Adaptive Control - A Survey, Automatica 19, pp.471-486.

Clarke D.W.,Gawthrop P.J. (1979): Self-tuning Control, Proc. IEE, 126, (6), pp.633-640.

Isermann R. (1982): Parameter Adaptive Control Algorithms - A Tutorial, Automatica 18, pp.513-528.

Kučera V. (1980): A Dead-beat Servo Problem, Int.J. of Control.

Kučera V., Šebek M. (1984): A Polynomial Solution to Regulation and Tracking, Part I,II, Kybernetika 20, pp.177-188, 257-282.

Kulhavý R., Kárný M. (1984): Tracking of Slowly Varying Parameters by Directional Forgetting, prepr. IFAC Congress Budapest, pp.78-83.

Maršík J. (1983): A New Conception of Digital Adaptive PSD Control, PCIT 12, pp.267-279.

Peterka V. (1986): Predictor Based Self-tuning Control, prepr. IFAC Congress, Washington.

Prokop R. (1986): Adaptive Control Based on Polynomial Approach (in Slovak), Automatizace 29, pp.3-7.

Šebek M., Kučera V. (1982): Polynomial Approach to Quadratic Tracking in Discrete Linear Systems, IEEE Trans. on Aut. Control, AC-27, No 6.

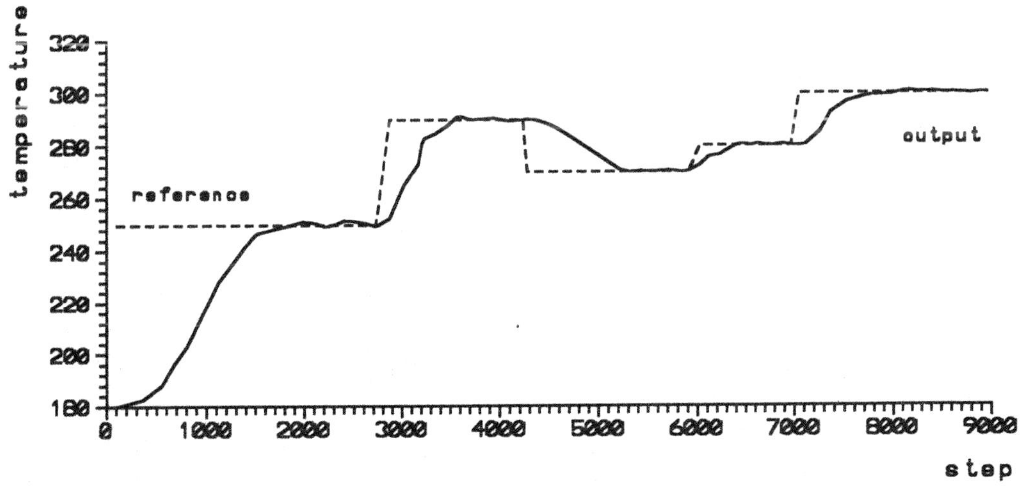

Fig.2: Self-tuning control of the temperature in the tin solder

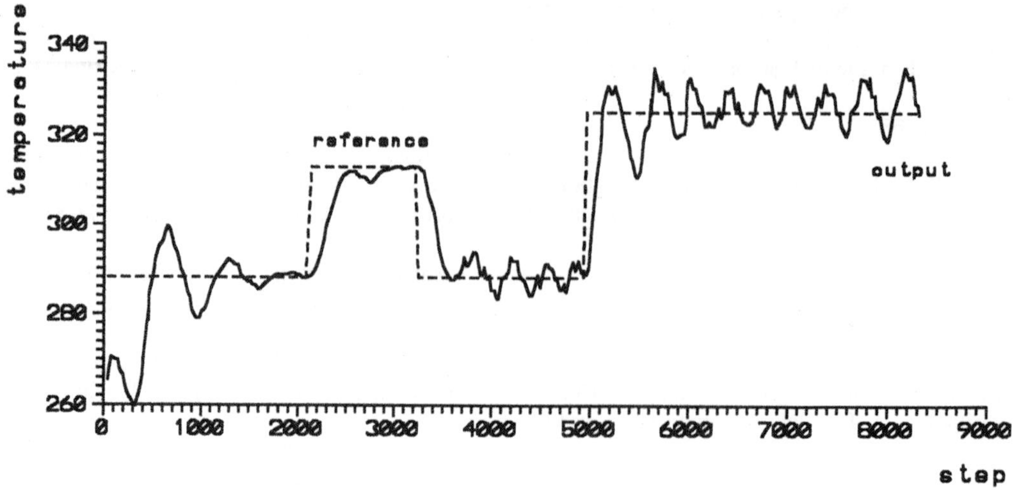

Fig.3: Adaptive PID control of the temperature in the tin solder

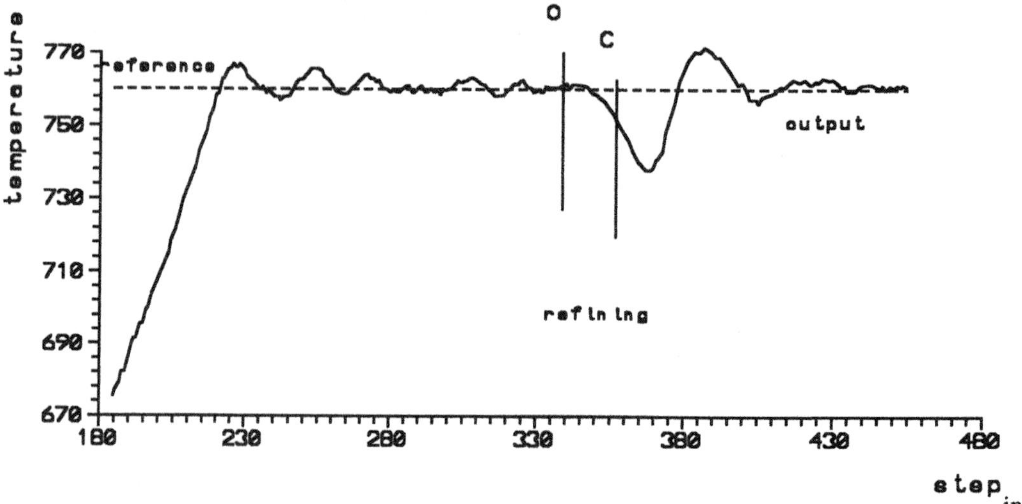

Fig.4: Controlled liquid alloy temperature profile
in the aluminium melting furnace

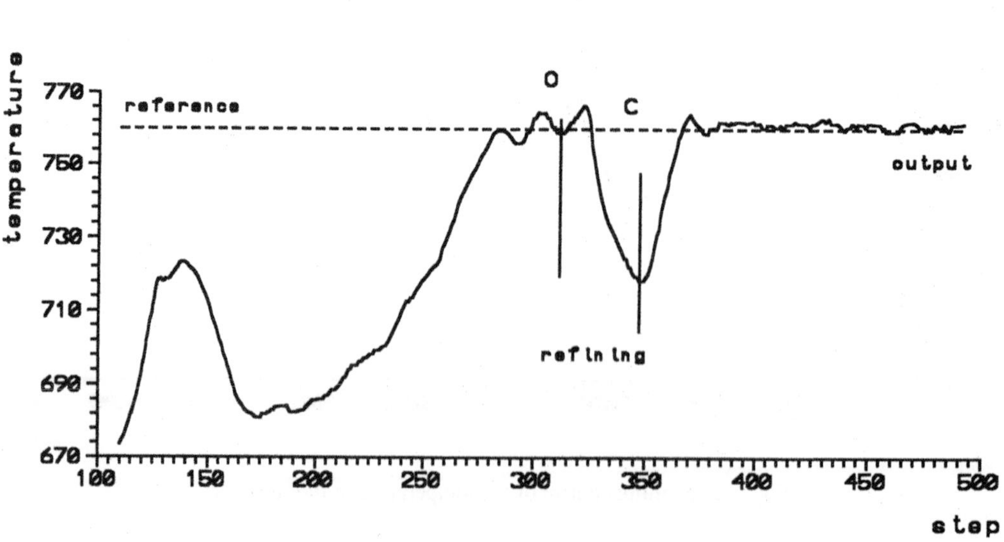

Fig.5: Controlled liquid alloy temperature profile

A PC-BASED REAL-TIME GENERALIZED PREDICTIVE CONTROLLER

A.P. de Madrid, F. Morilla and S. Dormido

Dpto de Informática, UNED, Avda Senda del Rey s/n, 28040 Madrid, Spain

Abstract. The Generalized Predictive Control (GPC) appears to be suitable as a general purpose algorithm for the stable control of the majority of real processes. Recent applications confirm the effort to take predictive control out of the laboratories and into industry. This work presents examples of simulation and real time control of a dc motor using the tool, for Personal Computers and based on the GPC basic algorithm, developed in our Department. These examples show that: it is possible to obtain good performance with prediction horizons less conservative than the 'default setting'; the developed tool is really useful to test different types of recursive parameter estimation; a new module should be added to GPC in order to get really adaptive control.

Keywords. Predictive control; adaptive control; process control; computer control, software tools.

INTRODUCTION

Predictive methods in adaptive control refer to a collection of control design formulations that pose control criteria at a given time explicitly in terms of predictions of future plant outputs and sometimes of future plant inputs, Bitmead and co-workers. Because such predictions become more difficult as they become more distant in time, these criteria are typically finite horizon optimal control criteria.

All these methods have certain features in common which distinguish them from previous design philosophies: the solution of a finite horizon optimization problem at each time instant implemented in a receding horizon way, the incorporation of plant output predictions, the provision of a small number of design parameters connected to various degrees with the closed loop dynamics.

The version which appears to have had the most acceptability is that derived by Clarke, Mohtadi and Tuffs (1987) and called GPC for Generalized Predictive Control. Because of its control parameters, it is possible to obtain several previous control strategies from it. This method is effective with a plant which is simultaneously nonminimum-phase and open-loop unstable with variable or unknown dead-time and whose model is overparameterized by the estimation scheme without special precautions being taken.

Predictive controllers are not usually employed in the industry; self-tuning PID controllers are the most frequently used. Our Department has been working in the development of self-tuning PID controllers for industrial application during the last ten years, Cruz (1984), Morilla (1987) . Nowadays, we are interested in the development of predictive controllers, for industrial application, and in the comparative study of GPC vs PID controllers. In this way, we have developed a simulation and real-time control tool using the GPC basic algorithm.

This work presents the GPC algorithm used, the tool developed and examples of simulation and real time control of a dc motor.

GPC: THE BASIC ALGORITHM

GPC is based on an ARIMAX process model. For a SISO system, it is given by the following form:

$$A(q^{-1})\, y_t = B(q^{-1})\, u_{t-k} + \frac{C(q^{-1})}{\Delta}\xi_t$$

where: u_t is the control input, y_t is the measured variable or output, ξ_t is an uncorrelated random sequence and Δ is the differencing operator. The control law is obtained by minimizing the cost function

$$J(u,t) = E\left\{ \sum_{j=N_1}^{N_2} (y_{t+j} - w_{t+j})^2 + \lambda \sum_{j=1}^{N_u} (\Delta u_{t+j-1})^2 \right\}$$

where: w_t is the set-point, N_1 is the minimum output horizon, N_2 is the maximum output horizon, N_u is the control horizon, λ is the control-weighting and $\Delta u_{t+j-1} = 0 \;\; j \geq N_u$. The equation that yields the future control increments for times t to $t+N_u-1$ is

$$\hat{u} = (G^T G + \lambda I)^{-1} G^T (w - f)$$

where (assuming that $N_1=1$) :

$$\hat{u} = [\Delta u_t, \; \Delta u_{t+1}, \; ..., \Delta u_{t+Nu-1}]^T$$

$$w = [w_{t+1}, \; w_{t+2}, \; ..., \; w_{t+N_2}]^T$$

$$f = [y_{t+1|t}, \; y_{t+2|t}, \; ..., \; y_{t+N_2|t}]^T$$

and the matrix G is composed of the impulse response parameters, g_i, of the plant model $B/A\Delta$,

$$G = \begin{bmatrix} g_0 & 0 & \cdots & 0 \\ g_1 & g_0 & \cdots & 0 \\ \cdots & \cdots & \cdots & \cdots \\ g_{N_2-1} & g_{N_2-2} & \cdots & g_{N_2-N_u} \end{bmatrix}$$

The controller parameters are chosen based on the type of process and the desired output. Different control strategies are obtained from different settings of the GPC parameters, but one of them the 'default settings', Lambert (1987), has proved to be adequate for a wide variety of applications:

$$N_1 = 1, \quad N_u=1,$$

$$N_2=10 \text{ (equal to the plant rise-time)}, \quad \lambda=10^{-6}$$

A recursive parameter estimator may be included, to obtain an adaptive control strategy. In this case, the control law has to be recalculated at each time instant.

PROGRAM DESCRIPTION

We have developed a simulation and real-time control tool for Personal Computers, based on the GPC basic algorithm. It was designed to allow the user to experiment with GPC in an interactive way and become more skilled in this control strategy, studying how the control parameters influence in the system response. It is configured for a Metrabyte DAS-16 board, but it can be easily modified for any other kind of data acquisition board. The user can access to all of its features by menus, and gets all the results in graphic format.

Next, we describe the four main menus, how the program works and how it has been implemented.

PROCESS: For the simulation case, the user defines the process to be controlled. As this process can change its dynamics in time, it is necessary to introduce the coefficients of polynomials A and B for every time instant. They are written in an ASCII file, and the user indicates the file name in this menu. For real time control, there is a real process, so this file (if it exists) is ignored.

The process model used to calculate the control law can be different (order and/or parameters) to the real process. If the parameters of this model are fixed, they are introduced in this menu, for both simulation and real time control.

GPC: This menu is used to introduce the GPC controller parameters: N_1, N_u, N_2, λ.

PARAMETERS: From this menu, the user can access to other menus, to set the sampling time, the set-point, the experiment duration and introduce white noise in the simulations (optional). For the adaptive controller, it is necessary to indicate the type of parameter estimation which will be used. A recursive least squares algorithm is selected, such as fixed exponential forgetting, variable exponential forgetting or fixed trace, and its parameters (initial forgetting factor and trace) are initialized. It is also necessary to select the process model order, and give the initial values of the estimated A and B polynomials.

For simulation, the user can decide if the control signal sent to the process will be just that one calculated by GPC, or if a control saturation must be included. This is explained in detail in the next section of this work.

RUN: From this menu, the user selects the type of experiment to do: simulation or real-time control, both in a fixed-parameters or adaptive way, or exit .

Once all the options and parameters are set up, the experiment starts. All the results are displayed in graphic mode, and some ASCII data files are generated.

If a fixed-parameter controller has been selected, a graphic screen with the closed-loop zeros and poles positions is shown. It is useful to decide whether the selected GPC parameters are correct or not, and discover if the system response is going to be unstable.

After this, in the simulation case, the set-point and the system output are displayed in a window, and the control signal applied to the process in another one. For real-time control, the set-point is shown just at the beginning, and the control and output signals are shown as they are obtained.

Finally, for the adaptive controllers, there is a screen for the estimated parameters, forgetting factor and trace evolution. This values are also written in an ASCII file.

Implementation

The main program was developed in Turbo Pascal. In order to get a flexible and easy-to-program graphic output, we first tried to use the Turbo Pascal Graphix Toolbox, but this tool was not able to display real-time data as they are being obtained. So, to solve this problem, we developed a graphic library in Turbo Pascal. It can handle 'real-time graphics', and it is quite similar to the graphic functions of Matlab: the screen can be split up in several windows, it is possible to display just a single point or some data vectors and, given a transfer function, it calculates and displays the poles and zeros in the z plane.

The library file TP4D16.PAS from Quinn-Curtis Metrabyte Turbo Pascal Data Adquisition and Control Tools (IPC-TP-018) has been used to communicate with the DAS-16 board but, in order to better handle the sampling time, we have modified some of its functions.

The 'pull-down' menu system has been developed in C, using the CSCAPE functions and the screen designer Look And Feel, both by Oakland Group, Inc. It allows the user to modify all the parameters and select the experiment to do in a friendly and interactive way.

APPLICATION EXAMPLES

We have chosen the Ball and Hoop equipment by TecQuipment to show how the program works. It is a quite easy system to control, but the influence of the GPC parameters can be studied. The speed of the dc electric motor was modelled as a first order linear system, unit gain and time constant of 1 second; a sampling time of 0.1 seconds is adequate. Thus, $A = 1 - 0.9048q^{-1}$ and $B = 0.0952q^{-1}$.

Next we introduce some representative experiments done with our program, and comment some of the results we have got. A sequence of set-point changes between 3 and 5 volts was provided with switching every 100 samples. The graphs display results over 500 samples of simulation or real-time control, using solid line for the set-point and dotted line for the process output and the control signal.

1) Effect of the horizons

Before starting the experiments, it is necessary to decide what horizons and λ are going to be used. We have studied the system response for prediction horizons smaller than 10. As we can see from the root-locus for N_2 (see Fig. 1), the position of the poles indicates that the output can be good for such an easy system with values of N_2 smaller than 10.

We have found the 'default settings' too 'conservative' for this process (see Fig. 2). Its is possible to obtain a good performance with a design less conservative than the 'default one', (see Fig. 3), at least for easy systems. And the smaller the horizons, the faster the computer can calculate the control signal. With a cheap - and slow - computer, the adaptive control of a fast process with a prediction horizon of 10 could be impossible.

We are studing the possibility of adding a new 'module' to GPC, to real-time modify the values of the horizons and λ, based on the system dynamics and the kind of output desired (see Fig. 4). This module would use both analytical and heuristic tools to modify the control parameters in two possible cases:

- The actual control parameters are too high, and it is possible to obtain the desired output with smaller values (faster computing).
- The system output is bad, because the horizons are no high enough (improving the performance).

2) Control signal saturation

GPC can obtain a really fast system response to changes in the set-point. But the cost is the high value of the control signal applied to it.

In general, the tension that can be sent from the board to the system has a minimum and a maximun value (in our case, we have configured the board for a range from 0 to 10 Volts). So, real processes cannot be controlled with different control values. If the control program tries to send a higher value to the system, it is saturated by the board, but in this case the algorithm is working with wrong tension values. To avoid this error, our program saturates the control signal before it is sent to the board, and includes this real value in all its future computations. What is more, this can be done in simulation too: the user indicates the control range he wants, which can be different to the board output range.

3) Adaptive control

Finally, the program has proved to be very useful to test different types of recursive least squares algorithms, allowing the user to study how they work and become more skilled in system identification and adaptive control.

The computer used for these experiments is a PC/AT (8 MHz) with math coprocessor, but it results too slow to real-time estimate the model parameters with a sampling time of 0.1 seconds. Thus, a sampling time of 0.3 seconds was used. In this case, $A = 1 - 0.7408q^{-1}$ and $B = 0.2592q^{-1}$. The plant rise-time is 3.3 sampling times, so we have chosen $N_2 = 4$.

In the example that we present, we have used a recursive least squares algorithm with fixed trace of 10. The initial estimated parameters are -0.7 and 0.2 (a bit different to those of the model).

The adaptation mechanism worked well and the system response improved in time (see Fig. 5) as the estimator converged to a model different to the initial one. In Fig. 6, the estimated parameters are displayed at the top left corner, the forgetting factor evolution at the bottom, and the trace (constant) at the top right corner.

CONCLUSIONS AND FURTHER WORK

The program we have developed has proved to be a good tool to study on a PC how GPC performs not only under ideal simulation conditions but when it is applied to real processes. GPC has proved to be a control algorithm as robust and versatile as it was expected.

It is our personal belief that a new feature should be added to GPC: the possibility of 'auto-tune' the horizons in real-time, based on the system observed dynamics. This could be done by using both analytical and heuristic criteria together. Nowadays, we are working in this way.

REFERENCES

Bitmead, R. B., M. Gevers and V. Wertz. Adaptive Optimal Control: The Thinking Man's GPC. Prentice Hall International. Series in Systems and Control Engineering.

Clarke, D. W., C. Mohtadi and P. S. Tuffs (1987a). Generalized Predictive Control. Part I. The Basic Algorithm. Automatica, Vol. 23, No. 2, pp. 137-148.

Clarke, D. W., C. Mohtadi and P. S. Tuffs (1987b). Generalized Predictive Control. Part II. Extensions and Interpretations. Automatica, Vol. 23, No. 2, pp. 149-160.

Cruz, J. M. de la (1984). Contribución al estudio y síntesis de reguladores autosintonizados. Doctoral Dissertation, Universidad Complutense.

Lamber, E. P. (1987). Process Control Applications of Long-Range Prediction. Report No OUEL 1715/87. Department of Engineering Science. University of Oxford.

Morilla, F. (1987). Contribución a los métodos de autosintonía de reguladores PID. Doctoral Dissertation, UNED.

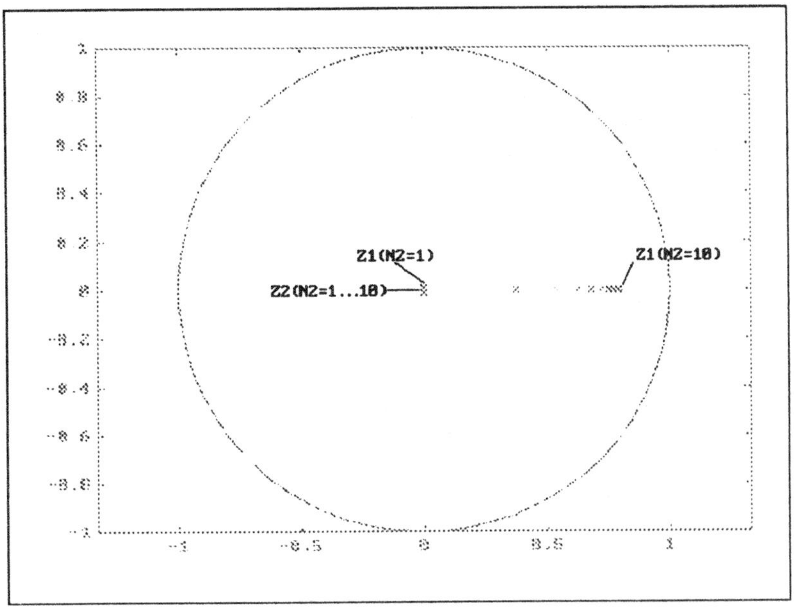

Fig.1 Root-locus for the dc motor with $N_1=N_u=1$, $\lambda=10^{-6}$ and N_2 variable from 1 to 10.

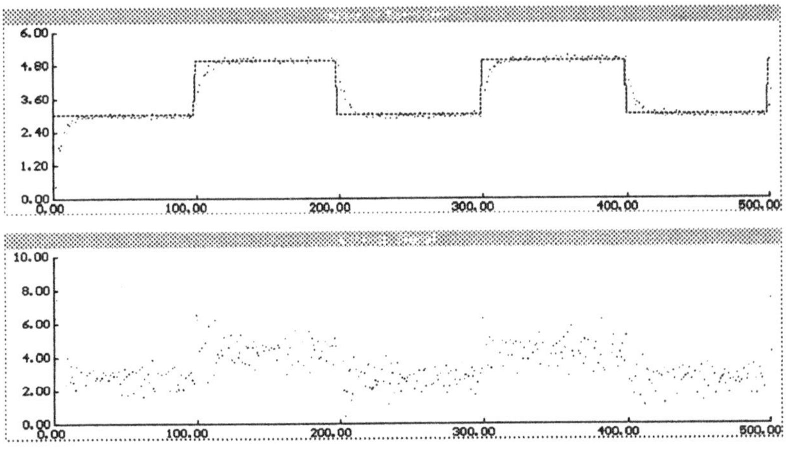

Fig.2 Speed control of the motor with $N_1=N_u=1$, $\lambda=10^{-6}$, $N_2=10$ and the fixed process model.

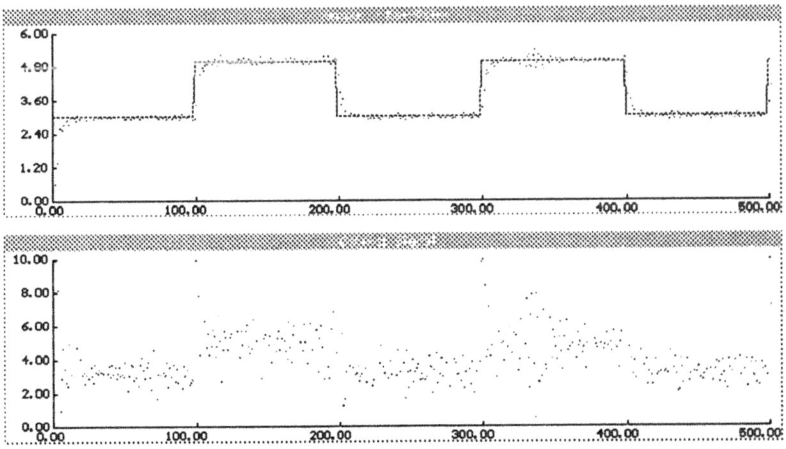

Fig.3 Speed control of the motor with $N_1=N_u=1$, $\lambda=10^{-6}$, $N_2=5$ and the fixed process model.

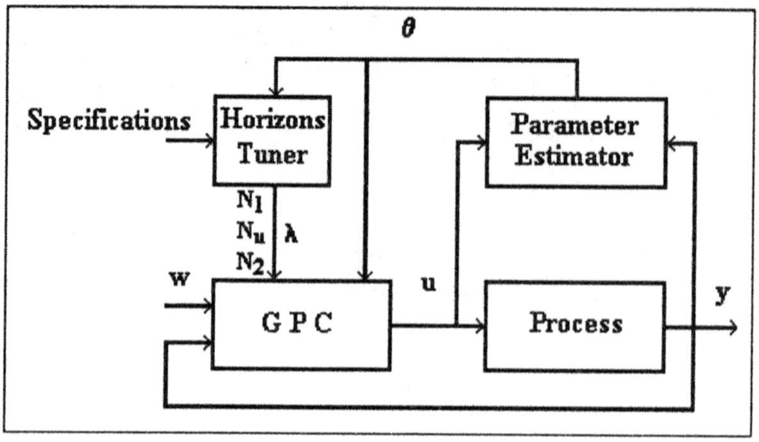

Fig.4 Scheme of a really Adaptive GPC.

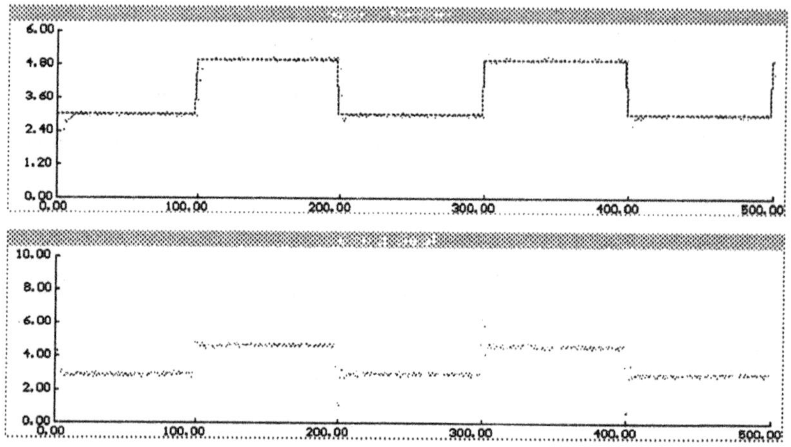

Fig.5 Speed control of the motor with $N_1=N_u=1$, $\lambda=10^{-6}$, $N_2=4$ and adaptive process model (sampling time 0.3 sec).

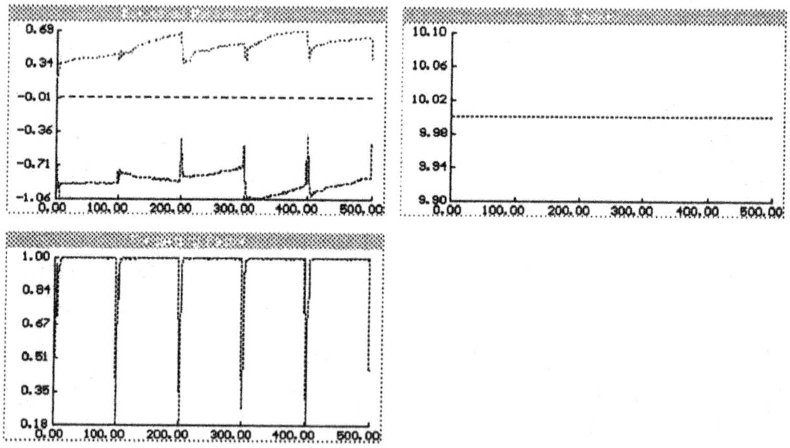

Fig.6 Speed control of the motor with $N_1=N_u=1$, $\lambda=10^{-6}$, $N_2=4$ and adaptive process model.

A PARALLEL ADAPTIVE CONTROLLER FOR TIME-VARYING PLANTS

M.B. Hadjiski*, A.K. Kordon** and V.C. Christov**

*Department of Automation, Technological University - Sofia, Kl, Ohridsky Bul. 8, 1156 Sofia, Bulgaria

**Paros Laboratory, P.O. Box 6, PTT-2, 9002, Varna, Bulgaria

ABSTRACT. A multi-functional adaptive regulator for a given class of time-varying plants with parameter variations influenced by a measurable scheduling variable is considered. The adaptive regulator has hierarchical structure with executive, adaptive, supervision and coordination levels. A satisfactory system performance in all possible operating points is established through switching of different appropriate concurrently acting controllers. An original procedure for generalised system state determination in a decision space with reduced dimension is developed. After an initial learning phase the system determines its generalised state through recognition of limited number of factors in a divided decision space. A procedure for bumpless transfer between the different regions is also defined. The proposed regulator is implemented by means of parallel algorithms on a PC-based transputer board. Simulation results for multi - functional control of a superheater of drum boiler are shown.

KEYWORDS. Parallel control, Adaptive control, Time-varying systems, Supervision methods, Gain scheduling.

1. INTRODUCTION

Recently, the problem with structure and parameter uncertainty is a dominating task for control theory. The problem becomes much more complex when the plant is time-varying. The standard algorithms with proved qualities (Astrom,1989) need additional improvements and limitations in non-stationary conditions. The current investigations are focused on linear time-varying systems with small in some sense parameter variations and their derivatives (Anderson,1983), (Middleton, 1988). A direct transfer of standard solutions and algorithms in case of fast parameter variations may cause to instability. The special investigations of time-varying plants with fast parameter changes are only for simple plants of low order (Annaswamy,1989). In order to establish a solution to the problem a modification and new development of the standard adaptive regulator structure is necessary.

Plants with structural parameter variations are considered in the paper. These variations are caused by variables connected with different operational conditions of the plant, usually called scheduling variables (Rugh,1990). When the parameter dependence is functional, the plants are successfuly controlled by gain - scheduling technique (Astrom,1989), (Hadjiski,1980). It is based on parameter freezing approach, but the time-dependence need some additional adaptive procedures. At the same time the influence of parameter deviation derivatives on the performance of gain scheduling system is still unclear from theoretical and implementation point of view.

The contemporary high quality requirements for process control impose the complex nonlinear time-varying system to work effectively in every possible technological mode. This additional complication requires a combination and development of adaptive control and gain-scheduling algorithms. A radical step toward this goal is the introduction of a hierarchical structure for adaptive control systems (Knapp,1990), (Kordon,1990b). It allows to cover all operarting conditions through a proper alternative regulator choice, coordination of identification procedures, criteria alternation etc. It is appropriate to implement such a complex hierarchical regulator on parallel processes (Minbashian,1990), (Hadjiski,1991). The transputer application of a parallel adaptive controller on a PC-based transputer board (Kordon,1990b), demonstrates the relatively low cost of such complex control system. The paper developes the ideas in (Hadjiski,1991) and focus the attention to the most difficult problems of the parallel adaptive controller for time-varying plants - the decision space set up and its reduction; the proper division of the decision space, recognition of the generalised system state and the bumpless transfer between the subspaces.

2. PLANT AND DISTURBANCE CHARACTERISTICS

2.1 PLANT MODEL CHARACTERISTICS

Time-varying plants with parameters dependent on a measurable variable λ, called a scheduling variable are considered (Hadjiski,1989), (Rugh,1990). A wide variety of technological processes could be described in this way: reheating furnaces, steam boilers, heat exchangers, plastic extruders, absorbers etc. (Hadjiski,1989). It is also posssible the existance of an unmeasaruble parametric dusturbance ξ.

The generalised multivariable time-varying plant is expressed as follows:

$$x = A(\lambda,\xi)x + B(\lambda,\xi)u + D(\lambda,\xi)\nu \qquad (1)$$

$$y = C(\lambda,\xi)x + F(\lambda,\xi)\nu, \qquad (2)$$

The generalised control could be of different type according to the selected control structure:

$u = u_1$ - only feedback control;

$u = (u_1,\lambda)^T$ - feedback and feed-forward control of the measurable scheduling variable λ;

$u = (u_1,\lambda,\mu)$ - feedback and feed-forward control of λ and μ.

The concrete contents of matrices in equations (1) and (2) depends on the control vector choice. Due to the dual character of λ (as a disturbance and as a scheduling variable), equation (2) is nonlinear (in most cases - bilinear).

Let λ and ξ be expressed as:

$$\lambda(t) = \lambda^o + \tilde{\lambda}(t) \qquad (3)$$

$$\xi(t) = \xi^o + \tilde{\xi}(t), \qquad (4)$$

where λ^o, ξ^o are the disturbance values in stationary mode, $\tilde{\lambda}$, $\tilde{\xi}$ are deviations. Depending on the intensity and frequency contents of the deviations $\tilde{\lambda}$ and $\tilde{\xi}$ three types of parametric time-varying plants are possible:

1. $\|\tilde{\xi}\| << \|\tilde{\lambda}\| \qquad (5)$

The plant is strictly parametric and its behavior could be exactly predicted because λ is measurable. A gain-scheduling technique is entirely applicable.

2. $\|\tilde{\xi}\| << q\|\tilde{\lambda}\| \qquad (6)$

The plant is with parameter uncertainty if q is small enough (q = 0.1-0.2), but gain - scheduling is still acceptable.

3. $\|\tilde{\xi}\| \approx \|\tilde{\lambda}\| \qquad (7)$

The plant is with unstructered parameter deviations and gain-scheduling is impossible.

2.2 DISTURBANCE CHARACTERISTICS

According to the type of the generalised scheduling variable vector

$$\zeta = (\lambda^o,\xi^o)^T + (\tilde{\lambda}(t),\tilde{\xi}(t))^T \qquad (8)$$

the plant behavior could be as follows:

a) Time invariant when $\zeta(t)$ is constant.

b) With structured parameter variations when conditions (5) - (6) are fulfilled. If λ is slow, the identification is still possible.

c) With unstructured parameter variations when condition (7) is valid. The identification is possible only if λ and ξ are slow.

The scheduling variable λ is measured with noise and the real signal is:

$$\psi(t) = \lambda(t) + \omega(t) \qquad (9)$$

The measurable scheduling variable $\psi(t)$ has the following characteristics: mean, covariance σ_λ, and first derivative covariance ϑ_λ.

Typical situations for system behavior on different kinds of disturbance action are shown in fig.2. As a decision criteria for possibilities to use identification and adaptive control algorithms the following relation is used:

$$\delta = \frac{\sigma_\theta(\lambda^o)}{\hat{\theta}(\lambda^o)} \qquad (10)$$

where $\sigma_\theta(\lambda^o)$ is the covariance of estimated parameter at the stationary operating point λ^o.

- If δ < 2-5% - the identification is possible (K - region);

- If 5% < δ < 15% - an approximate identification model is available and a robust adaptive control algorithm is necessary (L - region);

- If δ > 15% - the identification is very unaccurate and adaptive control is not recommended (M - region).

3. A HIERARCHICAL ADAPTIVE CONTROLLER FOR TIME-VARYING PLANTS

The structure of a hierarchical adaptive controller for time varying processes is shown in fig.1. It is simular to the structure proposed by (Knapp,1990), but the functions, coordination methods and decision - making procedures are considerably extended. The executive and adaptive levels are represented on fig.3 with the functional blocks CONTROLLER and TUNER.

The hierarchical adaptive controller could carry out different structures and control algorithms. The following possibilities are considered:

1. Feedback and/or feed-forward control.

2. Identifacation-based type of control (different adaptive control algorithms).

3. Gain scheduling control.

Table 1. List of alternative regulators.

	Regulator	Regulator parameters	Subspace Ω_i	Sign	Region
1.	Adaptive optimal	Exact identification	Ω_1	O	K
2.	Adaptive Smith	Exact identification	Ω_1	S	K
3.	Adaptive suboptimal	Approximate identification	Ω_2	O	L
4.	Adaptive suboptimal Smith	Approximate identification	Ω_2	S	L
5.	Adaptive PID	Exact identification	Ω_1	O	L'
6.	Parameterised regulator	Gain scheduling table	Ω_3	O	M
7.	Parameterised Smith	Gain scheduling table	Ω_3	S	M
8.	Parameterised PID	Gain scheduling table	Ω_3	O	M'
9.	Standard PID	Compromised parameters	Ω_4	O	N
10.	Standard P	Compromised parameters	Ω_4	O	N

4. Back-up control with standard PID regulators.

In every operating mode the supervisor determines the appropriate pair "tuner - executive controller" through switching the elements in the structure. The decision for a proper algorithm switching is taken by production rules from the decision space. The decisive factors for decision making are the covariance of the scheduling variable σ_λ, its derivative $\dot{\sigma}_\lambda$, and the relative identification error δ. The domain of the alternative regulators are systematized in Table 1, where Ω are the decision space regions shown in fig.3, and the symbol S means that the Smith controller is determined. The designation of regions K, L, M is according to fig.2, and N is a region with fast stochastic changes.

The requirements to the hierarchical adaptive controller for time-varying plants are the following:

- to assure robust stability and performance for all possible combinations of inputs r, g, μ, λ, ξ, ω.

- to guarantee a bumpless transfer between the alternative control algorithms.

- to maintain stability and minimal performance in failure conditions and sudden violation in the control system (parameter burst phenomena, signal disconnection)

In order to obtain the closed loop system descripion the controller equation is necessary. The control signal in the proposed structure is composed of a feedback control u_1, based on a measurable output y and of a feed-forward control u_2, based on a measurable disturbance $\zeta=(\lambda,\xi)$.

$$u_1 = -R(\rho,\hat{\lambda})(r-y) \tag{11}$$

$$u_2 = -K(\rho,\hat{\lambda})\zeta \tag{12}$$

$$u = u_1 + u_2 \tag{13}$$

where R is a parameterized feedback regulator, K is a parameterized feed-forward compensator, ρ is a structural factor, given by the coordinator.

For system analysis the decision-making algorithms has to be taken into account, too. The analytical solution of this complex system in general is impossible due to its strong nonlinear and time varying behavior.This analytical complexity imposes the necessity of learning procedures for determination the alternative control algorithms. They are based on analysis the parameter identification data (δ), the scheduling variable estimates and of performance criteria in this conditions. An approach for initial learning through off-line simulations is very appropriate.

There is a clear distinction between a coordinator and a supervisor in the proposed structure. The coordinator functions are as follows:

-Determination of the degree of plant's dependence on the ratio between λ and ξ.

-Data collection for factors and regions in decision space determination.

-Setting the width of the boundary layer between the regions Ω, as well as the weighting factors for the concurrently acting controllers in the boundary layer.

-Learning for decision rule set-up of regulator/compensator selection.

-Taking a proper decision for real-time selection of a "best" regulator.

-Filling the gain scheduling tables .

-Parallel process control.

-Supervisor control.

The supervisor functions are as follows:

-Identification coordination.

-Control algorithms synthesis.

-Regulator and compensator parameterization.

-Scheduling variable estimation.

-Calculation of scheduling variable characteristics.

-Realisation of bumpless transfer between the different control algorithms.

4. IMPLEMENTATION OF A PARALLEL ADAPTIVE CONTROLLER ON TRANSPUTER NETWOKS

The hierarchical adaptive controller for time-varying systems is suitable for implementation in parallel processes. The control of parallel processes is given to the coordinator. Its main task is to select the most appropriate regulator from a bank of concurrently activated controllers. The decision procedure is given in the previous section.

The different controllers used in the different operating areas and in the vicinity of every operating point require a condition for bumpless transfer between them. The parallel supervisor implements to the real plant the weighted control signal of both concurrently activated controllers R_i and R_j.

$$u = k_i(z)u_i(z) + k_j(z)u_j(z) \qquad (14)$$

where $u_i(z)$, $u_j(z)$ are the calculated control signals from regulators R_i and R_j for the corresponding subregions D_i and D_j; $k_i(z)$ and $k_j(z)$ are weighting factors. A boundary layer D_{ij} is defined . Outside it the independent action of the regulators $R_i$and R_j controls the plant. Inside the layer the weighted control signal (14) is applied.

Transputer networks are a very suitable hardware for implementation of complex control algorithms (IEE Coloquium,1992). The T805 transputer is claimed to perform 3.3 MFlops on 30 Mhz. It incorporates a 64 bits FPU, two timers, 4 KBytes of fast on-chip RAM and an external memory interface address a total of 4 Gigabytes external RAM. It has an interrupt entry called EVENT and 4 serial ports with data transfer of 20 Mbit/sec.

For parallel adaptive controller implementation the following hardware is used (Mark Ware Associate,1991):

- A TBX10D TRAnsputer Module (TRAM) mother board for the XT/AT bus with 10 TRAM sites which may be hardwired into pipeline. The board includes a T222 transputer and C004 link switch to provide software programmable link connections in any topology.

- Two TM 8922S-4 TRAMS with 32-bit T805 transputer running on 20 Mhz and 4 Mbyte of external memory.

- Dual channel 16-bit ADC TRAM with 50 Khz sampling rate per channel.

- Four channel 12-bit DAC TRAM with T400 transputer.

The software tool for parallel adaptive controller implementation on transputer networks is the DIstributed OperatinG ENvironment (DIODEN) (Paraskevov,1992). It supports multiple applications running same transputer networks. Each program consists of one or more tasks, which are independent modules and can be executed concurrently on every processor having the required resourses.

DIOGEN itself is a small kernel (less than 32 KBytes) residing on every processor. All applications run under its control and use its services. Functionally it can be divided into several parts: Communication Kernel, Memory Manager, Processor Manager and Graphics Server. The advantages of DIOGEN are as follows:

- independence of concrete network topology and processors' interaction;

- hiding system topology from the user;

- deadlock and livelock free message routing;

- fair distribution of workload between processes.

One of the benefits of parallel programming is the natural way of encapsulating the object structure into the programming code. The alternative controllers are implemented as transputer processes. The implementation of the supervisor is also distributed. It is based on a communication kernel and thus receives status from the supervisor at the lower level. The system of supervisors may be regarded as a tree network of intelligent multiplexors. In that case, their status is to manage the status from the lower level. Thus a supervisor at one level communicates with its network neithbours in a master-slave manner.

5. SIMULATION RESULTS

The hierarchical adaptive controller is used for simulation of outlet temperature control of a primary superheater of drum boiler. The control system structure is shown in fig.1, where y is the steam temperature, u - the spread water flow, λ - the relative steam flow through the superheater. The time-varying plant is simulated by its transfer functions obtained from real experiments for $\lambda = 0.5$, 0.7, 0.85 and 1.0. The transfer functions are of third order and a sampling rate of 0.1 sec. is used in MATLAB simulations. The identification algorithm is RLS with forgetting factor of 0.95.(Astrom,1989). The following alternative controllers are simulated: ageneralised minimum-variance (GMV) controller of Clarke and Gawthrop, an adaptive PID, a gain scheduling PID with adjusted parameters for λ=0.5, 0.7, 0.85,1, and a fixed robust PID controller with compromised tuned parameters for the middle of the range of scheduling variable variation - $\lambda = 0.65$.

In fig.4 the action of the parallel adaptive controller in an exponential decreasing of λ is shown. The classical adaptive control system without supervisor could not handle the parameter variations. The parallel controller detects the increasing identification error δ and the relatively slow variation of scheduling parameter first derivative. The coordinator takes the decision for suboptimal adaptive

PID selection at sample 150. After sample 200 the coordinator estimates the new stationary region in the decision space and at sample 212 it selects again the GMV controller.

Fig. 5 illustrates the systems behavior in case of drastical drop of λ from 1 to 0.7. The classical adaptive control system is unstable. The coordinator detects the increased δ, σ_{λ}, σ_{λ} and according to the decision rules in section 3, selects the gain scheduled PID controller. The combined action of both controllers in parallel stabilizes the system. The most severe case of scheduling parameter variation is shown in fig. 6. After an initial drop from 1 to 0.7, there is a combine action of an oscillation with amplitude of 0.25, a drift to λ=0.5 and a stochastic fluctuation with covariance of 0.02. Even in this case the parallel adaptive controller could stabilize the system with robust fixed PID.

6. CONCLUSIONS

A parallel adaptive regulator for time-varying plants is developed. It is based on hierarchical coordination, bumpless transfer between different concurrently acting control algorithms, learning and situation recognition in a decision space. The parallel adaptive controller is flexible and covers the whole range of plant's operating conditions as well as arbitrary scheduling variable variations. The proposed structure is suitable for transputer networks implementation and has a high algorithm complexity per hardware cost efficiency.

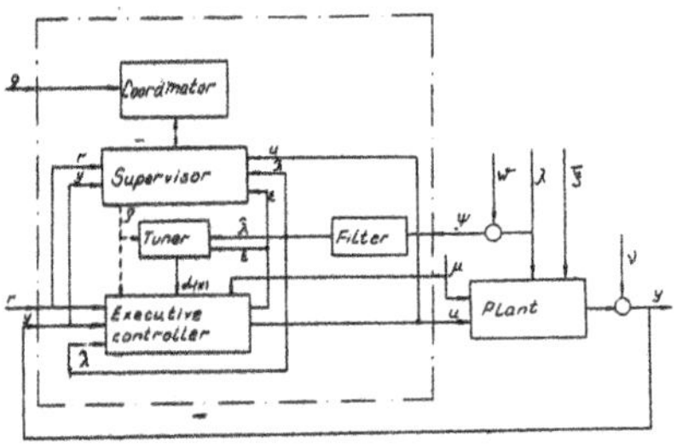

Fig.1. Hierarchical Adaptive Controller structure.

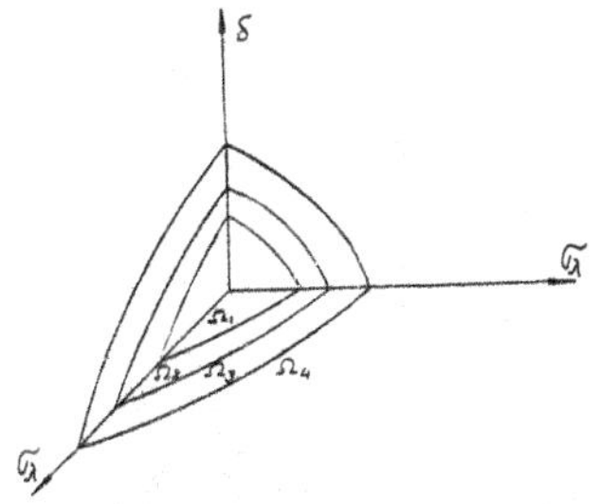

Fig.3. Decision space regions.

REFERENCES

Astrom K. and B. Wittenmark (1989). ADAPTIVE CONTROL, Addisson Wesley, Reading, MA.

Hadjiski M. (1989). AUTOMATIC CONTROL OF PROCESSES IN CHEMICAL AND METALURGICAL INDUSTRY, Technika, Sofia (in Bulgarian).

Hadjiski M. and A. Kordon (1991). Parallel Structure of the Supervision Level for Multivariable Adaptive Control of Time Varying Systems, PROC. OF THE CDC, Brighton, pp. 2946-2947.

IEE Colloquium "Generic Parallelisation of Algorithms for Control and Simulation" IEE 30/1992.

Isermann R. and others. (1991). ADAPTIVE CONTROL SYSTEMS, Prentice Hall, NY.

Knapp T. and R. Isermann. (1990). Supervision and Coordination of Parameter-Adaptive Controllers, PROC. OF THE ACC, Pittsburgh, pp. 1632-1637.

Kordon A. (1990a). The Structure of a Parallel Adaptive Controller and its Transputer Implementation, APLICATION OF TRANSPUTERS 2, IOS Press, Amsterdam, pp. 119-122.

Kordon A. (1990b). A Hierarchical Multivariable Self-Tuning Regulator, PHD DISSERTATION, Technical University, Sofia (in Bulgarian).

Kordon A. and others. (1991). Transputer Implementation of a Parallel Supervisor for Multivariable Adaptive Control of Time-Varying Systems, APPLICATION OF TRANSPUTERS 3, IOS Press, Amsterdam, pp.804-809.

Minbashian B. and K. Warwick. (1990). Flexible Parallel Control, PROC OF UK IT'90 CONFERENCE, pp.117-124.

Parashkevov A. and I. Nedeltchev. (1992). DIOGEN - The next step, accepted for PACTA'92, Barcelona

Rugh W. (1990). Analytical Framework for Gain Scheduling, PROC. OF ACC, Pittsburgh, pp. 1688-1694.

Shamma J. and M. Athans. (1991). Guaranteed Properties of Gain Scheduled Control for Linear Parameters, AUTOMATICA, 27.

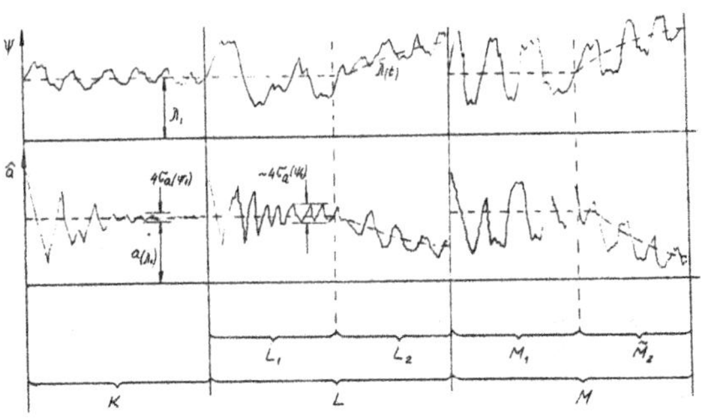

Fig.2. System behavior on different disturbances.

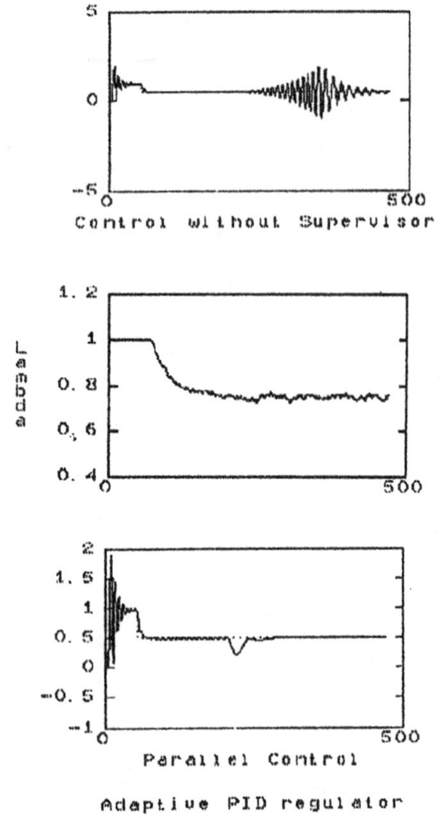

Fig.4. Parallel adaptive control in case of exponential scheduling parameter variation.

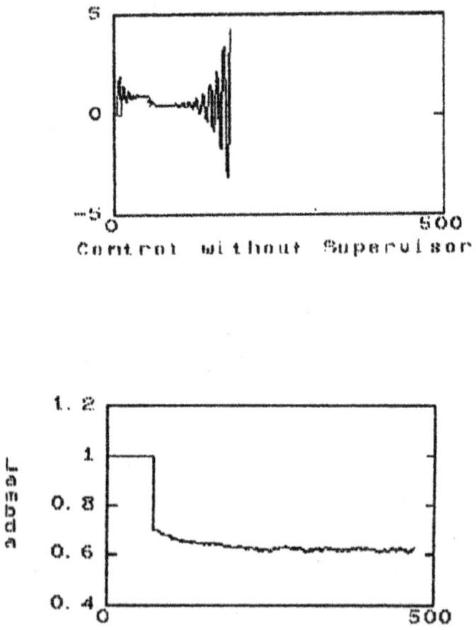

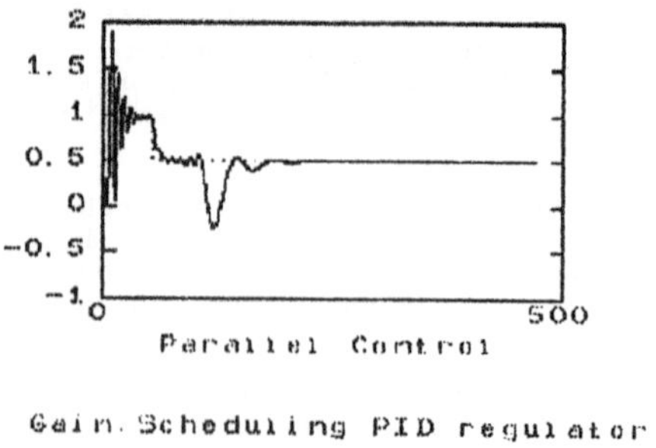

Fig.5. Parallel adaptive control in case of drastic scheduling parameter drop.

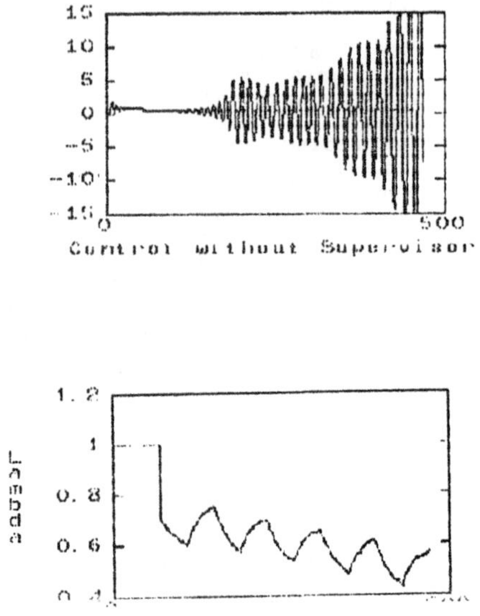

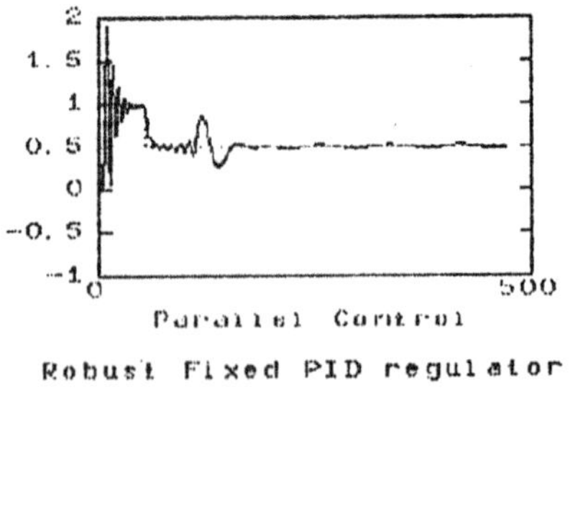

Fig.6. Parallel adaptive control in case of complex scheduling parameter variation (drop + drift + stochastic fluctuations).

CONTROL ALGORITHMS FOR SATURATING SYSTEMS

M. Huba, P. Bisták and M. Molnár

Department of Automation and Control, Faculty of Electrical Engineering, STU
CS-812 19 Bratislava, Czechoslovakia

Abstract: New type of predictive controllers for saturating systems is described. It relies on an analytical approximation of the controlled process by the I_2-model. The closed-loop behavior may be prescribed by means of reference trajectories defined as optimal braking trajectories of the actual model as well as by choosing a time horizon for matching with these trajectories. The method essentially improves features of classical relay minimum time systems. It enables to modify their behavior in a similar way as using the pole assignment method and so to achieve a desired degree of robustness against structural and state perturbation. Taking into account saturation limits and parasitic time lags makes the task of tuning controller transparent and reliable which is illustrated by the example of positional servo control.

Keywords. Minimum time control; linearization techniques; saturation; robustness; d.c. motors; position control.

INTRODUCTION

Saturation occurs in the vast majority of control application and dominantly influences the system dynamics. Its importance is also evident from large amount of literature available on this subject, where various approaches can be met. Here we will consider discrete (digital) controller for certain type of nonlinear and saturating systems with minimum settling time as the performance criterion. According to Feldbaum (1966), the first minimum time controller with quadratic feedback (which is typical for continuous optimal control of I_2-systems) was applied to the rolling mill control in 1935 and a bit later in the automatic potentiometer "Speedomax" produced by the firm "Leeds and Northrup". The main growth of minimum time systems came then in 50-ies and 60-ies. Nowadays, the academic perfectly tuned minimum time controllers (see e.g. Desoer and Wing, 1961; Athans and Falb, 1966; Feldbaum, 1966) are only rarely used - despite of the progress in implementation given by the currently available hardware and computers. As an important reason we can recognize their high sensitivity to fluctuations of all types. The aim of this paper is to show how some new solutions bringing essential simplification of discrete minimum time controllers (Buša and Huba, 1986) together with newly developed linearization technique (Huba and Pauer, 1991) and methods for compensation of different fluctuations change this traditional image of minimum time controllers and contribute to their revival attractivity.

SYSTEM'S LINEARIZATION

Let us consider digital t-optimal control of an idealized nonlinear 2nd-order system

$$\ddot{y} = Ku_R - f(\mathbf{y}); \qquad \mathbf{y} = (y, \dot{y})' \tag{1}$$

that must be brought from an arbitrary initial point $\mathbf{y}_0$ to the demanded state $\mathbf{w}=(w,0)'$ in the minimum time. The control signal is subject to amplitude constraints

$$U_{R1} \le u_R \le U_{R2} \tag{2}$$

Let K be a known gain of the system and $f(\mathbf{y})$ be a continuous feedback. Approximating it by zero-degree polynomial (Huba and Pauer, 1991) in the operating point $\mathbf{y}_p$ we come to the linear model

$$\ddot{y} = Ku_R - f(y_p) \quad (3)$$

After using the coordinate transformation

$$x = y - w; \quad \dot{x} = \dot{y}; \quad \ddot{x} = \ddot{y} \quad (4)$$

and the control signal transformation

$$u = Ku_R - f(y_p) \quad (5)$$

the minimum time control problem can be formulated in the commonly used form, i.e. the demanded state $x_w = 0$ and

$$\ddot{x} = u; \quad (6)$$

The transformed control signal u is now constrained between these new saturation limits

$$U_i = K_s U_{Ri} - f(y_p) \quad ; \quad i = 1,2 \quad (7)$$

and, at each sampling instant the computation of the control signal u_R consists of:

1) The computation of the control signal u in coordinates system (4) with constraints (7) using algorithms described in the next paragraph.

2) The expression of the control signal u_R using relation inverse to (5)

$$u_R = \left[u + f(y_p) \right] / K \quad (8)$$

In choosing y_p two elementary possibilities exists:

a) $y_p = w$ (classical linearization in the desired state);

b) $y_p = y$ (linearization in the actual state - the control algorithm (11) corresponds in this case to the well known nonlinearity compensation method (see e.g. Sommer, 1981).

These two limit situations can be continuously modified (Huba and Pauer, 1991) introducing new flexible operating point

$$y_p = mw + (1-m)y; \quad m \in <0;1> \quad (9)$$

T-OPTIMAL CONTROLLER FOR I_2-SYSTEM

Here we will briefly describe and interpret saturating minimum time controller for I_2-system (Buša and Huba, 1986; Huba and others, 1987; Huba, 1992). System (6) may also be described as

$$\frac{dx}{dt} = \begin{pmatrix} 0 & 1 \\ 0 & 0 \end{pmatrix} . x + \begin{pmatrix} 0 \\ 1 \end{pmatrix} . u = A_c x + b_c u \quad (10)$$

Under sampled data control u=const over sampling intervals T and the solution of (10) is

$$x(t) = e^{A_c t} x_0 + u \int_0^t e^{A_c \tau} b_c d\tau = A(t) x_0 + b(t) u \quad (11)$$

$$A(t) = \begin{pmatrix} 1 & t \\ 0 & 1 \end{pmatrix}; \quad b(t) = \begin{pmatrix} t^2/2 \\ t \end{pmatrix} \quad (12)$$

Denoting $x_0 = x_n$; t=T; $u = u_n$ and $x(T) = x_{n+1}$ we can derive the system's state transition equation as

$$x_{n+1} = A(T) x_n + b(T) u_n \quad (13)$$

Further we will also use its inverse form

$$x_n = A(-T) x_{n+1} - A(-T) b(T) u_n = $$
$$= A(-T) x_{n+1} + b(-T) u_n \quad (14)$$

In order to bring the representative point $x = (x, \dot{x})'$ to the demanded state x_w as quickly as possible, $\dot{x}$ must be maintained at maximum values that do not give rise to overshoot: The system must be accelerated as quickly and braked as late as possible. It is accomplished by the controller which attempts to achieve in one control step some of the optimal braking trajectories (OBT). In difference to the continuous minimum time systems with only two OBT, under sampled data control infinitely many OBT corresponding to the braking by one of the limit values U_i exist which differ by the value $u \in (0; U_i>$ in the last control step. Their points at sampling instants trace out polygonal "optimum switching curve" (OSC) with vertices X_N^i

$$X_N^i = A(-T) X_{N-1}^i + b(-T) U_i; \quad X_0^i = 0; \quad i=1,2 \quad N=1,2, \ldots \quad (15)$$

or simply by

$$X_N^i = b(-NT) U_i = \begin{bmatrix} N^2 T^2 / 2 \\ -NT \end{bmatrix} . U_i; \quad i=1,2 \quad (16)$$

Assuming constraints imposed on the control signal, OSC can be reached in one control step only from a region called "proportionality zone" (PZ) surrounding OSC and outlined by polygonal "shifted switching curve" (SSC) with vertices

$$Y_N^i = A(-T) X_{N-1}^i + b(-T) U_{3-i}; \quad i=1,2 \quad (17)$$

The third important polygonal curve called "critical curve" (CC) can be defined as the set of points x, from which OSC can be reached in one control step by u=0. By means of (14), vertices of such a curve can be expressed as

$$Z_N^i = A(-T) X_{N-1}^i; \quad i=1,2; \quad N=1,2,\ldots \quad (18)$$

If the value of the control signal computed for the achievement of OSC in one sampling interval T exceeds prescribed limit values (2), i.e. the representative point is outside of PZ, the system has to be accelerated or braked by the corresponding limit value U_i; i=1 or 2. Apart from full acceleration and braking, when $u = U_1$ or $u = U_2$, the value $u \in <U_1, U_2>$ can only occur in two sampling periods when the representative point reaches OSC and the demanded state. Optimal controller can be accomplished by a nonlinear PD-algorithm

$$u = sat\left[-(r_0 x + r_1 \dot{x} + r_c) \right] \quad (19)$$

with the state-dependant coefficients

$$\begin{aligned} r_0 &= 1/NT^2 \\ r_1 &= (2N+1)/(2NT) \\ r_c &= (N-1)U_i/2 \end{aligned} \quad (20)$$

where

$$i = [3 + sign(p)]/2 \quad (21)$$

$$p = 2x + T\dot{x} \quad (22)$$

$$N = int\left\{ 1/2 + \sqrt{1/4 + p/(U_i T^2)} \right\} \quad (23)$$

$$sat(u) = \begin{cases} U_2 \; ; & u > U_2 \\ u \; ; & U_1 \le u \le U_2 \\ U_1 \; ; & u < U_1 \end{cases} \quad (24)$$

As it has been already mentioned above, minimum time control systems are highly sensitive to all parasitic phenomenons occurring in the control loop (Fig.3). Thus, a relatively small parasitic time lag leads to overshoot and to oscillations in the neighborhood of demanded states. Therefore, in the next two paragraphs we will deal with the question how to overcome this essential lack of saturating controllers.

PREMATURE SAMPLING AND THE PREDICTION HORIZON

Assume at first, that the the representative point is just being in PZ and so the time optimal controller has produced a pulse $u \in \langle U_1, U_2 \rangle$ not exceeding the saturation limits. Then, in an idealized loop, the OSC will be reached in one sampling interval $T_p=T$. However, if we are not completely sure about the suitability of this control action, we can up-date the control yet before the originally planed sampling instant, say, at the moment $t=cT_p$, $c \in (0,1\rangle$. It is obvious that such a correction cannot lead to a failure. However, OSC will now be reached at the moment $t=T_p+cT_p$. If we apply such corrections in each control step, i.e. the controller will run with new sampling period

$$T = cT_p; \quad c \in (0;1\rangle \tag{25}$$

while the value $T_p>T$ is being set into Eqs. (11-14), the representative point will move in each sampling interval T toward OSC, but this motion will now be theoretically infinitely long: The transient from acceleration to braking will be softer and therefore not so much influenced by parasitic time lags.

To explain this procedure using terminology of predictive control (see e.g. Richalet and others, 1978) we can denote optimal braking trajectories as "reference trajectories" and the time horizon $T_p \geq T$ as the "prediction horizon" planed for the matching of OSC. Further we will introduce new parameter

$$\lambda = 1 - c = 1 - T/T_p \tag{26}$$

and rewrite the relation between T_p and T as

$$T_p = T/(1-\lambda) \tag{27}$$

Since in the predictive control of 1st-order systems with parasitic time lags (see Huba and Bisták, 1992) this parameter called "time margin" plays role identical with that of closed loop poles, for given parasitic time lags it is possible to derive appropriate values of the time horizon (or of the parameter λ) analytically. Thus, for the single integral + dead time T_d, when

$$G(s) = \frac{e^{-T_d s}}{s} \tag{28}$$

we get recommended values of the prediction

$$T_p = \left(\sqrt{T} + \sqrt{T_d}\right)^2; \quad T_d \leq T \tag{29}$$

$$T_p = (T + T_d)\left(1 + T/T_d\right)^{T_d/T}; \quad T_d > T \tag{30}$$

and for the single integral + time constant T_1

$$G(s) = 1/s(T_1 s+1) \tag{31}$$

$$T_p = \frac{\left(\sqrt{T} + \sqrt{T_1(1-e^{-T/T_1})}\right)^2}{1 - e^{-T/T_1}} \tag{32}$$

As we can interpret the positional control of the I_2-system as a velocity control of I_1-system with prescribed speed profile, we can also adopt these values in controlling I_2-systems.

SUBOPTIMAL CONTROLLERS WITH VARIABLE STRUCTURE

Here we will describe some modifications of the predictive controller that can (in applications with greater values of the prediction horizon T_p) improve the response quality as well as reasonably shorten the computation time.

Correction of the damping in the neighborhood of demanded states. Using standard technique of the pole assignment method (see e.g. Ackermann, 1972), for double real pole $z_{1,2}=\lambda$ of the prescribed characteristic polynom we get a control algorithm

$$u = -\left(\frac{(1-\lambda)^2}{T^2} x + \frac{3(1-\lambda)(1+\lambda/3)}{2T}\dot{x}\right) \tag{33}$$

Comparing this result with algorithm (19) used in the vicinity of demanded states (N=1) and modified by substituting T_p (27) instead of T into (20), when

$$u = -\left(\frac{(1-\lambda)^2}{T^2} x + \frac{3(1-\lambda)}{2T}\dot{x}\right) \tag{34}$$

we can see that the coefficient staying by $\dot{x}$ in the last algorithm is smaller than in (33). In controlling systems with values $\lambda \to 1$ by means of optimal controller (19) modified by (27) it causes small overshoot of the output variable. To improve this imperfection it suffice to use the control algorithm (33) for N=1.

Dual Mode Controller I. In the vast majority of practical applications it is desired to work with relatively small sampling periods, when the polygonal OSC (16) tends to a parabola

$$\mathbf{x}(t) = \mathbf{b}(-t)U_i; \quad t \in \langle 0;\infty) \tag{35}$$

After eliminating t and replacing x by x_1 this equation can be rearranged to the form

$$x_1 = F_x(\dot{x}) = \dot{x}^2/2U_i; \quad i=1,2 \tag{36}$$

The same also holds for SSC when, denoting x as x_2,

$$x_2 = F_y(\dot{x}) = (\dot{x}+U_i T)^2/2U_{3-i} - \dot{x}T - U_i T^2/2 \tag{37}$$

and for CC with x denoted as x_0

$$x_0 = F_z(\dot{x}) = \dot{x}^2/2U_i - \dot{x}T \tag{38}$$

The control signal u can now be computed to achieve in the time T_p the reference trajectory (35), when

$$\mathbf{A}(T_p)\mathbf{x} + \mathbf{b}(T_p)u = \mathbf{b}(-t)U_i \tag{39}$$

Eliminating t from the last equation and solving for u yields the control algorithm

$$u = \begin{cases} \mathrm{sat}\left(U_i/2 - \dot{x}/T_p + \sqrt{D}\right); & D \geq 0 \\ U_i; & D < 0 \end{cases} \tag{39}$$

where

$$D = U_i^2/4 + (2x + T_p\dot{x})U_i/T_p^2 \tag{40}$$

So, in comparison with algorithm given by Eqs. (19-23), we have reasonably reduced the number of mathematical operations. It is also suitable to omit the term $U_i^2/4$ in D and so to shift the reference trajectory in such a manner that it will tend the polygonal OSC from the inner side. So we can precede small overshoots arising due to the adopted simplifications.

The polygonal character of OSC must not be, however, neglected in the neighborhood of the demanded state. There it is recommended to use algorithm (33). Since the nonlinear algorithm (39) is inevitable only in PZ, i.e. only in the 2nd and the 4th quadrant of the phase plane, its

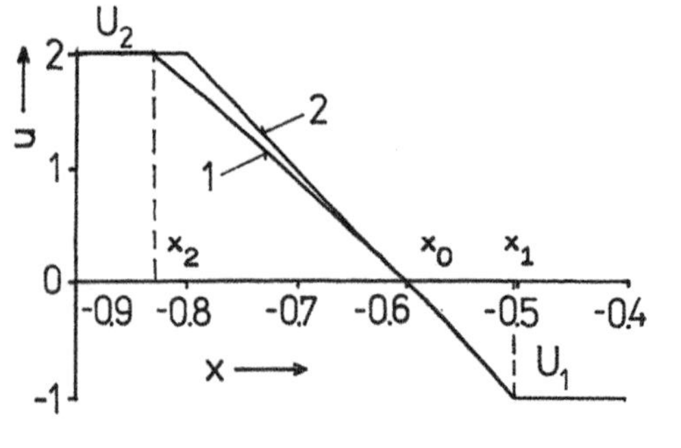

Fig. 1. Dependance (39) of the control signal u on x in a cross-section $\dot{x}=-1$ (curve 1) and linearized dependance (43) (curve 2); T=0.1.

application can be conditioned by testing

$$- \dot{x}.\mathrm{sign}(x) > \varepsilon \tag{41}$$

The constant ε should not be smaller than the coordinate $\dot{x}_1^i$ of the point X_1^i, i.e. $\varepsilon \geq |U_i|T$. To make the value of ε independent on the parameter i, it is recommended to choose

$$\varepsilon = (U_2-U_1)T_p = (U_{R2}-U_{R1})T/(1-\lambda) \tag{42}$$

Dual Mode Controller II. Further simplification can be achieved by replacing the nonlinear dependence u on x (39) in a cross-section of PZ defined by $\dot{x}$=const≠0 (Fig. 1) by a line crossing two of points $(x_1;U_i)$, $(x_0;0)$ and $(x_2;U_{3-i})$. The simplest algorithm corresponds to the combination of the first two points, when

$$u = \mathrm{sat}\left(\frac{x - x_0}{x_1 - x_0} U_i\right) = \mathrm{sat}\left(U_i(1+x/T\dot{x})-\dot{x}/2T \right) \tag{43}$$

EXAMPLE: POSITIONAL CONTROL OF A DC-DRIVE

Let us consider a positional control of the DC-drive (Fig. 2) described by equations

$$T_e \frac{dI}{dt} + I = -\frac{C_u}{R_m}\omega + \frac{1}{R_m}u_k \tag{44}$$

$$\frac{d\omega}{dt} = -\frac{1}{J}M_L + \frac{C_u}{J}I \tag{45}$$

$$\frac{d\varphi}{dt} = \frac{1}{i}\omega = \omega_2 \tag{46}$$

$$u_k = K_t u_R; \quad u_R \in <-10;10> \tag{47}$$

At first, in deriving controller equations the parasitic time constant T_e and the load M_L will be neglected, i.e. $T_e \overset{!}{=} 0$ and M_L=0. Then, denoting φ=y, $\omega_2=\dot{y}$, $a=C_u^2/JR_m$ and $K=C_uK_t/iJR_m$ we get the system equation in the form similar to (1)

$$\ddot{y} = -a\dot{y} + Ku_R$$

Introducing operating point $\dot{y}_p = (1-m)\dot{y}$ (provisionally let m=0) we can compute the control signal u_R using relations (4), (7), (10-15) and (8). According to possibilities of standard 8bit microprocessors the sampling period T=5ms was chosen. As we have neglected electrical time constant T_e and the time delay $T_d \doteq 1$ms caused by a finite speed of computer operation, the system shows oscillatory behavior (Fig. 3).

Taking these time lags into account by means of (29) and setting $T_p \doteq 2T$=10ms (λ=0.5), the system behavior reasonably improves (Fig. 4).

Now we can examine how the choice of the operating point $\dot{y}_p$ does influence the response character. It can be seen from Fig. 5 (T_p=2T) that an increase in the value of parameter m gradually slows down transient processes and the optimal damping corresponds to about m=0.5. The overshoot caused by the operating point choice can be substantially suppressed by an aditional increase of T_p (Fig. 6; T_p=5T=25ms). In the same manner we can also compensate other perturbations caused e.g. by the moment of inertia, or load variations, etc. In general, it is desired to work with as small sampling period as possible, to shorten the computer time consumption and so to have possibility to increase T_p without more recognizable lost of the response speed.

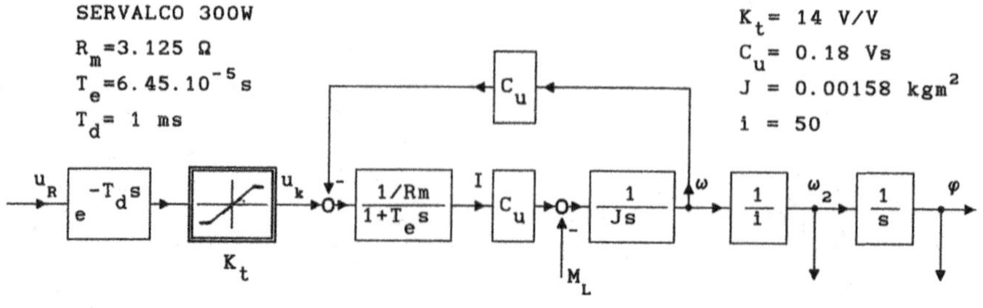

Fig. 2. Block diagram of DC-motor (T_d characterizes computer time consuption).

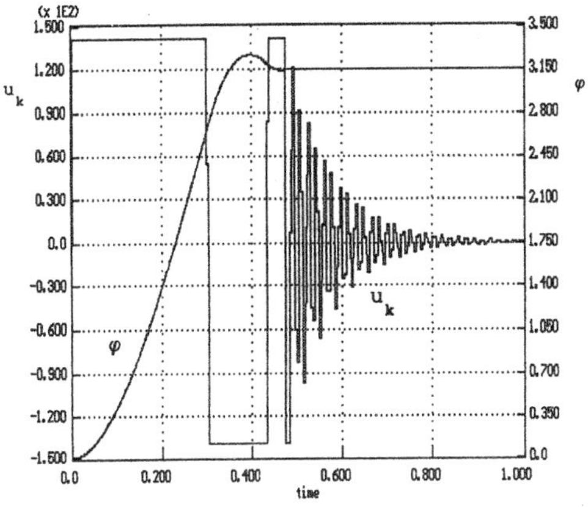

Fig. 3. Step responses corresponding to neglected parasitic time lags ($T=T_p=5ms$, $m=0$).

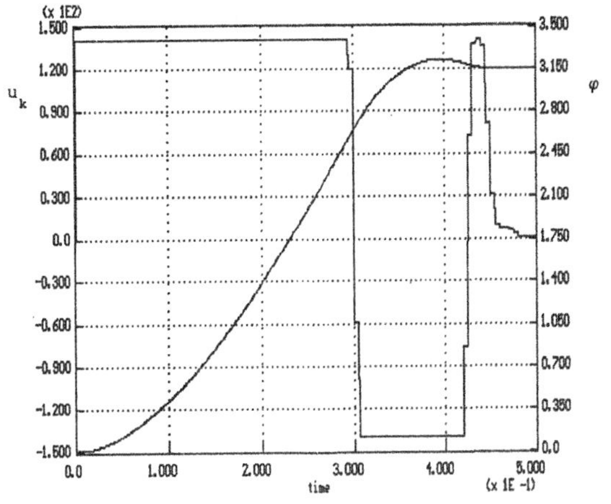

Fig. 4. Step responses for $T_p=2T=10ms$; $T=5ms$; $m=0$.

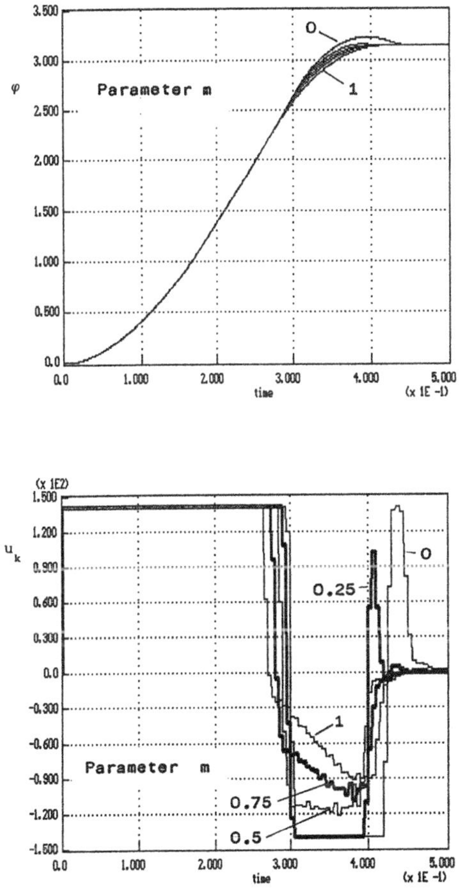

Fig. 5. Output signal φ and control signal u_k versus time for m={0, 0.25, 0.5, 0.75, 1}; $T_p=2T$; T=5ms

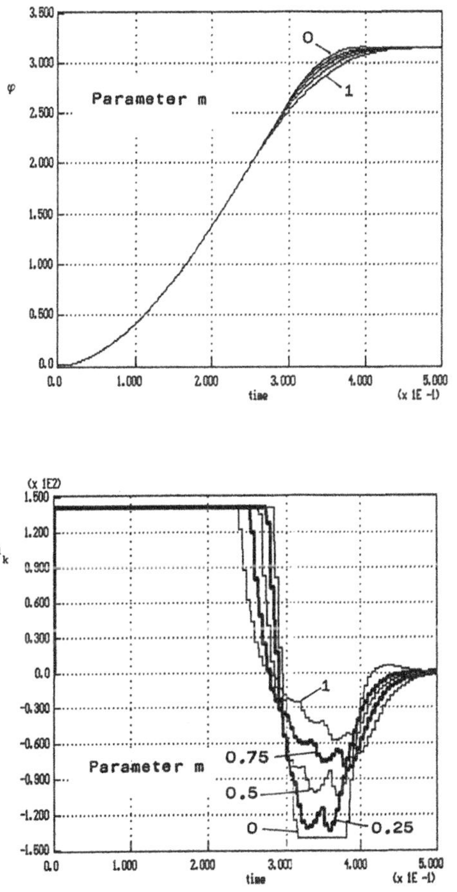

Fig. 6. Output signal φ and control signal u_k versus time for m={0, 0.25, 0.5, 0.75, 1}; $T_p=5T$; T=5ms

Finally, in Fig.7 step responses corresponding to the dual mode controller II for m=0.5 and T_p=2T are shown to illustrate a control quality very near to that achieved by the more complex optimal controller. In fact, according to smaller time consumption, they might be even better.
All simulations were carried out by means of program-package SIMULc developed in our department.

CONCLUSIONS

The paper has presented new predictive controller with several variants of different simplifications applicable for a broad class of saturating systems. The controller guarantees excellent behavior, is easy to implement by means of standard microprocessor technique as well as easy to tune.

REFERENCES

Ackermann, J. (1972). Abtastregelung. Springer, Berlin.

Athens, M. and P.L. Falb (1966) Optimal control. McGraw-Hill, N. York.

Buša, J. und M. Huba (1986). Entwurf von zeitoptimalen Abtastreglern für die Regelstrecke mit doppeltem integralen Verhalten. Automatisierungstechnik 34, 287-288.

Feldbaum, A.A. (1966). Osnovy teorii optimalnych avtomatitcheskich sistem. Nauka, Moscow.

Huba, M., Sovišová, D. and N. Spurná (1987). Digital time-optimal control of nonlinear second-order system. Preprints 10th IFAC World Congress Munich, 8, 29-34.

Huba, M. and A. Pauer (1991). Linearisierungsverfahren zur nichtlinearen Reglersynthese. Automatisierungstechnik 39, 35-36.

Huba, M. and P. Bisták (1992). Premature sampling - An analogy to the pole assignment for the saturating minimum time systems. Preprints 11th European Meeting on Cybernetics and Systems Research, Vienna, 1, 213-220.

Huba, M. (1992). Control algorithms for 2nd-order minimum time systems. Electrical Engineering Journal 43 (to appear in No.8)

Sommer, R. (1981). Synthese nichtlinearer, zeitvarianter Systeme mit Hilfe einer kanonischen Form. Fortschritt-Berichte der VDI Zeitschriften, Meß- Steuerungs- und Regelungstechnik 8/36.

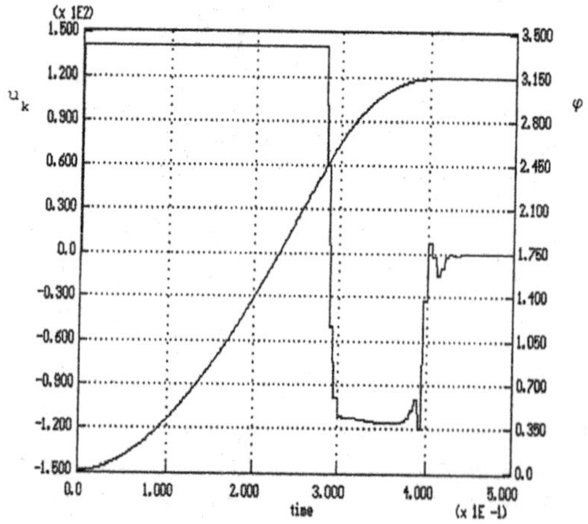

Fig.7. Dual Mode Controller II. m=0.5; T_p=2T; T=5ms

MULTICOMPONENT FORCE SENSING IN PRECISE MANIPULATION/ASSEMBLY TASKS

Š. Havlik* and G. Piller**

*Institute of Automation and Communication, Slovak Academy of Sciences, 97400 Banska Bystrica, Czechoslovakia

**Laboratory of Microengineering, Swiss Federal Institute of Technology, CH-1015 Lausanne, Switzerland

Abstract. This paper deals with the problem of the multicomponent force and torque sensing in robotic applications. It gives a brief but systematic method of approach to the design, solving and performance evaluation of the sensory devices. The design and realisation of a six-component capacitive sensor is presented. The integration of such a sensor into robot wrist provides an accommodation ability for the robot to be able to perform closely fiting assembly or precise manipulation tasks.

Key words. Force/torque sensors, robots, assembly, calibration, accommodation, factorial experiment

1. INTRODUCTION

One of the problems in advanced robotics is to perform precise operations where desired motion parameters cannot be known exactly a priori. How to ensure a high relative position accuracy of parts is an important area for research in robotic assembly. Frequently, the positional errors due to incorrect/exact unknown placements of the parts, parts tolerances and errors due to manipulation, which imply that the operation cannot be successfully executed without additional corrections of mutual position of mated parts at the moment of their first contact. Each misalignment of parts under contact results in the robot being complied and there exists a force/torque interaction.

The force and torque reactions that develop in a contact of parts are the important information for robot programming or feedback control.

Several sensing principles and many different constructions of sensors for this purpose have been developed during the past few years. A brief overview of the sensing principles is given in table of Fig. 1. The sensing devices are most frequently incorporated into robot wrist, between the arm flange and the effector, in the near vicinity of the contact.

The goal of this paper is to focus on some of the problems of the design, construction and performance evaluation of such sensory devices. An approach to the development of such a sensor, based on the capacitive displacement measurement technology is demonstrated.

2. SENSING THE FORCE AND TORQUE COMPONENTS

2.1 Problem, criteria and performance specification of a sensor

The sensing problem

The general task is to measure six components of the force and torque vector acting on the sensor $\mathbf{f} = [f_x \ \ f_y \ \ f_z \ \ m_x \ \ m_y \ \ m_z]^T$ in some given orthogonal sensor referential. The technological destination and

Physical parameter	Transducers
mechanical strain	**strain gauges** (Drake,1976; and al., about 50 different constructions have been developed)
displacement	**opto-electronic** -CCD (Hirzinger1987 et al.) -PSD (De Fazio,1980 ; Havlik 1991 et al.) **inductive (LVDT)** (Cutkosky 1982 et al.) **capacitive** (Piller1992 Wolffenbuttel 1990) **ultrasound** (Stauffer1987)
el.charge	**piezo** (Spescha 1970)
magnetic field	**magnetoelastic** (Vranish1982)
optical polarity	**fotoelastic**
conductivity	**current/voltage**

Fig.1. Sensing principles

the placement of such a sensor into robot wrist set some specific requirements that follow from its function of the transducer in information-control system and contemporary of the bearing member in the robot kinematics. The sensor in application is always a part of a production system and there are several criteria that a construction of such sensor should satisfy.

a) Application criteria.
- range of operation (values of the force, force/torque ratio, dimensions, etc...)
- interfaces (mechanical, signal, SW, HW)
- mode of operation (overload protection, vibration, performance of the robotic system, time response, compliance,..etc) (Van Brussel, 1986)
- environment

b) Technical parameters
- accuracy, dynamical range, resolution, etc...
- robustness
- reliability, etc...

c) Operational criteria
- maintenance, testing, diagnostics, calibration
- long term quality, etc...

d) Economical criteria
- cost, including the cost of calibration system
- flexibility, etc...

In reality the final product is always a compromise solution between the desired performance quality and the cost of the sensory system to be limited.

2.2 Functional parts of the sensor

Each sensor consists of three main functional parts: mechanics, transducers and data/signal processing.

Sensor mechanics. The role of the sensor mechanics is to produce measurable strains/displacements. The geometry of the elastic body and configuration of transducers is the main task for the design of sensor mechanics. Because of we always consider the function of the sensor mechanics within the range of linear characteristics we can therefore use the linear (static/dynamic) model representation.

Transducers. The function of the transducer depends on sensing principle. We shall see later that when one considers small displacements the linear model can also be used for capacitive transducers.

Then, for the large majority of sensors, we can consider the input-output static characteristics to be given by the relation

$$\mathcal{M}: \quad \mathbf{u} = \mathbf{S} \cdot \mathbf{K} \cdot \mathbf{f} \qquad (1)$$

where **u** is the vector of sensor readings and **K, S** are compliance and sensitivity matrices respectively.

Data processing. The main task of the data processing block in Fig. 2 is to perform the reconstruction of the force vector according to the model whose parameters have been estimated experimentally. With this principal function comes the need for a number of various other algorithms such as for digital filtering, force transformations and the elimination of the gravity or inertial effects .

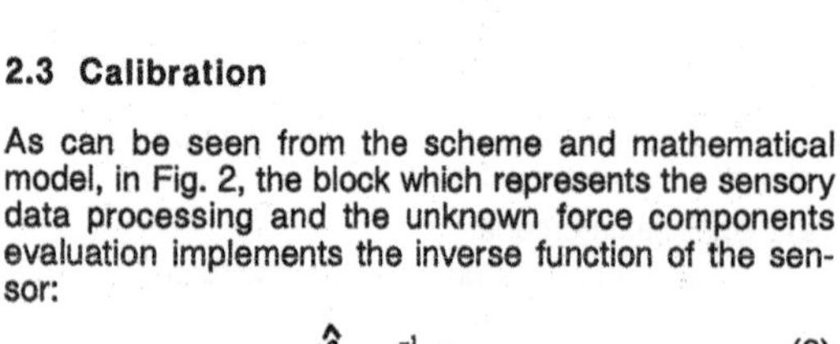

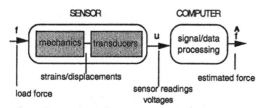

Fig. 2 Functional scheme of the sensor

2.3 Calibration

As can be seen from the scheme and mathematical model, in Fig. 2, the block which represents the sensory data processing and the unknown force components evaluation implements the inverse function of the sensor:

$$\hat{\mathbf{f}} = \mathcal{M}^{-1} \cdot \mathbf{u} \qquad (2)$$

In the case of the quasi-linear sensitivity characteristics we result in:

$$\hat{\mathbf{f}} = \mathbf{B}^{-1} \cdot (\mathbf{u} - \mathbf{u}_0) = \mathbf{C} \cdot (\mathbf{u} - \mathbf{u}_0) \qquad (3)$$

where **C** is the calibration matrix.

Because each realisation always differs from the ideal/designed case the real metrological characteristics of a particular device should be estimated individually by the calibration procedure. Also in order to keep a relatively good quality of measurements each sensor must be recalibrated from time to time. Calibration is then an experimental procedure for the estimation of the real parameters of the sensor given the structure of the model. For the majority of sensors and technological requirements necessary to an application it is sufficient to use a linear model representation and to estimate the coefficients of the calibration matrix **C**.

In general, there are two approaches to perform the calibration:
- To build a dedicated calibration stand and to make the experiments after dismounting the sensor. Although this procedure permits more complex measurements and gives the best results, the cost of such a laboratory equipment is prohibitive and experiments become too expensive.
- To use the orientation ability of the robotic system. The procedure is fast and cost effective but enables a limited variants of measurements what results in limited capability for the authentification of results.

Both these approaches exhibit some special advantages over the other. Let us define the requirements given on such a calibration system and related procedure:

a) Procedure should be fast and cost effective. It should be performed when there is any suspicion of sensor malfunction.

b) Accomplish the calibration on the robot without dismounting the sensor/effector.

c) Perform the calibration for all possible combinatios of force/torque components and or at least for several values (factorial plan of experiment as explained below).

Methodology of experimental measurements

There are several methods for preparing the plan of experiment and testing the multicomponent F/T sensors (Ferrero,1990). One of the general tests which gives the most valuable general information about characteristics of the device is the factorial method.

Denote by the symbols n - the number of components/factors that have to be measured and k - the number of force levels/values to which a particular component will be subjected. Then the total factorial plan which includes all possible combinations of factors and levels consists of k^n measurements. Thus for the six component sensor experimented on two levels we have 64 measurements.

According to the theory of experimentation (Schenck 1968) it is possible to create a fractional/reduced plan of measurements without the lost of much information. Such a reduced experiment for the sensor then consists of $k^{n-1} = 32$ measurements.

We denote by the index 1 the components of force having negative or zero values and by the index 2 the components with positive values. Then the fractional factorial plan 2^{6-1} can be described by the scheme in Fig. 3.

			mx1				mx2			
			my1		my2		my1		my2	
			mz1	mz2	mz1	mz2	mz1	mz2	mz1	mz2
fx1	fy1	fz1		X	X		X			X
		fz2	X			X		X	X	
	fy2	fz1	X			X		X	X	
		fz2		X	X		X			X
fx2	fy1	fz1	X			X		X	X	
		fz2		X	X		X			X
	fy2	fz1		X	X		X			X
		fz2	X			X		X	X	

Fig.3 Fractional factorial plan

Arrangement of experiments

In order to satisfy the above requirements on the calibration procedure we use the fact that the large majority of industrial robots nowadays exhibit sufficient accuracy to orient the end frame in any of three orthogonal directions (roll, pitch and yaw). We consider the sensor-mass configuration as is shown schematically in Fig. 4.

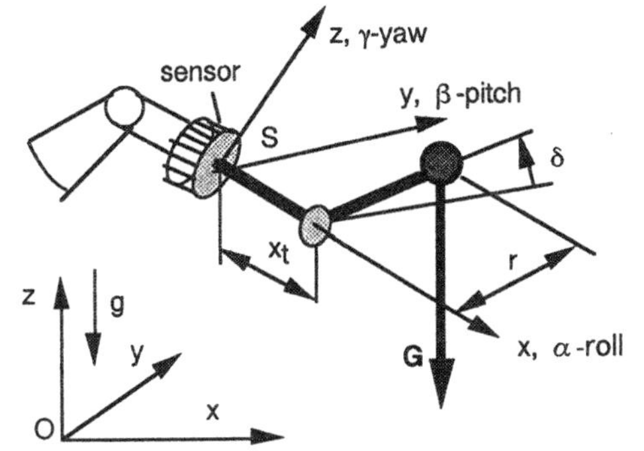

Fig.4 Configuration of the load

The load of the force sensor in the sensor reference system $S(x,y,z)$ depends on actual relative orientation of the robot end flange with respect to the gravity vector **G** in the base system $O(x_0,y_0,z_0)$.

We will change the orientation of the sensor and denote by the symbols: α-the angle of rotation about the x axis which corresponds to roll, β-the rotation about y axis which corresponds to pitch, and γ about the z axis which corresponds to yaw. We will also change the position of the mass by setting the angle δ.

The rotation of the base O relative to sensor reference system S, we can evaluate by sequentially pre-multiplying the three rotation matrices:

$$\mathbf{R}_{so} = \mathbf{R}(z,\gamma)\cdot\mathbf{R}(y,\beta)\cdot\mathbf{R}(x,\alpha) \tag{4}$$

where

$$\mathbf{R}(z,\gamma) = \begin{bmatrix} \cos\gamma & \sin\gamma & 0 \\ -\sin\gamma & \cos\gamma & 0 \\ 0 & 0 & 1 \end{bmatrix},\ \mathbf{R}(y,\beta) = \begin{bmatrix} \cos\beta & 0 & -\sin\beta \\ 0 & 1 & 0 \\ \sin\beta & 0 & \cos\beta \end{bmatrix}$$

$$\mathbf{R}(x,\alpha) = \begin{bmatrix} 1 & 0 & 0 \\ 0 & \cos\alpha & \sin\alpha \\ 0 & -\sin\alpha & \cos\alpha \end{bmatrix} \tag{5}$$

We note that these matrices represent the inverted transformations. See (Paul,1981).

The projection of the gravity force vector into sensor coordinates is then

$$\mathbf{f} = \begin{bmatrix} f_x \\ f_y \\ f_z \end{bmatrix} = \mathbf{R}_{so}\cdot\begin{bmatrix} 0 \\ 0 \\ -mg \end{bmatrix} \tag{6}$$

When we express the load of the sensor according to the mechanical configuration in Fig. 4, we have:

$$\begin{bmatrix} f_x \\ f_y \\ f_z \\ m_x \\ m_y \\ m_z \end{bmatrix} = \begin{bmatrix} \mathbf{R}_{so} & 0 \\ \mathbf{P}\cdot\mathbf{R}_{so} & 0 \end{bmatrix}\cdot\begin{bmatrix} 0 \\ 0 \\ -G \\ 0 \\ 0 \\ 0 \end{bmatrix} \tag{7}$$

where

$$\mathbf{P} = \begin{bmatrix} 0 & -r\cdot\sin\delta & r\cdot\cos\delta \\ r\cdot\sin\delta & 0 & -x_t \\ -r\cdot\cos\delta & x_t & 0 \end{bmatrix} \tag{8}$$

Performing the calibration experiment, according to the plan in Fig. 3, in order to load the sensor by a given combination of the force components we find a set of four angles α β γ δ.

Data evaluation

Experimental measurements give two sets of mutually corresponding vector pairs $\mathbf{f}_i$ and $\mathbf{u}_i$, $i = 1..32$. We estimate the parameters of the inverted model by performing the reconstruction of the **f** force-vectors according to the scheme in Fig. 5.

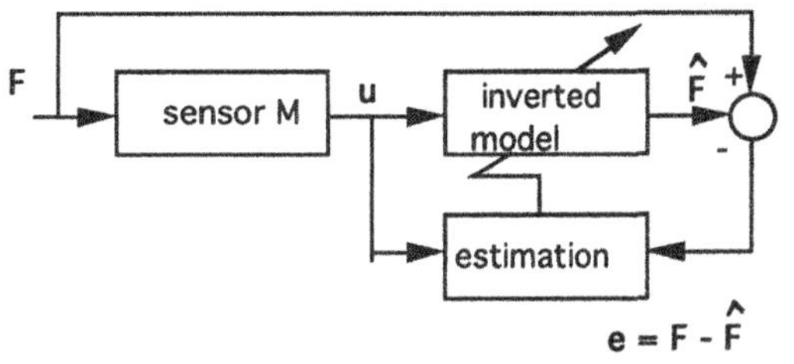

Fig. 5. The calibration scheme

For the estimation of the data processing model we minimize the cost function $\mathbf{E} = \sum \mathbf{e}_i^T\cdot\mathbf{N}\cdot\mathbf{e}_i$ for all independent force vectors in the n-component force space. Considering the linear representation we will select the multi-dimensional regression procedure.

3. THE SIX COMPONENT CAPACITIVE FORCE SENSOR

The proposed sensor consists mainly of three components (see Fig. 6):

- An elastic metallic tube 3. The elasticity of the tube is obtained by cutting out a number of *V*-shaped openings.
- A 6-D.O.F. capacitive small displacement transducer (6, 6', 6" and 7, 7', 7"), which consists of two printed circuit boards, mounted face to face with a small gap in between.

A processing board which is built inside the tube and directly plugged into the back face of the transducer. The circuitry for primary and post signal processing are fully digital. Two main components: a custom designed chip and a microcontroller are used for this purpose.

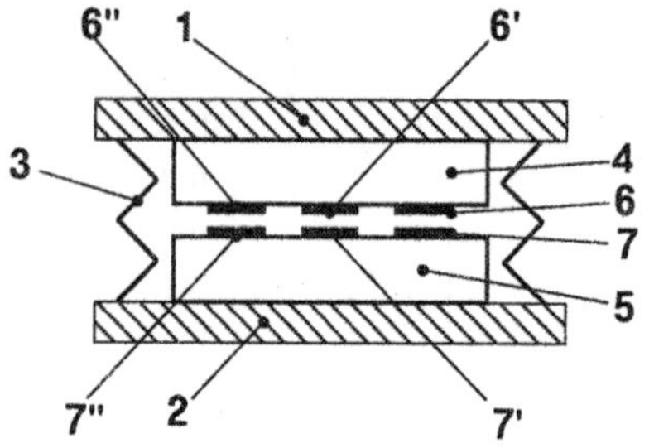

Fig. 6 Concept of the sensor

Two rigid flanges 1 and 2 are used for mechanical interfacing.

3.1 Sensing principle

Forces and moments applied to the sensor are evaluated through the measurement of displacements at given points inside the sensor body. The configuration of the electrodes, position and the direction of displacement measurements have been carefully chosen. Provided that the force sensor works within the elastic range of the material, there is a linear relation between the displacements and the forces/moments:

$$\mathbf{d} = \mathbf{D} \cdot \mathbf{f} \tag{9}$$

where **d** is a 6-dimensional displacement vector composed of the 6 measured displacements, **D** is an 6x6 displacement compliance matrix, and **f** is an 6-dimensional force/moment vector that contains the forces and the moments applied at the reference frame of the sensor.

The capacitive displacement transducer consists of two opposite electrode patterns as shown in Fig. 7 and Fig. 8 shows the displacements as they are measured by these electrodes.

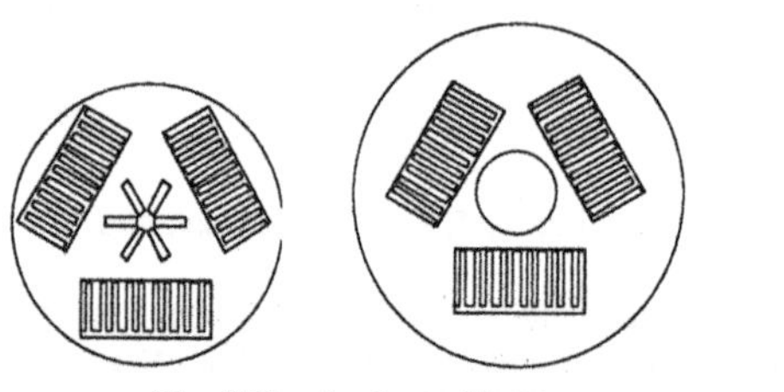

Fig. 7 Electrode patterns

Each of the three electrode patterns allows alternate measurement of either tangential displacements t_1, t_2 and t_3 or proximity displacements p_1, p_2 and p_3.

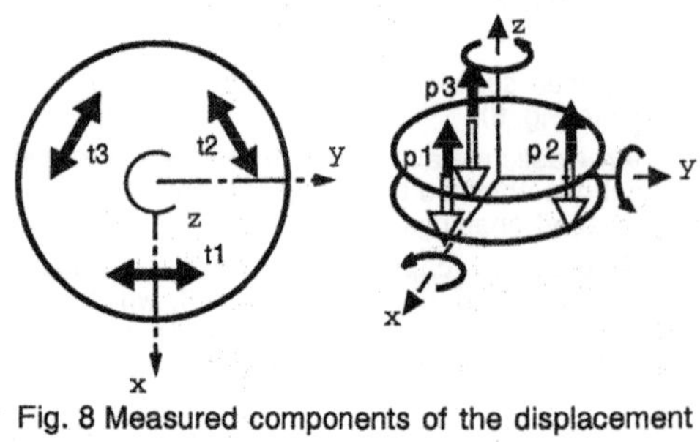

Fig. 8 Measured components of the displacement

The axial distances between electrodes and components of tangential displacements are given by (10,11).

$$\begin{bmatrix} p_1 \\ p_2 \\ p_3 \end{bmatrix} = a_p + b_p \cdot \begin{bmatrix} -1 & 0 & r \\ -1 & \frac{\sqrt{3}}{2}r & -\frac{r}{2} \\ -1 & -\frac{\sqrt{3}}{2}r & -\frac{r}{2} \end{bmatrix} \cdot \begin{bmatrix} d_z \\ \alpha_x \\ \alpha_y \end{bmatrix} \tag{10}$$

$$\begin{bmatrix} t_1 \\ t_2 \\ t_3 \end{bmatrix} = a_t + b_t \cdot \begin{bmatrix} 0 & 1 & r \\ \frac{\sqrt{3}}{2} & -\frac{1}{2} & r \\ -\frac{\sqrt{3}}{2} & -\frac{1}{2} & r \end{bmatrix} \cdot \begin{bmatrix} d_x \\ d_y \\ \alpha_z \end{bmatrix} \tag{11}$$

where the vector $\begin{bmatrix} d_x & d_y & d_z & \alpha_x & \alpha_y & \alpha_z \end{bmatrix}$ is the generalized displacement in a cartesian reference frame.

The basic configuration for measuring capacitances is shown in Fig. 9. Repeating this pattern we get the transducer more sensitive and appropriate to detect small displacements.

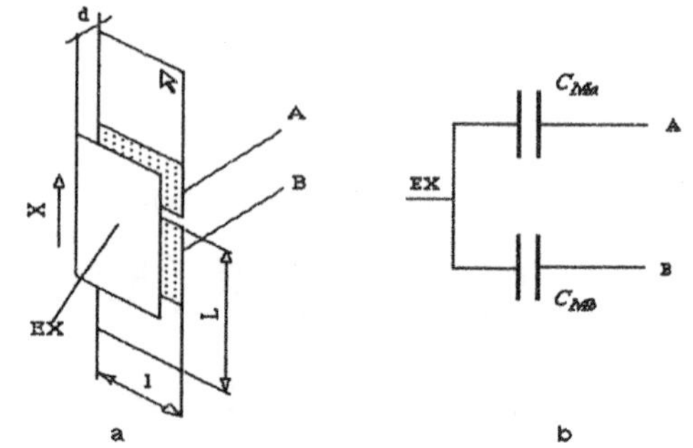

Fig. 9 Principle of the differential capacitive measurement

For the variable capacitances we can write the relations (12).

$$C_{Ma} = \varepsilon_0 \varepsilon_r \frac{lL}{2d}\left(1 \pm \frac{2\Delta x}{L}\right), \quad C_{Mb} = \varepsilon_0 \varepsilon_r \frac{lL}{2d}\left(1 \mp \frac{2\Delta x}{L}\right) \tag{12}$$

Using a digital switched capacitance bridge (see Fig. 10) we get a digital output signal, linearly proportional to displacement.

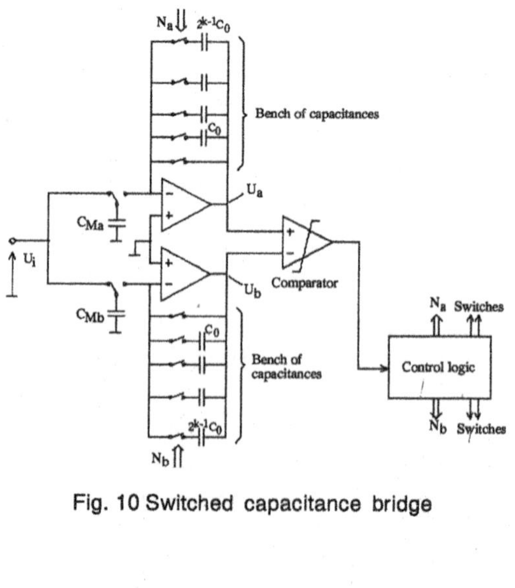

Fig. 10 Switched capacitance bridge

In order to ensure the balance during measurement there is a control logic which performs a successive approximation algorithm. The process ends when:

$$\frac{N_a}{N_b} = \frac{C_{Ma}}{C_{Mb}} \quad \text{with} \quad N_b = 2^k - N_a \tag{13}$$

We get a useful result from:

$$N_a = 2^{k-1}\left(1 \pm 2\frac{\Delta x}{L}\right) \tag{14}$$

After the first approximation, we assume that the dependance $N_a(\Delta x)$ is linear.

Relation (14) is valid for the tangential displacements measurement. For axial distance measurement, C_{Mb} is replaced by a fixed capacitance. The organization of the input circuitry enables one to obtain a number N_a proportional to the axial displacement.

3.2 Sensor construction

Fig. 11 shows one of the prototypes. The circular symmetry of the main mechanical pieces (including the transducers) results in a high precision sensor assembly. As can be seen on Fig. 11, there is a hollow cylinder form with several *V*-shaped openings made using laser cutting technology. Using this technology offers two main advantages: a) the laser beam is narrow so that a larger number of elastic bars could be obtained; and b) when machining no mechanical stress is generated.

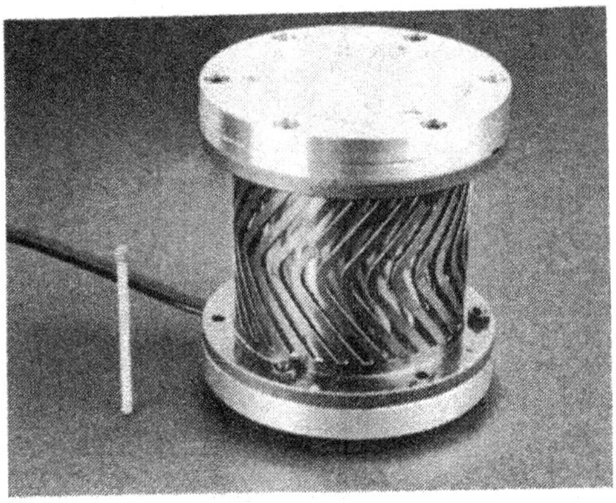

Fig. 11 Prototype of a capacitive six-axis force-torque sensor

3.3 Measurements

In order to evaluate the characteristics of the sensor two types of experiments (Fig.12) were made.

(i) Complex experimentation where the sensor is subjected, simultaneously, to all possible combinations of force components as given by the factorial plan.

(ii) The measurements for some chosen combinations that correspond to the intended application/assembly task. Then the sensor is calibrated for only the given application.

In both cases we considered the linear representation of the sensor function and estimate the calibration matrices using Moore-Penrose pseudo-inversion. As shown from the results of preliminary experimentation there is a small difference between the two approaches. Under complex experimentation (i) the sensor exhibits some small non-linearity effects if one changes the combination of the sensor load when in the full range of the force space. For a precise measurement, within the range of about 3-5 %, the non-linear model should be estimated and used. The sensor satisfies the performance criteria when it works within a given sector of the six-component force space (ii). Such a task oriented calibration better corresponds the main use of the sensor.

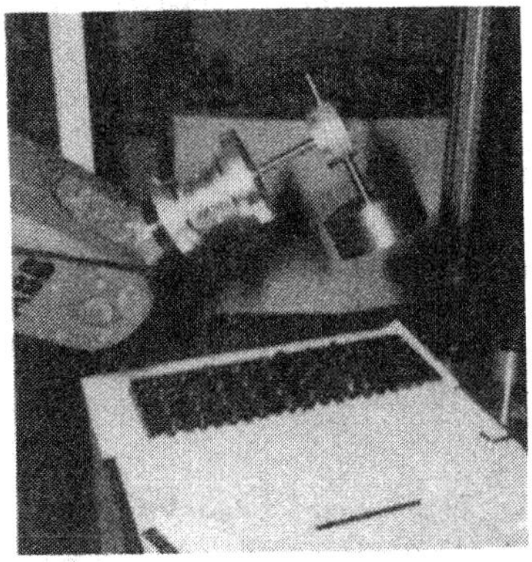

Fig.12 Arrangement of experimental measurement

4. CONCLUSION

Numerous studies, experimental measurements and practical results have confirmed the importance and real need of force/torque sensors for precise assembly tasks. But despite the potential ability of this concept it has not attained widespread practical use as was previously expected. The reasons for this are several:

- Sensors are still too expensive
- The integration of sensors and solving the sensor based accommodation presents numerous problems and imposes several constraints on performance that the majority of commercially available robots can hardly satisfy (fast time response, interfacing ability of HW, SW tools for programming/feedback control, etc...).

The development of the sensor at the Laboratory of Microengineering (Lausanne) showed that, using advanced fabrication and signal processing technology, it is possible to build a cost effective sensor with acceptable performance parameters. A principal application of this sensor is for sophisticated assembly as part of the configuration of the hand as shown Fig. 13.

Fig.13. End of arm configuration

Acknowledgements

This work was achieved entirely in the Laboratory of Microengineering (Institut de microtechnique de l'Ecole Polytechnique Fédérale de Lausanne). The authors would like to express their thanks to its Director, Professor Dr. C.W. Burckhardt and to F. Manzini for the use of his ABB robot.

References:

De Fazio, T.L., Selzer, D.S. and Whitney D.E.: (1984) The instrumented remote centre compliance. The Industrial Robot,Dec.1984, pp. 238-242

Ferrero, C.(1990): Multicomponent calibration systems to check force sensors. Proc. First.Int.Symp. on Measurement and Control in Robotics. June 20-22,1990, Houston, Texas (USA) pp. 5.1.1-5.1.10

Havlik, S. and Homola, D.: (1987) A force-moment sensing system for assembly robot. Proc. Artificial Intelligence and Information-Control Systems of Robots -87, Elsevier Sci.Publ. 1987, pp.235-240

Havlik, S.: (1990) The configuration and parameters of sensors for accommodation in robotic assembly. Proc. First Int. Symp. on Measurement and Control in Robotics. June 20-22, 1990, Huston, Texas (USA), pp.H1.1.1-H1.1.6

Hirzinger, G.: (1987) A new generation of robot sensors and their integration into sensor controlled robot system. Symp. Kommtech'87 May 1987, Essen

Paul, R.P.: (1981) Robot manipulators: mathematics, programming and control. MIT press.

Piller, G.:(1992) Aspects microtechniques de la construction d'un capteur d'effort multi-directionnel appliqué à la robotique. Doct. dissertation No 1054, EPFL-Lausanne,1992

Schenck, H.: (1968) Theory of engineering experimentation. New york,Mc Graw-Hill Book Comp.

Shimano, B.: (1978) The kinematic design and force control of computer controlled manipulators. Ph D dissertation, Stanford Univ.

Spescha, G.A.: (1970) Piezoelektrische Mehrkomponenten-Kraft und Momentmessung. (AT/M) Archiv fur technisches Messen, V131, July 1970(par1) pp.151-172,sept1970 (part2) pp-199-204

Stauffer, R.N.: (1987) Integrated sensors extend robotic system intelligence.Robotics Today, Aug 197, pp.11-16

Takada, R., and al.:(1988) An analysis of errors on 6-component force/moment calibration machines. Proc.Symp. IMECO 1988, pp.121-130

VanBrussel, H, and al.: (1986) Force sensing for advanced robot control. Robotics,2 (1986), pp.139-148

Vranish, J.M.,: (1982) Outstanding potential shown by magnetoelastic. Sensor Review, Oct.1982, No.4

Watson, P.C. and Drake,S.A.: (1978) Method and apparatus for six degree of freedom force sensing. Patent USA No.4,094,192, 1978

Wolffenbuttel, R.F., Mahmoud, K.M. and Regtien P.P.L.: (1990) Compliant capacitive wrist sensor for use in industrial robots. IEEE Trans. on Instrum. and Measurement, Vol.39, 6, Dec.1990, pp. 991-997

Yabuki, A.: (1990) Six-axis force/torque sensor for assembly robots. FUJITSU Sci. Tech. Journal, Vol.26,1, April 1990, pp. 41-47

A LOW COST PRODUCTION LINE INSPECTION SYSTEM

K. Warwick*, M.J. Usher* and R. Hagan**

**Department of Cybernetics, University of Reading, Berkshire, UK*
***SmithKline Beecham (Toiletries), Maidenhead, Berkshire, UK*

Abstract. The application of a computer based vision system in a high speed production line environment is described. Particular emphasis is placed on the need for a low cost solution which operates reliably in real-time, despite inherent implementation problems such as lighting. Comparisons are made between a number of different vision systems in terms of their performance and characteristics when trialled on the Brylcreem tub assembly, filling and packaging line at Maidenhead.

Keywords. Real-time computer systems; Cybernetics; Neural nets; Image processing; Visual inspection.

INTRODUCTION

Human labour is still used in a large number of situations for on-line production inspection, even though it has long been seen as a not very desirable state of affairs. The fact is however, that to replace a specific individual can be an extremely costly process, even when that individual's role appears to be simple and quite mundane. Recent years have, however, seen an increasing necessity for such an action, particularly because of high labour costs in the Western world.

One helpful recent factor in the use of machinery for on-line inspection, is the now more uniform approach to computer hardware and software design with stricter standards. This has also reduced the effect of purchasing out-of-date or outmoded computer systems, thereby increasing effective sales, reducing costs and hence allowing for many more implementations involving low-cost small systems. Also, because standard, fairly fast personal computers are now readily available, so hardware for visual data acquisition has also become more standardised, thus allowing for a greater use of such systems and hence improved efficiency.

THE PRODUCTION PROCESS

As an on-going exercise, SmithKline Beecham improve their production methods such that the manufacture and package of toiletry products, e.g. toothpaste, shampoo and hairspray, has changed from a fairly labour intensive exercise to one which is very machine dominant with a relatively small labour force acting largely as process supervisors.

Control of the production line is, almost exclusively, obtained through the use of Programmable Logic Controllers (PLCs), these being directly linked with simple detectors such as Photo-Electric Cells (PECs) for part-presence signalling, and finally a range of non-intrusive proximity switches. Brylcreem is produced on one of the most mechanised of the lines at Maidenhead, and is subsequently sold both in the UK and for export. The hair cream itself is batch produced and temporarily stored in vessels directly above the packaging line. On the line, white injection- moulded bases are filled with the fed-down cream, after which they are sealed, capped and collated into 12 tub packs. Although the entire packaging process is completely automatic, a number of process supervisors are employed firstly to detect, visually, defective components and secondly to remove components of poor quality, thereby ensuring both smooth production line operation and customer satisfaction with the product.

In practice three stations exist on the Brylcreem line as observation points, and a team of 3 operators/supervisors individually spend a period at each of the 3 sites, rotating every half hour. Despite this movement, however, mistakes are sometimes made, due to human error, no doubt partly boredom. Furthermore manual labour of this type is both fairly expensive and relatively scarce in the Maidenhead area, all of which makes the job a perfect candidate for mechanisation/ automation. It is of course also useful to release the staff so employed for jobs such as new product line development and short-run processes in which a more manual input is required.

But on-line labour is only one factor, and goes hand-in-hand with other long term considerations such as a move to Just-in-Time (JIT) manufacture, thereby removing the need for much expensive bulk storage and releasing an amount of precious space. Such a philosophy is particularly valid if varying, flexible production lines are to be employed with low production overheads. JIT manufacture generally allows for much less human intervention during the production process, which raises a number of issues regarding production quality monitoring and control.

INDUSTRIAL INSPECTION

Industrial inspection is possible by a number of means, and in order to investigate the potential use of such systems within SmithKline Beechams, the Brylcreem production line was selected as a pilot test bed, with a view to applying the techniques developed on to other packaging lines with the injection and blow-moulding facility following suit at a later date. Techniques such as laser-scanning and low-density photoelectric arrays were considered for the inspection process, however vision systems came out on top for a number of reasons.

Although it is not a hard rule, SmithKline Beecham wish to see a 2-year return on investment concerning machinery, which gave a ball park figure for project costing. One point on the packaging line was identified as being typical of problems within the plant and yet interesting in its own right. The specific task involved the detection of poorly positioned Brylcreem pot seals, which are made of highly reflective foil, and are placed on the pots and then glued down with a hot iron. When the foils are not well positioned the cap-tightening machine can jam up which will usually result in line down-time of several minutes and hence loss of production.

Specifications for the system were:

(a) identify problem components from a 2-D image;
(b) cost less than 20,000 fully installed;
(c) perform image capture and recognition in less than 0.2 secs;
(d) must be simple to operate and flexible;
(e) must run independently of a computer and be robust.

With this as a basis nearly 60 capable products were investigated, although the price range was from 2,000 to 100,000, which meant that several of the systems could be immediately discounted. The remainder were considered in terms of ability, cost, expandability, simplicity of operation, reliability, flexibility, support and availability.

These considerations easily reduced the list to a total of five products for further testing, and to this end a special purpose production test bed was designed in order to realise a comparative system.

TEST BED

The test bed consisted of an 'oval' conveyor system with two straight lengths, along with a Pulnix 460S CCD camera. A pair of 250W quartz halogen DC bulbs were used for lighting, these being high-stability and of long-life in order to provide a light source of high intensity and little or no variance. The set-up was such that the Brylcreem pots were viewed from the side with front lighting, also a through-beam proxistor was aligned with the camera as a triggering mechanism for frame-grabbing on component detection. Reflections proved to be a problem, partly due to the circularity of the pots and partly due to the reflectivity of the foil seal - such reflections were however reduced to a minimum by positioning the lighting.

As a benchmark standard, a neural network n-tuple hardware system (Aitken & co-workers, 1989; Bishop and co-workers, 1990) was employed firstly as a simple frame store for camera and lighting positioning, and secondly as a comparative system standard for testing other systems by. The neural network was used to select good and bad products without indicating any specific measurements or giving reasons for failure. Essentially the system was "taught" what an ideal Brylcreem pot looked like and as long as detected pot images came close to this the pots were passed as good, Warwick, Irwin and Hunt (1992). The network operates in parallel, enabling it to operate in real-time, realising image recognition at 50 Hz frame rate.

For testing further vision systems an incremental encoder digital tachometer was constructed with a PC interface, allowing for system testing at known conveyor speeds, both within and outside the real production line speeds. Each inspection system could then be thoroughly tested under similar conditions, something that would not be possible by simply regarding each supplier's sales demonstration.

SYSTEM SELECTION

Including the neural network standard, five inspection systems were extensively tested:

(a) Image Inspection;
(b) Allen-Bradley;
(c) T.A. Designs;
(d) Data Cell;
(e) Amerace.

The first two of these are based on matrix-scan data acquisition whereas the second two are based on line-scan, with the final one being the neural network system previously mentioned.

All of the systems were offered at a total installation price of less than 13,000, which was well within the total allowable.

As a primary consideration the Line-scan systems are not as easy to manipulate as matrix-scan, because only a cross-section of the image is considered at any instant, however the systems are usually more flexible as different camera resolutions are possible without the need to change hardware. This latter point is important for other than the Brylcreem line, when a higher degree of accuracy is necessary giving the line-scan devices a distinct advantage for high precision work.

The neural network system was set up in the first instance as an experimental arrangement in order to check on task viability. Good pots were shown in 20 orientations, all with the pots upright, and subsequently good and bad pots were fed in front of the system, and a confidence percentage was output. Typically with 90% confidence or above the pots were considered to be good, whereas with 85% confidence or below they were considered to be bad. The reason for such a high score for a bad pot was that the background, which formed part of the image, was unchanging and, with no windowing facility, was therefore a contributory factor. The high degree of repeatability and high success rate in bad pot detection did however indicate that the inspection task was both achievable and worthwhile.

The neural network system was very simple to use, and offered flexible input/output due to its STE bus structure. Further, the price is very competitive.

Of the other four systems, the T.A. Designs line-scan device, although it seemed suited to the application, was not made available to us despite numerous attempts. It was therefore considered immediately that they were unable to provide the required level of back-up support.

The other line scan device of Data Cell, contained hardware accelerator processing and a range of camera resolutions. Support for the system was good, although because the product was fairly new, a few 'teething' problems occurred.

Of the matrix scan devices, the Image Inspection Intelligent Camera is a PC-based upgrade, employing a frame-store and processing board all in a single casing. All of the Image Processing is carried out in software which naturally means a slower turn-around. However, despite this the coding was optimised sufficiently to meet the speed requirements of this problem. The supervision/instruction language is fairly simple to use, but a large point in favour of this system was that the support and interest from Image Inspection was far better than for any of the other systems.

The Allen-Bradley Vision Input Module (VIM), also a matrix scan device, was both simple and robust although it was designed to interface with Allen-Bradley PLCs which, unfortunately, were not compatible with those already used by SmithKline Beecham. The VIM could still be used in a stand-alone fashion, but this severely restricted input/output and data storage. Image processing was shared between hardware and software, allowing for recognition speeds around 30 Hz, i.e. faster than Image Inspection, but slower than the neural network. Finally, set-up time was fairly short.

PRODUCTION-LINE APPLICATION

Installation of each device on-line at the Maidenhead plant was shared between the supplier and Smith-Kline Beecham, with the lighting being down to the supplier and the component ejection apparatus down to SmithKline Beecham, Gelaky and co-workers (1991). The complete system had to be of a small enough size such that it could fit into existing production-line machine guards, although access had to be provided for maintenance and servicing. At this point it was decided that remote camera heads were necessary in order to allow for optimization of camera position.

In terms of performance the VIM system was better than the Image Inspection System with regard to speed in both operation and programming, the latter being partly due to a reduced instruction set. The Image Inspection device, however, was much more powerful, allowing for greater input/output capabilities, program saving and 'help' support. These features only become available on the further purchase of Allen-Bradley PLCs, thereby increasing the cost considerably.

Setting up the Image Inspection System took approximately two days in total, including fundamental experimentation, which included window arrangement for key image areas. Edge filters were applied for image enhancement, such that the resultant data could be checked against stored values. The input/output arrangement allowed for triggered frame-grabbing such that a pass/fail output could be given directly on the image collected. As basic add-ons, alarms can be signalled for such as high frequency or multi failures.

The Data Cell line scan system consisted of a 1024 pixel linear camera and an AT based half-size PC board, along with a Microsoft C library of 25 simple commands. The low level programming necessary increased the set-up time, but provided a great deal of flexibility. Programs were written to perform tests on the line-scanned data, these being thresholding and position testing against 'ideal' position data obtained from showing the system a perfect pot. A further program directed a gradient filter over the full grey-level image within a given window with an alarm condition being set at on/off dependent on the outcome of the comparison with the ideal data. General input/output facilities could be obtained by means of add-on PC cards.

COMMENTS

Input/output capability is high on the importance list for flexible production-line imaging systems, such that the systems should be able to interact with each other, with other sensory devices, with high level supervisors and also with PLCs. With this in mind, the Image Inspection device came out on top in terms of a basic work-horse system, both to solve the Brylcreem foil problem and also for potentially wider use, Gelaky and co-workers (1990). For some high accuracy requirements however, the Data Cell device could easily provide improved performance, whilst for high speed and simplicity of use, the Neural Network system was the easy winner. With matrix scan devices, data is converted from digital to analogue to different array size digital, such that positional confidence is less than with a line-scan linear array. Such jobs as moulding dimension checking are, therefore, perhaps best solved with a line- scan method.

Many potential commercial image processing systems exist, of a tried and tested nature, with simple to use instructions and of varying price. For the particular task of production-line inspection at high speed it was surprising therefore that the list of available systems was very quickly shortened to four or five potential products. In fact the major difficulties in setting up all of the systems are concerned with camera and lighting positioning, which is a very labour intensive task, and is one which is particularly application oriented.

Costs of around 10,000 for a full installation make such systems a viable, cost-effective alternative to human on-line inspection, and the consequent uptake is crucial to a modern, flexible production environment. The rapid move towards a more streamlined Just-In-Time manufacturing set-up will necessitate the requirement for more imaging systems of the type discussed, along with a variety of alternative fault detection and monitoring devices, such that greater efficiency, reliability and quality in the production process can be realised with only modest up-front investment.

ACKNOWLEDGEMENTS

The financial support of SmithKline Beecham, along with further input from the UK Science and Engineering Research Council and the Department of Trade and Industry is gratefully acknowledged. Further thanks go to Dr. R. Gelaky for his considerable involvement in the project.

REFERENCES

Aitken, D., Bishop, J.M, Pepper, S.A., and Mitchell, R.J. (1989). Pattern recognition in digital learning nets, Electronics Letters, 25, No. 11, 685-686.

Bishop, J.M., Minchinton, P.R., and Mitchell, R.J. (1990). Multi-class pattern association using neural networks, Proc. INNC'90, 926.

Gelaky, R., Warwick, K., and Usher, M.J. (1990). The implementation of a low-cost production line inspection system, Computer Aided Engineering Journal, 7, 180-184.

Gelaky, R., Warwick, K., and Usher, M.J. (1991). Industrial inspection system - the ease of use/ flexibility trade-off, IEE Colloq. on HCI Issues for the Factory, 5/1-5/3.

Warwick, K, Irwin, G.R., and Hunt, K.J. (Eds) (1992)., Neural networks for control and systems. Peter Peregrinus Ltd. pp.256.

INDUSTRIAL REFERENCES

Image Inspection Ltd., Unit 7, First Quarter, Blenheim Road, Epsom, KT19 9QN, UK.

Allen-Bradley Industrial Automation Products, Denbigh Road, Bletchley, Milton Keynes, MK1 1EP, UK.

TA Designs Ltd., Station Road Industrial Estate, Maiden Newton, Dorchester, Dorset, DT2 0EA, UK.

Data Cell Ltd., Data Cell House, 10 West End Road, Mortimer Common, Reading, Berkshire, RG7 3SY, UK.

Amerace Ltd., Votec House, Hambridge Lane, Newbury, Berkshire, RG14 5TN, UK.

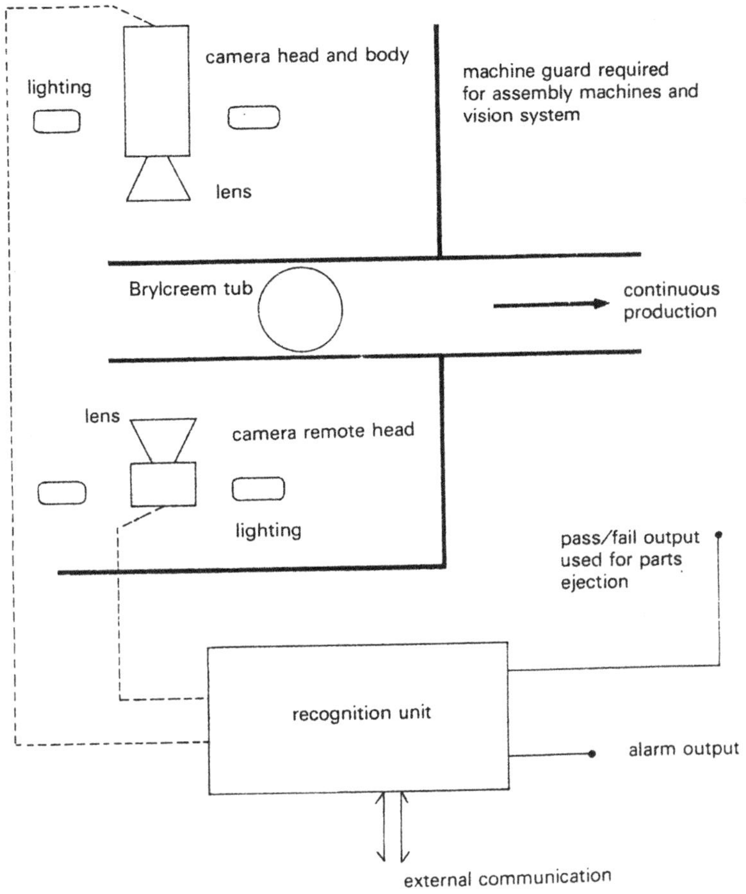

Figure 1. Top View of the Brylcreem production-line inspection system.

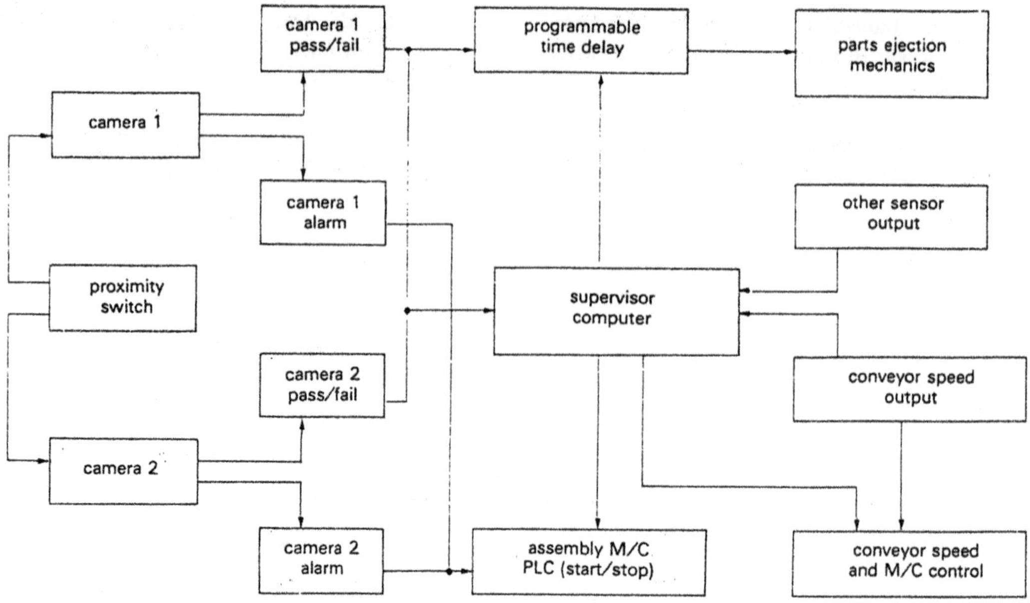

Figure 2. Production-line control.

PC IMPLEMENTATION OF OPTIMAL SAMPLED-DATA CONTROL FOR ROBOTIC MANIPULATORS

Z.G. Wei[1] and A. Johnson

Kramers Laboratory, Delft University of Technology, Prins Bernhardlaan 6, 2628 BW Delft, The Netherlands

Abstract. A stochastic optimal sampled-data tracking controller is implemented to control an industrial X-Y robot using a personal computer. The controller, incorporating a Kalman filter, takes account of the inter-sample behavior of the controlled system by using a continuous-time cost criterion, allowing the use of large sampling intervals. Most computations can be performed off-line. The experimental results show that the tracking accuracy and the sensitivity to the payload are good.

Keywords. Sampled-data system; real-time computer system; optimal control; robots; microcomputer-based control; stochastic control.

INTRODUCTION

The optimal sampled-data regulator and tracker for stochastic linear systems, which have been studied by Van Willigenburg (1991) in the context of robot control, do not make the usual approximation to the continuous-time cost criterion by a discrete-time formulation. Thus the inter-sample behavior of the system is explicitly considered. This allows the use of large sampling intervals and consequently a relatively simple, slow computer to realize real-time control. Even for robots with a large number of degrees of freedom, during the sampling interval many control calculations may be performed. The stochastic optimal sampled-data regulator and tracker constitute two solutions of the sampled-data LQG problem (Halyo and Caglayan, 1976), with a Kalman filter incorporated in the controller to filter away the state and measurement noise. These results permit the design and computation of a controller for stochastic linear systems that have to track reference trajectories.

The industrial X-Y robot we consider is modeled with both viscous and coulomb friction which is usually neglected (Asada and Slotine, 1986). This kind of robot is industrially used for various tasks including cutting, welding, spraying and moving loads. It is typically controlled using very fast sampling rates (Lin, 1987; Vukobratovic and Stokic, 1982) and therefore expensive, high performance computers. A minicomputer was used for real-time control of the robot with a deterministic optimal sampled-data tracking controller (Van Willigenburg, 1991). In this paper we present a procedure to design a stochastic optimal sampled-data tracking controller for such robots and show the experimental results from the implementation of such controller on a personal computer. From a practical point of view, this PC-based control system is attractive because of the low cost and the popularity of the personal computers.

The paper, first, presents the solutions of the sampled-data LQG problem for regulator and tracking problems and the robot dynamic model from the identification experiments. The optimal sampled-data tracker computation for the X-Y robot follows. Then the Kalman optimal filter design and a stochastic perturbation controller design are introduced. Finally, the PC implementation, the experimental results and conclusions are given.

[1] on leave from Wuhan University of Water Transportation Engineering, PRC

STOCHASTIC OPTIMAL SAMPLED-DATA CONTROLLER

Consider the stochastic continuous-time linear system in state-space form

$$\dot{x}(t)=Ax(t)+Bu(t)+w(t) \tag{1a}$$

where $x \in R^n$, $u \in R^m$, $A \in R^{n\times n}$, $B \in R^{n\times m}$. $\{w(t)\}$ is zero-mean, white, Gaussian process noise and

$$E\{x(0)\}=\bar{x}_0, \quad E\{x(0)x^T(0)\}=P_0, \quad P_0 \geq 0. \tag{1b}$$

where $\{w(t)\}$ is uncorrelated with $x(0)$. E denotes the expectation operation and T the matrix transposition.

The system is controlled by a digital computer, therefore measurements are taken at the sampling instants, i.e.

$$z(t_k)=Cx(t_k)+v(t_k) \tag{1c}$$

where t_k represents the sampling instants and $\{v(t)\}$ is a discrete-time zero-mean, white, Gaussian measurement noise uncorrelated with $\{w(t)\}$ or $x(0)$.

The control is piecewise constant, i.e.

$$u(t)=u(t_k), \qquad t \in [t_k, t_{k+1}), \tag{2}$$

The equivalent discrete-time system (EDS) of Eqs. (1) and (2) is (Johnson, 1985)

$$x_{k+1} = \Phi x_k + \Gamma u_k + w_k \tag{3a}$$

$$z_k = C x_k + v_k \tag{3b}$$

where x_k is the value of $x(t)$ at time t_k.

The finite-time stochastic optimal sampled-data regulator problem is to minimize

$$J=E\left\{\int_{t_0}^{t_f}\left[x^T(t)Qx(t)+u^T(t)Ru(t)\right]dt+x^T(t_f)Hx(t_f)\right\} \tag{4}$$

subject to Eqs.(1) and (2). Here $Q\geq 0$, $H\geq 0$, and $R\geq 0$. The sampling intervals are assumed to be equal, i.e.

$$t_{k+1}-t_k=T_s \tag{5a}$$

and the end time is assumed to coincide with the sampling time so that

$$t_f = t_N \tag{5b}$$

where N is a positive integer.

The solution to the problem is as follows (Halyo and Caglayan, 1976):

$$u_k=-(\tilde{K}_k+\tilde{R}^{-1}\tilde{M}^T)\hat{x}_{k/k-1}, \quad k=0,1,2,\ldots,N-1, \tag{6a}$$

$$\tilde{K}_k=(\Gamma^T S_{k+1}\Gamma+\tilde{R})^{-1}\Gamma^T S_{k+1}\tilde{\Phi}, \tag{6b}$$

$$S_k=(\tilde{\Phi}-\Gamma\tilde{K}_k)^T S_{k+1}(\tilde{\Phi}-\Gamma\tilde{K}_k)+\tilde{K}_k^T\tilde{R}\tilde{K}_k+G, \quad S_N=H, \tag{6c}$$

$$\tilde{\Phi}=\Phi-\Gamma\tilde{R}^{-1}\tilde{M}^T, \tag{6d}$$

$$G=\tilde{Q}-\tilde{M}\tilde{R}^{-1}\tilde{M}^T, \tag{6e}$$

where

$$\tilde{Q}=\int_0^{T_s}\Phi^T Q\Phi dt, \tag{7a}$$

$$\tilde{M}=\int_0^{T_s}\Phi^T Q\Gamma dt, \tag{7b}$$

$$\tilde{R}=\int_0^{T_s}(R+\Gamma^T Q\Gamma)dt, \tag{7c}$$

where $\hat{x}_{k/k-1}$ is generated by a discrete-time Kalman one-step predictor (Anderson and Moore, 1979) based on Eq. (3).

For the given system Eqs.(1), (2) and a given reference trajectory,

$$x_r(t), \quad t_0 \leq t \leq t_f \tag{8}$$

the finite-time stochastic optimal sampled-data tracking problem is to minimize

$$J=E\left\{\int_{t_0}^{t_f}\left\{\left[x(t)-x_r(t)\right]^T Q\left[x(t)-x_r(t)\right]+u^T(t)Ru(t)\right\}dt + \left[x(t_f)-x_r(t_f)\right]^T H\left[x(t_f)-x_r(t_f)\right]\right\} \tag{9}$$

where $Q \geq 0$, $H \geq 0$ and $R \geq 0$.

The solution to this problem is as follows (Van Willigenburg, 1991):

$$u_k=-(\tilde{K}_k+\tilde{R}^{-1}\tilde{M}^T)\hat{x}_{k/k-1}+K_k^1 V_{k+1}+K_k^2 T_k^T, \tag{10a}$$

$$\tilde{K}_k=(\Gamma^T S_{k+1}\Gamma+\tilde{R})^{-1}\Gamma^T S_{k+1}\tilde{\Phi}, \tag{10b}$$

$$K_k^1=(\tilde{R}+\Gamma^T S_{k+1}\Gamma)^{-1}\Gamma^T, \tag{10c}$$

$$K_k^2=(\tilde{R}+\Gamma^T S_{k+1}\Gamma)^{-1}, \tag{10d}$$

$$S_k=(\tilde{\Phi}-\Gamma\tilde{K}_k)^T S_{k+1}(\tilde{\Phi}-\Gamma\tilde{K}_k)+\tilde{K}_k^T\tilde{R}\tilde{K}_k+G, \quad S_N=H, \tag{10e}$$

$$V_k=(\tilde{\Phi}-\Gamma\tilde{K}_k)^T V_{k+1}-\tilde{K}_k^T T_k^T+\tilde{L}_k^T, \quad V_N=HX_r(t_f), \tag{10f}$$

$$\tilde{\Phi}=\Phi-\Gamma\tilde{R}^{-1}\tilde{M}^T, \tag{10g}$$

$$G=\tilde{Q}-\tilde{M}\tilde{R}^{-1}\tilde{M}^T, \tag{10h}$$

$$\tilde{L}_k=L_k-T_k\tilde{R}^{-1}\tilde{M}^T, \tag{10i}$$

where Eq. (7) holds and

$$L_k=\int_{t_k}^{t_{k+1}}x_r^T(t)Q\Phi dt \tag{11a}$$

$$T_k=\int_{t_k}^{t_{k+1}}x_r^T(t)Q\Gamma dt \tag{11b}$$

DYNAMIC MODEL OF THE X-Y ROBOT

The X-Y robot has two rigid links which move in perpendicular directions in the cartesian coordinate frame, so that the two links may be considered separately. The links of the X-Y robot are driven by current controlled DC servo motors and we assume that the transmission from the motors to the links is rigid and the motor current is proportional to the force applied to the links so that the motor currents are chosen as control variables. Furthermore, we assume each link suffers from both viscous and coulomb friction. With the above assumptions the dynamic model of the robot is (Van Willigenburg, 1991):

$$\begin{bmatrix}\dot{X}_p\\ \dot{Y}_p\\ \ddot{X}_p\\ \ddot{Y}_p\end{bmatrix}=\begin{bmatrix}0&0&1&0\\0&0&0&1\\0&0&-V_x&0\\0&0&0&-V_y\end{bmatrix}\begin{bmatrix}X_p\\ Y_p\\ \dot{X}_p\\ \dot{Y}_p\end{bmatrix}+\begin{bmatrix}0&0\\0&0\\b_x&0\\0&b_y\end{bmatrix}\begin{bmatrix}U_x\\ U_y\end{bmatrix}-\begin{bmatrix}0\\0\\C_x\text{sign}(\dot{X}_p)\\C_y\text{sign}(\dot{Y}_p)\end{bmatrix} \tag{12a}$$

$$\begin{bmatrix}X_p(0)\\ Y_p(0)\\ \dot{X}_p(0)\\ \dot{Y}_p(0)\end{bmatrix}=\begin{bmatrix}X_{pr}(0)\\ Y_{pr}(0)\\ \dot{X}_{pr}(0)\\ \dot{Y}_{pr}(0)\end{bmatrix} \tag{12b}$$

where X_p, Y_p are translations of links with respect to reference positions, V_x, V_y are viscous friction coefficients, C_x, C_y are coulomb friction coefficients, U_x, U_y are control variables, b_x, b_y are sensitivity coefficients of the control variables which depend on the mass of the link and the payload. $X_{pr}(t)$, $Y_{pr}(t)$ are reference trajectories and generally $\dot{X}_{pr}(0)$, $\dot{Y}_{pr}(0)$ equal zero.

For the X-Y robot, friction plays an important role in the robot dynamics. From identification experiments performed on a particular X-Y robot, we have:

$V_x=6.44$, $V_y=2.42$;

$b_x=83.12$, $b_y=64.29$;

$C_x=35.87$, $C_y=56.33$

We measure the speed and position of each link at the sampling instant. Without considering the measurement noise, complete state information of Eq. (12) is known at sampling instants. The output equation is

$$z(t)=C[X_p(t)\ Y_p(t)\ \dot{X}_p(t)\ \dot{Y}_p(t)]^T \tag{13}$$

where matrix C is a 4x4 identity matrix. Note that the robot dynamic model Eqs. (12) and (13) is not of the form Eq. (1), although shortly we will show how relevant the stochastic state space formulation is.

OPTIMAL SAMPLED-DATA TRACKER COMPUTATION

The deterministic model of a stochastic linear system can be used to determine the reference state trajectory (Athans, 1971). Therefore we use the deterministic model of the robot to compute the optimal sampled-data tracker which acts as a feedforward control in open-loop and derive the reference state trajectory.

The X-Y robot model Eq. (12) can be written as

$$\begin{bmatrix}\dot{X}p\\ \dot{Y}p\\ \ddot{X}p\\ \ddot{Y}p\end{bmatrix}=\begin{bmatrix}0&0&1&0\\0&0&0&0\\0&0&-Vx&0\\0&0&0&-Vy\end{bmatrix}\begin{bmatrix}Xp\\Yp\\ \dot{X}p\\ \dot{Y}p\end{bmatrix}+\begin{bmatrix}0&0\\0&0\\bx&0\\0&by\end{bmatrix}\begin{bmatrix}U'_x\\U'_y\end{bmatrix} \tag{14a}$$

where

$$U'_x=U_x-\mathrm{sign}(\dot{X}_p)C_x/b_x \tag{14b}$$

$$U'_y=U_y-\mathrm{sign}(\dot{Y}_p)C_y/b_y \tag{14c}$$

Equation (14a) is a linear state-space system, the deterministic version of Eq. (1a). If the links continue to move in the same direction during the sampling interval, the non-linear term in Eq. (12) which involves the coulomb friction could be compensated for by constant values of the control variables. Otherwise, the friction will not be compensated for. This compensation can be realized with a piecewise constant control.

We introduce the reference state trajectory :

$$x_r(t)=[x_{pr}(t)\ y_{pr}(t)\ \dot{x}_{pr}(t)\ \dot{y}_{pr}(t)]^T,\quad t_0\le t\le t_f \tag{15}$$

From Eqs. (14b), (14c), we have

$$U_x(t)=U'_x(t)+\mathrm{sign}(\dot{X}_{pr}(t))C_x/b_x \tag{16a}$$

$$U_y(t)=U'_y(t)+\mathrm{sign}(\dot{Y}_{pr}(t))C_y/b_y \tag{16b}$$

Given the compensation of the coulomb friction, in the sequel we can consider the linear system Eq. (14) with U_x, U_y as the control variables and compute the optimal sampled-data tracker Eq. (10) off-line from EDS of Eq. (14) once the design parameters, Q, R and H, are chosen for any given reference trajectory Eq. (15) and the initial state of Eq. (3) is given.

To equally weight deviations at all time from the reference trajectory, we choose Q to be time-invariant. To ensure all deviations from the reference state trajectory are a minimum, we choose

$$R = 0 \tag{17}$$

Any other choice of R would result in a compromise between the magnitude of the control and deviations from the reference state trajectory. Note that although continuous-time optimal trackers do not allow Eq. (17) because this may result in singular problems (Lewis, 1986a), the optimal sampled-data tracking problem does not give singularity problems (Van Willigenburg, 1991).

In the case of the X-Y robot movement, we are only interested in the deviations of the prescribed link positions so we choose Q as follows:

$$Q=\begin{bmatrix}q_1&0&0&0\\0&q_1&0&0\\0&0&0&0\\0&0&0&0\end{bmatrix} \tag{18}$$

where q_1 is a constant.

At the final time, we want the robot to stand still, so we choose

$$H=\begin{bmatrix}h_1&0&0&0\\0&h_1&0&0\\0&0&h_2&0\\0&0&0&h_2\end{bmatrix} \tag{19}$$

where h_1 and h_2 are constant.

To achieve better tracking accuracy, the values of Q and H should depend upon the sampling time and reference state trajectory. The choice of Q and H should be a compromise between reaching the final state and the tracking error.

For the two links of the X-Y robot, Eqs. (14),..., (19) hold and completely determine the optimal sampled-data tracking problem.

The reference state trajectory must be time scaled such that the final time is a multiple larger than the original final time.

KALMAN FILTER ALGORITHM DESIGN

The X-Y robot dynamics Eq. (14) is a linear system. Considering the modeling errors and inherent noise, we assume that it is disturbed by additive white noise. Although complete state information is available at the sampling time, the information may be corrupted by measurement errors and uncertainty which are assumed to be additive white measurement noise. The optimal linear minimum variance estimator is used to filter the noise from the measured state information. For the system Eq. (3), the discrete-time version of the Kalman filter is as follows (Anderson and Moore, 1979; Lewis, 1986b):

$$\hat{x}_{k+1/k+1}=(I-K_{k+1}C)\Phi\hat{x}_{k/k}+(I-K_{k+1}C)\Gamma u_k+K_{k+1}z_{k+1} \tag{20a}$$

$$K_{k+1}=P_{k+1/k}C^T(CP_{k+1/k}C^T+V)^{-1} \tag{20b}$$

$$P_{k+1/k+1}=(I-K_{k+1}C)P_{k+1/k}(I-K_{k+1}C)^T+K_{k+1}VK^T_{k+1} \tag{20c}$$

$$P_{k+1/k}=\Phi P_{k/k}\Phi^T+W \tag{20d}$$

with $\hat{x}_{0/0}=\bar{x}_0$, $P_{0/0}=P_0$ and I denotes identity matrix.

Providing V > 0 and conditions concerning stabilizability and detectability are met, there exists a stable optimal Kalman Filter (Johnson, 1985). For the robot system, after at most five sampling periods, both the error covariance and the Kalman filter gain converge to a steady-state value. This allows us to use the steady-state Kalman filter gain matrix to save memory. It also allows us to choose P_0 and $\bar{x}_0$ arbitrarily (Lewis, 1986b).

The process noise covariance W and the measurement noise covariance V are treated as design parameters which can be adjusted to produce an "acceptable" response from the Kalman filter. By varying the relative magnitudes of W and V, the Kalman filter is designed to provide what are considered proper weightings for the process model and the actual measurements in generating the filtered state information and to give the best tracking accuracy.

To model the uncertainty in the system Eq. (3), we choose W to be a diagonal matrix $w_x I$ for both the X and Y links.

To decrease the sensitivity to the modeling error, W is chosen larger than the possible true value according to the sampling time and the trajectory followed by the robot.

To model the measurement uncertainty, we choose $V=\begin{bmatrix}0.01&0\\0&9\end{bmatrix}$ for both links.

The Kalman filter gain terms are computed off-line.

STOCHASTIC OPTIMAL PERTURBATION CONTROLLER

In tracking, the optimal sampled-data tracker, which is open-loop, controls the robot moving along the given trajectory. A stochastic optimal perturbation controller, which acts in feedback, is needed to reduce the position and speed errors of the robot about the trajectory to zero. The design of the stochastic perturbation controller constitutes a stochastic optimal sampled-data regulator problem. In this case the choice of the cost matrices of the regulator in Eq. (4) determines the feedback gain. They can be chosen independently from those of the tracking problem.

The perturbation variables are:

$$\Delta x(t)=x(t)-x_d(t) \ , \qquad 0\le t<t_f \ , \tag{21a}$$

$$\Delta u(t)=u(t)-u_d(t) \ , \tag{21b}$$

where $u_d(t)$ is computed, given Eq. (15), from Eq. (10) and presented to the system in an open-loop fashion. $x_d(t)$ represents the state response of the system Eq. (14) to the optimal control $u_d(t)$. $\Delta u(t)$ is still constrained to be piecewise constant during the sampling interval. Hence, $u_d(t)$ is a feedforward component and $\Delta u(t)$ is a feedback component. The objective is to control $\Delta x(t)$ to zero by $\Delta u(t)$. This is a typical digital regulator problem.

The perturbation dynamics of the robot motion should be a linear equation in the vicinity of a desired trajectory. Therefore, according to Eq. (14), the stochastic linear dynamics of the perturbation variables is given as

$$\begin{bmatrix}\Delta\dot{X}_p\\ \Delta\dot{Y}_p\\ \Delta\ddot{X}_p\\ \Delta\ddot{Y}_p\end{bmatrix}=\begin{bmatrix}0&0&1&0\\0&0&0&1\\0&0&-V_x&0\\0&0&0&-V_y\end{bmatrix}\begin{bmatrix}\Delta X_p\\ \Delta Y_p\\ \Delta\dot{X}_p\\ \Delta\dot{Y}_p\end{bmatrix}+\begin{bmatrix}0&0\\0&0\\b_x&0\\0&b_y\end{bmatrix}\begin{bmatrix}\Delta U_x\\ \Delta U_y\end{bmatrix}+w(t) \tag{22}$$

According to Eq.(7) the feedback gain of the regulator is:

$$F_k = \tilde{K}_k + \tilde{R}^{-1}\tilde{M}^T \ , \qquad k=0,1,2,\ldots,N-1 \tag{23}$$

Therefore,

$$\Delta u(t_k) = -F_k\Delta\hat{x}_{k/k}, \qquad t_k\le t<t_{k+1} \ , \quad k=0,1,\ldots,N-1 \tag{24}$$

where $\Delta\hat{x}_{k/k} = \hat{x}_{k/k} - x_d(t_k)$, $\hat{x}_{k/k}$ is generated by the Kalman filter. Since the additive noise is assumed to be Gaussian, Eq. (24) constitutes a solution of the optimal sampled-data LQG regulator, given Eqs. (3), (20) and (22).

From Eq. (21), the stochastic optimal sampled-data tracking controller which consists of the optimal sampled-data tracker and the stochastic perturbation controller takes on the following form:

$$u(t)=u_d(t_k)-F_k\Delta\hat{x}_{k/k}, \qquad t_k\le t<t_{k+1} \ , \quad k=0,1,\ldots,N-1 \tag{25}$$

The stochastic perturbation controller counteracts not only the initial disturbance of the system states, but also the additive state and measurement noise during the tracking.

The cost matrices of the regulator problem are also chosen according to the sampling time and trajectory. Their basic forms are

$$Q=\begin{bmatrix}q_1&0&0&0\\0&q_1&0&0\\0&0&0&0\\0&0&0&0\end{bmatrix} \quad R=\begin{bmatrix}r_1&0\\0&r_2\end{bmatrix} \quad H=\begin{bmatrix}h_1&0&0&0\\0&h_1&0&0\\0&0&h_2&0\\0&0&0&h_2\end{bmatrix} \tag{26}$$

where q_1, r_1, r_2, h_1 and h_2 are constant. Q and H can be the same as in Eqs. (18), (19). When the sampling time is larger, r_1 and r_2 should be chosen smaller to guarantee large feedback gains to make the tracking error smaller.

Summarizing, the stochastic optimal sampled-data regulator, the optimal sampled-data tracker and the Kalman filter are implemented in the X-Y robot control. All calculations except those connected with the system states are performed off-line using PC-MATLAB.

PC IMPLEMENTATION OF THE OPTIMAL SAMPLED-DATA CONTROLLER

Because it is possible to use a large sampling time to control the robot, the optimal sampled-data controller described in the previous sections may be implemented in a PC. The personal computer used is compatible with an IBM PC-AT with a 80286 CPU, a 80287 co-processor and a 10 MHz clock, with only a standard PC add-on I/O card needed additionally. Most of time the PC is occupied with input/output, but we still have enough time to carry out good control.

Two 8-bit input channels of the I/O card are used to input 12-bit position and 10-bit speed data from the robot interface and two 8-bit output channels to output 12-bit control values and 4 bits control pulses to the interface. However, the I/O operation speed must match the speed of A/D and D/A conversion. The pulses for starting the conversion, reading data, addressing and resetting data channels should be long enough, since otherwise the computer will not get the correct data. Because every time only one byte can be input or output, any data more than 8 bits long has to be input or output separately in two bytes. Using the data latch for outputting data, it is possible to send 12 bit data to a D/A converter separately and start the conversion at the same time. After reading in the data, the two bytes should be combined into one word data.

Both monitoring and hardware interrupt methods are used to record the arrival of the sample instant. In the interrupt method, the output bit of the hardware clock offered by the on-board clock of the I/O card, is connected to the reserved bit of the programmable interrupt controller (PIC) of the PC. When the sampling instant is reached, the computer executes one control. If the floating point calculation is performed in the interrupt subroutine, the PIC's bit connected to the clock must be in a lower priority than that of the co-processor.

Unlike many other real-time applications, the control program is completely written in C. This makes programming and operation easy. Because the I/O functions of C operate on the low byte of a 16-bit integer variable for one byte operation, it should be shifted right 8 bits before outputting a high byte and left 8 bits after inputting and then combining it with the low byte.

Figure 1 displays the X-Y robot tracking a prescribed trajectory controlled by the control Eq. (25) with a sampling time of 30ms and the final time of 4s. The X-Y robot tracking with the deterministic optimal sampled-data tracking controller is shown in Fig. 2. We observe that the tracking accuracy in Fig. 1 is better than in Fig. 2. The sampling time can vary between 10 and 100ms and the tracking accuracy is not seriously affected (Wei and Johnson, 1991).

Figure 3 gives the robot tracking the prescribed trajectory controlled by a PD controller instead of an optimal perturbation controller. The control law is

$$u(t)=u_d+(K_p+K_d)e_{pk}-K_v e_{p,k-1}, \qquad t_k\le t<t_{k+1}, \tag{27}$$

where $e_{pk} = x_{pk} - x_{pr,k}$. K_d, K_p are constant derivative and proportional coefficients, respectively.

Table 1 gives the performances of the PD, the deterministic optimal and the stochastic optimal tracking controller. It is obvious that the tracking accuracy is improved by applying the stochastic optimal sampled-data regulator. The sensitivity of the controller to the payload is good.

CONCLUSIONS

The PC implementation of optimal sampled-data controller for a X-Y robot to track a given path in real time is presented. The experimental results show that the tracking performance with stochastic control is improved compared with the deterministic control. Moreover the performance is better with the optimal sampled-data controller then that with a PD controller. Because a great number of calculations are performed off-line, which reduces the burden of on-line computations on the computer, and large sampling intervals are used, the controller can also be realized on a personal computer to control a robot with more than two links.

REFERENCES

Anderson, B.D.O. and J.B. Moore (1979). *Optimal Filtering*. Prentice-Hall, Englewood Cliffs, New Jersey.

Asada, H. and J.-J.E. Slotine (1986). *Robot Analysis and Control*. John Wiley, New York.

Athans, M.(1971). The role and use of the stochastic linear-quadratic-gaussian problem in control system design. *IEEE Trans. Automat. Contr.*. 16, 529-552.

Halyo, N. and A.K. Caglayan (1976). A separation theorem for the stochastic sampled-data LQG problem. *Int. J. Control*. 23, 237-244.

Johnson, A. (1985). *Process Dynamics, Estimation and Control*. Peter Peregrinus, London.

Lewis, F.L. (1986a). *Optimal Control*. John Wiley, New York.

Lewis, F.L. (1986b). *Optimal Estimation*. John Wiley, New York.

Lin, S.K. (1987) Microprocessor implementation of the inverse dynamic system for industrial robot control. *Preprints 10th World Congress on Automat. Contr.*. 4, 332-339, Munich.

Van Willigenburg, L.G. (1991). *Digital Optimal Control of Rigid Manipulators*. Ph.D. dissertation, Delft University Press, The Netherlands.

Vukobratovic, M. And D. Stokic (1982). *Control of Manipulation Robot: Theory and Application*. Springer Verlag, Berlin.

Wei, Z.G. and A. Johnson (1991). Optimal sampled-data control of a cartesian robot using a personal computer. *PDR technical report no.73*. Delft University of Technology, The Netherlands.

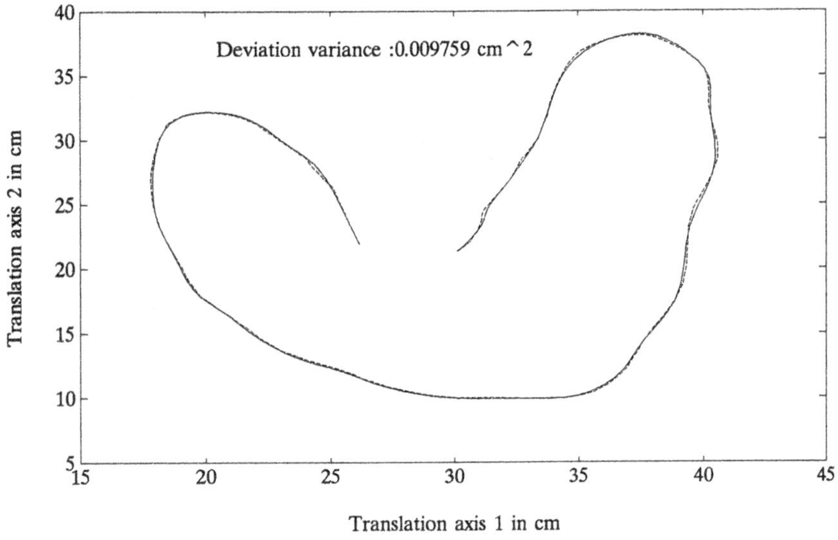

Fig. 1. Prescribed and actual (broken line) motion with stochastic optimal controller

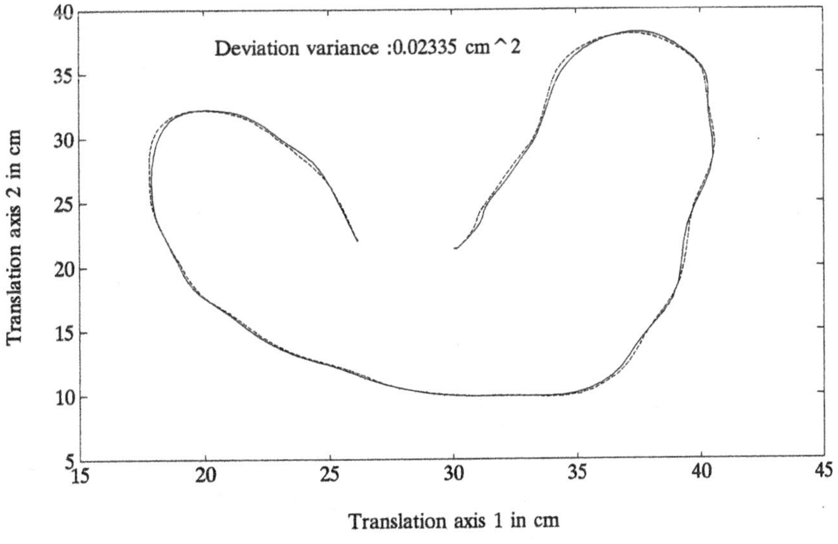

Fig. 2. Prescribed and actual (broken line) motion with deterministic optimal controller

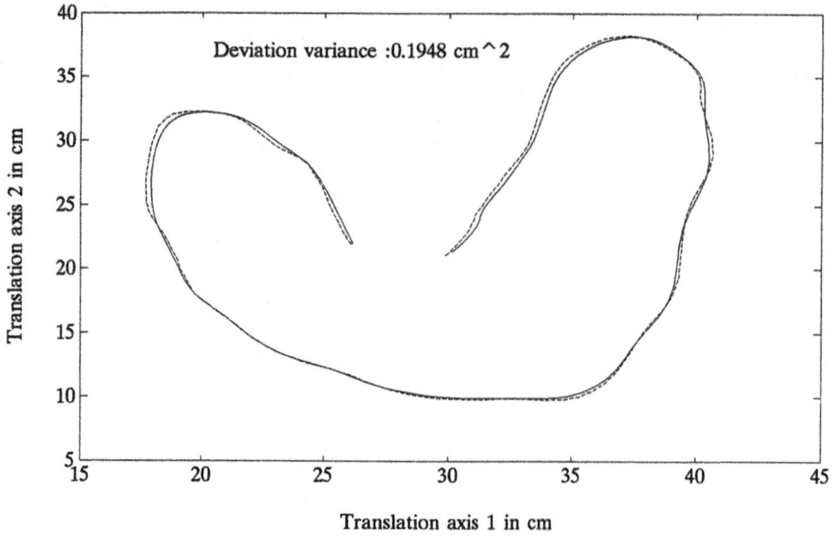

Fig. 3. Prescribed and actual (broken line) motion with PD controller

TABLE 1 Comparison of the Tracking Performances

		PD controller		Det.Opt.Controller		Stoch.Opt.Controller	
Load Condition	Link	Max.posi. error (cm)	Error variance (cm^2)	Max.posi. error (cm)	Error variance (cm^2)	Max.posi. error (cm)	Error variance (cm^2)
No load	X	0.4567	0.1948	0.2693	0.0234	0.1658	0.0098
	Y	0.8204		0.3555		0.2075	
10kg load	X	0.4913	0.2726	0.2761	0.0293	0.2131	0.0231
	Y	0.8485		0.4131		0.3114	

ROBOT PROGRAMMING BASED ON COMMERCIAL CAD SYSTEMS

E. Vendrell, M. Mellado and J. Tornero

Departamento de Ingeniería de Sistemas, Computadores y Automática (DISCA), Universidad Politécnica de Valencia, P.O.B. 22012, E-46071 Valencia, Spain

Abstract. Cost reduction in commercial CAD systems makes them appropriate to be used as a support tool for robot programming. The use of these systems allows a graphical, intuitive tool to be acquired for supporting robot programmers. The paper describes the development of a robot programming system based on a low cost commercial CAD software. This programming system admit robots to be handled in an interactive way from a personal computer. Programs are made by modifying the robot configuration graphically on the computer screen. The developed tool is for general use, and accept any robot modelled in the CAD system to be programmed. In the paper, an education laboratory for training robot programmers using this software is described.

Keywords. Robot Programming, Robot Simulation, CAD Tools, Modeling

INTRODUCTION

Programming is one of the most relevant topics within throughput of robotic systems. A perfect programming means a right understood in man/machine interface and provides improvement of performance in production phase, where is included.

Traditional robot programming by *guided*, from the use of a joystick while the robot is following these movements, is the most extended solution (Angulo 1985). Several steps for movement teaching, characteristic point recording and stored movement repeating are done, until robot working is run at correct velocity. Some operations, such as gripper openinig/closing or welding system activation, are also specified in some points. The result is a sequence of movement points together with signal activations.

Nevertheless, this method presents a set of drawbacks for optimal robot programing, such as:

- Low flexibility, because little changes in a program can represent a repetition of a big part of teaching process.
- Program edition and mistake corection is usually uncomfortable.
- Error detection is oftenly difficult.
- Program possibilities from the point of view of control flow and data structure are null.

Also, robot programming is made directly in the working cell robot is located, so its work must be stopped to receive guided to new positions for storing. Therefore, new drawbacks appear:

- Lost of time in production.
- Inconveniences for the programmer, such as environment polution or external noises.
- Programming during overtimes, with its evident cost.

In this type of programming appears lack of verification for critic situations such as configurations out of working-area, warning approaches to objects, visual comprobation of distances, so on. All these situations only can be avoided by means of a high skin and know-how of the programmer.

Other traditional method in robot programming is the use of *textual languages*, from programs or sequences of commands. Robot programming is made in an external computer with text editors, obtaining text files which, after compiling and linking, will be transmited to the robot and run by it.

Most of textual languages are based on robot oriented programming, specifying sequences of movements corresponding to robot actions. There are also object oriented languages for programming robots, where actions are described in terms of objects to be manipulated instead of robot movements.

Using textual languages avoids robot halt during programming task, with time saving, but movement checking can not be verified until the program is executed. In this way, critic situations listed above are impossible to be checked previously.

On the other hand, this programming method force the programmer to learn the language syntax. To make easy this learning, simple robot oriented languages were developed but with working capabilities quite limited. In contrast, high capability languages were performed, but they result difficult to be learnt, and also programmers can hardly take the best of them.

This paper describes the obtaining of a graphic and interactive robot programming system based on a low cost commercial CAD system. The system presents a great improvement of characteristics because robot programming is done from a CAD software, in an intuitive way. Also, it gives a friendly interface environment with the use of input devices such as mouse. All its characteristics makes the system very easy to learn.

Programming can be made externally to the working chain of the robot, modifying directly the robot configurations of its model on a computer. Simultaneously to robot programming, all movements are simulated and verified, allowing to program and check the robot without stopping robot tasks.

Graphical programming of the robot on a computer avoids programmer to learn a language syntax, because robot actions are selected by options in menus with a mouse and introducing required data through keyboard.

Using a commercial CAD software as the core of the system takes advantage of all the facilities that this software incorporates, making possible not only to model the robot, but also the objects within the robot work-space. Some functions, such as zooms on conflictive zones, distance computation between robot and objects, different perspectives and so on, enables the robot to be programmed more conveniently using the software described.

This programming tool admits a library with several robot models to be managed. Each one of the robot-arms available within the library have been modelled previously on the CAD system. Once the robot to be programmed is selected, its model appears on the screen, as well as menus appropriate to the chosen robot type.

To program the active robot model, inverse and direct kinematic problem resolution is applied, displaying the movement of the robot on the screen, with simulation and verification of every configuration. The system can be used in two ways: i) step by step, by programming movements of the robot, or ii) saving a sequence of movements and actions to form a trajectory or complete robot task, with the possibility of later runnings. The basic funtion of the former way is to check validity of configurations, while the objective of the latter is robot program generation.

Trajectory generation accepts macros to be made to define complete robot tasks for latter execution, as, for example, picking-up or placing objects on robot work-space, drilling cycles in parts modeled in the CAD system, etc.

The use of CAD systems as a support in robotic has been a research topic for several years. For example, it has been applied to work-space automatic determination (Borrel 1985), obstacle avoidance (Liegeois 1984) and time optimal trajectory planning (Dubowsky 1986). Related to the use of CAD systems as a robot programming tool, Dombre (1986) presents the CARO system for teleoperation tasks while Faverjon (1986) focusses more on object level programming of industrial robots.

Some commercially available CAD/CAM systems include robotic applications, such as CATIA system (Dassault), ROBOGRAPHICS (Computer Vision) and GRASP (Univerity of Nottingham). Nevertheless, the main limitation of these systems is their high cost, both for hardware and software, making it difficult to extended implantation to small and medium enterprises.

For this work, AutoCAD v10.0 for personal computers has been considered as the Computer Aided Design software. AutoCAD, with its AutoLISP programming system, has enough power and flexibility to implement the proposed robot programming support tool. Moreover, its very low cost, as well as its easy and confortable use, together with the high number of personal computers installed nowday, makes of AutoCAD the most extended CAD software available on the market.

ROBOT MODELING

3D modeling in a CAD system is usually done starting from a set of primitive objects in such a way that as combination of them, parts with great detail can be defined. For example, solid modelers use to include boolean operations as union, intersection and diference over primitive objects to generate complex structures.

In a simple wireframe modeler, as the one used in this work for the reasons mentioned above, the facility incorporated within the system consists on constructing blocks as union of different objects. Afterwards, these blocks will be handled as an independent object. In this way, each element of the articulated chain which forms the robot-arm is modeled as a block (Fig. 1).

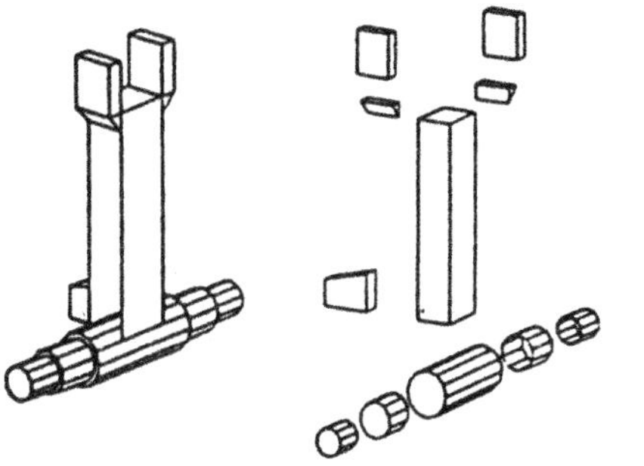

Figure 1: Arm of ASEA IRB6 with Component Objects

For example, to model the ASEA IRB6 robot, the following elements have been considered: base, arm, forearm, hand and gripper (as articulated elements) and beam, counterpoise and two motors (as other elements to complete the robot definition). Fig. 2 shows all the elements which compound the ASEA IRB6 robot. Each one of these elements has its own local coordinate system solidarily linked to it, which shall be used to asemble the elements and to form the robot-arm. This technique allows the independent movement of the elements on the screen in function of rotation angles of each configuration.

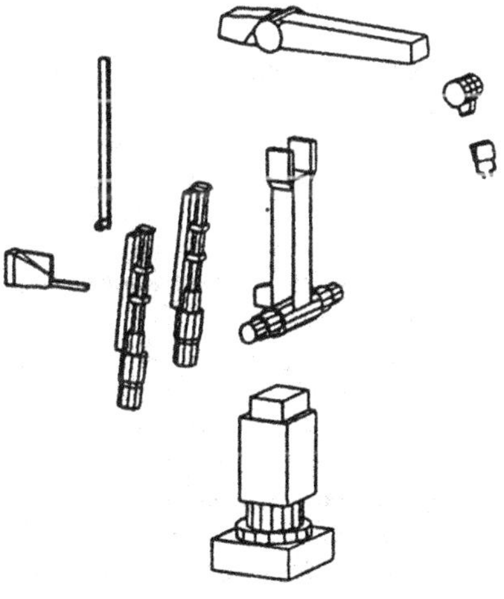

Figure 2: Elements of ASEA IRB6 robot

The link of different elements and their relative position is defined by means of homogeneous transformation matrices which relate corresponding local coordinate systems. Denavit-Hartenberg parameters (1955) are used to obtain those transformation matrices.

In Fig. 3 are shown the coordinate systems linked to each of the elements that form the articulated chain of ASEA IRB6 robot, together with an end-effector system. The coordinate systems of the figure are for the synchronism position of the robot.

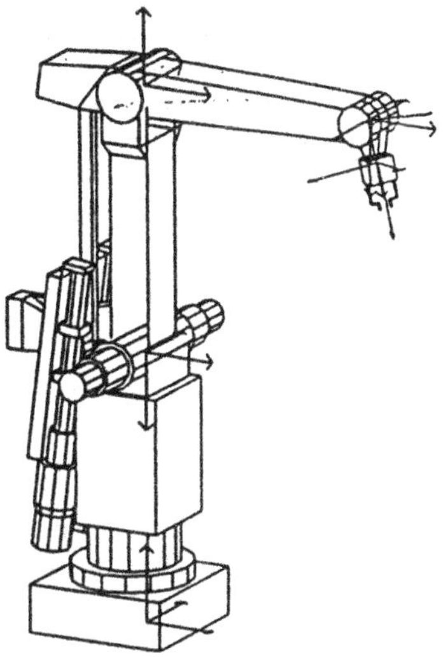

Figure 3: Denavit-Hartenberg Representation for ASEA IRB6

For this position, the values of Denavit-Hartenberg parameters and the joint variable ranges are shown in Table 1. These parameters allow computing homogeneous transformation matrices to convert from coordinate system of element $i-1$ to coordinate system of adjacent element i.

Joint	θ_i	α_i	a_i	d_i	Range
1	0	-90	0	0	-170 to 170
2	-90	0	690	0	-130 to -40
3	90	0	670	0	65 to 130
4	-13	-90	0	0	-90 to 90
5	42	0	0	95	0 to 360

Table 1: Denavit-Hartenberg Parameters of ASEA IRB6

ROBOT PROGRAMMING

Direct kinematic model can be obtained directly multiplying the matrices commented above (Fu 1988). Solving direct kinematic problem makes possible to represent the robot on the screen for any configurations. By this means, it is possible to program robot movements introducing the joint angles by computer input devices, updating the configuration on the screen.

Obtaining the inverse kinematic model is a problem whose complexity depends on the kind of robot to program. For its resolution, some of the already existing techniques can be used, for example, Pieper (1968) or Paul (1981) methods. This solution admits to program the robot introducing end-effector position and orientation.

Solving direct and inverse kinematic problems admits to display robot movements on the screen, verifying at the same time, all of joint ranges. In this way, it can be automaticaly guaranteed, transparent to the user, that generated movements will present no problem about configurations out of working-area.

There are two basic ways of working with the developed system:

- Programming robot movements by configurations, where for each step it is possible: i) to change independently one of the joint variables or ii) to introduce an end-effector position and orientation to be reached. Therefore, performing approaches to objects to be manipulated by the robot is facilitated.
- Generating trajectories, from certain movements performed by the preceding method. In order to generate one valid trajectory, robot momevements are made by configurations, storing those suitable for the robot task. Accesory orders, as velocity or kind of movement, can be indicated to be stored in the trajectory.

Trajectory generation accepts to make macros to define tasks or full working robot cycles for latter use, i.e., picking-up and placing objects in the robot working-area, or performing drilling cycles on pieces modeled in the CAD system.

The system has an option to activate the connection to the robot, in such a way that the robot is forced to follow the movements of its model as programmed on the computer. Then, initialy, the robot is moved only on the screen to simulate programmed movements. When the movements have been validated, activating the connection to the robot will convert the system into an actual simulation tool.

Moreover, it is possible to execute orders which imply no movement, as activate a TCP (Tool Center Point), change movement velocity and so on. These orders, as well as movements orders, has to be included in a trajectory only when the programmer has checked and validated them.

Full displaying of internal robot state and data helps more, if possible, programming task, improving programmer work.

For the specific case performed in the work, the connection to the ASEA IRB6 robot, the computer (a 386-PC) is connected to the Computer Link board in the ASEA Control Unit S2 via RS232. Communication Tools library was used to implement the software of the system for robot connection.

ROBOT PROGRAMMING SYSTEM INTERFACE

The acquired interface permits using all facilities of the CAD system, such as zooms, scaling, desplacements, framing, perspectives, etc. To get this, the interface was developed programming a set of functions in AutoLISP integrated in a personalized menu file which substitutes the own AutoCAD menu. Therefore, it is possible to accede proper functions of AutoCAD, typing them through the command line.

Once the execution of the system has been started, a model of the robot to be programmed is displayed on the screen in two windows with different perspectives. At the top of the screen there is a top-down menu with the following options:

- <u>Movement Menu.</u> Movement options allow to modify robot configuration to reach an end-effector position and orientation as desired. This can be done by two basic ways:
  - By joint variables: With this option the robot configuration can be modified joint by joint. For example, for the ASEA IRB6 model, the menu shown in Fig. 4 gives leave modifying rotation angles for the elements base, arm, forearm, wrist and hand, as well as open or close the gripper. Auxiliar orders to cancel the action or undo last action are also provided.

    These actions allow the movement of the robot similarly to the movement with the joystick of the *teach-pendant*. Once the joint to be modified has been selected, one of the windows displays the robot from a point of view located in the axe of the joint. Also, to make easier the programmer work, the cursor is activated in this window to let the introduction, in a dynamic way with the mouse, of the new position of the element to be moved.

    Thus, the programmer, with the use of the mouse and in a dynamic and interactive way, can rotate the element with the desired angle (when great precision is required, the angular value can be typed in). Fig. 5 shows how the movement of the forearm of ASEA IRB6 robot is specified with the mouse.

    Once again, to make easier programmer task, the introduced angles are related to an horizontal axe, with internal transformation into joint variables of Denavit-

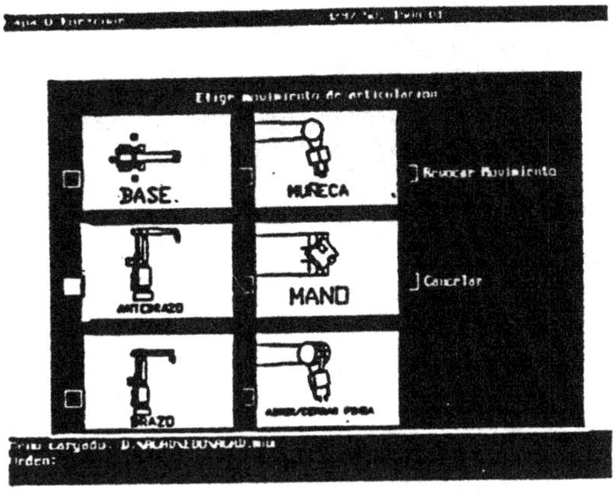

Figure 4: Movement Menu by Joints for ASEA IRB6

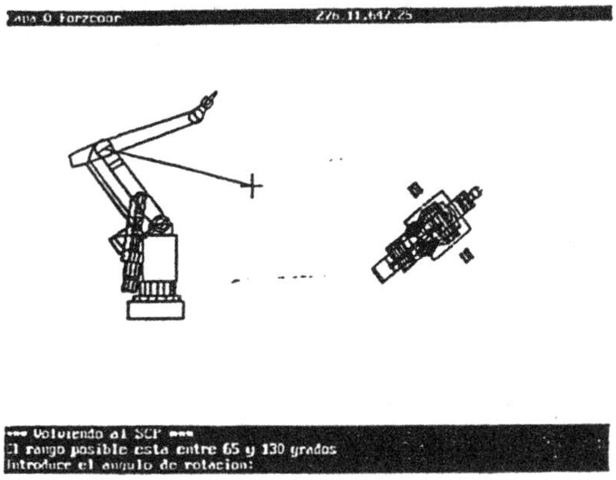

Figure 5: Movement of the Forearm of ASEA IRB6

Hartenberg. Also, the range of movement for each element is displayed on the screen. If programmer makes a mistake lying out of these ranges, an error message is displayed, the action is cancelled and previous configuration is recovered.

- By end-effector coordinates: This option admits the introduction of end-effector position and orientation by means of the coordinate of a point in the space and Euler angles. An internal resolution of the inverse kinematic problem will give joint variables for adequate configuration for such a position and orientation, with all variable ranges checked and verified.

Other options included in this menu are:

- Connection to robot: This option let activate or unactivate the connection to the robot in such a way that only when connection is active, the robot follows accurately the movements of its model on the screen as guided by the programmer.

  This allows to simulate robot model movements on the screen without moving the robot to be programmed or stopping the production process the robot could be developing at this moment. Programs will be safer if they are firstly simulated and verified on the screen and then activate the connection to the robot to execute the program.

- Initializing Robot: Initiate the robot to be programmed, moving it to a synchronism position and giving initial values to all its parameters.
- Synchronism: Move the robot to synchronism position.

• Utility Menu. Several utility options give facilities to:

  - Help trajectory generation, such as activation of TCP, activation of a reference coordinate system, kind of coordinate for movement (robot coordinate with free movements or rectangular coordinate with TCP moving in a straight line) or velocity definition (maximum, basic and programmed velocity).
  - Inform internal state of the robot, as well as visualize graphic representation of Denavit-Hartenberg representation.

• Trajectory Menu. Its options accept the generation and execution of trajectories as sequences of robot actions. A trajectory can be generated and saved into a file.

  The programmer can simulate several configurations of the robot on the screen until the correct one is found out. Then, this one can be stored in the trajectory file, just selecting an option. Utility orders from the previous menu can also be stored in the trajectory file. When all the trajectory is defined, the file has to be closed.

  The execution of the trajectory can be made latter just typing the name of the file. If robot connection is active, the robot will follow this trajectory, else only the model on the screen will do.

- Auxiliar Menu. Several auxiliar orders are included in this menu, such as undo last action, save current configuration into a file or recover it.

CONCLUSIONS

The system described provides a tool for industrial robot programming which is easy to learn and use. The process consists on a general, simple and intuitive methodology of robot programming, based on the graphic facilities within commercial Computer Aided Design software, whilst also providing a friendly man-machine interface for the programmer's work.

Robot programming is performed in an interactive way through the CAD system, working directly on the screen with a 3D model of the robot to be programmed. The system automatically checks every configuration to guarantee validity of robot program generated for latter execution.

REFERENCES

Angulo, J.M. & Aviles, R. (1985)
Curso de Robotica.
Paraninfo.

Borrel, P., Liegeois, A. & Tanner, P. (1985)
"Automatic Modelling of the Work Area of a Robot Arm with Reference to its Various Configurations". SIAM Conf on Geometric Modelling and Robotics, Albany (NY).

Denavit, J., & Hartenberg, R.S. (1955)
"A Kinematic Notation of Lower-Pair Mechanisms Based on Matrices".
J. App. Mech. Vol. 77, pp 215-221.

Dombre, E., Fournier, A., Quaro, C. & Borrel, P. (1986)
"Trends in CAD/CAM Systems for Robotics". IEEE Int. Conf. on Robotics and Automation, San Francisco, California.

Dubowsky, S., Norris, M.A. & Shiller, Z. (1986)
"Time Optimal Trajectory Planning for Robotic Manipulators with Obstacle Avoidance: A CAD Approach". IEEE Int. Conf. on Robotics and Automation, San Francisco, California.

Faverjon, D. (1986)
"Object Level Programming of Industrial Robots". IEEE Int. Conf. on Robotics and Automation, San Francisco, California.

Fu, K.S., Gonzalez, R.C. & Lee, C.S.G. (1988)
Robotica: Control, Deteccion, Vision e Inteligencia. McGraw-Hill.

Liegeois, A., Borrel, P. & Dombre, E. (1984)
"Programming, Simulating and Evaluating Robot Actions". 2nd Int. Symp. on Robotics Research, Kyoto.

Paul, R.P. (1981)
"Robot Manipulator: Mathematics, Programming and Control". MIT Press, Cambridge, Mass.

Pieper, D.L. (1968)
"The Kinematics of Manipulators under Computer Control". Artificial Intelligence Project Memo No 72, Stanford University, Palo Alto, California.

A PC-BASED CAD/CAM SYSTEM FOR AUTOMATIC LASER PROCESSING OF MATERIALS

R. Sanz and A.R. Nores

Dpt. System Engineering, E.T.S.I.I., University of Vigo, P.O. Box 62, 36200 Vigo, Spain

Abstract. In this paper, we present a system consisting of a low-power carbon-dioxide (CO_2) laser, a small moving table and a low cost personal computer. This system is mainly used for laser cutting process automation although doping and deposition operations can be also performed. Hardware and software developments have allowed direct manufacturing based on the geometric model of the piece, generating the moving table motions, and controlling its speed and the laser's intensity. Experimental results have also shown that the use of computer control has improved design for manufacturing.

Keywords. CAD; CAM; computer-aided design; manufacturing processes; microcomputer-based control.

INTRODUCTION

Laser is an active area of research and development, having a great number of applications in industrial fields. Actually, laser manufacturing applications such as material processing are of practical interest. In particular, CO_2 lasers are used for cutting, welding and heat treatments. Experiences in manufacturing industry have shown the flexibility of such systems (Johnson, 1986). Additionally, the use of CAD/CAM reduces design time modifications and improves the manufacturing process.

Based on a simple laser system, we have developed a complete system to control a laser cutting process (Nores, 1991). The principal aim of this work is to supply an easy-to-use tool for all phases involved, from the piece design to its mechanization. Three major issues must be addresed in the development of such a system: providing simple and low-cost automatic operation, using existing equipment and improving the cutting process quality.

The original system was composed of a 50 w. CO_2 laser; and a moving table managed by simple numerical control (NC).

Although this system works correctly, it presents very important limitations, such as:

- Semiautomatic operation;
- Tedious handling;
- Low memory capacity, that does not allow the automatic manufacture of complex pieces;
- Low cutting accuracy, specially in curve sections, because of rudimentary moving table control;
- Absence of laser intensity adjustement.

Since all these factors have a negative effect on final products, we considered the necesity of carrying out a system to improve the cutting process.

SYSTEM STRUCTURE

A laser machine is generally composed of a laser, a beam delivery, a drive mechanism and a motion control system. Figure 1 shows a block diagram of the piece processing system.

The workpiece to be manufactured is placed on a moving table. Both x-y movements allow to track

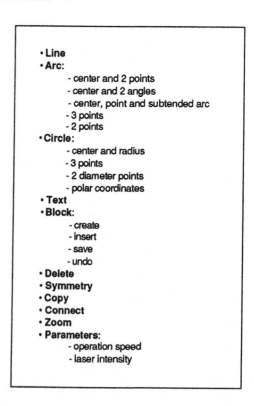

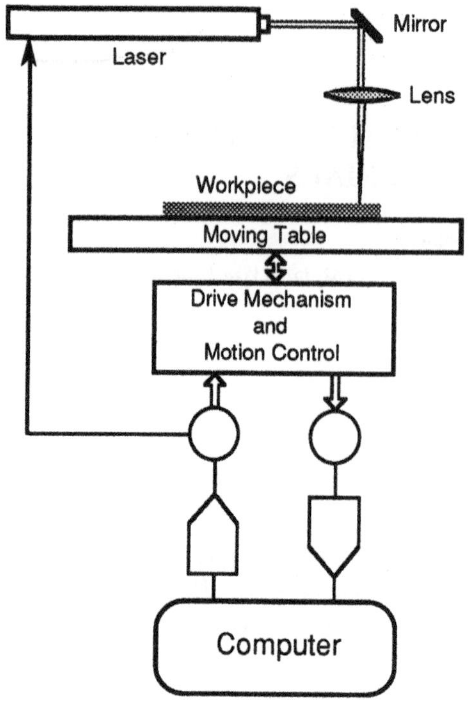

Fig. 1. Schematic diagram of the laser machine.

the cutting trajectory. These movements are drived by two d.c. servomotors which are controlled by conventional analog regulators and a NC. The motor speeds are proportional to two reference signals that are determined by a control program.

The 50 w. CO_2 laser provides a vertical polarization output whose intensity is adjustable by means of an analog reference signal. A copper surface reflecting mirror is used to redirect the laser beam. To attain sufficient power density to cut, the laser beam must be focused on a small spot by using a simple lens made of Potasium Cloride (KCl). Since the working distance from the lens to the workpiece is small, it is necessary to protect the lens from damage. This is done by the use of a fan that generates an air current.

The early moving table system was handled by a small numerical control (NC). In order to improve the process automation, an integrated computer control system has been implemented. A low cost personal computer is used to draw the workpiece and to perform direct processing. The NC is attached to the personal computer through two adaptation cards which have been developed in order to have suited I/O and signal adaptation. In addition, a communication card, that holds analog and digital input/output, has been developed.

The overall system includes a number of protections such as against high tension discharges.

DESIGNING THE PIECE

The software that we have developed consists of a CAD program used to design the workpiece and a CAM program that performs the cutting process from a previous design. For the automatic processing of pieces, the design and mechanisation phases have been integrated in the same software package, in such a way that the operator must not change the application during the manufacturing process. All functions and operating modes are accessible by means of menus. Data input can be performed by using either a keyboard or a mouse.

The CAD program, named LASERCAD, has been implemented with the aim of offering a complete tool for laser processing. Besides usual working options, LASERCAD provides extended design capabilities, adding new sorts of basic design elements and allowing direct piece processing from the geometric model. The main drawing options are recorded in Fig. 2. The desired cutting speed and laser light intensity are included for each element of the model.

- **Line**
- **Arc:**
  - center and 2 points
  - center and 2 angles
  - center, point and subtended arc
  - 3 points
  - 2 points
- **Circle:**
  - center and radius
  - 3 points
  - 2 diameter points
  - polar coordinates
- **Text**
- **Block:**
  - create
  - insert
  - save
  - undo
- **Delete**
- **Symmetry**
- **Copy**
- **Connect**
- **Zoom**
- **Parameters:**
  - operation speed
  - laser intensity

Fig. 2. Listing of drawing elements

Moreover, the designs can be printed or plotted, and they can also be imported and exported by means of the DXF standard file interchange format.

PROCESSING THE PIECE

Besides the cutting operation, the CAM program offers two other mechanization modes: doping and deposition.

Cutting is used to engrave or cut the flat surfaces of a wide range of materials. Deposition can be carried out in two different forms: by repetition and cyclically. In repetion mode, the system returns to the starting point once the final trajectory point has been reached, repeating this process as many times as specified in the design. In cyclical mode, each time the final point is reached, the inverse trajectory to the starting point is made. This cyclical trajectory is repeated as many times as indicated by the user. Doping is used to carry out doping over surfaces of a variety of materials.

The type of processing must be indicated before introducing the design and cannot be later modified.

As the drawing elements are introduced by using the LASERCAD program, they are coded internally as nodes of a linked list structure. This information is converted into a B-tree representation in order to be correctly interpreted by the CAM program. Users can choose any starting point but automatic selection is also possible. The element that contains this point is assumed to be the root node. Each branch is constituted by contiguous open elements (lines and arcs) that will be procesed in such an order. If there are three or more concurrent elements in the same point, the extra elements are placed in the other branch of the tree. The same is valid for non open elements. It is interesting to notice that, however complex the piece may be, this structure guarantees a unique binary tree representation. Another feature is that continous trajectories will be carried out regardless of element introduction order.

From the generated tree structure, each geometric element is converted in a set of points corresponding to the trajectory. Finally, orders are given to the moving table that executes the trajectory. This process is performed according to the speed defined for each element. The laser intensity is proportionally adjusted to this speed.

The implemented control scheme is a simplified version of the proposed in Aström and Wittenmark (1984), in which the observer has been eliminated (see Fig. 3). This is done to satisfy time constraints.

A simulation option is included to monitor the manufacturing process and to verify the order in which piece elements are processed. The processing simulation speed is proportional to the real implementation.

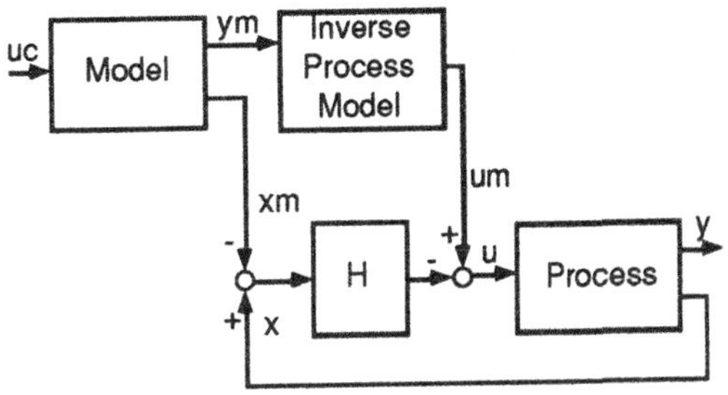

Fig. 3. Block diagram of the motor control systems.

EXPERIMENTAL RESULTS

Different experiences with a number of pieces having complex outlines have been carrying out. These show that the proposed structured for internal information seems to work very well.

However several drawbacks arise with trajectory following due to mechanical flaws observed in a d.c. motor actuator. These malfunctions can be solved by using an observer as it is proposed in Aström and Wittenmark (1984). An accurate behaviour with a modified control algorithm can be expected.

Futhermore, experimental results have shown that intensity adjustment must be improved in order to achieve better cutting quality. The necessity of implementing a more sophisticated intensity control is being considered. In particular, a fuzzy control system is under design.

In any case, the use of a more powerfull computer system is an essential requirement for optimum processing results.

CONCLUSIONS

This paper has reviewed a CAD/CAM system for material processing that has been implemented using existing equiment and a low-cost personal computer.

In particular, we have performed the following developments:

- A suited CAD program, that provides geometric design of pieces for later manufacture;
- A CAM program, that can automatically process a piece designed by the CAD program, controlling the intensity of the laser light;
- A communication card, for a personal computer, that holds analog and digital input/output;
- Two adaptation cards, that isolate and arrange signals between the computer and the moving table's numerical control system.

The software package run on a PC-microcomputer with graphic terminal and two floppy disks, although the employ of a hard disk is recommended.

Even if the system is simple, effective and highly flexible, problems in trajectory control have been observed .

REFERENCES

Aström, K.J. and B. Wittenmark (1984). Computer Controlled Systems. Prentice-Hall, Englewood Cliffs, N.J. pp. 214-217.

Johnson, T.A. (1986). Flexible laser manufacturing systems. In W.W. Duley and R. Weeks (Ed.), Laser Processing: Fundamentals, Applications, and Systems Engineering, Proc. of the SPIE, Vol. 668, pp. 45-52.

Nores, A.R. (1991). Computer control of a moving table and a laser, for automatic cutting of pieces (in spanish). Master Dissertation. University of Vigo.

--- (1982). Model 570 CO_2 Laser System Instruction Manual. Apollo Laser Inc.

FURTHER EVALUATION OF AN AUTO-CALIBRATION METHOD FOR PID CONTROLLERS

A. Voda and I.D. Landau

Laboratoire d'Automatique de Grenoble and GR "Automatique", ENSIEG - BP 46, 38402 Saint-Martin d'Hères Cedex, France

Abstract. The paper presents further results on the evaluation of a method for the auto-calibration of PI or PID controllers when a point of the frequency characteristics of the plant to be controlled (gain and phase) has been identified, using for example, a closed loop "oscillation" experiment. The underlying design methodology is inspired by Kessler's "Symmetrical Optimum" [1958]. The method takes in account both robustness aspects and closed loop characteristics. In the case of a PID, the proposed method can be interpreted as a refinement of the Ziegler-Nichols second method rules for which a theoretical base is given. Simulations on various examples and experiments on a real plant conclude the paper.

Keywords : PID Control, Automatic Calibration

I - INTRODUCTION

PI and PID controllers represent still the majority of controllers used in the industrial control systems. It is therefore interesting to develop an appropriate methodology for the automatic calibration of the parameters of these controllers following the general approach of modern control

1- Identification of the plant

2 - Computation of the controller parameters based on identified model and desired closed loop performances.

The practical requirements for a "good" method of auto-calibration of PID controllers are:

* Simple computation of the parameters (table).
* Test on the plant with small signals.
* Existence of a theoretical basis for the comprehension of the tuning.
* Easy interpretation of the tuning.
* Clear indications for "accelerating" and "slowing down".
* Large field of application (guaranted preformances).
* Good robustness margins in the (theoretical) field of application
* Acceptable performances beyond the theoretical field of application (stability first).

The paper is organized as follows. In section II are briefly presented the plant models for which "symmetrical optimum" tuning rules are applied. Section III deals with the auto-calibration of the PI controller. In section IV two methods for the auto-calibration of PID controllers are presented. A comparison with the Ziegler Nichols (second method) is done in section V, since the second method can be interpreted as an improvement of the original Ziegler Nichols second method. Simulation results and experimental results on an air heater are given in section VI.

II - THE KESSLER'S "SYMMETRISCHE OPTIMUM"

We will briefly recall the background of the "symmetrical optimum"

2.1. PI controller

The plant to be controlled is assumed to be of the form :

$$H_p(s) = \frac{G_0}{(1 + T_1 s) \prod_{1}^{n-1} (1 + s T_j)\, e^{sT_n}} \tag{2.1}$$

where T_1 correspond to a large (compensable) time constant with respect to the sum of the "parasitics" time constants and time delay i.e.

$$T_1 >> \sum_{1}^{n} T_j = T_\Sigma \tag{2.2}$$

Therefore for the frequencies below $\omega \leq 1/T_\Sigma$ the plant transfer function can be approximated by :

$$H_p(s) = \frac{G_0}{(1 + sT_1)\,(1 + s\,T_\Sigma)} \tag{2.3}$$

For this plant a PI controller is considered :

$$H_{PI}(s) = \frac{1 + s\,\tau}{s\,\tau_i} = K_p\left(1 + \frac{1}{s\,T_i}\right) \tag{2.4}$$

The tuning rules of symmetrical optimum lead to the Bode diagram of the open loop (controller + plant) shown in Figure 2.1.

As it can be observed the PI assures that the crossover frequency is $w = 1/(2\,T_\Sigma)$ and a slope of 20 db/dec is assured one octave at the left and at the right of the crossover frequency.The phase margin is larger than 60° and it depend on the ratio $T_1/(4\,T_\Sigma)$ (it improves when T_1 approaches $4\,T_\Sigma$).

The two structures of PI controllers which we use in simulations and experiments are shown in Figure 2.2.

The responses to load disturbances are given by the sensitivity function, which for the two extreme cases (integrator plant and $T_1 = 4\,T_\Sigma$) are shown in the figure 2.3.

The advantages of this approach for the tuning of PI(D) controllers are the followings :

1 - the ultimate performances of the closed loop are defined by the neglected (uncompensable) dynamics.

2 - the tuning of the controller depends upon the neglected dynamics and the ratio process gain / time constant. (i.e. two informations).

3 - it assures a good phase and gain margin by forcing a 20 db/dec slope around the crossover frequency.

4 - it clearly shows how to incorporate tolerances in the tuning of the controller [Kessler, 1958], [Landau 1968].

2.2. PID controller

Using the same considerations about the "parasitics" as in section 2.1, the plant transfer function is of the form :

$$H_p(s) = \frac{G_0}{(1+sT_1)(1+sT_2)(1+sT_\Sigma)} \qquad (2.5)$$

where it is assumed that T_Σ represents the sum of uncompensable time constants and delays and that T_1 and T_2 the compensable plant constant have the property T_1, $T_2 >> T_\Sigma$.

The transfer function of the PID in series and parallel form is given by :

$$H_{PID}(s) = \frac{(1+s\tau_1)(1+s\tau_2)}{s\tau_i} = K_p\left(1 + \frac{1}{sT_i} + sT_d\right)$$

(2.6)

The Bode diagram for the PID is shown in Figure 2.1. As one can see the fact that a 60 db/dec slope appears now in the low frequency domain requires that the lead effect of the two PID zeros occurs at a twice lower frequency ($1/(8\,T_\Sigma)$ instead of $1/(4\,T_\Sigma)$) than in the PI case when the slope in the low frequency domain was 40 db/dec.

III - AUTO-CALIBRATION OF PI-CONTROLLERS

For a plant having a transfer function of the form (2.3), where T_Σ may represent either the sum of uncompensable small time constants and time delays or the small time constant which will define the closed loop band pass, one needs to know for tuning a PI :

1) the value of the time constant T_Σ
2) the ratio (G_0/T_1) between the process gain G_0 and the large (compensable) time constant T_1.

Depending on the ratio T_1/T_Σ it results that for $\angle\phi \approx -135°$, one has $\omega_{135} = (\alpha/T_\Sigma)$ where the coefficient α varies between $\alpha = 1$ in the case $T_1 >> T_\Sigma$ (integrator) and $\alpha = 1.28$ for $T_1 = 4\,T_\Sigma$.

Therefore making a feedback experiment with a relay with hysteresis in the loop one can obtain oscillations at the frequency $\omega = \omega_{135}$ where the phase lag of the plant is $\angle\phi \approx -135°$. From this experiment one obtains :

$$T_\Sigma = \frac{\alpha}{\omega_{135}} \qquad (3.1)$$

and the gain at the frequency ω_{135} denoted by $G(\omega_{135})$.

From the knowledge of the process gain at $\omega = \omega_{135} = \alpha/T_\Sigma$ one can compute the ratio (G_0/T_1) which is necessary for tuning the PI controller.
The tuning rules (KLV/PI) are summarized in table 3.1., for $\alpha = 1.15$, which is the mean value between its limits, and corresponds to $T_1 = 10\,T_\Sigma$

Table 3.1. KLV/PI Tuning rules

PI "series" form
$\tau = \frac{4\alpha}{\omega_{135}} = \frac{4.6}{\omega_{135}}$
$\tau_i = 8\,\alpha^2\sqrt{1+\alpha^2}\,\frac{G(\omega_{135})}{\omega_{135}} = 16\,\frac{G(\omega_{135})}{\omega_{135}}$
PI "parallel" form
$T_i = \tau = \frac{4.6}{\omega_{135}}$; $K_p = \frac{1}{3.5\,G(\omega_{135})}$

IV - AUTO-CALIBRATION OF PID CONTROLLER

The PID controller has 3 parameters to be tuned while a "feedback relay experiment" or "oscillation experiment" gives two infomations upon the plant. Therefore an additional "reasonable" hypothesis has to be added for tuning a PID. (Further improvements of the methods will be possible whith two "feedback relay experiments", for different phase lags of the plant).

Two methods will be presented, one which uses a "feedback relay experiment" for finding the gain and frequency for a plant phase lag $\angle\phi = -135°$ and a second one which uses a "feedback relay experiment" or the "Ziegler-Nichols oscillation experiment" for a plant phase lag $\angle\phi = -180°$. In particular the second method allows a comparison on analytic grounds with the Ziegler-Nichols tuning rules. This second method can be also interpreted as an improvement of the Ziegler-Nichols tuning rules.

4.1. Method 1

The method 1 has as objective to accelerate the closed-loop response with respect to the use of a PI tuned according to the method presented in section III. One assumes that the "feedback relay experiment" for $\angle\phi = -135°$ has been done. To accelerate the closed loop response in the spirit of the "symmetrical optimum" means that we assume that the "parasitics" time constant is in fact lower than the one estimated and that the estimated "high frequency" time constant can be compensated.

Therefore it is assumed that the plant model is of the form :

$$H_p(s) = \frac{G_0}{(1+sT_1)(1+sT_2)(1+sT_\Sigma)} \qquad (4.1)$$

where now

$$T_2 = \frac{1}{\omega_{135}} \qquad (4.2)$$

and

$$T_\Sigma = \frac{T_2}{\beta} = \frac{1}{\beta\,\omega_{135}} \quad (\beta = \text{acceleration factor}) \qquad (4.3)$$

If $(t_r)_{PI}$ is the rise time obtained with a PI, the β factor will depend upon the desired response time expressed as a function of t_r.

$$(t_r)_{desired} = (\frac{1}{\beta})\ (t_r)_{PI} \qquad 1 < \beta \le 2 \qquad (4.4)$$

We will use a PID of the form :

$$H_{PID}(s) = \frac{(1 + s\tau_1)(1 + s\tau_2)}{s\tau_i} \qquad (4.5)$$

Since T_2 is known, choosing :

$$\tau_1 = T_2 = \frac{1}{\omega_{135}} \qquad (4.6)$$

one brings back the problem to the tuning of a PI for the plant transfer function :

$$H'_p(s) = \frac{G_0}{(1 + sT_1)(1 + sT_\Sigma)} \qquad (4.7)$$

with T_Σ given by Eq. (4.3) and (G_0/T_1) calculated from ω_{135} and $G(\omega_{135})$.
The tuning rules are given in table 4.1.

Table 4.1 : KLV/PID 1 Tuning Rules

PID "series" form
$\tau_1 = \frac{1}{\omega_{135}}$ $\tau_2 = 4T_\Sigma = \frac{4T_2}{\beta} = \frac{4}{\beta\omega_{135}}$; $1 < \beta \le 2$ $\tau_i = 2(\frac{G_0}{T_1})\,4T_\Sigma^2 = \frac{8\sqrt{2}}{\beta^2} \bullet \frac{G(\omega_{135})}{\omega_{135}}$
PID "Parallel" form
$T_i = \left(\frac{4+\beta}{\beta}\right)\frac{1}{\omega_{135}}$; $T_d = \left(\frac{4}{4+\beta}\right)\frac{1}{\omega_{135}}$ $K_p = \frac{T_i}{\tau_i} = \left(\frac{4+\beta}{4}\right) \bullet \frac{\beta}{2\sqrt{2}\,G(\omega_{135})}$

Remark :

(1) the implementation of the derivative term under the form : $D(s) = (s\,T_d) / (1 + s\,T_d/N)$ can be easily handled, by incorporating (T_d/N) in T_Σ or by choosing $(T_d/N) << T_\Sigma$

4.2. Method II

One assumes a plant transfer function of the form given in Eq. (2.5). One assumes that $T_1, T_2 >> T_\Sigma$.

A "feedback relay experiment" or a "Ziegler-Nichols experiment" is done for finding the critical frequency ω_{180} for which the plant phase lag is $\angle\varphi = -180°$.

From the assumption that $T_1, T_2 >> T_\Sigma$ it results that in the frequency region around ω_{180} the plant transfer function can be approximated by $(T_1 = T_2 = T >> T_\Sigma)$:

$$H'_p(s) = \frac{G_0}{(1 + sT)^2(1 + sT_\Sigma)} \qquad (4.8)$$

At ω_{180} , the phase lag of - 180° is essentially determined by the term $1/(1+ s\,T)^2$ plus a small amount coming from $1/1 + sT_\Sigma$ (T_Σ is smaller than $1/\omega_{180}$).

To be specific, assumes that $\omega_{180} = 10/T$. At this frequency the term $1/(1+ s\,T)^2$ introduces a phase lag of 170°. This means that at $\omega_{180} = 10/T$ the contribution of the term $1/1 = sT_\Sigma$ is an additional phase lag of $\angle\Delta\phi = -10°$. But using the normalized gain-phase characteristics of a first order transfer function it results that :

$$T_\Sigma = \frac{1}{5\,\omega_{180}} \qquad (\omega_{180} = \frac{10}{T}) \qquad (4.9)$$

Similar results are obtained if we assume that $\omega_{180} = 8/T$ in which case $T_\Sigma = 1/4\,\omega_{180}$.

Therefore from the $\angle\phi = -180°$ experiment one estimates the "parasitics" time constant as :

$$T_\Sigma \approx \frac{1}{(4 \text{ to } 5)\,\omega_{180}} \qquad (4.10)$$

If one uses the "symmetrical optimum" rules for the plant transfer function given in Eq. (4.8), what is needed in addition to T_Σ to end up the tuning is the computation of G_0/T^2 from the measurement of $G(\omega_{180})$.

The Table 4.2 summarizes our second method for tuning PID and the Ziegler-Nichols (second method) tuning rules [0gata 1990] are given for reference.

Table 4.2 : KLV/PID2 Tuning Rules

	KLV/PID2 (a)	KLV/PID2 (b)	Ziegler-Nichols Method 2
$\tau_1 = \tau_2$	$\frac{2}{\omega_{180}}$	$\frac{1.6}{\omega_{180}}$	$\frac{1.57}{\omega_{180}}$
τ_i	$\frac{2.1\,G(\omega_{180})}{\omega_{180}}$	$\frac{1.05\,G(\omega_{180})}{\omega_{180}}$	$\frac{5.23\,G(\omega_{180})}{\omega_{180}}$
T_i	$\frac{4}{\omega_{180}}$	$\frac{3.2}{\omega_{180}}$	$\frac{3.14}{\omega_{180}}$
T_d	$\frac{1}{\omega_{180}}$	$\frac{0.8}{\omega_{180}}$	$\frac{0.785}{\omega_{180}}$
K_p	$\frac{1.9}{G(\omega_{180})}$	$\frac{3}{G(\omega_{180})}$	$\frac{0.6}{G(\omega_{180})}$

V - COMPARISON WITH ZIEGLER-NICHOLS TUNING RULES

Figure 5.1 shows Bode diagram for the compensated system (PID + plant) when the PID is tuned using :

a) Ziegler-Nichols 'second method
b) KLV second method, with the choice given in table 4.2, column (b),
for a plant characterized by the following transfer function :

$$H_p(s) = 1/(1+ s\,50)^2\,(1 + s) \qquad (5.1)$$

Note that the zeros of the PID are (almost) identically in this two cases but the KLV second method gives a much larger K_p.

Comparison of the time responses for these two cases are shown in figure 5.2 showing clearly better performances for the KLV second method with respect to the Ziegler-Nichols second method.

The fact that higher K_p (or equivalently a smaller τ_i) given by the KLV second method provides better results is easily explained by looking to the Bode diagram of figure 5.1. For the Ziegler-Nichols second method, the crossover frequency is in a region characterized by a slope of 60 db/dec close to the breaking point from 60 db/dec to 20 db/dec. The KLV second method gives a K_p which pushes further the crossover frequency in a region where the compensated system has a slope of 20 db/dec at the left and and the right of the crossover frequency.

Table 5.1 gives more detailed results since it includes also the results for the KLV second method, column (a) and the exact "symmetrical optimum".

Table 5.1 : Comparison of calibration methods

Method	Ziegler-Nichols	KLV2 (a)	KLV2 (b)	Exact Sym. Opt.
Phase Margin (°)	18.8	45.8	40	40
Delay Margin (s)	2.01	1.93	1.36	1.45
Gain Margin (db)	10.7	>>	25.6	25
Max Sensitivity function (db)	8.98	3.09	4.04	3.96
Crossover Frequency (r/s)	0.16	0.414	0.512	0.48

Note that the KLV method 2 (table 4.2, column a) which gives larger zeros provides better results in terms of gain,phase and magnitude margin (the inverse of the maximum sensitivity function) with respect to the values given by column b.

VI - SIMULATIONS AND EXPERIMENTAL RESULTS

We will illustrate in the following some of the results obtained with the "auto-calibration" method presented in the previous sections.

6.1. PI controller

Table 6.1 summarizes the results obtained in simulation for a plant transfer function of the form :

$$H_p(s) = \frac{1}{(1 + s\,T)(1 + s)} \qquad (6.1)$$

for different values of T. Two types of PI controller are considered.

a) P + I actions on the error
b) I action on the error and P action on the measurement

Figure 6.1 shows the time responses for T = 8 s in case (a) and (b), and Figure 6.2 responses to load disturbances, for the PI tuned with the exact symmetrical optimum rules and with the auto-calibration method ω_{135}).

In Table 6.1 t_r represents the rise time (at 90 %) and M the percentage of maximum overshoot.

Table 6.1. PI Controller

	Time constant T (s)	4	6	8	10
P + I on error	t_r (s)	2.16	2.29	2.40	2.53
	M (%)	23	28	30	31
I on error	t_r (s)	5.75	5.80	5.96	6.23
P on measur.	M (%)	0	0	0	0

Figure 6.3 shows the results obtained on an air heater. The frequency for a phase lag of 135° has been found using a closed loop relay experiment (ω_{135} = 0.56; G(135) = 0.08). The step responses with the two types of PI controllers are illustrated.

6.2. PID controller

Method 1

To illustrate the method 1 we have considered the following plant transfer function :

$$H_p(s) = \frac{1}{(1 + s\;10)(1 + s)(1 + 0.5\,s)}$$

Table 6.2. summarizes the result for different β in the case of a PID acting on the error without and with filtering the reference. One can see the improvement of performances with respect to a PI controller.

Table 6.2. PID / Method 1

	acceleration factor β	1,5	2	4	PI
without filter	t_r (s)	2.78	1.95	1.01	4.78
	M (%)	9.8	13.5	35.1	18.6
with filter	t_r (s)	8.58	6.07	2.52	12.46
	M (%)	0.8	1.3	1.1	0

The experiments on the air heater with a PID, using accelerating factors β = 1.5, 2 are shown in Figure 6.4.

Interesting results are obtained on an industrial plant, at Heating Company of Grenoble.The controlled loop is the network differential pressure and the functional diagram is shown in Figure 6.5.

A closed loop relay experiment was realised and the values of ω_{135} = 0.765 r/s; G(135) = - 5.8 db were founded.

The step response of the closed loop with a PID using an accelerating factor β = 2 is shown in Figure 6.6.

Method 2 :

In addition to the results given in section 5, Table 6.3 summarizes the results obtained witht the two adjustements indicated in table 4.2 (columns a and b) without filtering the reference for the same plant considered in section 5.

Table 6.3. PID / Method 2

Adj.	$\angle G$	$\angle \Phi$	t_r (s)	M (%)
KLV/PID2 (a)	>>	45.8	3.38	30
KLV/PID2 (b)	25.6	40	2.74	41

VII - CONCLUSIONS

An auto-calibration method for PI and PID controllers based on the knowledge of a point of the plant frequency characteristics (gain and phase) has been presented. The underlying design method is based on the "symmetrical optimum" introduced by Kessler [1958].

The main advantages of this method are the followings :

1) It is a simple method of the same complexity as the Ziegler-Nichols' second method.

2) It gives better results than the Ziegler-Nichols' second method.

3) The closed-loop characteristics (time, response, overshoot) can be computed easily.

4) Gives clear indications for "accelerating" or "slowing down" (T_Σ, α, β).

5) It has a large fiel dof application (ratio $T/T_\Sigma > 4$ (2)).

6) It assures good robustness margin.

7) The auto-calibration method can be easily extended to incorporate plant parameter variation.

8) The test signals on real plants are small (0.2 V).

REFERENCES

Kessler C. (1958). "Das Symmetrische Optimum", Regelungstetechnik 6, n° 11, pp. 395-400, n°12, pp. 432-436.

Ogata K. (1990). "Modern Control Engineering", Prentice Hall, Englewood Cliffs.

Ziegler J.G., Nichols N.B (1942). "Optimum Settings for Automatic Controllers", ASME Trans. 64, pp. 759-768.

Aström K.J., Hagglund T. (1988). "Automatic Tuning of PID Regulators", I.S.A., Research Triangle Parc.

Landau I.D., Grossu A.L., Gavat St. (1968). "L'utilisation de la commande adaptative d'un modèle dans la régulation des moteurs électriques", Automatisme 13, n°4, pp. 146-152.

Aström K.J., Hagglund T. (1984). "Automatic Tuning of Simple Regulators with Specifications on Phase and Amplitude Margins", Automatica 20, pp. 645-650.

Persson P., Aström K.J. (1992). "Dominant Pole Design - A Unified View of PID Controller Tuning", IFAC-ACASP 92 Symposium, Grenoble July, 1-3.

Landau I.D., Voda A. (1992). "The "Symmetrische Optimum" and the Auto-Calibration of PID Controllers", IFAC-ACASP 92 Symposium, Grenoble July, 1-3.

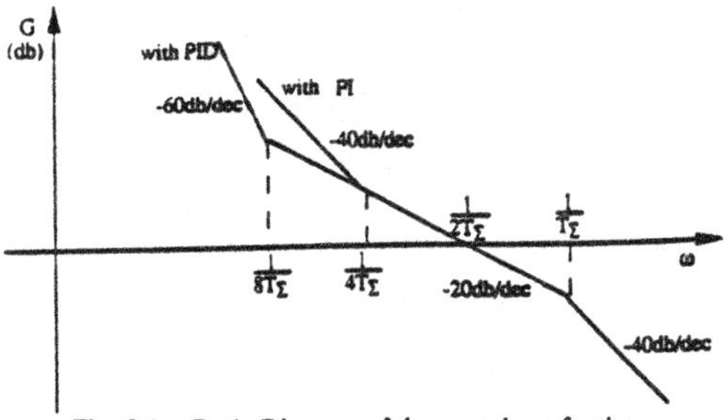

Fig. 2.1 Bode Diagram of the open loop for the "symmetrical optimum"

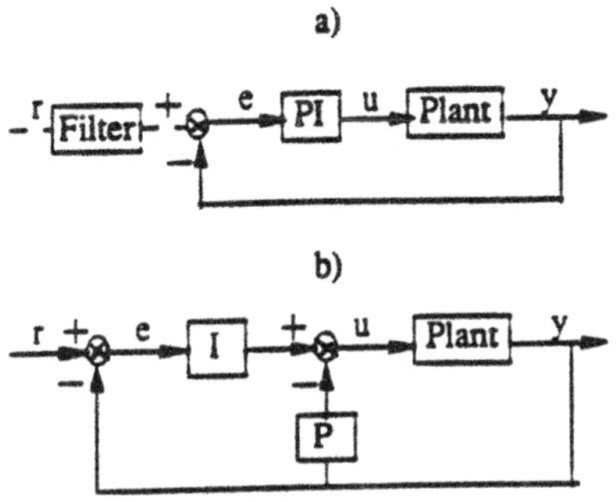

Fig. 2.2 PI Controller:
a) PI on the error
b) PI with I action on the error and P action on the mesurement

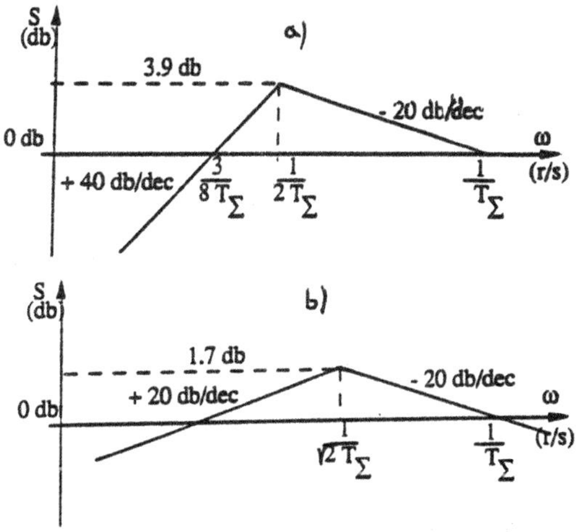

Fig. 2.3. Sensitivity Bode diagram for closed loop with PI controller :
a) integrator plant
b) $T_1 = 4 T_\Sigma$

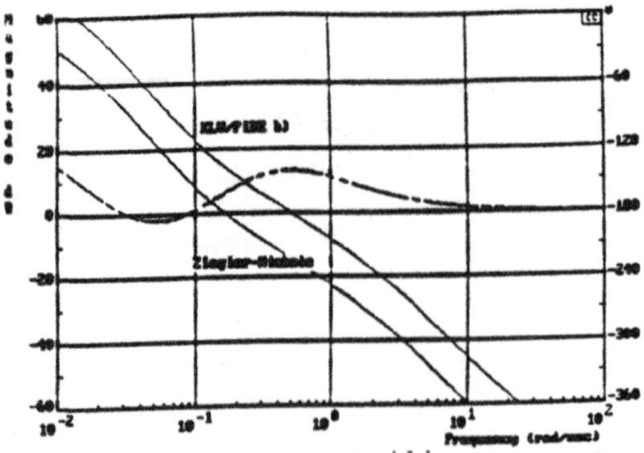

Fig. 5.1 Bode diagram for Ziegler-Nichols 2nd method and KLV2 method

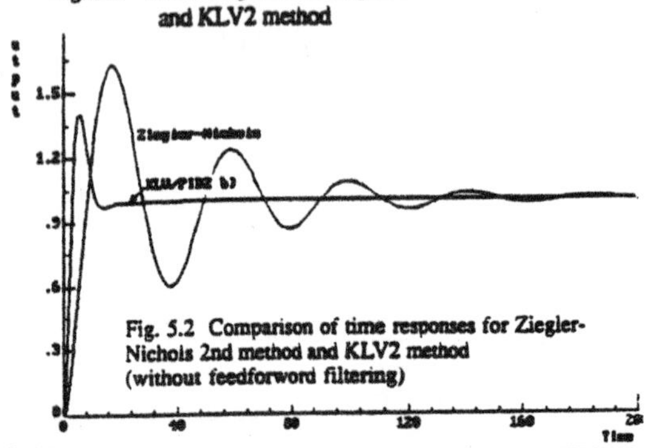

Fig. 5.2 Comparison of time responses for Ziegler-Nichols 2nd method and KLV2 method (without feedforword filtering)

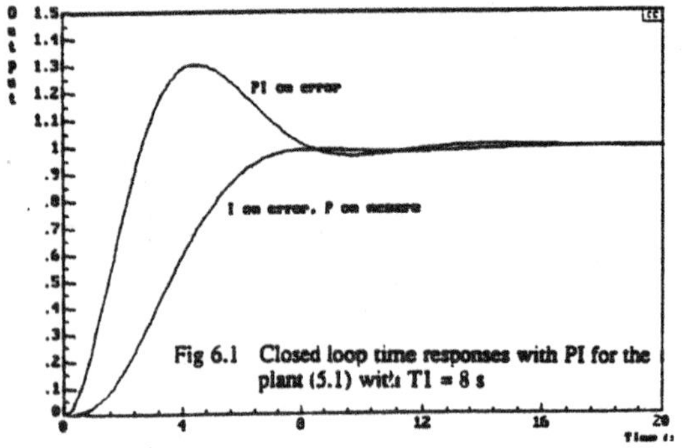

Fig 6.1 Closed loop time responses with PI for the plant (5.1) with T1 = 8 s

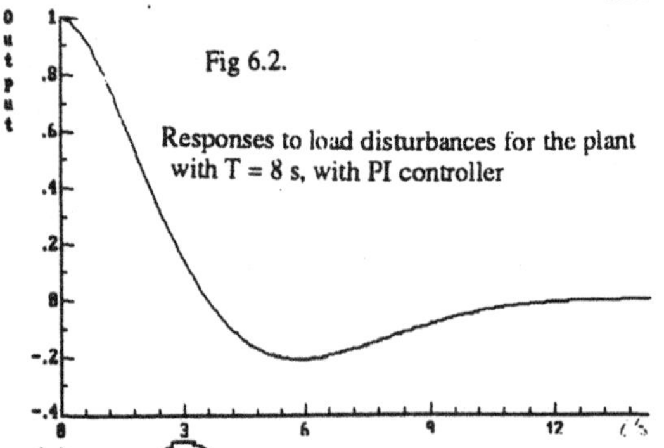

Fig 6.2.

Responses to load disturbances for the plant with T = 8 s, with PI controller

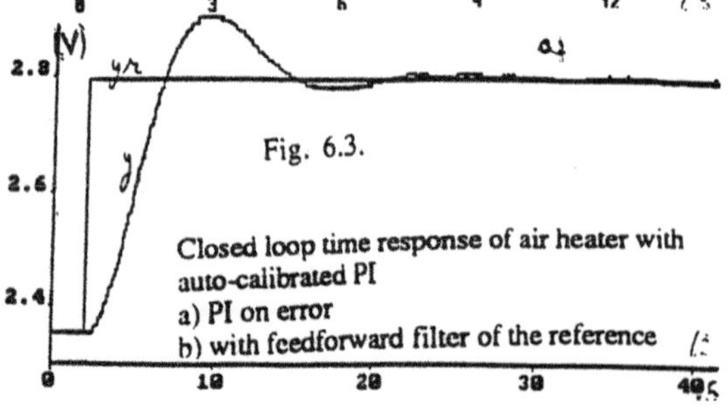

Fig. 6.3.

Closed loop time response of air heater with auto-calibrated PI
a) PI on error
b) with feedforward filter of the reference

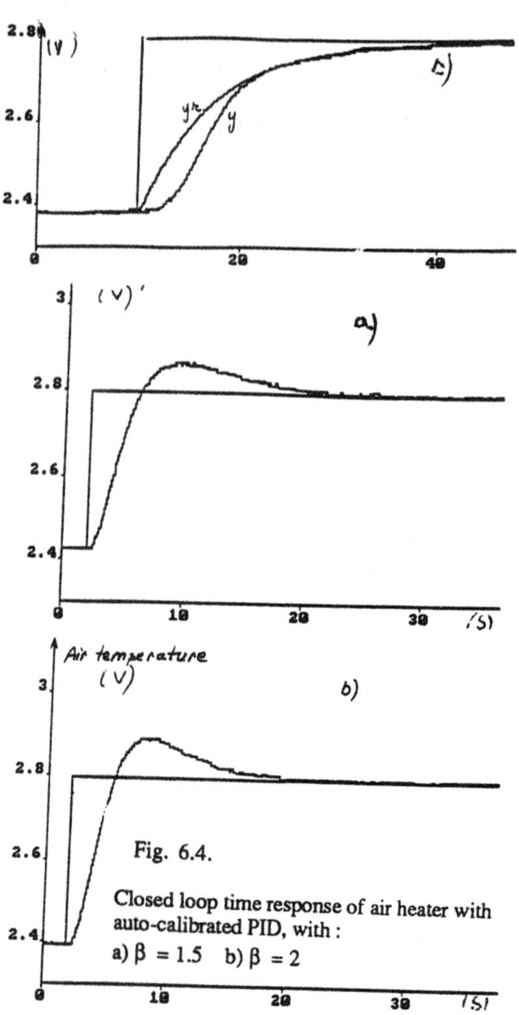

Fig. 6.4.

Closed loop time response of air heater with auto-calibrated PID, with :
a) $\beta = 1.5$ b) $\beta = 2$

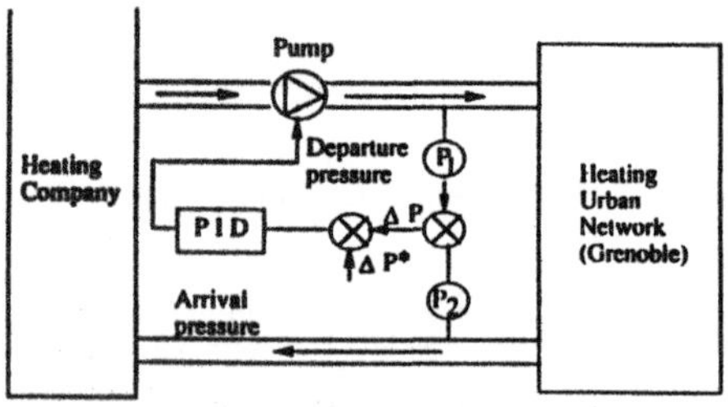

Fig. 6.5. Functional diagram of the network differential pressure from the Heating Company

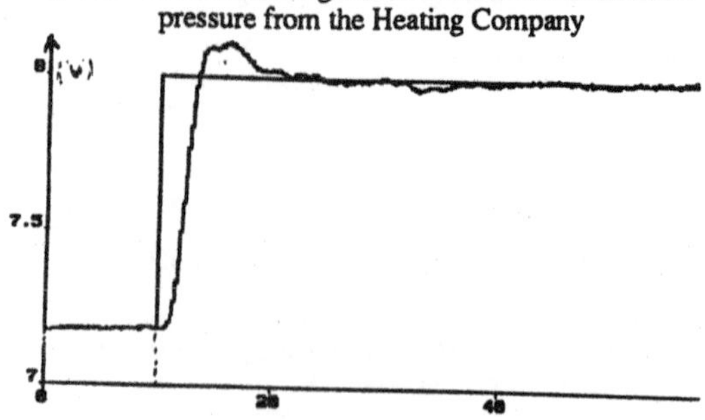

Fig. 6.6. Step response of the closed loop with auto calibrated PID, for $\beta = 2$

STEAM CONSUMPTION CONTROL SYSTEM FOR BATCH PULP COOKING

J. Petrovčič, A. Bitenc and S. Strmčnik

Department of Computer Automation and Control, J. Stefan Institute, Ljubljana, Slovenia

Abstract In batch pulp cooking the heating steam is used only in some phases of the technological procedure. This causes very irregular total steam consumption. High steam flow-rate variations impose a high dynamic load at the steam generation plant which reduces the efficiency of the steam generation. In this contribution a system for steam consumption control is presented, which is based on the three multiloop microprocessor-based controllers. The implemented control structure consists of a set of temperature controllers, a set of "smooth" temperature set-point generators, a set of state-machine algorithms for sequential control, a steam distribution algorithm, a simple steam flow-rate estimation and a steam flow-rate controller. The control structure is simple enough to allow the implementation as a stand-alone small-size control system. The low-cost solution is provided also by elimination of the need for a separate steam flow-rate measuring system for each digester's heat exchanger. After the installation a reduction of 70% in steam flow-rate variance was measured. The reduction may be adjusted by the operator as a compromise between the acceptable dynamic loading of the steam boiler and the requested pulp production rate.

Keywords Flow control, Microcomputer based control, Pulp industry

1. INTRODUCTION

Batch pulp cooking process consists of several technological phases. Among them the most important are filling the digesters with wood chips and acid, pre-heating and impregnation, heating, cooking, relieving and emptying. The steam flow is usually used as the energy source, which is needed only during the pre-heating and heating phases. Usually several digester systems are sequenced to achieve a near-continuous production rate and also to level the total steam consumption. However, only in the optimal sequencing of digesters, which is in practice very seldom, may approximately constant total steam flow be achieved. In normal operation the variations of total steam flow are highly irregular. The high capacity steam boilers are therefore essential to cover the consumption peaks but their average load is low. Due to this their efficiency is also low and dynamic loading very high. Fig. 1 shows a hypothetical example of building up the total steam flow (the sequence of only three digester systems are shown).

To decrease the variations in total steam flow-rate several approaches are known:

- implementation of the appropriate scheduling system, which defines the optimum time relations between the technological phases of all digesters in the sequence,
- limiting the velocity of the control of steam flow valves,
- generation of "smooth" temperature set-point profiles which should be well tracked by temperature controllers,
- elimination of any need for manual intervention by steam flow valves during the normal cooking procedure,
- closed-loop control of the total steam flow by redistribution of the steam consumption from pre-heating to heating phase of the technological procedure.

The enumerated measures are usually implemented on relatively large process computers in the pulp cooking control systems. They require a few tenths of measured process variables per digester and the cost of the system is very high. Based on some experiments and on analysis of the operation of a middle-sized pulp cooking plant, we have realized, that it would be possible to implement a stand-alone steam consumption control system, which would not require so many process measurements and large process computer for the implementation, but would provide a satisfactory leveling of the total steam consumption. The main idea was to separate the implementation of the optimal scheduling system and the steam flow-control system. In this respect the highest quality of the steam flow control could not be obtained, but at the expense of a little lower effectiveness the steam flow leveling system becomes much simpler, easier to understand and to operate, cheaper and uses significantly less measuring equipment.

In this paper we would like to present the design of a low-cost steam-leveling system, which was implemented on three in-house developed microcomputer-based multiloop controllers, which uses only one additional measurement of the total steam consumption beside the normal measurement of digester temperatures, and controls the total steam flow very efficiently. The positive effects of the steam flow control were measured, recorded and compared to non-control operation and are given at the end of the paper.

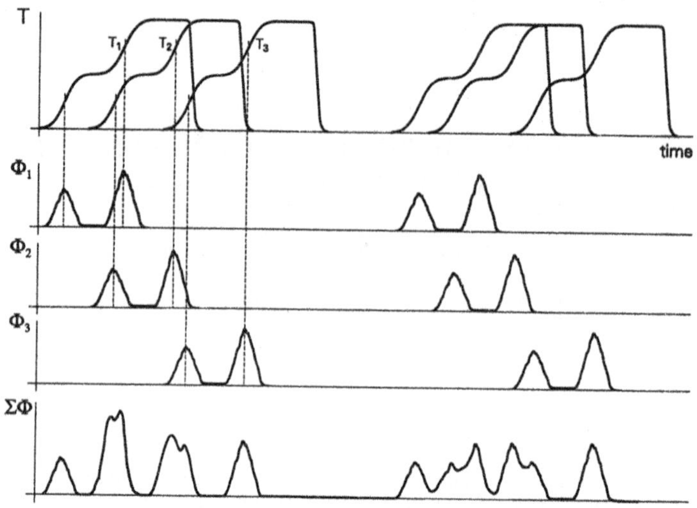

Fig. 1. Hypothetical example of summation of individual steam flow consumptions. Batch pulp cooking produces very high variance of the total steam flow.

2. DESIGN AND OPERATION OF THE STEAM CONSUMPTION CONTROL SYSTEM

The possibility for the steam consumption control is based on specific properties of the batch pulp cooking. During the pre-heating phase of the pulp cooking, the chemical reactions are practically not yet started, only the impregnation of the wood chips occur. In the heating phase, chemical reactions start and the technological process should be controlled very accurately to meet the desired quality of the product. Low sensitivity of the product quality on temperature variations in the pre-heating phase gives us the possibility to redistribute the steam consumption from the digesters in the pre-heating phase to digesters in the heating phase. With this an effective leveling of the consumption could be achieved. When there are more digesters in the heating phase, steam flow should be transferred from digesters in the pre-heating phase and when there is not enough consumption for heating, the steam flow should be used to speed up digesters in the pre-heating phase.

Fig. 2 explains the influence of steam flow manipulation on the temperature of the circulating acid of each digester by showing a hypothetical course of temperatures and steam flow versus time. The set-point temperature profile is generated by means of a second order set-point generator which is designed to provide smooth temperature transitions and consequently smooth steam flow transitions basically desired. In the heating phase the digester temperature tracks the set-point profile very accurately by means of a precise temperature control and by using the steam as needed. Here the steam flow should not be limited due to the technological restrictions as indicated by the 100% limiting value of the temperature controller. In the cooking phase the reaction is exothermic and normally no steam is used.

In the pre-heating phase the fastest temperature set-point transition is defined by means of the set-point generator. The maximum speed of the temperature transition from the starting point to the temperature of impregnation (T_{IMP}) is limited accordingly to the restrictions of the technological process. The set-point would be perfectly tracked in case of full availability of steam flow. However if we limit the available steam flow, the digester temperature would rise more slowly, but at the end of the pre-heating phase it should reach the temperature of impregnation. This could be obtained by using a temperature control loop which consists of a temperature set-point generator and a single-loop controller to manipulate the steam value of the heat exchanger. Also the controller should have a variable output limiter, by which the steam valve manipulation could be limited by means of a steam control system. This solution excludes the need for separate steam flow controllers in cascade with temperature controllers. The costs are lower by omitting separate steam flow measuring systems and cascade controllers.

2.1. Temperature control algorithm with variable output limiter

To achieve parallel operation of temperature and steam flow control we had to adopt a classical PID (discrete) algorithm for this especial purpose as indicated in Fig. 3. As a main distinction a variable limiter with the additional signal input is exchanged for a classical parameter defined output limiter. The limiter is included in the "bumpless feedback" loop of the control algorithm to provide smooth manual-to-automatic and automatic-to-manual transitions and protects against "wind-up" of the integrator.

The linearization module was added to the control algorithm to approximately compensate the static nonlinearity of the steam valves and their backlash behavior. A relatively successful linearization gives us an opportunity to take the signal in front of the linearization block as a pre-estimated steam flow signal. In this respect we may call the variable limiter input signal as an "available steam flow" signal. Functions in front of the variable limiter prepare the "available steam flow" signal as follows. The main switch (programmable block) distinguishes between the heating and the pre-heating phase. In the heating phase maximum steam flow is available (100%). The temperature controller operates without any limiting and keeps the digester temperature tracking perfectly its set-point profile. In the pre-heating phase the "available steam flow" signal is given by a steam flow control system by means of two different priority signals Φ_{P1} and Φ_{P2}. Switching between "high priority steam flow" signal Φ_{P1}, "low priority steam flow" signal Φ_{P2} or zero is performed by means of a priority signal. The selected signal then passes through a MIN-MAX structure which defines its minimum and maximum value. The minimum value Φ_{MIN} determines the lowest possible steam flow which

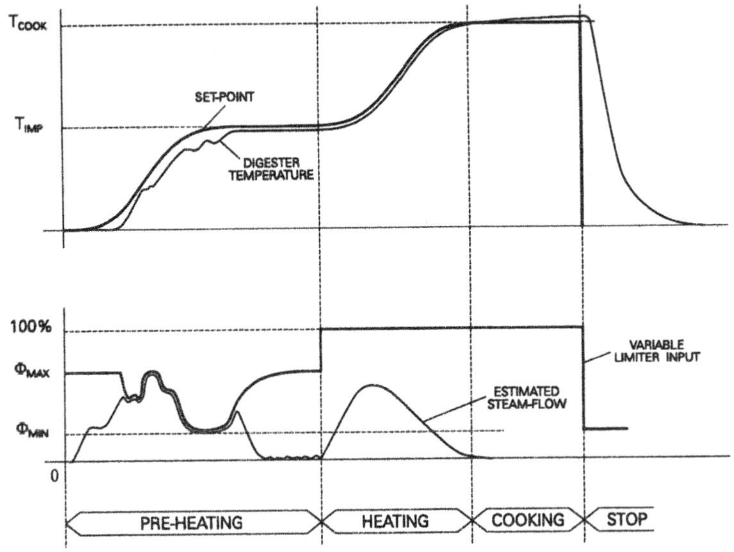

Fig. 2. Limitation of the steam flow in the pre-heating phase. Steam flow limiting in the pre-heating phase causes a deviation between a set-point and measured digester temperature. This has no negative effects on the chemical process of cooking.

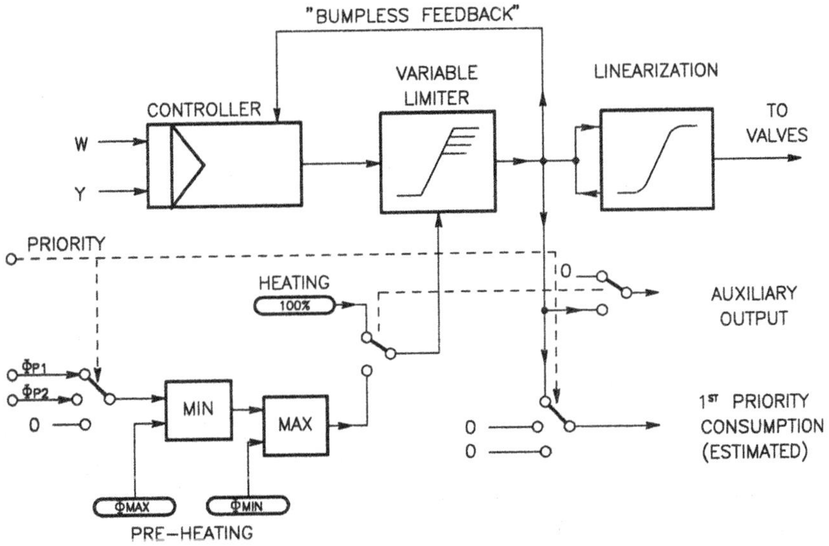

Fig. 3. Block diagram of temperature controller together with variable limiter and linearization. Limiting signal is provided by steam flow control system.

prevents a partially pre-heated digester to cool down in case of extreme deficiency of steam flow. On the other hand, maximum value Φ_{MAX} limits the steam flow in case of a great difference between the set-point and current temperature after a period of long steam flow deficiency, when a sudden increase in steam flow availability occurs.

The built-in PI(D) control algorithm also includes an output signal velocity limiter. It limits the change of steam valve position per time unit to prevent fast changes in steam flow due to noise in temperature or flow measuring signals, amplified by the proportional gain of the controllers. Using velocity limiters, the deterioration of dynamic response of temperature control loops is less noticeable than using ordinary low-pass filtering.

2.2. Steam flow control

The main idea of the solution presented is in using only one measurement of steam flow (total steam flow) and approximating steam flow of each digester by using valve linearization functions. Steam flow control is realized as shown in Fig. 4. Each digester has its own set-point generator, temperature controller with variable limiter and linearization block. Individual set-point generators are triggered by a simple finite state machine, which is driven manually or by means of a main computer via serial link. This state machine is responsible for proper sequencing of technological phases of each digester and not for the general sequencing of all digesters.

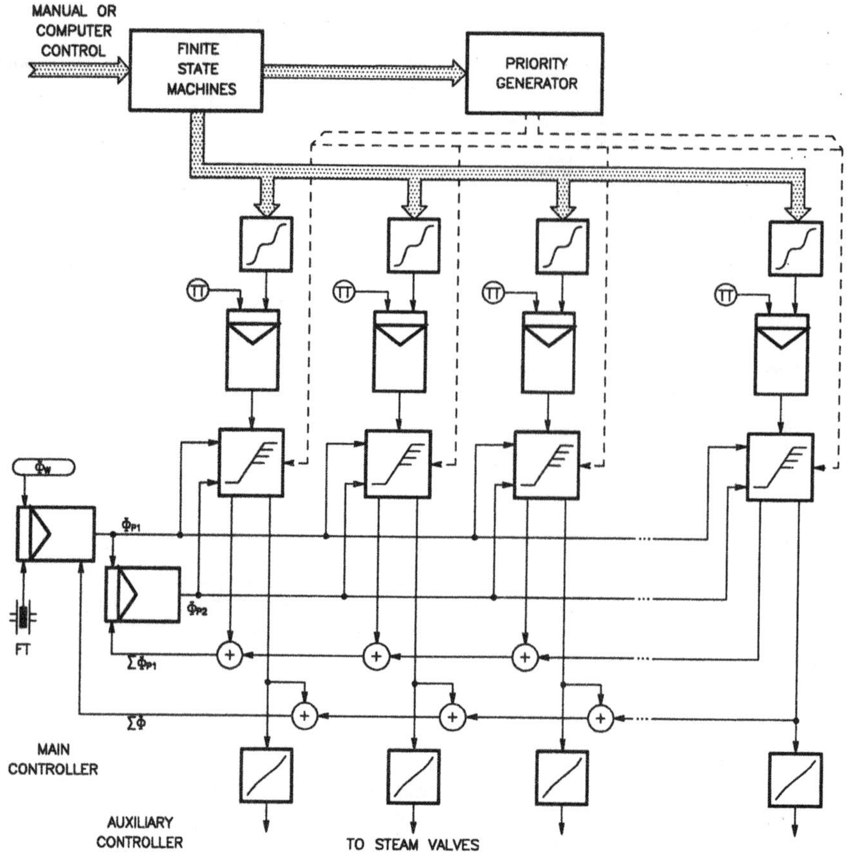

Fig. 4. Simplified block diagram of steam flow leveling system.

Steam flow control is processed by two cascaded controllers. They influence only those temperature control loops, which belong to digesters in the pre-heating phase. The main controller compares the measured total steam flow to desired set-point value Φ_W. Based on their difference it adjusts the value of "available steam flow" Φ_{P1} and sends the signal to variable limiters of all digesters. Only one of the temperature control loops is assigned to first priority on consuming the steam flow for pre-heating. Depending on the temperature in this loop, the temperature controller may take all of the "available steam flow" Φ_{P1} or less. The amount of steam taken is signaled back on the "1st priority consumption" output (see Fig. 3) and returned on a summing line to the auxiliary controller ($\Sigma\Phi_{P1}$). The auxiliary controller calculates a difference between "available steam flow" Φ_{P1} and used flow-approximation ($\Sigma\Phi_{P1}$). The difference is "available steam flow" Φ_{P2} of the second priority and is also sent to all variable limiters. Again only one of the variable limiters of digesters in the pre-heating phase is assigned to the second priority and is allowed to consume the rest of the available steam flow. Other digesters in the pre-heating phase have no priority and may take only the minimum amount of steam flow to keep them warm. All of the approximated steam consumptions of digesters in the pre-heating phase are then summed up ($\Sigma\Phi$) and returned back to main controller. This feed back signal is then increased or decreased by means of the PI algorithm in accordance with the current difference between the desired and measured steam flow. The PI algorithm of the main steam flow controller is designed as a "pure" discrete algorithm with slow sampling rate (2 sec.). The slow sampling rate allows enough time for summed signals to be returned back to the main or auxiliary controller because temperature control loops operate asynchronously with different sampling rates. The feedback signal ($\Sigma\Phi$) is a part of the main controller's integrator and is used to provide "bump-less" transfer from manual-to-automatic mode or vice-versa and to prevent "wind-up" of the integrator. The integrating nature of the main controller substitutes for imperfect linearization of the characteristics of steam valves, so the parameters of linearization functions need not be very accurately adjusted.

The auxiliary controller operates as a simple P-type controller with unity gain but as a controller gives the additional possibility of direct manual control if necessary.

Priorities of steam flow consumption in pre-heating phases are directed by a simple priority generator which is driven by the central finite state machine. The digester which enters into the pre-heating phase first, gets the 1st priority, the next one is assigned to 2nd priority. When a new digester comes into the pre-heating phase, it is attached to the end of the priority queue. When the digester leaves the pre-heating phase, it is removed from the priority queue and the queue is shifted forward.

This system of flow-control could be expanded to more priority levels by using more auxiliary controllers in cascade. By observing normal operation of a 7-digester pulp cooking plant it was found that using only two priority levels (two steam flow distribution levels) has no decreasing influence on the production rate.

2.3. Temperature set-point generators

Smooth temperature transitions from start to end temperatures are essential for keeping steam flow variations basically low. To achieve this we designed a dedicated set-point generator which is presented in Fig. 5. It generates parabola-shaped transitions as used in other application (Petrovčič and Strmčnik, 1988). It should be noted that a classical approach (ramp generator + non-linear function block) is not appropriate for this purpose due to variable starting temperature values of the process.

The set-point generator consists of two cascaded integrators which are driven by a simple finite state machine through "initial condition - operation" (IC/OP) digital inputs. The first integrator I1 determines the slope of temperature transition and saturates at MAXSLOPE or -MAXSLOPE values. It always starts with zero initial condition. The second integrator always starts with the value of measured temperature T_y as an initial condition to prevent sudden steam valve transitions at the beginning of each phase. Integrator I2 after starting, integrates up to the saturating value of the desired end temperature T_{MAX}. The saturation is soft due to early warning by a built-in comparator.

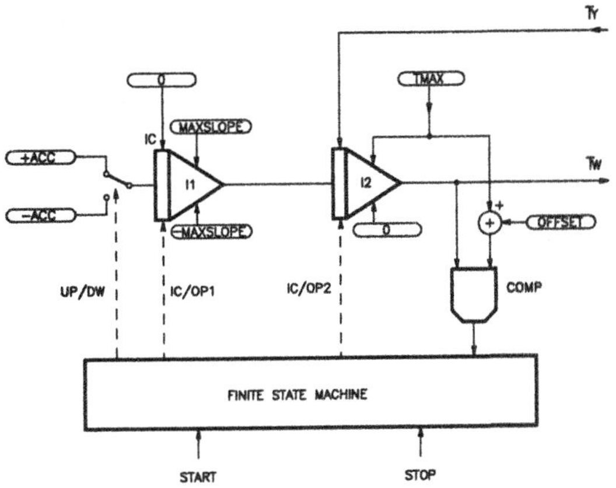

Fig. 5. Second-order temperature set-point. It always starts with the last value of measured temperature.

3. IMPLEMENTATION OF THE SYSTEM AND RESULTING EFFECTS

The system could be implemented on a large process computer but the design goal was to use smaller and cheaper process control components. Also standard programmable single-loop controllers together with programmable logic controllers were not adequate due to the need for special set-point generators and "pure discrete" nature of the steam flow controllers. Modern programmable logic controllers with analog inputs and outputs with the possibility of free programming of some special functions would now be appropriate for the implementation of the presented system.

We have implemented the system of steam consumption control on three in-house developed microcomputer-based multiloop controllers MMC-90 (Bitenc and coworkers, 1992) which are connected together by a simple communication loop. They perform steam flow control as described in this paper, seven temperature control loops and also seven pressure control loops for seven pulp cooking digesters at the Videm Pulp and Paper Factory in Krško, Slovenia. The effectiveness of the system may be presented by diagrams in Fig. 6, where total steam flow consumption were recordered without and with an implemented system. The reduction of steam flow variance is very high and it was measured as approximately 70%. This reduction significantly reduces the loading of the steam boilers. Their effectiveness increases. It was estimated by the user that at this means that the pay-off time of the system implemented would be only 3 to 4 months.

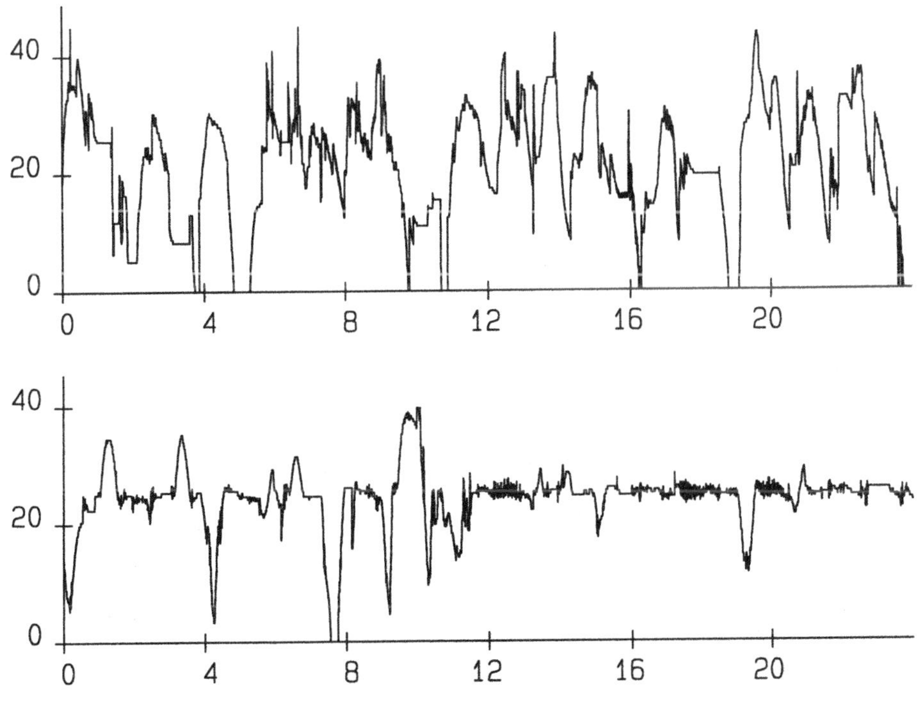

Fig. 6. Total steam consumption (24 hour recording); upper diagram: without steam flow control; lower diagram: after the application of the steam flow consumption control system.

Setting the optimal steam flow set-point is not a trivial task for the operator. This setting influences not only the average steam flow level, but also the steam flow variance and the production rate. At optimum setting the variance is minimum. Increasing or decreasing the set-point value from the optimum increases the variance of the steam flow. On the other hand a lower set-point means lower production-rate and vice-versa. The operator gets the opportunity to optimize the production rate in respect of the quality of steam consumption.

4. CONCLUSION

The system presented was designed as a low-cost solution. This was achieved by omitting separate steam flow control for each digester of the pulp cooking plant and using approximate linearization of steam valve characteristic. A low cost was gained also by excluding the system for digester sequencing which normally requires a great amount of additional process sensors and actuators. It was shown that the system of steam flow control alone may be very effective and gives significant reduction of steam flow variance.

Of course, steam flow control would be even better if temperature controllers were cascaded also to individual steam flow controllers. In this case less periodical adjustments of the parameters are expected. However, during two years of operation only after major maintenance of steam valves some small changes on system parameters (linearization curves and control parameters) were needed.

The system designed is not dedicated to pulp cooking plants only. By small modifications it would be appropriate also to similar batch-wise technological processes.

REFERENCES

Bitenc, A., J. Čretnik, J. Petrovčič and S. Strmčnik (1992). Design and application of an industrial controller. *IEE Computing & Control Engineering Journal*, 3, 29-34.

Hvala, N., S. Strmčnik, J. Černetič (1991). Scheduling of batch chemical reactors on parallel production lines. *Preprints of IEEE Mediteranean El. Conf. MELECON 91, Ljubljana*, B. Zajc and F. Solina (Ed), 2, pp. 880-883.

Juričić, Đ., S. Strmčnik and J. Petrovčič (1986). Compensator for static nonlinearities - design and application. *Electronic letters*, 22, 532-534.

Petrovčič, J., S. Strmčnik (1988). A microcomputer-based speed controller for lift drives. *IEEE Transactions on industry applications*, 24, 487-498.

ON-LINE IDENTIFICATION OF A THERMAL CONDITIONING SYSTEM

V. Manni*, C. Minelli*, S. Vecchiotti*, C.M. Ciaramelletti** and A. De Carli**

**Elasis Scpa, Research Centre for Control and Telecommunication Systems, Via Salaria Km. 91, (Alcatel-Telettra), 02015 Cittaducale (RI), Italy*

***Department of Computers and Systems Sciences, University of Roma, "La Sapienza", Via Eudossiana 18, 00184 Roma, Italy*

Abstract A Low-Cost approach to improve the performances of a Computer Integrated Manufactory System by means of on-line identification of industrial plants is discussed. The paper shows the experiences gathered in applying to a thermal conditioning system the theoretical approach to identification and adaptive control. Off-line and on-line identification has been carried out, with a particular effort for determining the best data generation inputs for the system. The GPC adaptive control algorithm has been tested in this application. Some critical points have been pointed out in this phase and some solutions have been proposed.

Keywords process parameter estimation, identification, predictive control, adaptive control, control applications

INTRODUCTION

In many applications the updating of the already installed supervisory device realizes the Low Cost Automation of an industrial plant, since the hardware structure of the supervisory device can in general support the programs for the on-line control.

Following this trend, the more relevant problem is to work out the programs for the on-line control according to a control strategy getting appreciable improvements to the behaviour of the plant. Since customers of control systems are generally unable to provide good information on the static and dynamic characteristics of the plant, the designer of the control strategy should first define the model of the plant. To obtain a feasible model, it is necessary to match the value of some parameters to the real operating conditions of the plant. The processing of a suitable set of measurements allows to determine the value of the uncertain parameters and to apply an adaptive control strategy for attaining the performance specifications in the whole range of the operating conditions.

According to this trend, the research centre of ELASIS Company developed some years ago a Computer Integrated Manufacturing System, labelled PRODAS (Process Optimization and Data Acquisition System). The first realizations of this device are now mainly devoted to problems connected to the management of energetic flows (Mattiacci and Franceschini, 1990). The hardware structure of PRODAS system allows its application to many different plants.

The Level 1 of PRODAS system manages the processing of the Input/Output variables and implements the control of plant according to a PLC procedure. The extension to the on-line control requires the implementation of the software for an adaptive control.

The aim of this paper is to present the experiences effected by ELASIS Company to attain the above mentioned results. The procedure for deducing the structure of model and for the off-line computation of the parameters will first be presented. Subsequently, moving from the off-line identification procedure, the on-line identification programs have been implemented on the device. An adaptive controller is finally proposed and some validation tests are presented. Since the power involved in real plants is very high, all the experiments have been carried out by means of a laboratory model. It has the same instrumentation of a real plant and a quite similar control of the thermal flow.

PROBLEM APPROACH

The most general approach to the problem of modelling and identification is to consider the plant as a "black box" and to deduce the static and dynamic characteristic by applying suitable waveforms of the input variables for stimulating the plants. The quicker way to attain the parameter identification is to use the existing software libraries in an MS-DOS environment instead of modify the firmware. The computer utilized is a portable PC-AT, the same one is used as development system for PRODAS Level 1.

Since the main target of the control is the tracking of the temperature set-point, the first step is to find out the static characteristic of the plant and to linearize it if strictly necessary. The second step is to identify the fundamental dynamics by applying an off-line procedure. The third step is to work out an on-line procedure suitable for the application of an adaptive control strategy, that realize the target of the on-line control.

Adaptive control strategy overcomes the different plant characteristics (e.g. actuators, rooms dimensions, distance between actuators and transducers...) and variations in endogenous and exogenous heats.

The Generalised Predictive Control (GPC) (Clarke, Mohtadi and Tuffs, 1987) has been implemented so as to obtain the insensitivity to over-parameterization and "dead time" variations. Moreover GPC procedure is very indicated since the synthesised controller has an inherent integral action.

EXPERIMENTAL SUPPORT

Laboratory Model

The laboratory model has been realized so as to simulate the heat propagation in a real thermal system. An insulated box, equipped by a power resistors set and a constant speed fan, has simulated the thermal conditioning system. The control of the heat flow has been effected by acting on the voltage supply of a power amplifier feeding the resistors. The temperature inside the box has been measured by a linearized industrial PT-100 transducer.

Data Processing

A Metrabyte's DAS-8/AO interface board has been used to convert the Analog Inputs into a 12 bits format. The program developed for the identification effects the real time monitoring of the I/O variables and sets the parameters of the anti-aliasing filter and the sampling rate. This latter should be fixed by considering also the effects on the plant of the stepwise variable used for its control (see Modified PRBS).

The analysis of the data has fundamentally been effected by Matlab Identification Toolbox implemented on a PC-AT (Ljung, 1985). The functions for the parametric identification got reliable results. On the contrary, SPA function for the non-parametric identification has not given quite satisfactory results. Therefore a dedicated function has been developed on the basis of FFT Matlab function.

IDENTIFICATION

System Non-Linearities

The first experiments has been aimed at obtaining the static characteristic and some hints about the dynamic of the plant. A set of increasing amplitude steps has been imposed and the steady state amplitude of the output variable has been measured. The rise time has been also deduced.

A minimum threshold level and an approximately quadratic static characteristic has been pointed out. The first one is due to the amplifier and the second one to the resistors behaviour.

The rise time has shown no sensible variation inside the range of the step inputs. A linearized model can therefore describe the plant dynamics. Only the gain should be linearized for matching the operating conditions.

Test Signals

A Pseudo Random Binary Sequence, PRBS (Landau, 1988), has been used for stimulating the plant. Fig. 1 illustrates some periods of a PRBS waveform.

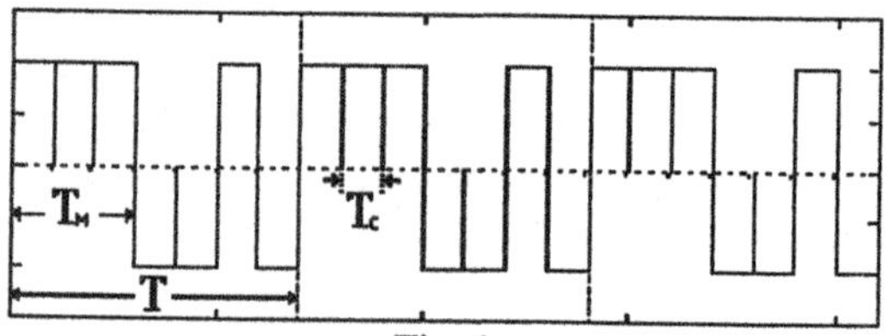

Fig. 1

The PRBS is generated by a dedicated instruction implementing a binary shift register with feedback. The number of bits of the register and the period T of PRBS are strictly related. If N indicates the number of bits and T_c the clock period of the shift register, the PRBS period results:

$$T = nT_c = (2^N\text{-}1)T_c$$

Clock period T_c coincides with the width of the minimum duration pulse, and should be chosen so as $1/T_c$ is bigger than the bandwidth of the plant. The width of the maximum duration pulse T_M should be determined so as so attain the steady state. The number N of the bits, used for producing PRBS, is equal to the rate between the maximum and minimum width.

The harmonic response of the plant realize the non-parametric model used for the identification. It has been computed in terms of the FFT of input and output variables. Reliable results are obtained by neglecting the starting up response and by effecting the FFT in an integer number of PRBS periods. Fig. 2 illustrates the harmonic response obtained by this approach and demonstrated that the fundamental dynamics of the plant is very well evidenced.

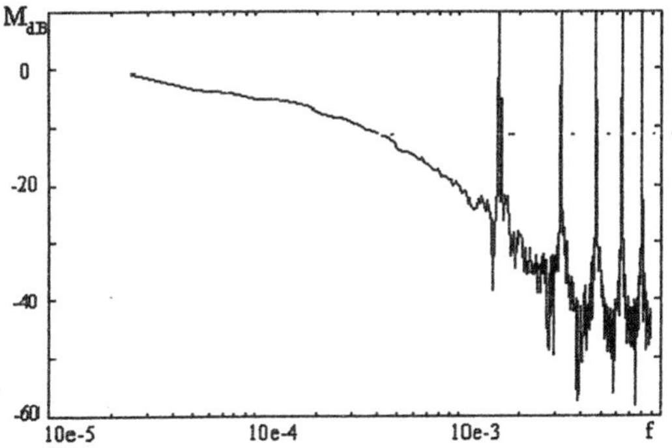

Fig. 2

The clearly visible peaks in the harmonic response in correspondence to the $1/T_c$ frequency (and its multiples) are due to the division for the FFT of the PRBS, which goes to zero at that frequencies (Fig. 4).

Modified PRBS

Unfortunately PRBS inputs could produce abnormal solicitations of the plant and possible damages. The reduction of the amplitude makes more difficult the signal processing for the identification due to the superimposed noise. The PRBS waveform should be therefore modified in order to avoid the injury of the plant. The most simple way to obtain this results, without relevant modification to the harmonic content, consists in linear shaping of the initial and final edges of the pulses as shown in Fig. 3.

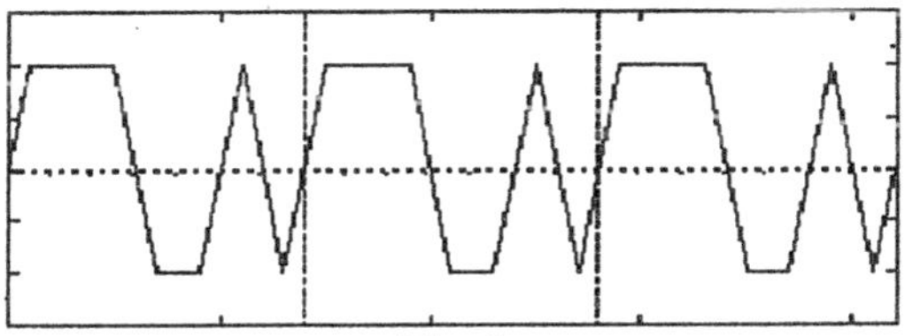

Fig. 3

The harmonic content of the modified PRBS (slashed one) and the original one (continuos line) are represented in Fig. 4.
The comparison shows a faster decay in the higher frequency range and a small amplitude reduction in frequency range lower than the cross-over frequency. Improvements are obtained by shaping the pulse edges with a sinusoidal waveform (this could require an increase of the sampling rate of the digital control).

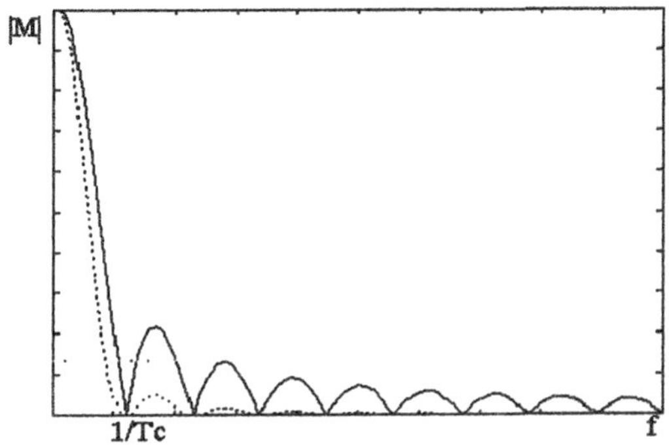

Fig. 4

Off-Line Identification

This step is pointed to the fitting of the best structure for the dynamic model and the estimation of the parameters. The identification procedures already available in the Matlab Identification Toolbox have been used. Moreover the harmonic response of the model has been compared with the harmonic response obtained by processing the real data set. This latter was recorded by using a sampling rate of .04 Hz and a fourth order Bessel filter with a cutoff frequency of .02 Hz.
The range of the data set has been normalized between 0 and 1 values, subtracting their minimum value and dividing for the difference between maximum and minimum value.

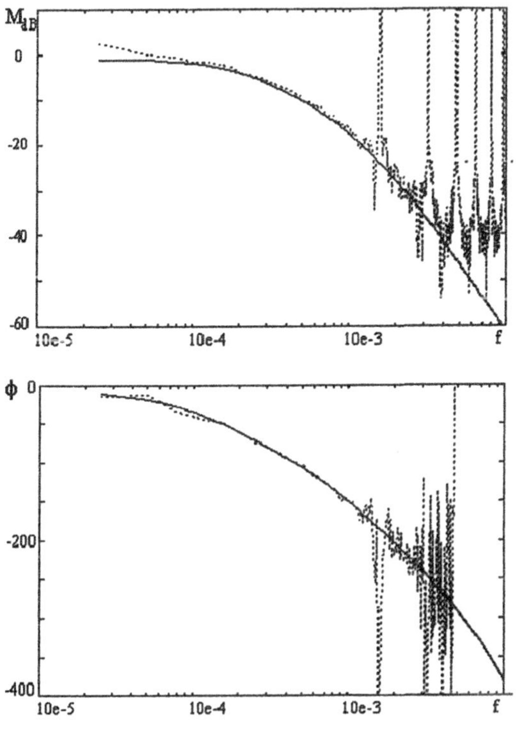

Fig. 5

The Matlab identification functions have been used and the results have been compared. The criterion already present in Matlab (Ljung, 1985) have been used to validate the identified model (structure and values of parameters), together with the comparison between the frequency analysis of both results and recorded data. The ARMAX procedure gives the most feasible model. It results

$$G(z)=\frac{b_1z^{-4}+b_2z^{-5}+b_3z^{-6}}{1+a_1z^{-1}+a_2z^{-2}+a_3z^{-3}+a_4z^{-4}+a_5z^{-5}}$$

Fig. 5 compares the amplitude and the phase of the model with the values worked out by processing the recorded data.

On-Line Identification

The on-line identification has been effected by accepting the structure of the model determined by means of the off-line identification procedure. Improvements of accuracy can be obtained by taking into account the prediction error. This latter can be easily estimated in an on-line procedure.
The effects of a possible bias have been avoided by a parameter p independent of other variables. Therefore the structure of the model has resulted:

$$\begin{aligned} y(k) = {} & a_1y(k-1) + a_2y(k-2) + a_3y(k-3) + a_4y(k-4) + \\ & a_5y(k-5) + b_1u(k-4) + b_2u(k-5) + b_3u(k-6) + \\ & c_1e(k-1) + c_2e(k-2) + c_3e(k-3) + c_4e(k-4) + \\ & c_5e(k-5) + p \end{aligned}$$

Variations of the time delay are easily taken into account by acting on the number of the b_i coefficients.
The recursive identification procedure chosen (ELS) allows to update the value of all the coefficients. Both ARX and ARMAX approach have been tested. Similar results have been obtained as concerns accuracy, while ARX needs less computing time and memory.
Fig. 6 compares the harmonic response obtained by the on-line parametric identification procedure with the non-parametric one. By neglecting the bias coefficient p a discrepancy in the lower frequency range appears as shown in Fig. 7.

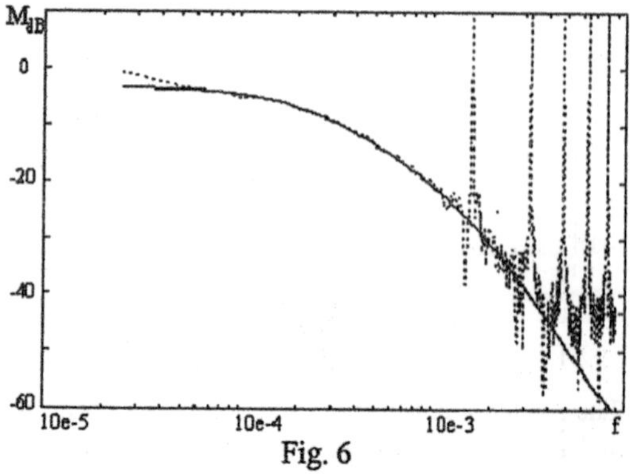

Fig. 6

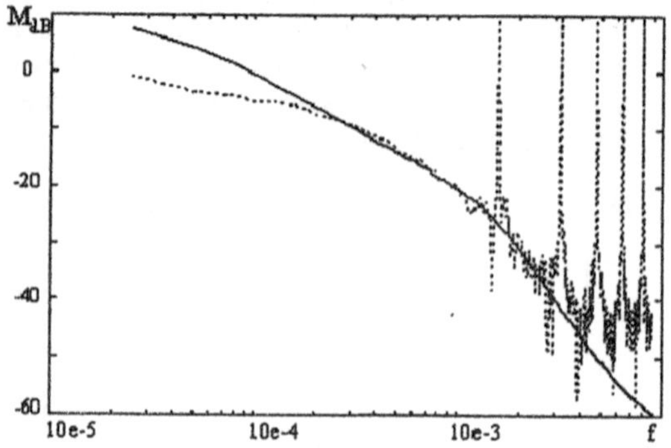

Fig. 7

ADAPTIVE CONTROL

To obtain adaptive control, the GPC control algorithm has been joined to the on-line identification procedure. The main problem is to tune the on-line identification procedure to the real operating conditions, in order to avoid critical situations in the adaptive control.
In fact the target of adaptive control is to attain a good regulation by reducing the transients due to tracking of the set point. The steady state operating conditions is therefore the target of the regulation, but in this situation the identification procedure gives unfeasible values of the parameters. A low quality performance of the adaptive control system results if suitable expedients are not used.

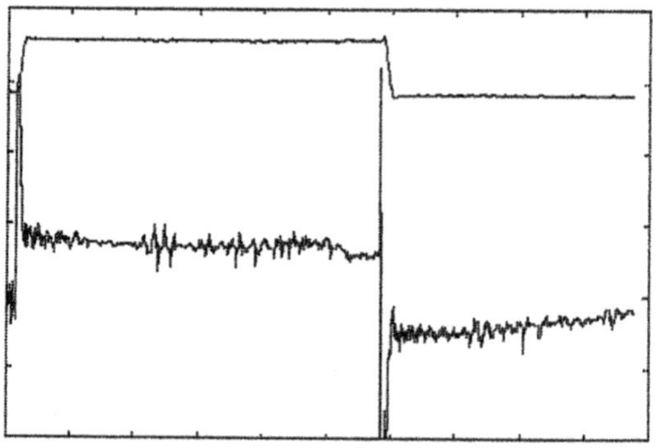
Fig. 8

Fig. 8 represents two step transients for an adaptive control implemented without introducing suitable expedients. The upper bound shows the waveform of the output variable and the lower bound the waveform of the control action. It could be seen as, leaving the set-point unchanged for a certain time, the output variable gives no information to the identification procedure. Consequently, the identification procedure started to drift, leading to wrong control action, clearly visible in the second step transient.
The expedient for this problem consists to get the identification procedure in a sleeping condition when the plant operates practically in steady state.

An other problem, related more strictly to the GPC procedure, is the existence of a relevant control activity during the time interval that should characterize the steady state operations (see again Fig. 8).

Relevant improvements have been obtained by assigning a variable weighting factor to the predicted control action, in order to avoid the sensitivity of the control action to the quantization errors when a steady state operating condition is attained. The actual value of the weighting coefficient $\lambda(t)$ depends on the distance between the actual output variable and the steady state one, fixed by means of the set point. It is computed by means of the following relationship:

$$\lambda(t) = \lambda_{\min} + (\lambda_{\max} - \lambda_{\min}) \cdot e^{-\eta(t)}$$

where

$$\eta(t) = \rho \cdot \eta(t-1) + | y(t) - y^* | \cdot (1-\rho)$$

$\lambda_{\min}$, $\lambda_{\max}$ are respectively the minimum and maximum desired value of λ;
ρ is a forgetting factor ranged [0..1];
$y(t)$ is the controlled variable;
y^* is the actual value of the controlled variable corresponding to the set point.

This relationship is structured so as to obtain reduced values of the control action when the controlled variable is close to the steady state operation.

Fig. 9 reproduces the same transitory conditions of Fig. 8 and shows the obtained improvements. In fact, in correspondence of the steady state, the control action is very smooth and assumes small amplitude in transitory conditions too.

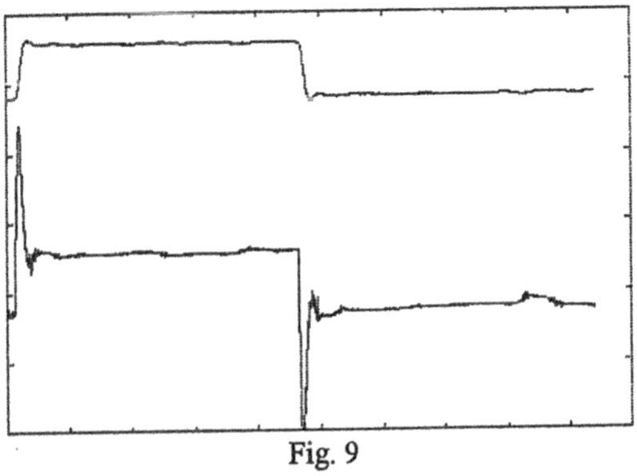

Fig. 9

CONCLUSIONS

This paper describes a low-cost strategy to improve the performances of control systems.

Some theoretical approaches (Landau and Ljung) to identification and adaptive control have been tested in a MS-DOS environment (a PC AT equipped with an I/O card).

Regarding the identification theoretical approach, it has been pointed out that the PRBS signals can be modified to become less stressing for the plant, without loosing information content.

A systematic procedure has been described, that could be successfully followed to face the identification of an industrial plant, starting from total lack of information.

To pass from off-line to on-line identification, some modifications to the models structure should be applied, in order to consider the non-zero mean disturbances.

The obtained on-line identificator could be easily used to implement an adaptive controller. The result, especially suitable in industrial contexts, could be an all-purpose controller, to apply to a wide number of plants without any tuning.

The theoretical approach followed in the implementation of the adaptive controller has been the GPC. The experiences collected during the identification phase have been used to prevent controller misbehaving due to the identification drift in steady state operating conditions.

A strategy has been suggested to reduce the quantization errors effects on the control action, dynamically varying the GPC "control weight".

REFERENCES

Brigham, E.O. (1972). The Fast Fourier Transform, Prentice-Hall Inc., Englewood Cliffs, New Jersey.

Ciaramelletti, C.M. (1991). Metodi di identificazione e tecniche di controllo per sistemi di produzione e distribuzione di fluidi energetici, degree thesis, Dipartimento di Informatica e Sistemistica Universita' "La Sapienza", Roma.

Clarke, D.W., C. Mohtadi, and P.S. Tuffs (1987). Generalized predictive control part I & II. Automatica, 23, N.2, 137-160.

Formichetti, L. (1991). Controllo adattativo di processi industriali con metodi predittivi (GPC), degree thesis, Dipartimento di Informatica e Sistemistica Universita' "La Sapienza", Roma.

Landau, I.D. (1988). System identification and control design using P.I.M. + software, Prentice-Hall Inc., Englewood Cliffs, New Jersey.

Ljung, L. (1985). System identification Toolbox for use with PC Matlab: user's guide, Version 2.0, The Mathworks Inc., Portola Valley, California.

Mattiacci, T., and F. Franceschini (1990). PRODAS process optimization and data acquisition system: un sistema per la gestione integrata dei servizi energetici. L'Elettrotecnica AEI 77, 945-952.

APPLICATION OF VARIABLE-STRUCTURE CONTROL TO AIR-CONDITIONING SYSTEMS

N.E. Gough*, M.S. Srai*, M.J. Leach, G.M. Dimirovski**, O.L. Iliev** and Z.A. Icev**

*Control Systems Group, Faculty of Science & Technology, Wolverhampton Polytechnic, Wolverhampton WV1 1SB, UK

**Laboratory of ASE, Electrotechnical Faculty, P.O. Box 574, St. Cyril and Methodius University, y-91000 Skopje, Republic of Macedonia

Abstract. A modern discontinuous control strategy, based on variable-structure systems control,is proven to be vastly more efficient than the commonly used (optimal) on-off control without the need for additional hardware. It may be acheived, however, on the expence of a higher actuator switching frequency. A wider application in energy control and management systems is suggested.

Keywords. Energy control, variable structure systems and control, energy management systems, robustness.

INTRODUCTION

Communities of our contemporary world have faced the challenge of rational use of energy by memeans of various methods and technologies (*Gough & coworkers, 1988,1992)*, following the two great energy crises diring the seventies. In particular, a concept refered to as energy conservation through control has attracted considerable research effort *Clark (1983), Ashley (1984), Dimirovski (1987), Gough & others (1988,1992),Dimirovski & Tomsic (1989)* One of most important recent trends in the control and management energy systems (EMS) is based on the argument that future EMS control systems will use two/three level controller-based computing systems implementing sophisticated but simple controls, designed on the grounds of approximate but phenomena efficienly representing models, rather than nowdays typical computer process control systems *Ashley (19 84), Dimirovski & Tomsic (1989), Gough & coworkers (1988)*.

The research reported in here is devoted to the EMS type of control problems associated with energy con trol in the air-conditioning technologocal instalations. Our current study has provided a closer insight in the adventages of the application of variable-structure control theory, as well as a promissing VSSM control system design solution *(Srai & others, 1991 a,b)*.

The discontinuous feedback control strategy known as variable-structure, sliding-mode (VSSM) control has originated in Russia by works of Emelyanov (19 67) and his collaborators on SISO systems case. It has been already extensively described in the literature, and by Itkis (1976) and Utkin (1977, 1987) in particular. Further theoretical results of interest in the present study are those of White (1983, 1986).

CONCEPTUALIZATION OF A VARIABLE-STRUCTURE CONTROL

An Theoretical Overview

In its simplest form,the basic theory of VS control provides for results which make the system (A,B) to be constrained to move along the hypersurface $C(x)==0$ by the feedback switching law $u=-Kx$ ensuring

$$C(x)dc(x)/dt<0 \qquad (1)$$

by setting

$$K_i > A^tC/b_i^tC \quad \text{if} \quad b_i^tC>0 \qquad (2)$$

subject to the orientation x with respect to the hyperplane. The inequalities are reversed with the inner product inequality

$$C(x)x_i > 0 \qquad (3)$$

describing this orientation. The form of u is then discontinuous but it can be made symmetric by using Taran's Transformation (Itkis,1976) without loss of generality. Gough & coworkers (1988 a,b, 1992) have shown that this symmetry has a beneficial effect on the initial transient. The control is then described by

$$u= -\sum_i K_i |x|_i \ \mathrm{sgn}(C) \qquad (4)$$

This control can be implemented by using the op-amp circuit as presented in Fig. 1.

The literature describes how uncertainties in the system model and external disturbances can be in most cases completelly overcome using VS control. In particular, Gough & others (1988a, 1992) shows how the time delay, inherent in temperature control systems, can be compensated by using sufficiently high values K for delays of up to ten percent of the object system inertial time constant.

In relation to the ON/OFF control used in common energy management systems, the control gain K represents a more sophisticated form of switching without the need for extra hardware.

A Control Law for Room Temperature Regulation

Figure 2 shows a simple model for the air-conditioning of an isolated room, which may be regarded as a syb-system of a multizone building. The controls k1 and k2 control the humidity and temperature,respectively, and have values of unity and zero representing on and off status in the conventional commonly used systems.

It is shown in the next section that this system may be represented by differential equations of the form

$$\begin{vmatrix} T_1 \\ T_2 \end{vmatrix} = \begin{vmatrix} a_{11} & a_{12} \\ a_{21} & a_{22} \end{vmatrix} \begin{vmatrix} T_1 \\ T_2 \end{vmatrix} + \begin{vmatrix} b_{11} & b_{12} \\ b_{21} & b_{22} \end{vmatrix} \begin{vmatrix} T_1 & T_0 \\ T_4 & T_2 \end{vmatrix} \begin{vmatrix} K_1 \\ K_2 \end{vmatrix} \qquad (5)$$

This is a special case of the form with non-linear f,g,h

$$x = f(x) + g(x)\ u$$
$$y = h(x) \quad (6)$$

used by Fernandez and Hedrick (1987). For this particular application case $f(x) = A\ x$ and $g(x) = g\ x + g0$, where A, g, and g0 are constant matrices, and $h(x) = x_1 = T_1$ for controlling the temperature T_1 in the room. If the hummidity control K_1 is the con stant preset fresh air cycle, then it may be combined in g(x), and the system becomes single-input-single-output (SISO).

For such a system, the control strategy is to apply VS control on the error between the actual output y the desired output, defined as follows

$$e = y - y(\text{desired}) \quad (7)$$

A sliding surface is calculated as a linear combination of time derivatives of e, with the number of terms equal to the linearisability index of the non-linear system (Fernandez & Hedrick, 1987; White 1983). Intuitivelly, the model is a second order system with states T_1, T_2 so that sliding surface

$$S = e + ke = 0$$

is effectivelly the surface

$$S = T_1 + kT_1 = 0. \quad (8)$$

To ensure sliding mode along this surface the control u is designed from (1) to be

$$u = g\ (x)\ \{u + n\ \text{sgn}(S)\} \quad (9)$$

where bar indicates nominal values. The following control law satisfies equation

$$K_2 = 0 \text{ when } T_1 - 19 + k\ T_1 > 0$$
$$K_2 = 1 \text{ when } T_1 - 19 + k\ T_1 < 0 \quad (10)$$

for a desired temperature of 19 degrees centigrade. These switching conditions are illustrated on a phase plane for k = 1 in Fig. 3.

A Simple Object System Model

In the Appendix attached, a description of variables involved is given in detail. Via associating the time derivative of temperature with heat flow, the following deffierential equations are derived:

$$T_1 = \frac{U_2A_2(T_2-(1-K_1)T_1-K_1\ T_0)}{V_1R_1C_1} + \frac{U_1A_1(T_1 - T_0)}{V_1R_1C_1} + \frac{K_1F_1(T_1-T_0)}{V_1} \quad (11)$$

$$T_2 = \frac{K_2F_2R_2C_2(T_4-T_4)}{V_2R_2C_2} + \frac{U_2A_2(T_2-(1-K_1)T_1-K_1\ T_0)}{V_2R_2C_2} \quad (12)$$

The separation of linear terms in the variables and the product terms between the states and inputs obtains the form (5). Nominal values of the flow and transfer coefficients are as follows:

$U_2A_2 = U_1A_1 = 1;$

$F_1R_1C_1 = 0.8;$ $\quad V_1R_1C_1 = 80;$

$F_2R_2C_2 = 1.3;$ $\quad V_2R_2C_2 = 20.$

A CAD TECHNIQUE FOR VSSM CONTROL DESIGN

Some General Remarks on the CAD Technique

Here, a package aimed at CAD of VS control systems, which stands alone in terms of library routines and graphics packages, designed for a PC environment is presented as well as its application in the design of the system considered.

The algorithms for VS design have been developed in Pascal. Their furter development relies on prototyp ing algorithms using the 386 matlab package written in C. Eventually, these functions can be duplicated to form an integral part of the package. A database written in MS C will hold system models of the form shown in Fig. 4, and hold data about the system in a form to facilitate perturbation analysis. The gra phics facilities required need to display time responses, phase plane trajectories and sub-spaces up to dimension 3 and surfaces within them. Fig.4 show the software environment used and the functions ser ved by the individual components.

User Interface and Database

The user interface represents a series of drop-down menus to the user, and these are shown in sequence in Tables 1 to 6. An example will serve to illustra te the kind of analysis required in VS system control design - the multizone building of Fig. 5.

Database involves storing the system model in a separate file in the database writen in C and accessable from Pascal via ASCII representation. The model is created with function SS_create which returns a pointer to structure state_space_sys.

```
Struct State_Space_Sys  *STATESP
      {    unsigned shar flag
           shar name [SS_name_lenght]
           int type
           double dtime
           MATRIX mat_a, mat_b, mat_c, mat_d;
      }
```

```
STATESP SS_create (char *name), uint nins, uint
nouts, uint order, int type, double dtime);
```

```
/* initialises structure*/
```

```
STATESP SS_load (char *filename) /* initialises
SS_create with the file data */
```

```
STATESP SS_save (STATESP SS, char *fiLename)
/* writes *SS data to the file */
```

```
STATESP SS_make (char *name, MATRIX amat,....,
MATRIX dmat, int type, double dtime)
/* initialises directly using parameters */
```

SS_create creates memory for the model in terms of its size - inputs, outputs, states, etc. The model is acctually initialised by SS_make which is passed the actual matrices.

Results and Graphics

Tables etc. can be used to record data but the following essential visual design aids need to be offered for display:

- Display of the trajectories of x.
- Display of 3D sub-spaces of x and 2D surfaces within them.
- Drawing schematics/transfer function diagrams of applications, e.g.multizone buildings Fig.5, etc.

Mathematical Routines and Transformation

System Transfomations to controllable, observable, companion and discrete; forms-use of Mathlab and/or the C library. Singular value decompositions using SVD and USV functions. Solutions of Lyapunov and Riccati matrix equations.

Similarity transformations of the form

$$z = T\ x,$$

where T is a nonsingular matrix, are required to make a study of systems to controllable, observable, companion etc. forms and to discrete forms. The fun ction SIM_TRANS,BAL_TRANS achieve this operation in conjunction with the controllability/observability matrices returned by CONTORL and OBSV.

System Representation, Design and Simulation

The general multivariable system is described by the following equations:

$$x = (A + E)\, x + B\, u$$

where E is required for singular perturbation analysis; and

$$\sigma = C\, x$$

the huperplane required for VS design. Note that B may be of the form B(x).

The discrete model is simulated using the discrete transition form, or its approximation given by

$$x(K+1) = (hA + I)x(K) + Bu(K).$$

Discontinuous VS SISO and NINO algorithms with system simulation coded in Pascal.Inorporating the digital simulation algorithm given above functions VS_SIM for autonomous system, and VS_SIMU for excited systems perrforms this task. IT_SIM allows iteration of parameters during simulations.

Arbitrary selection or users choice for C. Optimise

$$J = \frac{1}{2}\int_0^\infty x_Q^2 + u_R^2\, dt$$

with C solving Riccati equations from Matlab. Eigen value assignment method for C. Allowing for SISO, MIMO or decentralized situations, based on the following algorithms, respectively for canonical transformations x' = Mx; overviw of these algorithms is given bellow.

Algorithm 1:

Step 1: Find M so that it spans B and B nonsingular.

Step 2: Find C_1 to set eigen-values of $A_{11} - A_{12}C_1$ where A_{ij} are portions of $M^{-1}AM^{11}$ and C_1 is the hyperplane.

Step 3: $C = [C_1 \mid I_m][M]$

Algorithm 2:

Step 1: Find M

Step 2: Solve for P: $0 = PA^1 + A^1P - PAQAP + DD$ where $A^1 = A_{11} - A_{12}Q_{22}^{-1}Q_{12}^T$ and Q_{ij} are partitions of Q

Step 3: $C = [\ -Q_{22}^{-1}A_{12}P + Q_{12}^T \mid I_m][M]$

Algorithm 3:

Step 1: Set $C_2 = I_m$ $T = C_1A_{11} - A_{22}C_1 - C_1A_{12}C_1 + A_{21}$
$S = B^{-1}RB^{-1}$
$W = Q_{11}^2 = T^1ST - 2Q_{12}C_1 + C_1^TQ_{22}C_1$
making sure $Q_{11} > 0$.
$Q_{12} = Q$ in choosing M

Step 2: Solve Lyapunov equations
$K(A_{11} - A_{12}C_1) + (A_{11} - A_{12}C_1)^TK = -W$
$L(A_{11} - A_{12}C_1) + (A_{11} - A_{12}C_1)^TL = -I$
for K and L.

Step 3: Solve for C which satisfies
$\delta I/\delta C_i = (A_{12}^T C_1^T - A_{22}^T)STL - STL(C_2^T A_{12}^T + A_{11}^T) + (Q_{22}C_1 - Q_{12}^T)L - A_{12}^T KL = 0$

Step 4: Solve for C possibly using gradient method. Then $C = [C_1 \mid I_m][M]$

Singular values of equivalent control feedback matrix provide for robustness analysis alternative to eigen assignement; use of Matlab; solution of matrix Lyapunov and Riccati eqations.

Finally, hierarchical concept makes use of various penalty functions in decomposition techniques. Interaction balance and iteraction prediction design and simulation; use of Matlab routines in decomposition and optimal VS design methods.

Algorithm 3 above requires a gradient method to find the optimum hyperplane. This can be difficult for large systems while it is adequate for medium sized systems. Hence a hierachical decomposition is required. The cost function during optimisation is modified for the two level algorithm

$$J_1 = \int_0^\infty (x_{iQ}^2 + u_{iR}^2 + \lambda_i(z_i - \Sigma_j L_{ij}x_j) + F_ix_i)dt$$

Essentially, this is the same J as the last two terms disappear at the optimum. X_i, Y_i and F_i are calculated using VS control at the lower level and λ set at the higer level. It is obtained by conjugate search of the error vector $e_i = z_i - \Sigma_j L_{ij}x_j$, while, simultaneosly, F can be obtained from algorithm 4.

Algorithm 4:

Step 0: Find F : $A - BF_iC$ stable. Set i=k=0.

Step 1: Find $J_i(F_i)$ from solution of $KA_{qe} + A_{eq}K + CF\,RFC = 0$ and $J = 1/2\ trK$.

Step 2: Find $g_i = col(J_{iFi})$. F_i from $J_1 = 0$ and solution of $LA_{eq} + A_{eq}L + X_0 = 0$. If $||g_i||$ sufficiently small step, else continue.

Step 3: If K = 0 define $H_i = H_{i-1} +$

$$+ \frac{||z_i - z_{i-1}||}{(z_i - z_{i-1}) \pm (g_i - g_{i-1})} + \frac{H_i||g_i - g_{i-1}||\,H_{i-1}}{(g_i - g_{i-1})^TH(g_i - g_{i-1})}$$

Step 4: Perform a one-dimensional minimisation $J_1(z + \alpha_i s) = \min_\alpha J\ (z + \alpha_i s) > 0$. Let $z_{i+1} \to z_i + \alpha_i s_i$, $i \to i+1$. If k=2dim(z) set k=0, else k->k+1.

Step 5: Go to step 2.

However, the above constitutes the unfaeasible method. The feasible method is obtained from partitioning the control into local and global terms, neutralising the global control by means of the interconnection constraint.

SIMULATION OF THE DESIGNED SYSTEM

The system response simulated for 400 seconds under on/off control is shown in Figure 6. The humidity control K_1 is presset for the simulation cycle as in figure 6(a) serving two purposes.

Firstly, the system becomes SISO so that only room temperature is regulated, illustrating the relative merits of the controllers. Secondly, by switching during the simulation run it has the room of a disturbance input. Under on/off control the 'dead band' can be set to allow regulation to have a nominal percentage error.The smaller value of this the greater the number of switching required of this heater. Figures 6(b) and 6(c) show the variation of the state temperatures. The corresponding rate for switching is shown in 6(d).

By contrast, VS control produces virtually zero steady state error in room temperature and reduced variation in water temperature (Fig. 7). The signal levels are much reduced at the expense of a much higher switching rate. It can also be seen that there is little change afected by the "disturbance" K1 compared with on/off control. The heat supplied in the two cases is aproximatelly the same but the wast reduction in the signal energy is illustrated by Figure 8.

Figure 9 shows the variation of switching frequency against dead band for on/off control and "slope" of hypersurface for VS control. The rapid switching under VS control is not rudimental to performance but may cause additional wear on the actuator. A smoothing of this phenomenon, if it were possible, would be desirable. In fact, this is possible.

CONCLUSIONS

The result presented in here show that temeperature regulation is vastly improved using VS control law described in a non-linear disturbance affected enviroment. Our investigations have shown how even considerable time-delays, typical in such environments, can be controlled. On the other hand, on/off systems must often be designed with power failure consideratins, for example, uppermost.

The theory suggests that the VS control law described forces tracking of the desired response from anywhere within the state-space. This indicates a built in safety and also has implications for temperature prediction model following.

The much reduced signal variations show that quadratic cost functions if the inputs and states will have low values comparable to optimal LQG design *(Gough & Coworkers, 1988,1992; Srai & coworkers 1992 a,b)*. These are important for energy management.

While the room temperature control example may be a simple one, especially by making it SISO one, it can be extended to many rooms or multi-zone buildings. Similar decentralised strategies have been improved upon using VS control in the area of water systems. An alternative strtegy is to design MIMO systems by extending the number of states including humidity, lighting, water supply and other controls for building energy management.

Appendix. The following is a list of the symbols used in the model of the system in figure 2:

- U_1 - heat transfer coefficient between room and exterior;
- U_2 - heat transfer coefficient between heating coil and air;
- A_1 - surface area of room;
- A_2 - surface area of heating coil;
- V_1 - volume of room;
- V_2 - volume of heating coil;
- R_1 - density of air;
- R_2 - density of water;
- C_1 - specific heat of air;
- C_2 - specific heat of water;
- T_0 - temperature of exterior;
- T_1 - temperature of air in room;
- T_2 - temperature of water in coil;
- T_4 - temperature of hot water;
- F_1 - volumetric flow of air;
- F_2 - volumetric flow of water.

REFERNCES

Ashley, S. (1984), "What next in energy management?". Building Services, 60-61.

Clark, W.E. (1983), "Energy management and control system selection: Critical components". ASHRAE J., 25, 42-44.

Dimirovski, G.M. (1987/1990), "Control of energy and energy conservation through control" (in Macedonian). *In J.Pop-Jordanov, N.Uzunov & B.Andrejevski (Eds.)*, Long Term Development of Energetics in Rep. of Macedonia, Macedonian Academy, Skopje, 117-122.

Dimirovski, G.M. & M.Tomsic (1989), "Control and optimization of energy consumption: The impact of information technologies on energetics" (in Serbo-Croat). Energy and Development, Scientific Forum Monograph, Beograd, 487-493.

Emelyanov, S.V. (1967), Variable Structure Control Systems (in Russian). Nauka Pub., Moscow.

Fernandez H.G. and R.Hedrick (1987), "Control of multivariable nonlinear systems by the sliding method". Int. J. Control, 46, 3.

Gough, N.E., Z.M.Ismail & R.E.King (1984), "Analysis of variable structure systems with sliding modes". Int. J. Systems Sci., 15, 401-409.

Gough,N.E., M.J.Leach & M.S.Srai (1988a),"Use of variable structure controls in energy management". Proc. 6th Int. Conf. on Systems Enegineering, Coventry Polytechnic, Coventry, U.K.

Gough, N.E., G.M.Dimirovski, B. Abul-Huda, M.J. Leach & M.S.Srai (1988b),Decentralized Multivariable Variable-Structure Control Systems with Sliding Modes. Res.Rep. SCET/R102,Faculty of Sceince & Technology, Wolverhampton Polytechnic, Wolverhampton.

Gough,N.E., G.M.Dimirovski, M.S.Srai, M.J.Leach, Z.A.Icev & T.D.Kolemisevska (1992), "Application of variable-structure control in energy management". Prepr. 1st IFAC Workshop on Automatic Control for Quality and Productivity (A.T.Dinibutun & A.Kuzucu, Eds.), Istanbul, TR, 2, 424-431.

Itkis, U. (1976), Control Systems of Variable Structure. J.Wiley, New York.

Srai, M.S., N.E.Gough, G.M.Dimirovski & Z.A.Icev (1991a), A Study of Robustness and Sensitivity in Variable Structure Systems. Tech.Rep. No.121, Faculty of Science & Technology, Wolverhampton Polytechnic, Woverhampton, U.K.

Srai, M.S., N.E.Gough, G.M.Dimirovski & Z.A.Icev (1991b), A Portable Software Environment for CAD of Variable Structure Control Systems. Tech. Rep. No. 132, Faculty of Science & Technology, Wolverhampton Polytechnic, Woverhampton, U.K.

Srai, M.S., N.E.Gough, G.M.Dimirovski & Z.A.Icev (1992a), "On robustness of EMS variable-structure sliding-mode control systems". Prepr. 1st IFAC Workshop on Automatic Control for Quality and Productivity (A.T.Dinibutun & A.Kuzucu, Eds.), Istanbul, TR 2, 377-384.

Srai, M.S., N.E.Gough, G.M.Dimirovski, B.A.Huda, V.P.Deskov & Z.A.Icev (1992b),"On multivariable and decentralized variable-structure sliding-mode controls in applications". Prepr. 1st IFAC Workshop on Automatic Control for Quality and Productivity(A.T. Dinibutun & A.Kuzucu,Eds.), Istanbul,TR, 2,369-376.

Utkin, V.I. (1977), "Variable structure systems with sliding modes". IEEE Trans., AC-22, 212-222.

Utkin, V.I. (1987/88), "Discontinuous control systems: State of the art in theory and applications" Invited Plenary Paper. Prepr. 10th IFAC World Congress (R.Iseramnn, Ed.), Munich, FRG, also in the Proceedings, The IFAC & Pergamon Press, Oxford, 1, 75-94.

White, B.A. (1983),"Reduced order switching functions in variable-structure control systems". IEE Proc., vol. 130, Pt. D, no.2, pp. 33-40.

White, B.A. (1986),"Range-space dynamics of scalar variable-structure control systems". IEE Proc., vol. 133, Pt. D, no. 1, pp. 35-41.

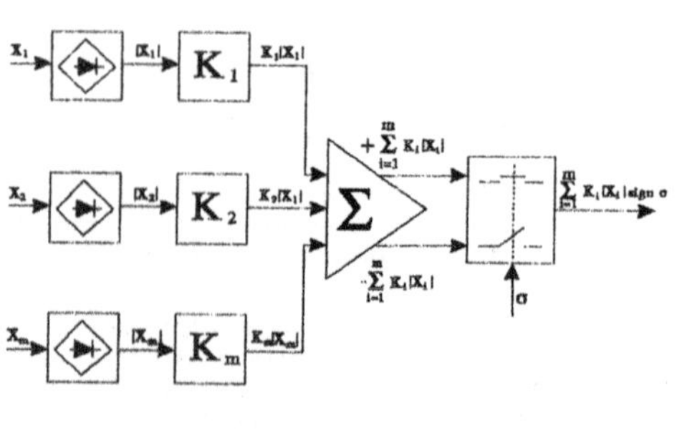

Figure 1. The Op-Amp Implementation

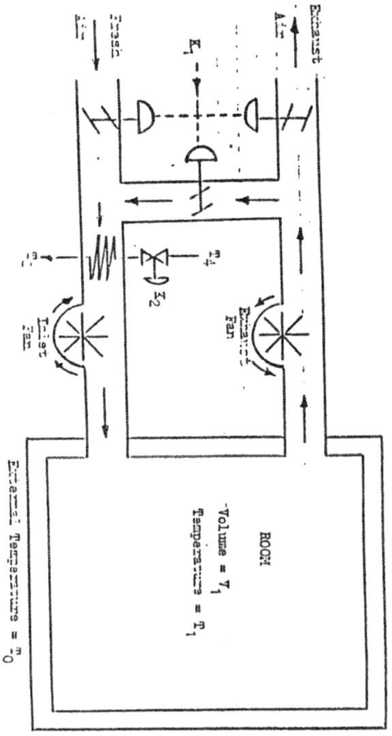

Figure 2. A Schematic of Controlled Object

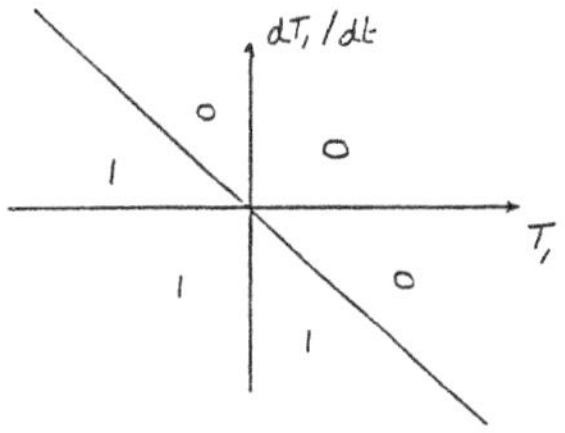

Figure 3. The Hyepersurface for K=1

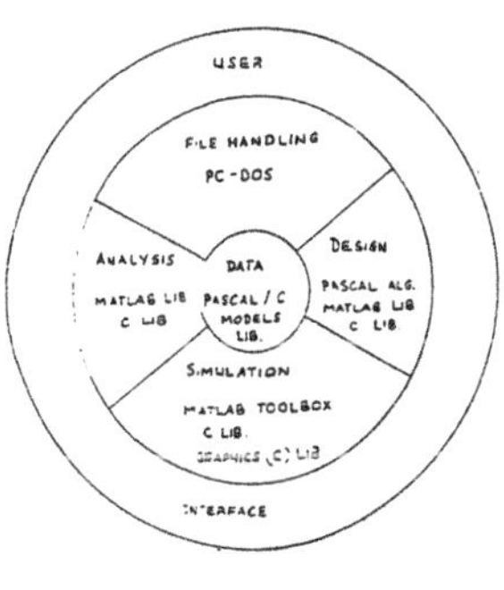

Figure 4. CAD Package Overwiew

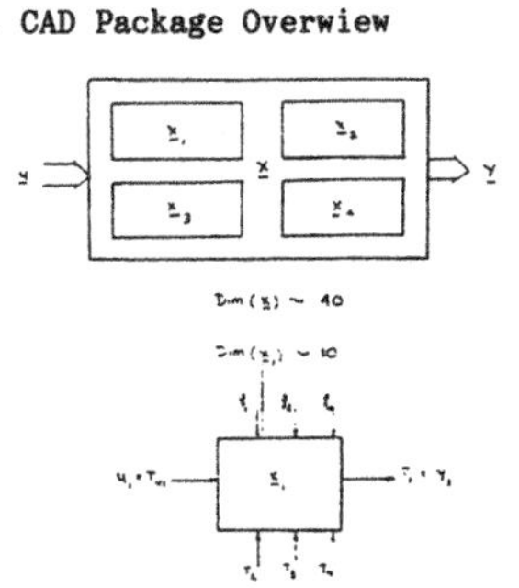

Figure 5. Multi-Zone Building and its Decomoposition

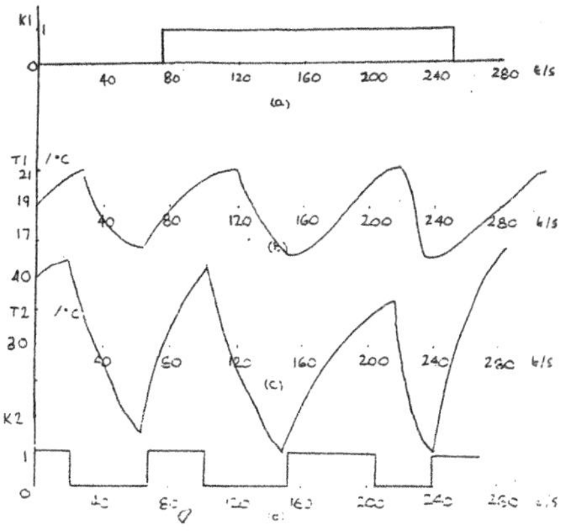

Figure 6. ON/OFF Control

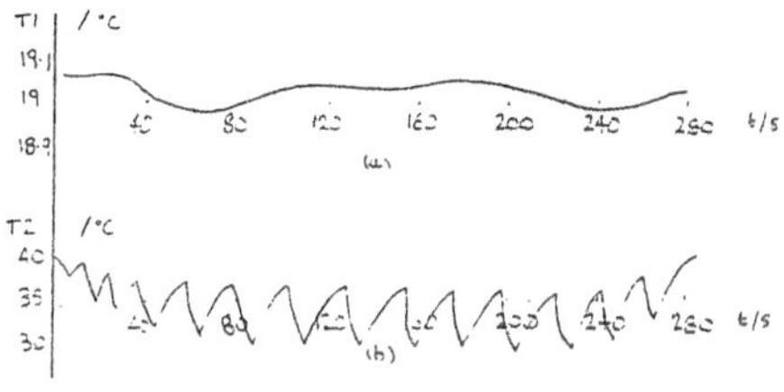

Figure 7. VS Control

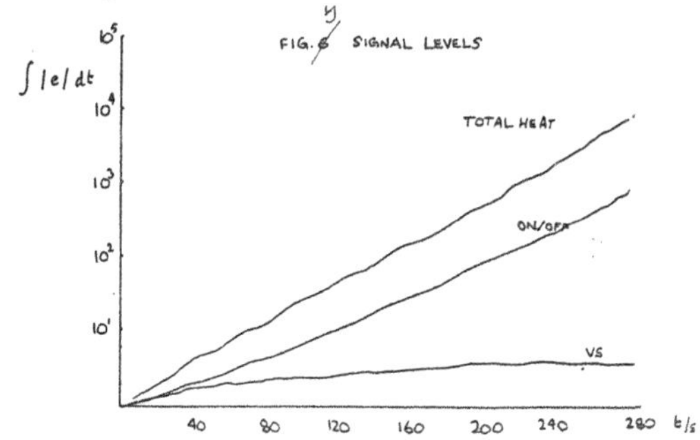

Figure 8. Signal Levels

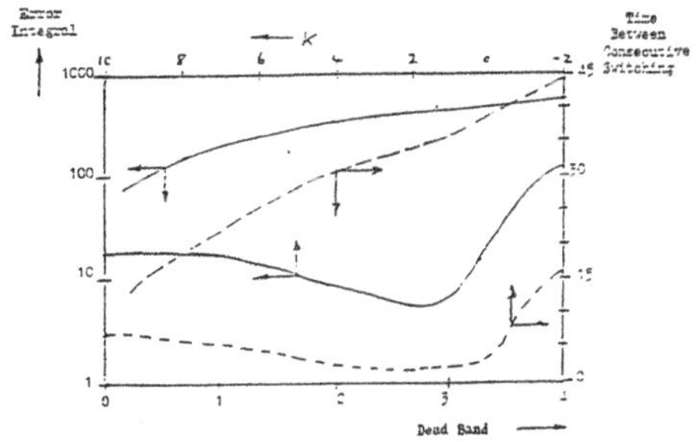

Figure 9. Variation of switching frequency against dead band for on/off control and "slope" of hypersurface for VS control.

DISTRIBUTED AND SELF-TUNING CONTROL SYSTEM FOR DIESEL GENERATORS

H. Rasmussen* and J. Aagaard**

*Institute of Electronic Systems, Aalborg University, Fr. Bajers Vej 7, DK-9220 Aalborg, Denmark
**A/S DEIF, Daadyrvej 2, DK-7800 Skive, Denmark

Abstract. The paper describes a distributed control system for synchronous generating sets. Self-tuning regulators for synchronizing and load sharing are described. Pole placement is used as controller design method and robustness against unknown system dynamics is achieved by proper selection of filters in connection with Least Square system identification. Simplifications leading to a single tuning knob for a controller with PID structure are shown. The methods are tested on a programmable simulator. Measurements on a 1MW MAN B&W diesel generator have given a set of parameters for the simulator and verified the simulator model.

Keywords. Distributed control, Self-tuning regulators, Robust control, Power management, Diesel generator control.

INTRODUCTION

Generating sets and associated control systems are normally made by different manufactures. To facilitate the installation of a control system for a specific system of gen sets the requirements to the control system are that it must be flexible and modular to make it fit easily into the system in question. Enhanced control of large diesel engines may be obtained by using an improved model in the controller design (Blanke, 1986). As the dynamics of the diesel generators forming part of the total system may not be known in advance by the manufacturer of the control system, a need for self-tuning controllers arises. A/S DEIF, which produces distributed control systems, and Aalborg University have therefore co-operated in developing robust self-tuning regulators for synchronization and load sharing. Apart from performance improvements it has been emphasised that a user briefly experienced in conventional PID-control should be able to install the system. This has been achieved by using a robust control design giving a controller with PI(D)-structure.

DISTRIBUTED CONTROL SYSTEM

By the end of this year, A/S DEIF will introduce a new generation of low cost control system named Delomatic-3 to the European market. This system is specifically designed for control of generating sets (DEIF, 1992). Delomatic-3 forms the target system for the self-tuning regulators described in this paper. The following serves as a short presentation of the system.

The Delomatic-3 makes up a distributed control system based on the principle shown in Fig. 1. Each diesel generator is equipped with a separate controller, communicating with the other controllers in the system using a Local Area Network (LAN). Each Delomatic-3 controller consists of one control panel for Man Machine Interface (MMI) and one control module, implemented in a 19" rack frame, for interfacing and control of the diesel generator. Due to the design of the control module, a highly flexible system has been obtained, in which each logical function is implemented in a separate module. The following modules are available:

- Power supply
- Logical controller and LAN interface
- Input (16 analog or digital inputs)
- Output (16 relay outputs)
- AC measuring and synchronizing module

The AC measuring and synchronizing module uses Digital Signal Processing (DSP) to calculate active power, reactive power, apparent power, RMS values of current and voltage,

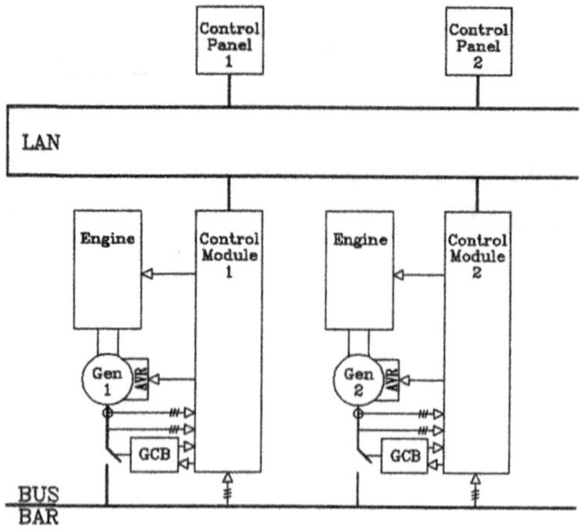

Figure 1: *Distributed Control System*

power factor and finally the frequency of the generator as well as of the bus bar. The module also controls the Generator Circuit Breaker (GCB) and gives a reference output to the speed governor on the diesel engine and to the Automatic Voltage Regulator (AVR) on the generator.

The software for each controller of the Delomatic-3 system has been divided into a number of separate modules forming the total structure illustrated in Fig. 2. As shown in the figure, the self-tuning regulators are implemented in the lowest layer of the structure. This is done in consideration of the performance of the regulators as well as the tuning process.

OPERATION MODES AND CONTROL PRINCIPLES

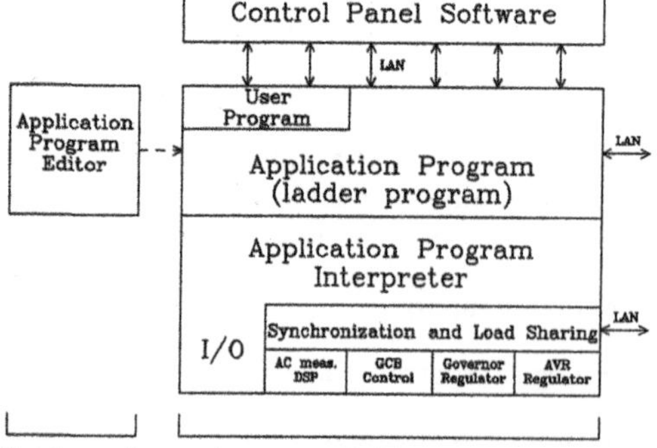

Figure 2: *Software structure of the Delomatic-3 system*

The most essential control inputs to the diesel generator are the reference value to the speed governor and the voltage reference to the AVR. Depending on the actual mode of operation, these control inputs are used to control various parameters in the power plant.

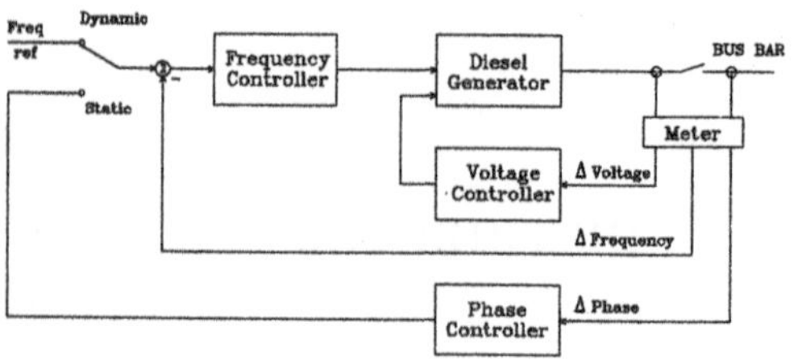

Figure 3: *Concept for static and dynamic synchronization*

Synchronization
During synchronization of a generator to the bus bar, the frequency and phase are controlled using the speed reference input of the diesel engine governor. The generator output voltage is controlled, giving a set point to the AVR. The concept for synchronization of the diesel generator is shown in Fig. 3.

The input to the voltage controller shown in the figure is the difference between the amplitude of the bus bar voltage and the generator output voltage. During the synchronization procedure this voltage difference must not exceed a specified value. The frequency controller gives the reference signal for the speed governor. The difference between measured generator frequency and bus bar frequency is subtracted from a frequency reference value, giving the input signal for the frequency controller.

Synchronization of diesel generators to the bus bar is usually performed in either static or dynamic mode. In static mode the phase of the generator is controlled, ensuring that the phase shift across the circuit breaker stays below a set value, before the GCB is activated. In static synchronization mode the time delay of the circuit breaker may be measured. Dynamic synchronization is obtained by giving the generator a frequency lead within the range of 0.5Hz to 2Hz. The synchronizing signal is then transmitted, the time delay of the circuit breaker, before phase accordance. When a dynamic synchronization has been performed, the generator exports power immediately after connection to the bus bar.

Load Sharing
When generators are connected to the bus bar, the operation mode for the controllers changes entirely. The generating set operates either

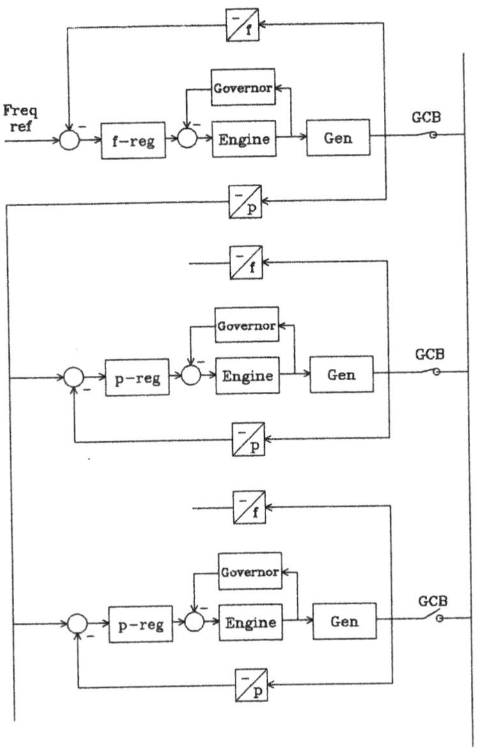

Figure 4: *Concept for load sharing*

in island mode or in parallel with the mains. In island mode (DEIF, 1989) the controllers must maintain the frequency and voltage despite change of load, and share the load among the generating sets according to specified conditions. In parallel operation with the mains, the controllers must only share the load among the individual generating sets, as the frequency and voltage are determined by the mains. Furthermore the controllers must maintain a specific import/export of active and reactive power between the generator plant and the mains.

If a generator is connected in parallel with the mains, the reference to the speed governor controls export of active power from the generator. In a similar way the reference to the AVR, due to the inductor in the output of the generator, controls the import/export of reactive power. These relationships are used to maintain a specific import/export of active and reactive power.

Concepts for load sharing are divided into the following basic principles:

- Droop
- Isochronous
- Master/slave isochronous

Due to the equivalent nature of active power and frequency, and reactive power and voltage respectively, the following only describes the basic principles for control of active power and frequency.

Speed droop can be used for balancing active power. Speed droop is obtained by giving all the engines in a generator plant a frequency versus active power characteristic with a slope within the range of 1 - 5%. In such a system no information interchange takes place among the generating sets. Speed droop is identified by its static frequency error. Droop characteristic is often implemented in the speed governors, to ensure a safety load sharing system.

The term "isochronous" refers to load independent frequency control. In a isochronous load sharing system each of the gen sets are equipped with a feedback controller for active power. In island mode the reference input to these controllers is the output from one frequency regulator common to the generating plant.

A variant of the isochronous system is the master/slave concept shown in Fig. 4. During island operation one generator, equipped with a frequency regulator, performs the task of maintaining a constant bus bar frequency. Control of the remaining generating sets in the plant is performed by active power controllers. The measured value of active power from the constant frequency controlled gen set is used as reference input for all power regulators. In both isochronous modes, the controllers use the LAN for information interchange.

In practical applications a number of logical functions, e.g. load depending start/stop, asymmetrical load sharing and control of base power gen sets, are normally applied to the load sharing system. Further information on these functions is available in DEIF (1989, 1992).

SELF-TUNING REGULATORS

The modules of the self-tuner system are, as shown in Fig. 5, divided into a 3-level hierarchy: The supervision level, the design level and the interrupt level (Isermann and Lachmann, 1985). The supervision level is divided into tasks for interface to the operator, control of the interaction between the algorithms in the lower levels and security logic concerning proper operation of the basic algorithms for system identification and design of the control parameters. The selection of algorithms is essential in order to ensure the overall robustness of the control sys-

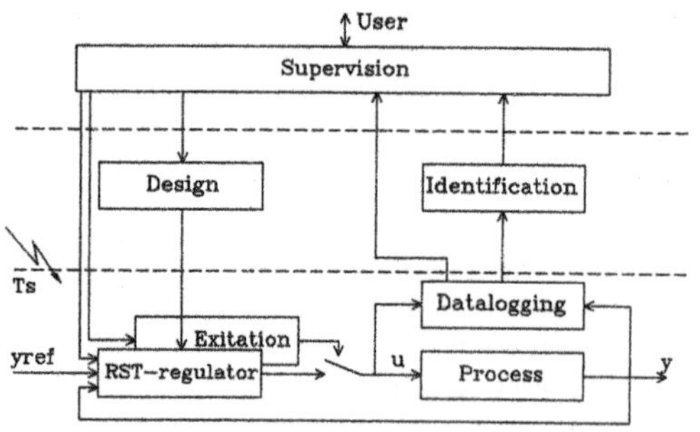

Figure 5: *Modules of the self-tuner*

tem. Polynomials A and B describing the system dynamic are estimated using least square algorithm based on filtered values of input and output. Robustness of the system identification is achieved by including a design polynomial C in the model giving a best estimate in the frequency region near the cross-over frequency. Controller design is based on a pole placement method. The method is analyzed with regard to robustness, and parameters giving simple tuning are selected. For the true system is assumed

$$y_t = Pu_t + v_t \tag{1}$$

We shall use a parametrized model set of the form

$$y_t = \frac{B(\theta)}{A(\theta)}u_t + \frac{C}{A(\theta)\Delta}e_t \tag{2}$$

where $\Delta = 1 - q^{-1}$. The one step ahead prediction error is for white noise e_t

$$\epsilon_t(\theta) = \frac{A(\theta)\Delta}{C}[(P - \frac{B(\theta)}{A(\theta)})u_t + v_t] \tag{3}$$

If C is used as a design polynomium a Least Square identification gives

$$\theta_N = arg \min_{\theta \in D} \frac{1}{N}\sum_{t=1}^{N} \epsilon_t(\theta)^2 \tag{4}$$

The estimate converges asymptotically (Ljung, 1987) to

$$\begin{aligned} \theta^* &= arg \min_{\theta \in D} \lim_{N \to \infty} E\frac{1}{N}\sum_{t=1}^{N} \epsilon_t(\theta)^2 \qquad (5) \\ &= arg \min_{\theta \in D} \int_{-\pi}^{\pi} \Phi_\epsilon(\theta,\omega)d\omega \\ \Phi_\epsilon(\theta,\omega) &= [| P - \frac{B(\theta)}{A(\theta)} |^2 \Phi_u + \Phi_v] \mid \frac{A(\theta)\Delta}{C} |^2 \end{aligned}$$

The asymptotic model can therefore be seen as a compromise between fitting the input-output transfer function B/A , to the true transfer function P in a frequency weighting norm. How to select C depends of the controller to be designed.

As a design method for the controller is used pole placement

$$\begin{aligned} AR'\Delta + q^{-1}BS &= A_0 A_m \qquad (6) \\ T &= \frac{S(1)}{A_0(1)}A_0 \end{aligned}$$

giving $R = R'\Delta$, S and T when A_0 and $A_m$is specified.

When the plant and the model have the same number of unstable poles and the designed closed loop is stable, Gevers (1991) shows that stability is implied by

$$| P - \frac{B}{A} || \frac{AS}{A_m A_0} |< 1 \tag{7}$$

Selection of C as

$$C = A_0 A_m \tag{8}$$

gives for $\Phi_v \ll \Phi_u$

$$\Phi_\epsilon(\theta,\omega) = \{| P - \frac{B}{A} || \frac{A\Delta}{A_0 A_m} |\}^2 \Phi_u \tag{9}$$

Comparing Eq. (7) and Eq. (9) a Φ_u approximating $| \Delta |^2 \Phi_u =| S |^2$ gives a proper frequency weight.

Defining

$$\begin{aligned} u_t^f &= \frac{\Delta}{A_0 A_m}u_t \\ y_t^f &= \frac{\Delta}{A_0 A_m}y_t \qquad (10) \end{aligned}$$

means that the parametrized model may be written as

$$A(\theta)y_t^f = B(\theta)u_t^f + e_t \tag{11}$$

and least square system identification then gives A and B.

Many process control problems can be adequately and routinely solved by conventional PID control strategies. The overriding reason that the PID controller is so widely accepted is its simple structure which has proved to be very robust with regard to many commonly met process control problems as for instance disturbances and non-linearities. Tuning of the PID settings is quite a subjective procedure, relying heavily on the knowledge and skill of the control engineer or even plant operator. Although tuning guidelines are available the process of controller tuning can still be time consuming with the result that many plant control loops are often poorly tuned and full potential of the control system is not achieved. One way of deriving a self-tuning control law with the structure of a PID controller is to choose the order of the parameterized model to $n = 2$ and the degree of A_m and A_0 to n. The parameters in a PID-controller

$$K_P\{1+\frac{T_s q^{-1}}{T_I(1-q^{-1})}+\frac{T_D(1-q^{-1})}{T_s(1-gq^{-1})}\}=\frac{S}{R} \quad (12)$$

are then given by

$$\begin{aligned}
g &= r_2\\
\frac{T_s}{T_I} &= (1-g)(\frac{s_2}{s_0}+\frac{s_1}{s_0}+1)/N\\
\frac{T_D}{T_s} &= (g(g+\frac{s_1}{s_0})+\frac{s_2}{s_0})/N\\
K_P &= s_0/(1+\frac{T_D}{T_s})\\
N &= 1-g(\frac{s_1}{s_0}+2)-\frac{s_2}{s_0}
\end{aligned}$$

If further

$$A_0 = A_m = (1-\gamma q^{-1})^n \quad (13)$$

the only tuning knob (fast/slow) is γ.

If A and B have common roots a first order model n=1 is used and the design then gives a PI-controller.

TEST OF THE CONTROL SYSTEM

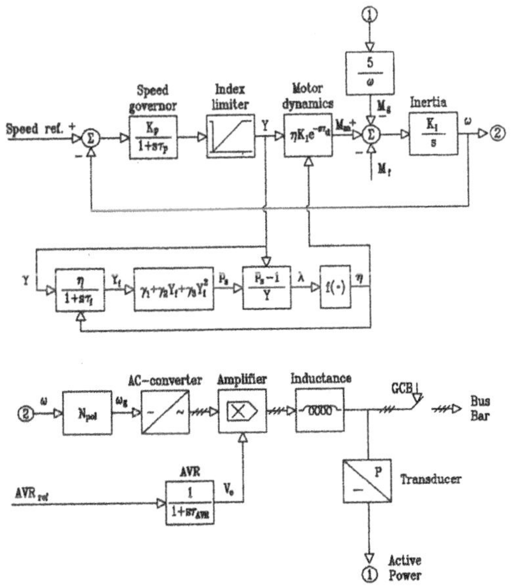

Figure 6: *Schematic drawing of the Diesel Generator Simulator*

The system has been tested on a programmable simulator having the same electrical interface as a diesel generator set. The simulator can

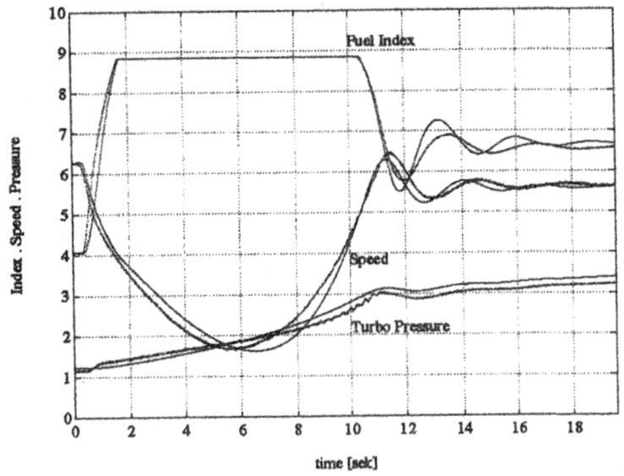

Figure 7: *Simulated and measured values of speed, fuel index and turbo pressure for a load step from 0 to 500kW*

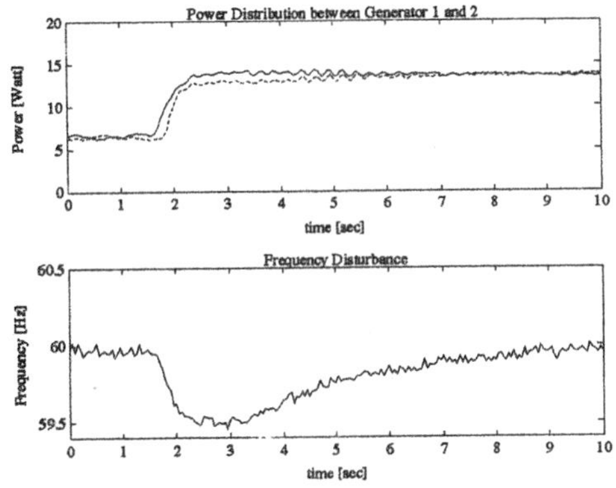

Figure 8: *Power distribution between generators and disturbance of frequency for a load step*

be set to many different types of diesel generators within the range 100kW to 5MW. One of the settings of the simulator has in co-operation with MAN B&W Diesel Holeby been found by measurement on a 1MW diesel generator. Figure 6 shows the hardware/software modules of the Diesel Generator Simulator.

Simulator input: Speed reference, Voltage reference AVR_{ref} and Generator Circuit Breaker (GCB) on/off.

Simulator output: The tree phase voltages and a signal indicating the GCB state.

The generator output is modelled as an inductance scaled to 10V output phase voltage and 30W maximal output power. A non-linear model of the shaft speed dynamics is derived from the conservation of energy for the diesel engine as shown in Blanke (1986). The crank shaft's angular speed ω is a function of indicated engine moment M_m, frictional moment M_f and load moment M_g. The mass of fuel injected into a cylinder in each cycle is controlled by the fuel pump index Y. The scavenging air pressure p_s is modelled as shown in Fig. 6. The dynamic model of the combustion process in Blanke (1986) is simplified by using a mean value engine torque for each cylinder firing. The thermal efficiency curve $\eta = f(\lambda)$ is essential and using the curve Arbøl (1965) a model giving a fairly close match for load step in all working points is obtained. A set of parameters for the model Fig. 6 are found in Christiansen and co-workers (1992) for a 1MW B&W diesel generator, where simulated and measured response for 0-25%, 25-50%, 50-75%, 75-100%, 100-0% and 0-50% load step are showed too. Figure 7 shows the simulated and measured values of speed, fuel index and turbo pressure for the 0-50% load step. The values are shown as 0-10volt transducer outputs. Transformation to standard units is given in Christiansen and co-workers (1992).

The load sharing control concept shown in Fig. 4 has been tested for different settings of the tuning knob γ in Eq. (13) in the power and frequency controllers. Figure 8 shows a load step response for fast closed loop dynamics for the power controller and relatively slow closed loop dynamics of the frequency controller. A reduction of the disturbance can be achieved by tuning the frequency controller to a faster closed loop response (less γ). The trade off for this reduction is a lower robustness against variation of the dynamics of the total system.

CONCLUSION

The status of the co-operation between A/S DEIF and AUC concerning development of a distributed control system for synchronous generating sets having facilities for automatic tuning of the controllers has been described. Robust self-tuning regulators for synchronizing and load sharing are described, and simplifications leading to controllers with PID structure with a single knob for tuning are shown. The methods are tested on a programmable real time simulator. Measurements on a 1MW MAN B&W diesel generator have given a set of parameters for the simulator and verified the simulator model.

ACKNOWLEDGEMENT

The authors acknowledge the assistance of the staff of MAN B & W Diesel Holeby, Peter Frederiksen in particular, for providing the measurements used for verifying the simulator for the diesel generator.

REFERENCES

Arbøl S. (1965). Forbrændingsmotorens arbejdsproces. *Teknisk Forlag*, København, Denmark.

A/S DEIF. (1989). Standard principles for island operation. *Fax +4597524720.* Skive, Denmark.

A/S DEIF. (1992). Delomatic-3 power management system for generator plants. *Fax +4597524720.* Skive, Denmark.

Blanke M. (1986). Requirement of adaptive techniques for enhanced control of large diesel engines. *Proceedings of the 2nd IFAC Workshop.* Lund, Sweden.

Christiansen T., L. Jørgensen and T. Pedersen (1992). Distribueret proceskontrolsystem. *Master's thesis, University of Aalborg.* Aalborg, Denmark

Gevers M. (1991). Connecting identification and robust control: a new challenge. *IFAC proceedings.* Budapest.

Isermann R. and Karl-H. Lachmann (1985). Parameter-adaptive control with configuration aids and supervision functions. *Automatica, 21(6).*

Ljung L. (1987). System Identification: Theory for User. *Englewood Cliffs NJ. Prentice-Hall.*

AUTOMATIC HIGH VOLTAGE CONTINUITY TEST FOR WINDING WIRES

J. Vehí and F. Coll

Dept. Enginyeria de Sistemes, Automàtica i Informàtica Industrial, Universitat Politècnica de Catalunya, Spain

Abstract The paper describes a whole quality control automation for a manufacturing plant of winding wires. Every section contains several production machines which produces from 8 to 24 wires simultaneously. In a first step, we develope a high voltage continuity test apparatus connected to a microprocessor board. The board's function is to save the results coming from eight test apparatuses and preprocess them for further transmission. From 1 to 3 boards may be placed into a 24" rack which has a keyboard and an alfanumerical display for on-machine programming and visualization. This device also provides visual and acoustic alarms. The control unit receives data from the microprocessor boards by RS-485 links (up to 16 microprocessor boards for every link) and generates a single label with the quality parameters for every coil. From the control unit or from the on-machine device it is possible to change many parameters such as test voltage or alarms configuration.

Keywords Automatic testing, Communications control applications, Data transmission, Quality control, Industrial production systems, Real-time computer systems.

INTRODUCTION

Actually, the manufacturing companies must constantly review not only profitability, operating efficiency, market share, etc, but also quality performance.

The benefits of an appropriate quality improvement program included:

- great reductions in product defects;
- improvement in product relibiality;
- substantial decreases in company quality costs;(scrap, rework, warranty, inspection costs, etc.)

and as a result:

* increased market share;
* increased operating efficiency and profitability;
* reduced inventory costs;
* reduces lead/delivery times.

The application field is a manufacturing plant that has three sections with several machines for section. Every machine is able to produce simultaneously from 8 to 24 wires, with nominal diameters which can vary from 0.25 mm to 1.6 mm (this is the range where the test is applicable). The range of quality classes and diameters has forced us to build a system able to make different types of test.

The current requirements for high quality winding wires prove that is not enough a statistical quality control but it is necessary a continuity test along the entire wire. This kind of test allows a better grading of the winding wires produced according to international standards.

SYSTEM DESCRIPTION

The system must perform the integration of some testing subsystems on an industrial network linked to a central unit that processes the information for further statistical analysis and monitoring. The control unit also produces quality labels for every wire coil.

The conception of the system is modular and easily re-configurable and extendible and allows the posibility to work with a single machine or several machines. Figure 1 shows the configuration of the global system.

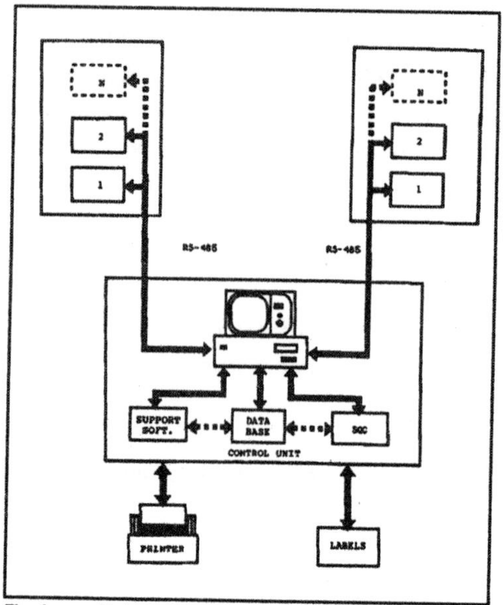

Fig 1: Global system

The system has several subsystems that may be configurated according to some specific characteristics of every machine and section:

- *Fault detection system*. Provide digital signals for fault counting.

- *Microprocessor board*. Pre-process data information and performs the communication.

- *Remote unit*. Designed for the man-machine interface and alarm generation.

- *Control unit*. Centralize all the information, works as the master of the communications system and makes the real-time monitoring. Also, the control unit post-process data for further statistical applications and generates hardcopies and labels.

There are two main performances requested by the manufacturer: a complete monitoring in a centralized control unit, and a man-machine interface for every machine. The first one is obtained by the control unit, and man-machine interface is provided by the remote unit that integrates a keyboard and an alfanumerical display for up to three microprocessor boards.

FAULT DETECTION SYSTEM

The International Electrotechnical Commission in their Technical Committee N° 55 publishes a revision of the norm IEC-251 that describes the procedure of high voltage continuity test.
The fault detection is made by pulling the wire specimen with the conductor earthed over a "V" grooved electrode at a constant speed. The test voltage is connected to the electrode. Defects in the insulation are detected and registered on a counter. The results are listed in terms of faults per 30 m according to the IEC - 251. Figure 2 illustrate the test apparatus.

The sensitivity of the fault detection circuit shall be such that the threshold fault current shall be as shown in Table 1 with a tolerance of 10%

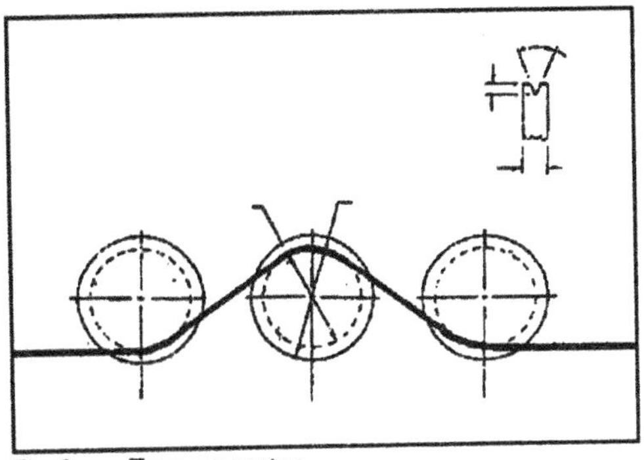

Fig 2: Test apparatus.

TABLE 1

Test Voltage d.c. (V)	Threshold fault current (μA)
3000	16
2500	14
2000	12
1500	10
1000	8
750	7
500	6
350	4

To obtain a low cost product, the test voltage is common in every group of four wires and may be changed from the control unit or from the remote unit. This is not a handicap because every machine manufactures similar kinds of wires.

The fault detection is independent for every wire, and provides a high-level TTL compatible signal for every fault in 25 mm. The test apparatus is used also for speed measurement and, in addition, generates an alarm when the wire is broken.

MICROPROCESSOR BOARD

This board is based on MCS-51 family 80C32 microcontroller witch support high level programming (with PL/M-51) and it has embedded one full-duplex serial port as well as others common capabilities.

Digital signals coming from two separated detection boards are jointed in a microprocessor board. This board acquires data from PPI ports and evaluate some rates as speed of every wire, number of faults per 30m, number of faults per 100m, metre with more faults, etc. The microcontroller stores data grouping the evaluate data of every 100 m. At the request of the central unit, this board sends to the central unit this pre-processed information.

REMOTE UNIT

The remote unit is designed for the man-machine interface. This system is able to integrate from one to three microprocessor boards in a remote unit. It has built-in some input-output functions that check the microprocessor boards and allow man-machine interface. It is possible to modifier voltage test and other parameters by means of a keyboard and to generate alarms. An alfanumerical display provides visualization of the status of the machine and information about speed of every wire, number of faults per 30 m., alarm recording, etc.

COMMUNICATIONS SYSTEM

To obtain a communication system at lowest cost as possible, we are choice the RS-485 standard. This choice is advisable that we are several devices to be connected in a multipoint link and the range of distances (from 20 to 300m).

The communication mode is half-duplex with a personal computer a master and a configurable number of devices as slaves. This is possible because we store data into the microprocessor board memory and we only need to do a data transfer when the devices are asked by the host computer.

Hardware. For the prototype system, we are mounted into the host computer a double channel RS-485 communications board but it is able to support eight RS-485 channels for further extends. That is to say, is able to support 128 remote devices (the number of wires we may supervise simultaneously are up to 1024).

The communication of the remote devices is performed by the microprocessor boards which are embedded one full-duplex serial port.

Sotfware There are three classes of messages to transfer:

* Measurements of the detection board
* Alarms of wire broken
* Orders for voltage test change

The two first data transfers are from microprocessor boards to the main computer, and the other is on the contrary. The communication software of the AT-286 computer is implemented in turbo C. The transmission speed is 38400 bauds.

The communication procedure is very simple. The host computer asks sequencially all the microprocessor boards. Each device is asked one time for 30 seconds. The voltage test orders are sent in this moment. When it is asked, the microprocessor board sends to the host computer the information of the last 100 m. of wire or a message if 100 m are not completed. Alarm recording is made in this moment.

CONTROL UNIT

The control unit is intended to deal with all the available information to produce a real-time monitoring of the production status.

The control unit sends to the printer periodically reports of the main parameters. When one coil is finished, a single label is generated with the quality parameters. Also, the main quality and production information are stored in data files arranged for further statistical analysis.

CONCLUSIONS

We are developed an open control system that allows a wide range of configuration levels, i.e., from a single production machine to several machines distributed in separated sections without waste of performance.

The centralized quality control assures the detection of malfunctions and provides a better knowledge about the production system that will result on a general improvement.

REFERENCES

[1] J. Boettcher, H.R. Traenkler. Trends in Intelligent Instrumentation. IFAC Low Cost Automation. Milan 1989

[2] P. MsKeown. Implementing Quality Improvement Programes. Proc. IFIP TC5/WG 5.3 Working Conference on Computer Integrated Quality System in CIM Systems. Belgrade 1989

[3] G.C. Barney. Intelligent Instrumentation. Prentice Hall International. London 1985

Copyright © IFAC Low Cost Automation,
Vienna, Austria, 1992

THE APPLICATION OF INTELLIGENT CONTROLLERS IN THE AUTOMATION OF A SMALL PLANT FOR RECYCLING OF SECONDARY POLYMERS

M. Hadjiiski*, E. Panchev*, Z. Zlatev* and S. Patsoff**

**Sofia Technological University, Bulgaria*
***"Podemplast", Podem, Bulgaria*

Abstract. An intelligent system for integral management of a small plant for recycling of secondary polymers has been developed. The system uses intelligent regulators with wide functional potentialities that can meet the peculiarities of the this type of enterprises, such as vast and frequent changes of operative conditions. The developed regulators include functions on short-term production planning, steady-state optimization of the separate units, optimization of the nonstationary regimes of the most important technological parameters. Combination of model-based and fuzzy reasoning-based methods for choosing optimal static regimes and gain-sheduling technique in controlling modified standard regulators has been accomplished. Data for experimentally obtained characteristics of the basic units of a small plant and the results of the simulation research work are adduced. The peculiarities of the realizations of the basis of PC and industrial controllers are studied.

Keywords. Gain control, fuzzy control, intelligent machine, PID control, polymer recycling, production control, small plant.

INTRODUCTION

A remarkable development of the theoretical methods and the practical application of the management of large production systems in the chemical industry (Konig and Stokhansen, 1990) has been observed during the last few years. There are not many research works carried out into application of intelligent computer system for management of small factories. Obviously, the direct transfer of results from a Computer Integrated Production System to small factories is ineffective. The main reasons of this are: (i) small factories have various structures, production, organization, size; (ii) external economic disturbances are very fast and relatively great (market, energy and raw materials costs, legislation, etc.); (iii) serious investment and innovation activities are bounded. The efforts of the experts in the different fields, united in the trend of Low Cost Automation (Albertos, 1989, De Carly, 1989) lead to the effective realization of the benefits now available with present day computer, information and control techniques and technologies. The experience in the development of intelligent systems by using basic control techniques, relatively cheap and available components (Aracil, Ollero and Morant, 1990, Navaro, Albertos et al, 1990) are suitable for a wide range of processes. The trend LCA has large potentialities, but a great deal of theoretic and practical problems remain unsolved (Astrom, 1991, Wittenmark, 1991).

This article presents the development and defining of the ideas set first in (Hadjiiski, Patsov, Zlatev and Panchev, 1992) for designing of an intelligent control systems of a small factory for recycling of plastics. It combines some large complex control systems features: hierarchical coordination, management and production integration, advanced control algorithms with LCA achievements: relatively cheap basic hardware and software, easy operation and maintenance by low qualified personnel, high degree of flexibility.

PROBLEM FORMULATION

The economical structure of the small plant for recycling of plastics is presented at Fig. 1. The following symbols are accepted: C denotes storehouse; M- a waste product grinding mill; W1 - a grinding store, first reservoir; W2 - a washing reservoir; Z - a centrifuge; S - a drier; P - an electric heater for hot air; B1, B2 - hoppers for ground polymer; EL - two extruders line, each one consisting of two cascade extruders; BG - a final production hopper; EF - a foil extruder line; TCH - foil products; MR - an extruder line for other extruder products; SPI - an injector products line; SK - storehouse for final production; R - realization of final production. The production process may be divided into four technological stages:

1) Preliminary preparation of the plastic by-product. This stage includes the elements C, M, W1, in which the secondary polymer is ground and washed. The gauge of the foil of the secondary polymer and the state of the cutting tools of the mill are main technological problems;
2) Drying of the ground products consecutively in the elements W2, Z, S, B1. The size of the ground products and the loading are main technological problems;
3) Obtaining of regranulated products in the extruder line EL. Here, the inlet moisture, the size of the ground product, the composition of the polymer, and the temperature of the environment are the main problems;
4) The manufacture of the final product includes extruder and injector for foil and miscellaneous products. The feed rate of the production line has a main technological effect.

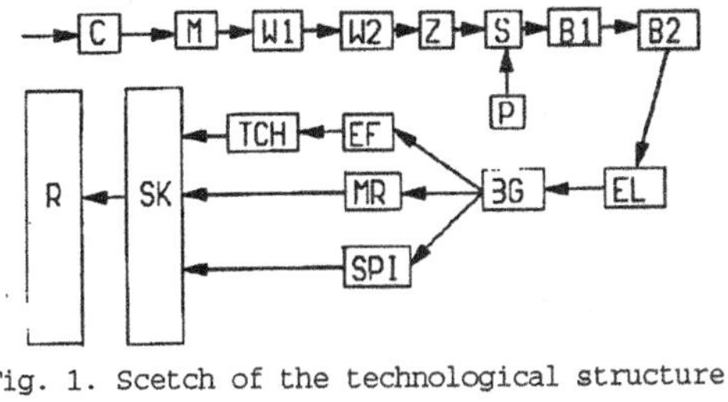

Fig. 1. Scetch of the technological structure.

There are some other problems besides the above mentioned. Such are the planned and unforeseen interruptions because of current repair and damage, and also as a result of changeability of the economic factors - the amount and the rhythm of the orders, cost of the energy consumed, etc. There are many difficulties in the productions af secondary polymers compared with the traditional production of plastics: (i) - the level of the internal disturbances is much higher. Main the kind of the secondary polymer, the gauge and the size of the ground products, and the moisture of the by-product appear to be main problems; (ii) - the grid of the end extruder needs frequent changing because of plugging; (iii) - some difficulties occur in synchronizing the work of the separate elements on the scheme on Fig. 1. Unfortunately, most of problems can not be registered immeasurable, and the judgement of the effect of their influence is based on indirect or linguistic information. The main tasks of the system for control of a small factory are: (i) - minimum production cost; (ii) - guarantee for the quality demanded by the client. The attainment of these reguirements under all possible operational conditions is realized by a control system in the three hierarchical levels - Fig. 2. The first level comprises measurement, instruments, actuators, basic controllers like PID, ''On-off'', PLC. The second level manipulates the set-points of the first level controllers, and realizes all functions of intelligent steady state and dynamic control. The third level determines optimum productivity scheduling (production rate, planned interruption). This paper deals mainly with the second and third level. The coordination among the hierarchical levels is paid special attention to. The coordination is determined in a broad and usable way by application of measurable variables (production rates and technological parameters) as coordinating variables, and optimization is executed at all levels. The communication among the subsystem is a prerequisite for the achievement of good synchronization. The intelligent decentralized regulators of the second hierarchical level possess the following properties: (i) - they may be provide solution of the optimization problems in steady state and dynamic behavior; (ii) - they are flexible under permanent changes of the conditions: change in the properties of the secondary polymer, damage, organizational reasons, costs of the electric power; (iii) - they extend the kind of the information that is used - interference approach, expert knowledge, linguistic information - because of a lack of possibility of taking direct measures of the most important parameters (indices of the raw material, quality of the products); (iv) - they posses properties for adaptation that are determined by gain scheduling technique and model-based off-line optimization; (v) - they incorporate know-how and operator's experience in the algorithms; (vi) - they combine advanced control methods with the peculiarities of the production units; (vii) - they are realized an basic computer devices with a relatively low price.

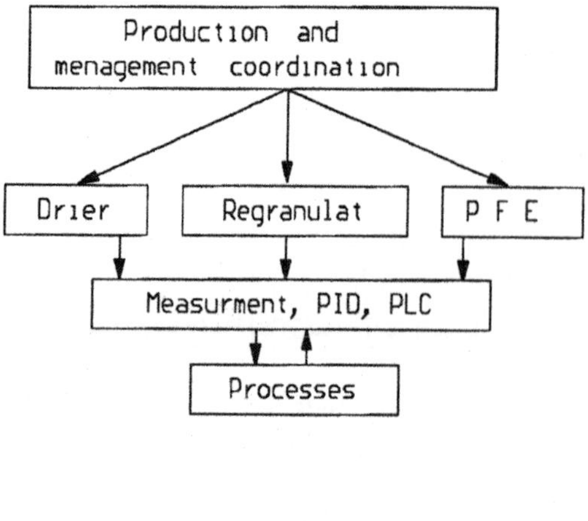

Fig. 2. Three level hieararchical control system.

PRODUCTION PLANNING

A PC-based planning system for coordination of the manufacturing processes of the secondary polymer, transportation and storage processes has been worked out. The optimum hour feed rate the technological units from the first three technological levels in Fig. 1 is determined by the means of this system. The electric power is the main variable of the production costs. The relationship of the price during the night, day, and peak hours is as 1:2:4. The energy power characteristics of all units have been examined in detail and stated in (Hadjiiski, Patsov et al, 1992). They show that everywhere the relative consumption of electric power is minimum when production rate is maximum. That is why the following approach is introduced in the planning programme: (i) all the equipment, with the exception of the extruder line EL work under ''on-off'' regime with nominal productivity. (ii) the planning outlook is 24 hors, and the interval of discretization is 1 hour. (iii) the necessary average hour productivity is attained by verios correlations of the ''on-off'' intervals. (iv) the extruders EL for secondary granulated products cannot work with zero consumption of electric power; during short interruption they are being warmed up to keep their thermal state. When production rate is below maximum, the extruder works constantly, but its average hour production rate is determined by an equivalent ''on-off'' regime. The criterion for optimality is minimum consumption of electric power for a given twenty-hour production Q:

$$J=\sum_{j=3}^{4}\sum_{i=1}^{24}X_{ij}N_jC_i+\sum_{j=3}^{4}\sum_{i=1}^{24}(1-X_{ij})N_j^{*}C_i \quad \text{->min} \qquad (1)$$

where:

X_{ij} - relative time for work of the units from group j in i hour, (Fig. 1), when: (i) j=1 the units from the stage of preliminary preparation of the secondary polymer; (ii) j=2 washing and drying; (iii) j=3,4 - the extruder couples E_{12} and E_{34} from the extruder line EL;

N_j - nominal electric power of the j - group;

C_i - tariff price of electric power for i hour (i=1:24)

N_j^* - electric power of the heaters of the EL extruders during idle running.

The problem (1) is solved within the following limits:

(a) the twenty-four-hour productivity Q of a small factory is given :

$$\sum_{i=1}^{24}X_{ij}P_j=Q \quad (j=1,4) \qquad (2)$$

where:

P_j - nominal productivity of the j group (P_1=250 kg/h; P_2=230 kg/h; $P_3=P_4$=100 kg/h);

(b) the units of the groups work under ''on-off'' regime:

$$0 \leq X_{ij} \leq 1 \qquad (3)$$

(c) the intermediate storages have limited capacity:

$$\sum_{i=1}^{24}X_{ij}P_j-X_{i,j+1}P_j+1 \leq S_{jH} \qquad (4a)$$

$$\sum_{i=1}^{24}X_{i,j}P_j-X_{i,j+1}P_{j+1}\geq S_{jL} \qquad (4b)$$

where j=1 refers to washing reservoir S_{1H}=1800 kg, S_{1L}=200kg; j=2 refers to intermediate storage S_{2H}=1200kg, S_{2L}=200kg.

The problem (1) is solved by the simplex method of linear programming within these limitation (2), (3), and (4). The solution is characterized by the following peculiarities: (i) The power of the extruder line EL as a component of N_3 and N_4 is accepted for a given productivity according to the average statistic data; (ii)

The solution of problem (1) is valid for a given configuration of working units and planned interruptions (respectively $X_{ij}=0$). When unforeseen interruptions occur (for example, a damage), the programme is restarted. Depending on the damage, it is possible: (a) to meet the requirements for the given twenty-four-hour productivity Q by increasing the relative consumption of electric power $(J/Q)_1 > (J/Q)_2$; (b) to decrease the given productivity and solve problem (1) when $Q_1<Q_0$; (c) to introduce admissible relative consumption of electric power $J/Q \geq (J/Q)_a$ as a limit and find the maximum possible productivity under these conditions. The programme defines 96 unknown variables at the initial starting for twenty-four hours. Fig. 3 shows the twenty-four-hour work-schedule for grinding mill (a), and the improved schedule in case of damage in 12 hour (b).

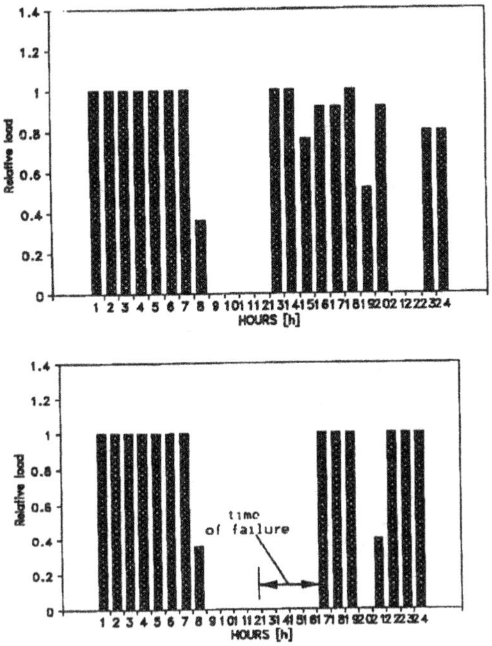

Fig. 3. Load of grinding mill.

SYSTEM COORDINATION

The coordination in the production of a secondary polymer is accomplished by the hour product rate Q(k) between the third and the second level, and by the assignment $r^0(k)$ of the local regulators and controllers between the second and the first level. The mechanisms for coordination and control are shown in TABLE 1. The mechanic and transportation processes are most important for units from the first stage of production (Fig. 4). That is why, the hour production rate Q(k) is a direct coordination factor.

Table 1 Sctech of levels of the system

Level	Function	Approach
Third level ↓ Q(k)	Optimal planning	Linear Programming
Second level ↓ $r^0(k)$	Steady - State Optimization	Model - based fuzzy reasoning
First level	Dynamic control and optimization	Feedback control Feedback-feedforward Gain scheduling Fuzzy control

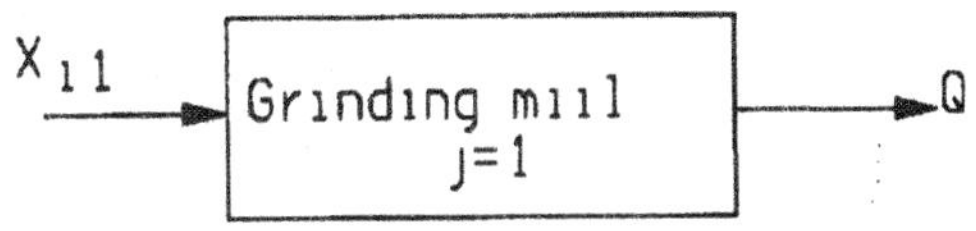

Fig.4 Scheme of the first subsystem.

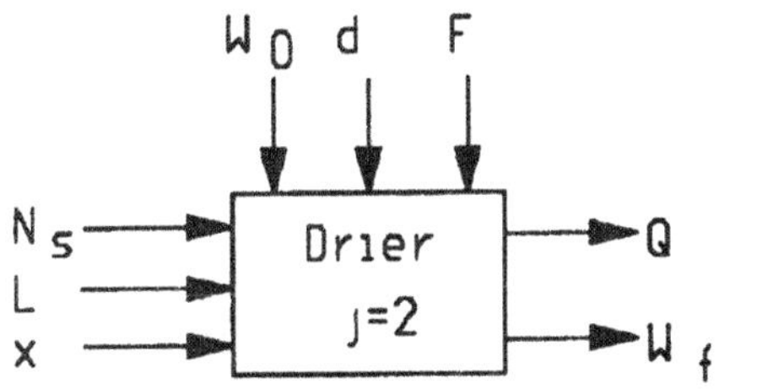

Fig. 5. Scheme of the second subsystem.

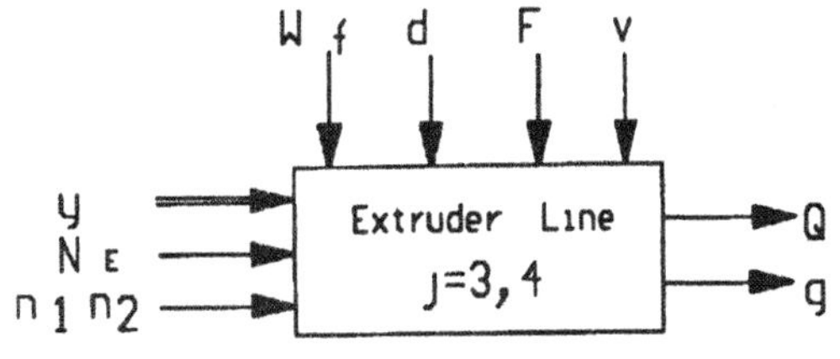

Fig. 6 Scheme of the third subsystem.

The productivity Q and the determined moisture of the ground products W_f are the basic initial variables of the second subsystem Fig. 5 (washing and drying). They are maintained at the determined minimum values by the local regulators of electric power of the drier H, the quantity of drying air L and the degree of recirculation x. The initial moisture W_0, the average size F, and the gauge d of the ground products affect the process from the outside. A local steady-state optimization for minimizing the consumption of electric power is introduced with preset $Q=Q^0$ and $W_f=W_f^0$, and with in certain limits. In this way Q is the coordination variable for the drier.

$$N_s \dashrightarrow \min \qquad (5)$$

The productivity Q and the quality of the secondary polymer q are the basic initial variables of the third subsystem (extruder for granules) Fig. 6. The temperatures in the heating zones y, the power N=N1+N2 of the two extruders E_1, E_2 and also their relationspeed n_1, n_2 influence the process mainly while the moisture W_f, the size F and the gauge of the ground products d cause disturbances. A local criterion for steady-state optimization, minimizing the consumption of electric similar power (5) has been introduced:

$$N_E \rightarrow \min \qquad (6)$$

And in that case, provided the local criteria (6) and the defined reguirements for the quality g and production rate characteristics q, the assignments of the regulators y^0, N_E^0 and n^0 are determined uniquely as a result of the use of an optimization procedure. The hour production rate Q plays the role of a coordinating variable.

STEADY-STATE OPTIMIZATION

Stead-state optimization solves problems (5) and (6) and at the same time is a coordinating procedure for defining the assignments of the local regulators in the subsystem for drying (j=2) and obtaining the secondary granulated product (j=3,4). As it has been already proved, (Hadjiiski, Patsov, Zlatev, Panchev, 1992) the

main external influence in Fig. 5 and Fig. 6 - mousture W^0, W^f, and size F and gauge d of the ground products cannot be measured precisely. This is the reason for using a combination of model-based and fuzzy-based approaches for steady-state optimization. The combination of these two approaches proves to be very effective (Nevins et al, 1990, Hadjiiski, Spasov and Filev, 1991, Kaemmerer and Shristoferson, 1985). Quantitative models have been used to estimate mainly conclusion parts of the membership functions. Expert knowledge has been used both for condition and conclusion part of the membership functions. Because of parameter estimation uncertainties received by skilled operators in the cognitive part of investigation the following scales have been accepted:
(i) for the condition part - three linguistic variables : Big (B), Medium (M), and Small (S);
(ii) for conclusion part - two types of scales :
a) five linguistic variables: High (H), Nearly High (NH), Normal (N), Nearly Low (NL), and Low (L);
b) seven linguistic variables, H, NH, Slightly High (SH), N, Slightly Low (SL), NL, and L.
A uniform scheme of fuzzy reasoning is presented at Fig. 7 (Sugeno, 1985).

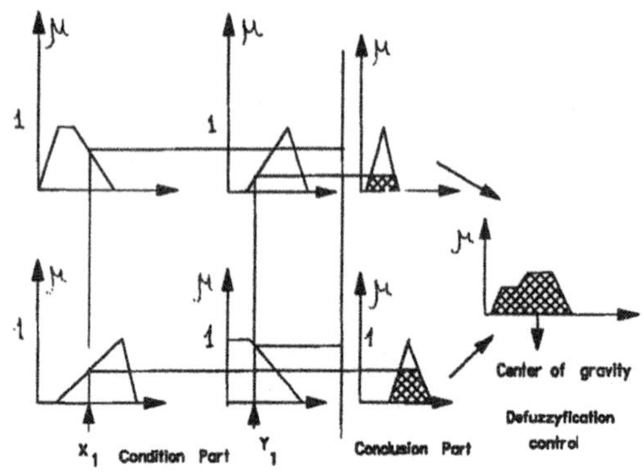

Fig. 7. Fuzzy reazoning.

DRIER FUZZY CONTROL

In conformity with Fig. 5 the initial moisture W^0 and the gauge of the ground products d are accepted as entrance data in the optimizational problem together with the productivity Q and the desired initial moisture. A quantative balance model for the drying process has been designed. It defines the optimum values of the electric heater and the recirculation of the drying air x. It is combined with a fuzzy reasoning approach (Hadjiiski, Patsoff, et al, 1992). The consumption of drying aqend L (which is a result of independing investigation) (Shinskey, 1978) is considered to be constant.

EXTRUDER LINE STEADY-STATE OPTIMIZATION AND CONTROL

The steady-state optimization of the extruder line EL is much more complex that the ones of the other subsystem. First of all, each of two branches of EL is comprised of two extruders E_1 and E_2 (Fig. 8), which have independent parameters and are synchronized only by the equality of the productivities $Q_1=Q_2$. Due to this peculiarity the problem for steady-state optimization of the coupled extruders E_1 and E_2 is divided into two separate problems. The processes of initial melting of the ground products in the extruder E_1 are very complex and are not studied very deeply (Fenner, 1979). The extruder undergoes the effects of moisture W_f, gauge d and size F of the ground products. The preset values Q_i^0 (i=1,4) of the temperature controllers in the four heating areas of the extruder have the control actions. The optimum values of Q_i^0 depend mainly on the type of the polymer and on the input moisture W_f, the decisions are taken on the basis of fuzzy reasoning (Fig. 9). The experimental studies showed that under determined temperatures y_i^0, the productivity Q is affected by the revolutions n and the gauge d of the ground products in a relationship shown in Fig. 10. At preset values of productivity and rational speed controllers n_0^0 are defined directly. Ignoring of the relationships in Fig. 10 would result in disruption of the continuity of the material flow in the coupled extruders. If linguistic data is used to characterize the thickness, fuzzy reasoning may be applied. The experimental results show that regardless of b, at constant temperature, electric power is in linear relationship with the production rate. In this way the adjustment of the power controller is determined uniquely by the rate of production. Due to the fact that the main extruder E_2 processes secondary polymer, the resistance grid in the head of the extruder is being gradually plugged, which leads to a decrease in the productivity Q_2 and an increase of the consumed power N_2. The experimental results are shown in Fig. 11. The relative energy consumption for the production of a single of quantity is also increases (Fig. 12). That is why, the local optimizational problem is formulated as a problem for minimizing the relative variable production costs for electric power and for change of the grid:

$$J=S_1+S_2 \quad \text{-->} \quad \min \tag{7}$$

$$S_1=C_E*N/Q=C_E*f_1(t_c,n) \tag{8}$$

$$S_2=C_g/(t_c*Q)=C_g*f_2(t_c,n) \tag{9}$$

where S_1 and S_2 stands for financial expendititures for electric power and for the change of the grid for the production of 1 kg. of granulated secondary polymers; C_E stands for rate of electric power during night, day, and peak hours; C_g stands for price of the grid; t_c stands for time between two consecutive changes of the grid; N stands for energy consumption.

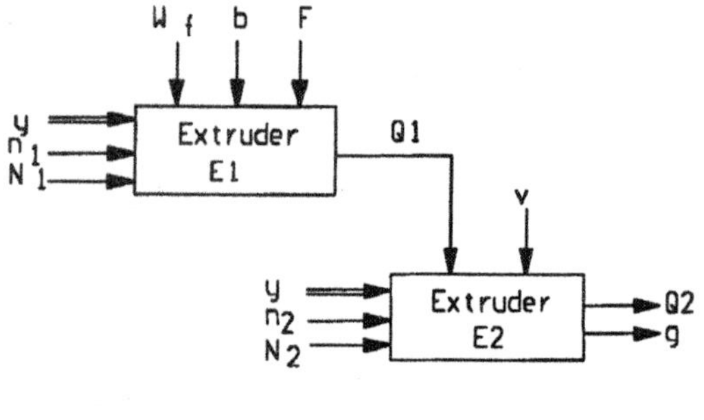

Fig. 8. Scheme of the extruder line.

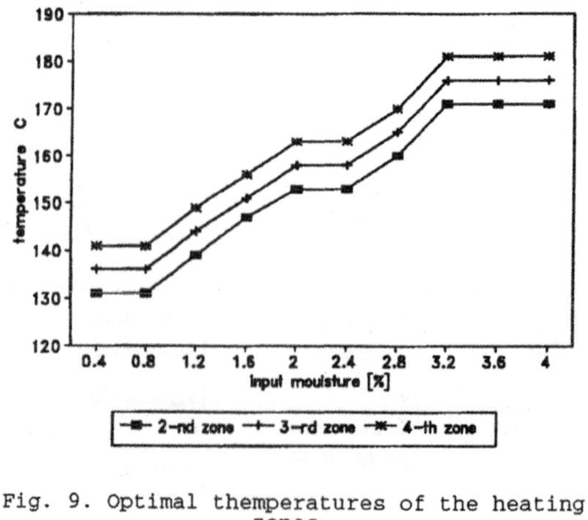

Fig. 9. Optimal themperatures of the heating zones.

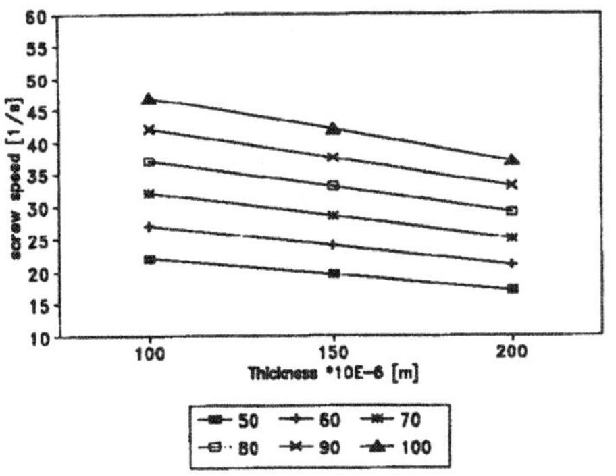

Fig. 10. The screw speed of E_1 according to Q and b.

According to Fig. 11 and Fig. 12, the components S_1 and S_2 of J may be presented as dividing variables by n and t_c. For the criterion J we get:

$$J=C_E*a_1(n)+C_E*b_1(n)t_c+C_g/(t_c*(a_2(n)-b_2(n)*t_c)) \quad (10)$$

From the necessary condition for minimization of J:

$$dJ/dt_c=0 \quad (11)$$

we obtain an algebraic equation of the type:

$$\sum_{i=1}^{4} h_i(n,C_E)t_c^{\,i}=0 \quad (12)$$

From the numerical solution of equation (12) we obtain a clear view of the relationship between the revolutions of the screw n and time interval of the grid changing t_c when the cost of the electric power is a parameter $C_E(i)$:

$$t_c=f_1(n,C_E) \quad (13)$$

According to Fig. 11, the productivity Q may be expressed in the following way:

$$Q=f_2(n,t_{cm}) \quad (14)$$

Since Q is well known from the system of production planning and is constant within 1 hour, and $t_{cm} << 1$, the system of equation (13) and (14) can be solved numerically with parameter $C_E=(C_{E1},C_{E2},C_{E3})$ for the verios price rates during the twenty-four-hours of the day:

$$n=f_3(Q,C_E) \quad (15)$$

In this way, the local optimizational problem (7) is solved: a) the optimum adjustment of the speed controller is obtained from (15); b) the optimum time for changing of the grid is obtained from (13) after the substitution of n from (15); c) the optimum power N for electric operation is obtained from the experimentally achieved relationship in Fig.13:

$$N=f_4(n,t_c) \quad (16)$$

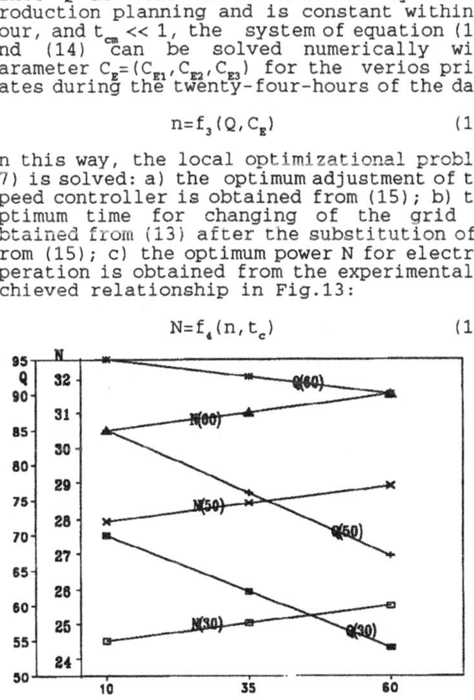

Fiig. 11. Q and N of E_2 depending on t_c and n.

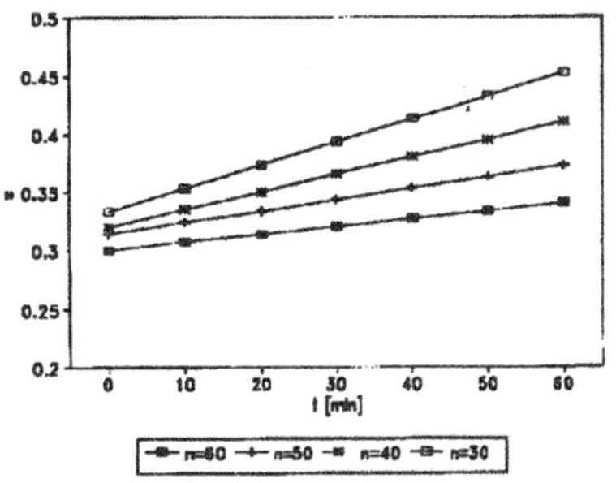

Fig. 12. The relative energy consumption depending on t_c and n.

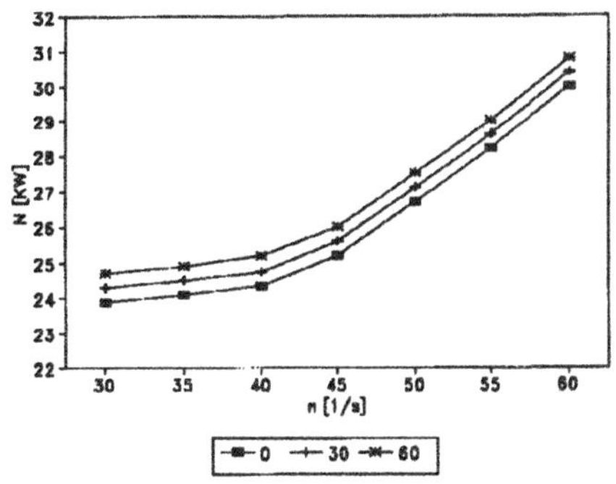

Fig. 13. N depending on n and t_c.

GAIN SCHEDULING DYNAMIC CONTROL

The temperature regime of the extruders, which causes great influence on the quality of the secondary granulated product g (Fig. 8), is very interesting for the dynamic optimization. The experimental studies of all the eight areas of the extruders E_1 and E_2 show great parametric relationship between the coefficients of the mathematic models and the revolutions of the screw. A generalized parametric model of the following form has been accepted :

$$W(p,l)=\frac{K(l)*e^{-td(l)*p}}{T(l)p+1} \quad (17)$$

where $l=n/n^0$ stands for relative screw speed; n, n^0 stands for current and nominal screw speed. The above mentioned results of the production planning and the local steady-state optimization of the extruders show that the screw speeds are an important optimizational parameter and vary within wide ranges according to the productional conditions and disturbances. That is why, a new system for controlling the temperature regime of the extruders has been designed. Its purpose is to complete the existing scheme where the electric heaters work under ''on-off'' regime. The changing of the relaying type of control with a thyristor control proved to be economically ineffective. Because of this, it has been accepted that a new functional element called Puls Duration Modulator (PDM) should be installed after the standard numerical PID-algorithm. The former determines the duration of ''on'' and ''off'' position of the switches of the electric resistance heaters (Kersic and Isenberg, 1985). A gain scheduling control principle is realized in the scheme in Fig. 14. The parameters of the PID - controller are constantly changing according to the scheduling variable l which, in this case, stands for the screw speed. They are determined from the system for local optimization: for extruder E_1 from (15) and for extruder E_2 - according to Fig. 10. The methods of designing a gain scheduling control system are well-developed (Hadjiiski, 1989) and during the few years an increasing interest in them has been registered (Astrom and Wittenmark, 1989, Rugh, 1990). The parameters for adjustment of the PID controllers according to l are obtained by applying the method of the frozen parameters:

$$x(l)=(K_p(l),\ T_i(l),\ T_d(l)) \tag{18}$$

Gain scheduling table is made in the free programming memory of the used controller ISOMATIC _ 1001. Science the constants in the project model change 2-3 times in the interval of the alternation of the revolutions, the effect of the introduction of gain scheduling control technique is considerable. Here, the transitional processes are extremely unsatisfactory and the system may become unstable with a PID regulator of constant coefficients.

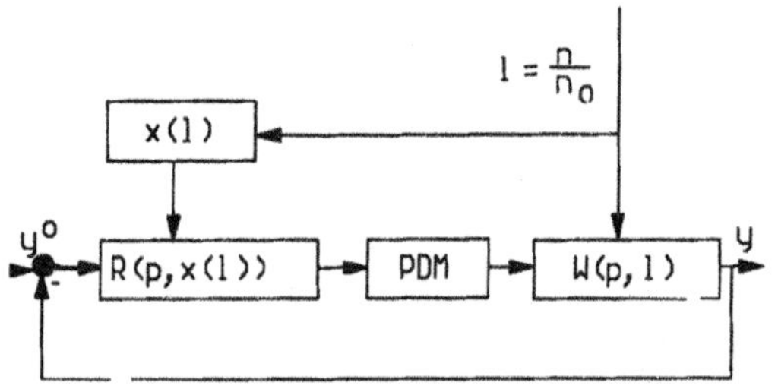

Fig.14. Gain scheduling control system

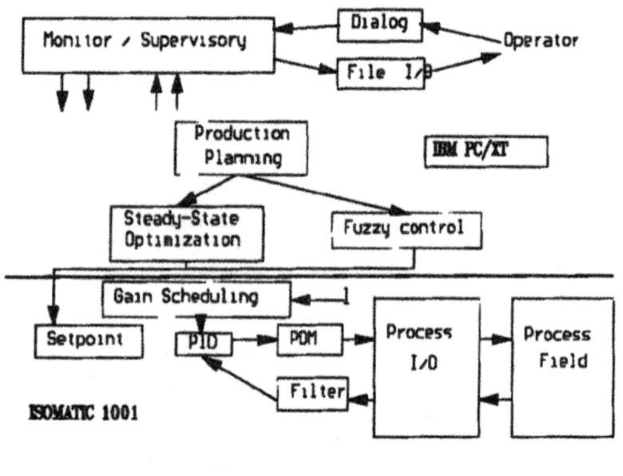

Fig. 15 United functional scheme of the intelligent controller.

CONTROL SYSTEM REALIZATION

The general functional scheme of the proposed intelligent decentralized regulator is shown in Fig. 15. The potentialities of available and cheap industrial controllers with a sufficient free programmable memory (32-64 K) for connection with the object of control are combined with the potentials of personal computers (PC) for interface with the operator and more complex calculations. The applied programmes are in FORTRAN77. Science the programme language of the industrial controller ISOMATIC - 1001, used in this particular case, was FORT, it was necessary to use an intermediate interface language. In view of the fact that the system for control of a small plant comprises about 20 controllers, the general scheme for intelligent control in Fig. 15 comprises of different modules for each of them. A multifunctional intelligent controller has been recently designed. It uses suitable strategies for control according to the particular economic conditions, the production developments and the existing disturbances, Due to this fact, a uniform approach for automation of the various production units may be applied for small factories, taking into full consideration peculiarities. The optimization of the regimes in steady-state and dynamic operation conditions is attained by minimum changes in the existing measurement and regulating effects and by working out a computer-based control system. The application of discussed approach in the case of a small plants for recycling of secondary polymers shows hopeful prospects for a wider usage as a principle and practice.

REFERENCES

Albertos, P. (1989). Low Cost Automation Issues. Proc. of the IFAC Symp. Low Cost Automation, Milan, Italy.

Aracil, J., A. Ollero and F. Morant (1990). Knowledge Acquisition and dynamic behavior by using Intelligent Controllers, Proc. 11 IFAC World Congress, Tallin, Estonia, vol.8,pp. 136-146.

Astrom, K. and B. Wittenmark (1989). Adaptive control. Addisson Wesley, Heading MA.

Astrom, K.(1991). Directions in Intelligent Control. Proc. of IFAC Symposium on Intelligent Tuning and Adaptive Control, Singapore.

De Carly, A. (1989). Proc. of the IFAC Symp. Low Cost Automation, Milan, Italy.

Fenner, R. (1979). Principles of Polymer Processing. Mc Milan Pres, London.

Kaemmerer, W. and P. Shristopherson (1985). Using Process Models with Expert Systems to Aid Process Control Operations, Proc. Amer. Contr. Conf.

Kersic, R. and R. Isenberg (1985). A Concept of Industrial Adaptive Control Based on Robust Design Principle and Multi-Model Approach, Proc. of IFAC Adaptive Control of Chemical Processes, Frankfurt am Main.

Koing, J. and W. Stokhansen (1990). An Information System for Production and Process Control in the Chemical Industry, Proc. of 11 World Congress, Tallin, Estonia.

Hadjiiski, M., K. Spasov and D. Filev (1991). Knowledge-Based Model of Thermal State of Metallurgical Ladle, Proc. of IFAC Workshop on Expert Systems in Mineral and Metal Processing, Helsinki, Finland.

Hadjiiski, M., S. Patsov, Z. Zlatev, E. Panchev (1992). Control of the Technological Processes of a Small Factory for Recycling of Plastics, Preprints of the IFAC Workshop on Automatic Control for Quality and Productivity, Istanbul, Turkey.

Hadjiiski, M. (1989). Automatic Control of Processes in Chemical and Metallurgical Industry, Technica, Sofia, (in bulgarian).

Navairo J.N., P. Albertos, M. Martinez and F. Morant (1990). Intelligent Industrial Control, Proc. of 11 IFAC World Congress, Tallin, Estonia.

Nevins, J., D. Whitney and A. Edsall (1990). Intelligent Systems in Manufacturing, Proc. of 11 IFAC World Congress, Tallin, Estonia.

Rugh, W. (1990). Analytical Framework for Gain Scheduling, Proc. of the ACC'90, pp.1688-1694.

Shinskey, F. (1978). Energy Conservation through Control. Academic Press, New York.

Sugeno, F. (1985). Industrial Applications of Fuzzy Control, Academic Press, New York.

Wittenmark, B. (1989). Theory as a part of Low Cost Automation Control, Proc. of the IFAC Low Cost Automation, Milan, Italy.

SELF-ORGANIZING FUZZY CONTROLLER WITH NEURAL NETWORK

L. Jurišica and M. Sedláček

Department of Automation and Control, Faculty of Electrical Engineering, 3 Ilkovičova St., Bratislava 812 19, Czechoslovakia

Abstract: The method for adjustment of fuzzy controller parameters is presented. In fuzzy control the control signals are generated by an appropriate inference mechanism from the linguistically expressed rule base. To refine the parameters of the rules it is necessary to have either a thorough insight into the dynamics of the overall system or to develop an appropriate adaptation mechanism. Such mechanism with Kohonen neural network like structure is presented and documented on the example of motion control system.

Keywords: Fuzzy control; Neural networks; Learning systems; Motor control; Servomechanisms

INTRODUCTION

Fuzzy logic controllers (FLC) are useful when the processes are interpreted inexactly or uncertainly. We have recognized them suitable for the motion control systems in robotic applications.

The performance of FLCs, however, is dependent on the availability of good linguistic control plan which is not sometimes easy to formulate. The usual solution is the long refinement - adjustment of FLC parameters for the given plant. An attractive solution for this problem can be the incorporation of learning strategy into the FLC.

One possible solution is to refine the controller by means of self-organization. Some learning algorithms for the fuzzy logic controllers have been already proposed (Morita, 1990). They can either modify the expressions in "Then" clauses and shift membership functions or incorporate neural network structure as learning element. Neither in first nor in second case the problem of proper functioning when the membership functions are not properly described is succesfully solved.

In this paper we present FLC with rule base expressed in the form of neural network-like architecture with learning capabilities. In this structure the Kohonen neural network like architecture adapts the values to reach desired performance. After the learning, the membership functions are closer to the optimal values and can be easily (if necessary) refined. Thus the time-consuming stage of refinement can be avoided.

The proposed structure is documented on the design of FLC for motion control system.

DESCRIPTION OF THE PROBLEM

Foundations of Fuzzy Logic Theory

For the clarity we shall briefly introduce some basic facts of the fuzzy set theory and fuzzy logic, which will be needed (Lee, 1990).

Definitions:
Fuzzy set F in an universe of discourse U is characterized by a membership function μ_F which takes values in the interval [0,1]. Thus F in U can be represented as a set of ordered pairs of element u and its membership function:

$F=\{(u, \mu_F(u)) | u \in U\}$.

Let A and B are two fuzzy sets in U.
Union: The membership function μ_C of the union $C=A \cup B$ is pointwise defined for all $u \in U$ by: $\mu_C(u) = \max\{\mu_A(u), \mu_B(u)\}$.
Intersection: The membership function μ_D of the intersection $D=A \cap B$ is pointwise defined for all $u \in U$ by:

$\mu_D(u) = \min\{\mu_A(u), \mu_B(u)\}$

Complement: The membership function μ_E of the complement of set A is pointwise defined for all $u \in U$ by: $\mu_E(u) = 1 - \mu_A(u)$

Foundations of Fuzzy Logic Controllers

In this part we present the main ideas, essential for the FLC. Figure 1 shows the basic configuration of a FLC. The fuzzifier relates real values of inputs to fuzzy sets, according to the input membership functions in the knowledge base. Then the fuzzy reasoning upon the rules is performed. Rule base provides control rules for the process of fuzzy reasoning. Defuzzifier transforms the outcome of fuzzy inference into the real value for the output.

Motion Control Problem

In motion control systems for robotics the problem of control of uncertain systems is crucial. Time varying components come from the robot's manipulation with payload (step changes of load torque T_L) and changes in robot configuration (changes of the moment of inertia J). The time varying J is an important varying parameter. This fact greatly affects the robot servosystem performance.

The typical d.c. velocity servosystem is presented in Fig.2. The objective is to design the robust FLC for the case of varying J and step changes of T_L.

In the Fig.2:
K_R - gain of the velocity transducer
Te - electromagnetic time constant
Rm - armature resistance
Cu - motor torque coefficient
Note that J represents the sum of the motor and reflected load inertias through the axis of rotation.

FUZZY CONTROLLER FOR THE MOTION CONTROL SYSTEM

We present the robust FLC for the velocity d.c. servosystem, dealing with the effect of the nonstationary moment of inertia and variable load torque.

Designed FLC has two inputs:
error e(k) = r(k)-y(k) and the rate of error de = abs(e(k))-abs(e(k-1)), where r(k) is the desired value in the kth sampling instant, y(k) is the actual value and abs(*) is the absolute value. The first input is denoted a_1 and:
$a_1(k)=q_1 e(k)$, where q_1 is scaling factor.
The second input is denoted a_2 and:
$a_2(k)= q_2 de(k)$, where q_2 is scaling factor.

The considered membership functions are trapezoidal and have the form from the Fig.3. The membership functions for the defuzzification are documented in Fig.4, where c_1, c_2, c_3 - knowledge base parameters
Together with the knowledge base design the crucial point is the rule base. The proposed rule base is in Table 1, where N, Z, P, NZ, PZ is negative, zero, positive, negative zero and positive zero.

TABLE 1 Rule Base of the FLC

de ↓ / e →	N	NZ	PZ	P
N	NZ	PZ	NZ	PZ
Z	N	NZ	PZ	P
P	N	NZ	PZ	P

Unfortunately, the systems performance rapidly decreases with the unproper choice of parameters in knowledge base. The performance with the initial parameters is documented in the Figure 5. Before the refinement the performance of the system was not sufficient. y(t) is a transient response of velocity with a step change of T_L at time 0.5s.

The way to automatically refine the parameters in the knowledge base is the incorporation of learning procedure, as will be described in the next paragraph.

LEARNING FUZZY LOGIC CONTROLLER WITH NEURAL NETWORK

Introduction

A neural network type fuzzy logic controller has been proposed by Morita (1990). It differs from the conventional fuzzy model in the fact, that the input and output values of each fuzzy rule are multiplied by weighting factor. Since the weights of the proposed model correspond to the synaptic weights of a neural network, such model is called a neural network type. This scheme is more sophisticated than usual ones, but still solves only the problem of shifting membership functions and changing their width. It does not solve the problem of bad choice of shape and parameters of the initial set of membership functions.

In our approach, first , a knowledge base is created by describing an operators knowledge in the form of fuzzy rules and membership functions. This is the same as in conventional fuzzy logic controller design procedure.

Then the knowledge base is refined iteratively by varying the neural network parameters so that the difference between the desired and actual value becomes smaller with number of iterations.

Controller Architecture

The fuzzy logic controller has its usual structure as described above. The only exception is the implementation of the knowledge base. The knowledge base is creating the structure similar to Kohonen neural network. In this structure the learning algorithm is used.

The structure of the knowledge base is shown in Fig.6. It performs the mapping $V: R^2 ---> R^6$. The inputs of the two input layers are the error and rate of error. Then the winning elements in each input layer are found. The winning elements are the elements v_i and w_m, for which holds:

$$\min \|v_j(k) - e(k)\| \quad \text{for } j=1..N$$

$$\min \|w_n(k) - de(k)\| \quad \text{for } n=1..N$$

where N is a number of cells in the input layer and i and m are the positions of the winning elements in the input layers and $\|*\|$ is a suitable norm.

Each of the two input layers has three output layers with the associated values of membership functions. In the first output layer for e are the values of membership function for the set Error Negative - μ_{eN}, in the second output layer for the set Error Zero - μ_{eZ}, and in the third output for the set Error Positive - μ_{eP}. The knowledge base of the rate of error de is similar.

When the values of e(k) and de(k) are fed into the first two layers, the winning elements are found. Then proper values of membership functions for the actual operating point are associated and further processed by the FLC. In the next sampling instant the adaptation procedure occurs. The goal is to adjust the associated values of membership functions in the ith and m^{th} position thus, that the selected criterion is decreased.

Let us denote p as the position i of the winning element in the input layer for error e and q as the position of the winning element in the input layer for rate of error de. Then the associated actual values of membership functions are:

$\mu^{p}_{1,r}(k)$ for the pth cell of the rth output layer of error in the kth sampling instant.

$\mu^{q}_{2,s}(k)$ for the qth cell of the sth output layer of the rate of error in the kth sampling instant.

In this sense

$\mu^{p}_{1,1}(k)$ is the pth value of $\mu_{eN}(k)$,

$\mu^{p}_{1,2}(k)$ is the pth value of $\mu_{eZ}(k)$,

$\mu^{q}_{2,1}(k)$ is the qth value of $\mu_{deN}(k)$ etc.

Note: It is reasonable to have values of error and the rate of error monothonically increasing in the input layers. Thus the scheme becomes more computationally efficient, because it is not necessary to calculate the norm $\|*\|$ for all N elements in the input layer. Some of the sophisticated searching methods can be used. It can be shown, that the maximal number of norm calculations is n+2, where:

$N/(2^{n}) < 1 \le N/(2^{n+1})$.

Learning Algorithm

The algorithm adjusts the membership values so that the performance index is decreased. The learning process can be divided into two phases:

1st stage of learning algorithm

Let L(k) is the performance index of an error in the kth iteration:

$$L(k)= \frac{1}{2} * \left(r(k)-y(k)\right)^{2}$$

The value of membership function in the kth instant in the actual operating point is modified in the $(k+1)^{th}$ sampling instant so that the criterion L(k+1) is decreased:

$$\mu(k+1) = \mu(k) + \Delta\mu(k)$$

$$\Delta\mu(k) = - \gamma \frac{\delta L(k)}{\delta \mu(k)}$$

where γ is a small step value.

Note: For clarity the sub- and superscripts have been omitted. In this paragraph the adaption is performed with the value with position p and q in the output layer. Let the max rule, used in defuzzification, has the number t from Table 1, then the adapted membership values are those, from which the rule number t has been evaluated:

$$\min \{ \mu^{p}_{1,r}(k), \mu^{q}_{2,s}(k) \} = \mu_{t}(k)$$

$$\mu_{t}(k) \ge \mu_{j}(k) \qquad j=1..9, \ j \ne t$$

In the following this will be omitted.

The problem which occurs is similar to the neural network problem of learning with distal teacher, because:

$$\Delta\mu(k) = - \gamma \{ r(k)-y(k) \} \frac{\delta y(k)}{\delta \mu(k)}$$

and the sensitivity function: $S(k) = (\delta y(k))/(\delta \mu(k))$ is not known. We are using rough approximation of Saerens, (1990), where ν is a small step:

$$S(k) = \frac{\delta y(k)}{\delta u(k)} \frac{\delta u(k)}{\delta \mu(k)} =$$

$$= \nu \operatorname{sign}\left(\frac{\delta y(k)}{\delta u(k)} \right) \left(\frac{\delta u(k)}{\delta \mu(k)} \right)$$

Then the resulting adjustment is:

$$\mu(k+1)=\mu(k) + \alpha e(k) \operatorname{sign} \left(\frac{\delta y(k)}{\delta u(k)} \right) H(k)$$

where sign(*) is a signum function, α is a learning step, error $e(k)=r(k)-y(k)$ and $H(k) = (\delta u(k))/(\delta \mu(k))$ is dependent on the actual rule in defuzzyfication procedure (simple for the standard max version of defuzzification procedure). For example, if the actual rule for defuzzyfication is the rule in the 1st column and 2nd raw of the Table 1 then $\mu_{a} = \min(\mu_{eN}, \mu_{deZ})$ and H(k) can be derived from the defuzzyfication:

$$q_{3}a_{3}(k) = - c_{3}\mu_{a}(k)$$

Note: The signum approximation is valid only if the characteristics of the plant is monothonic. In our case of linear time varying system this assumption holds.

2nd stage of learning algorithm

In this stage the Kohonen learning mechanism is applied for the values in the neighborhood of winning element. The underlying idea is to extend the learning action. The Kohonen self organizing procedure has the following form:

$$\mu^{p-1}_{1,r}(k+1)=\mu^{p-1}_{1,r}(k) + \beta \left(\mu^{p}_{1,r}(k) - \mu^{p-1}_{1,r}(k)\right)$$

$$\mu^{p+1}_{1,r}(k+1)=\mu^{p+1}_{1,r}(k) + \beta \left(\mu^{p}_{1,r}(k) - \mu^{p+1}_{1,r}(k)\right)$$

for the case of adapting the pth element in the rth output layer. β is a coefficient, depending on time and the distance from the centre p. It is about 0.4.

RESULTS

Number of simulations was performed to verify the validity of the proposed learning FLC.

In this paper we deal with the effect of varying inertia moment J and load torque T_L and the ways of their elimination in velocity d.c. servosystem. The results were verified by considering the d.c. disk motor SERVALCO 300 W with following parameters: (For description of the parameters see Fig.2)

C_u = 0,18 Vs, R_m = 3,1 Ω, Te= $6,45.10^{-5}$s,

J_o = 0,00158 kgm^{2}, K_R = 0,032 $V/rad.s^{-1}$

T_{nom} = 96 Ncm.

where T_{nom} is the nominal motor torque.

Figure 5 shows the transient response of system with the initial set of parameters. The same experiment is documented in the Fig.7 for the FLC with the neural network. The coefficient α was 0.01. The transient response for the system with adjusted parameters and coefficient α equal to 0.1 is shown in Fig.8.

After the process of automatic learning the membership functions can be further manually tuned, if necessary. In this case only the small corrections instead of time consuming refinement were necessary. The resulting control strategy can be plotted in three-dimensional space, as illustrated in Fig.9.

SUMMARY

In the paper the fuzzy logic control for the dc velocity servosystem was designed. It was shown, that unproper choice of parameters of the controller deteriorates the system performance. The method of automatic refinement of membership functions in knowledge base was developed. The method is based on the Kohonen neural network like structure. The performance of the system after the process of learning increased.

REFERENCES

Lee, Ch. Ch. (1990). Fuzzy logic in control systems: fuzzy logic controller. *IEEE Transactions on systems, man, and cybernetics*, 20, 404-417.

Jurišica, L., and M. Sedláček (1991). Neural networks for robotics and fuzzy control. In *Discrete processes control*, Bratislava, ČSVTS, 114-115.

Morita, A. (1990). A neural - net type fuzzy model. *Mitsubishi Electronic Advances*, 51, 12-14.

De Silva, C. W. (1991). An analytical framework for knowledge based tuning of servo controllers. *Engineering applications of artificial intelligence*, 4, 177-189.

Buckley, J. J., Ying, H., and W. Siler (1990). Fuzzy control theory: a nonlinear case. *Automatica*, 26, 513-520.

Chen, Y. Y., and T. Ch. Tsao (1989). A description of the dynamical behavior of fuzzy systems. *IEE transactions on systems, man, and cybernetics*, 19, 745-755

Saerens, M., and A. Soquet (1991). Neural controller based on backpropagation algorithm. *IEE Proceeding-F*, 138, 55-62.

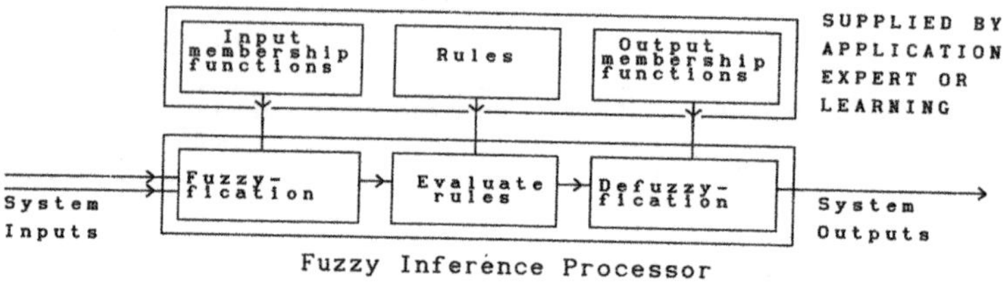

Fig.1. General scheme of the fuzzy logic controller

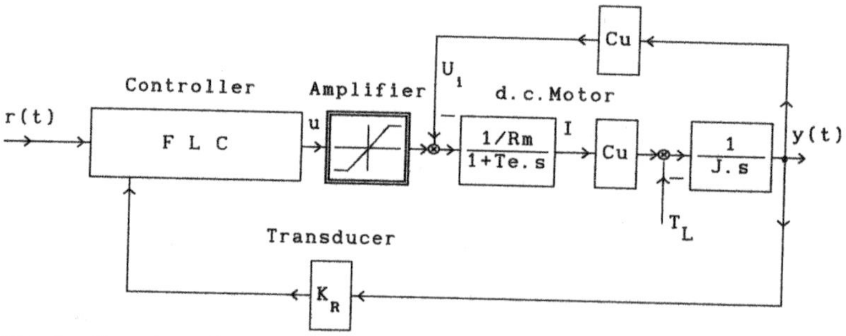

Fig.2. Fuzzy controlled velocity d.c. servosystem

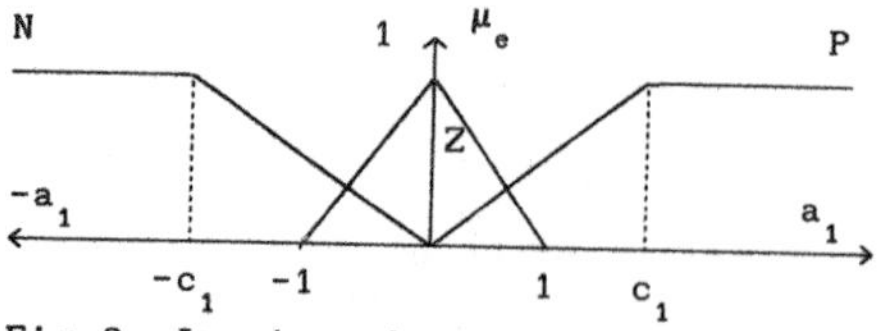

Fig.3. Input membership functions

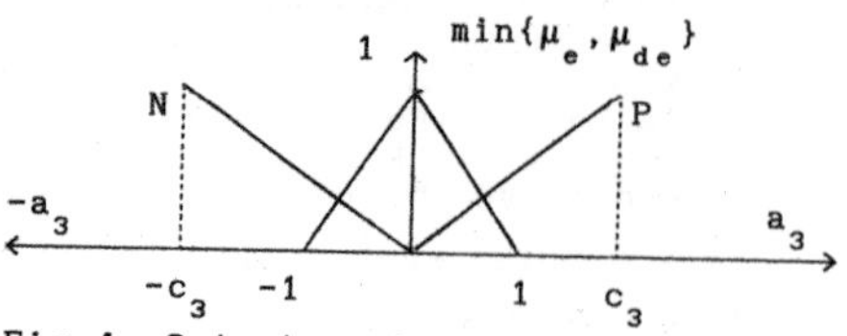

Fig.4. Output membership functions

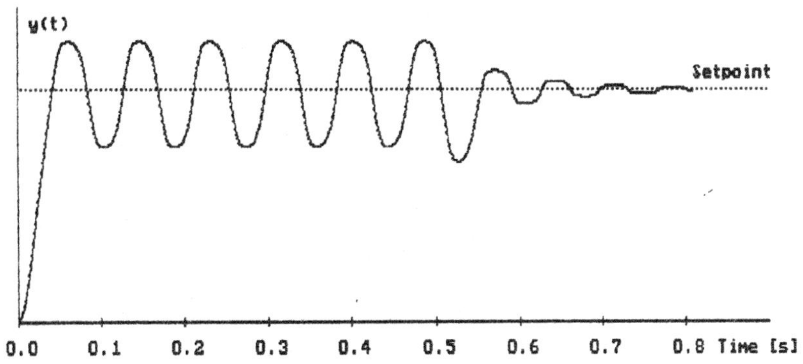

Fig.5. Performance of the system with initial parameters

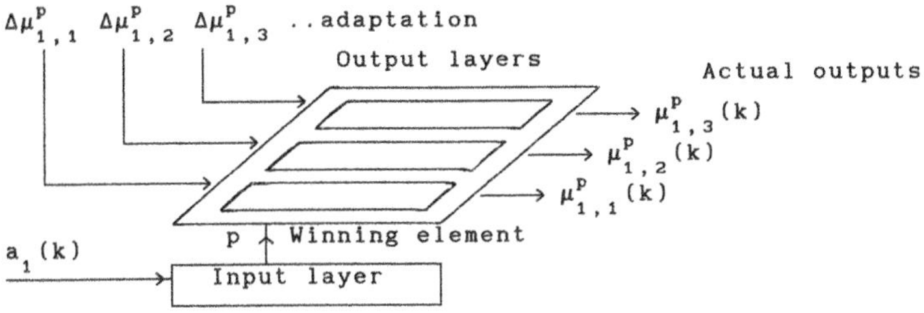

Fig.6. Structure of the knowledge base for the $a_1(k)=q_1\ e(k)$

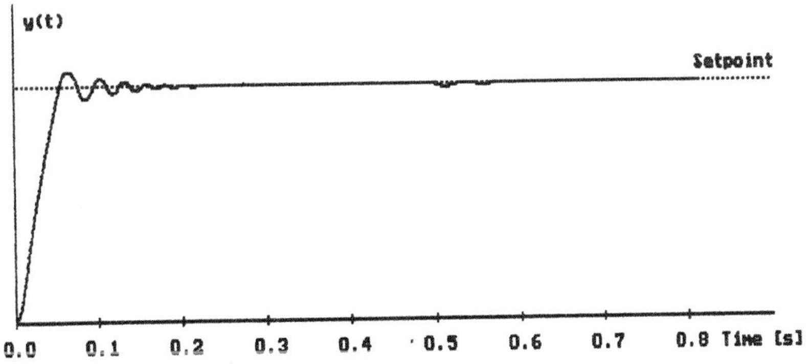

Fig.7. Performance of the system with learning

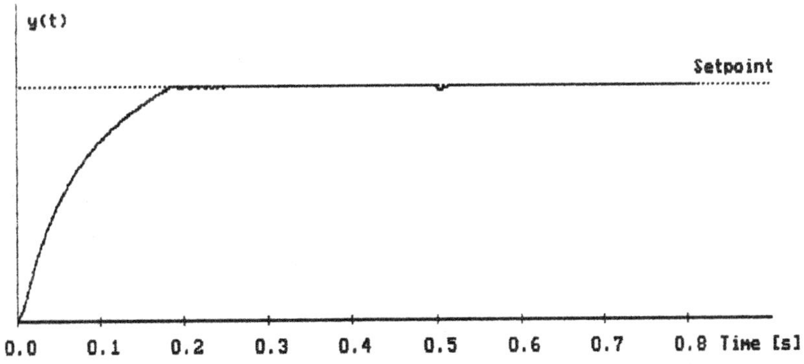

Fig.8. Effect of varying learning step

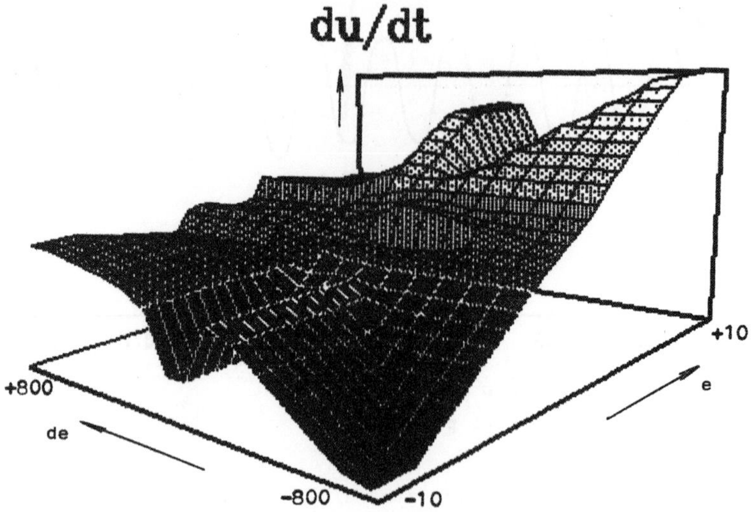

Fig.9. Resulting control strategy of the FLC

REAL-TIME SUPERVISION BY A GRAFCET ALGORITHM

R.F. Garcia

Universidad de La Coruña, E.S. Marina Civil, Paseo de Ronda 51, 15011 La Coruña, Spain

Abstract This paper describes a low cost method useful to be applied on industrial process control as a mean to increase system safety and reliability. Such method consist in a control supervisor capable for evaluating in any stage of the control process, the condition of the sensors and/or detectors involved on to previous and present stage of the sequential control path , and consequently decides which action to take in order to avoid system damage or system failure. The method is based in a grafcet language algorithm to be implemented on any programable logic controller (PLC) for fault finding and decision-making even in hybrid sequential processes, that is (acquired data is a combination of logic and analog information), operating simultaneously with the logic control sequence.
Keywords decision-making, diagnosis, failure detection, grafcet, supervisory control

INTRODUCTION

During the last decade the grafcet language have been used as one of the most important means for designing, programming and describing logic sequential systems (Silva. M, 1982, 1985). Recently, (Telemecanique, 1989) developed some software based-computer tools to control sequential processes in which logic and analog events take place at same time, furthermore such process may be described and programed under grafcet language description.
Supervising a sequential process implies commonly the use of an expert or rule-based system which performs tasks like monitoring, diagnostics and decision-making. On the other hand when a sequential system is controlled by Grafcet language, the task of supervision through diagnosis, is centered on finding an effective method of representing and organising the domain knowledge as well as a suitable problem-solving strategy for the diagnistic task which depends of the human experience, or on the human knowledge about that particular process.
The task of supervising a sequential control process implies commonly the use of an expert system capable for perform tasks like monitoring, diagnostics decision-making and the consequent actuation. Knowledge in expert systems can be expressed by one or both of two ways, the one who is based in the if-then rules and the second which is based in object oriented structures. In any case a real-time computer with special software and input/output hardware is needed.
In the task of process supervision, under any technique, the human expert is responsible for implement a safety method for a correct diagnostic and consequent actuation. Grafcet description has revealed as a powerfull tool to solve the problem when programmed on a PLC because of its capacity of control both, the system logic sequence and the reliability of acquired data.

PROBLEM FORMULATION

In a rule-based expert system to be applied in an industrial plant for the evaluation of any diagnostic task or a malfunction diagnosis (Russo 1987, Shum 1988), it is frequently the evaluation of rules such that:

Rule No 34
IF starter command of pump No.1 is on. (b1)
THEN start pumo No 1 (A1)

Rule No 35
IF starter command of pump No.1 is on (b1)
AND running time is more than 10 seconds. (b2)
AND its speed is less than 500 rpm. (b3)
THEN it has failed. (A4-b4)
ELSE open valve No.15 (A3)

Rule No 36
IF pump No 1 has failed (A4-b4) THEN start pump No 2. (A2)

Rule No 37
IF starter command of pumo No.2 is on (b4)
AND time is more than 10 seconds. (b5)
AND the speed of second pump is less than 500 rpm.(b6)
THEN stop the plant. (A5-b7)
ELSE open valve No. 15 (A3)

Last rules expressed as logic values for any codition gives:

IF b1 THEN A1
IF b1 AND b2 AND b3 THEN A4 ELSE A3.
IF b4 THEN A2
IF b4 AND b5 AND b6 THEN A5 ELSE A3.

Such rules translated into a grafcet representation are showed at figure 1.
being easy to apply and implement as well as simple to understand.

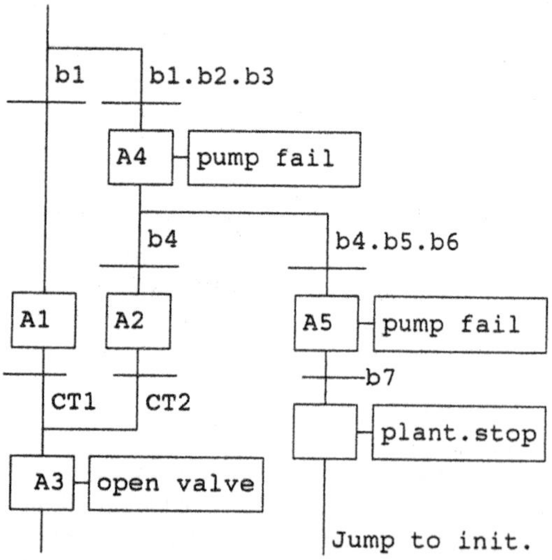

CT1=b1.b2.b3
CT2=b4.b5.b6

Fig. 1 Grafcet of a diagnostic task

In similar way it is possible to implement by grafcet a rule-based system to validate the acquired data as the inputs to any logic control process by means of an extended grafcet to be added to the initial grafcet of the process control. To do that, it is assumed that supervision of acquired data and concluded actions depends on the past and present data. Let's consider part of a flow control sequence described by means of a grafcet representation where CT(i) is the condition for transition to stage S(i). In any situation or stage (i) of a control sequence, the evidence of a fault in an activity is the time T(i) the action takes to reach the next condition for a normal transition. Other analog magnitudes helps a lot for fault finding. With the data available, the problem of data validation or process supervision can be formulated by means of a true table showed at table 1.

TABLE 1 True table of supervision algorithm

T i	CT i	CT i+1	possible fault in
1	0	0	S(i), CT(i+1)
1	0	1	slow actuation
1	1	0	S(i)
1	1	1	CT(i)

Decision-making through the conclusions given by the proposed true table, is to be implemented by grafcet depending on the desired action to perform to be programmed by an human expert in the particular problem-solving task under the criteria of avoiding critical or ambiguous situations.
The mentioned part of a flow control sequence is showed at figure 2.

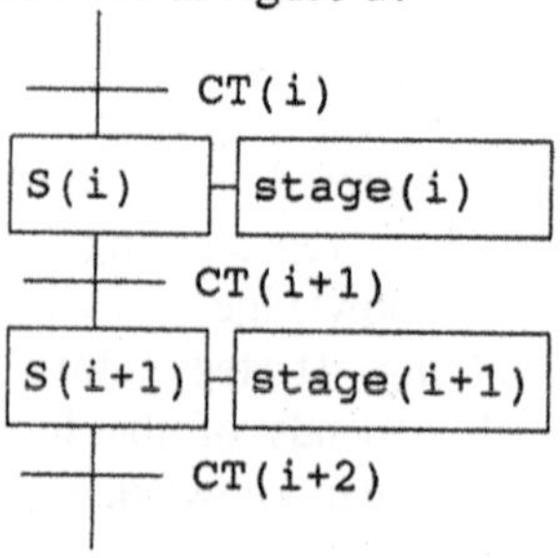

Fig. 2. Grafcet flow control sequence.

The conclusions to find or detect some fault are interpreted directly from the true table, where for any abnormal situation, it must be generated the proper action to be performed as an output according the human expert programmer, who is based in his experience about malfunction criteria for every stage of the plant.
From mentioned true table, there are four functions that identifies failure of one or more of the parts of the plant (detectors, actuators or both). The definition of any function is as follows:

$$A = T(i).\underline{CT}(i).\underline{CT}(i+1) \Longrightarrow S(i), CT(i+1) \quad (1)$$

$$B = T(i).\underline{CT}(i).CT(i+1) \Longrightarrow \text{slow actuation.} \quad (2)$$

$$C = T(i).CT(i).\underline{CT}(i+1) \Longrightarrow S(i) \quad (3)$$

$$D = T(i).CT(i).CT(i+1) \Longrightarrow CT(i) \quad (4)$$

The variable of reference to detect abnormal situations is in this case the time T(i) the action takes to reach next transition under normal condition. Another analog variable, depending on the process model could be useful also.
Expressions (1), (2), (3) and (4) are programmed by means of some extended grafcet which must be added to to the flow control sequence described at figure 2.
Figure 3 shows the outputs produced by the extended grafcet for an stage supervision.

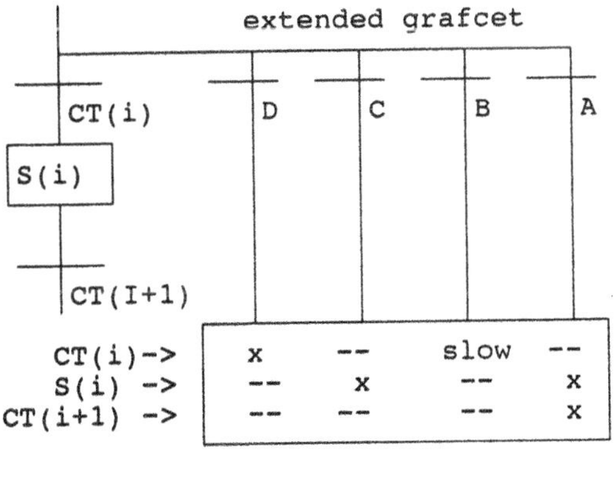

Fig. 3 Spervision task by extended grafcet

The task of identification the part of the plant that fails requires some extra-effort and it may concluded with ambiguous results due to the internal coupling between an activity and the detectors responsisble for generate the condition for transition. According the conclussions from true table 1, after a fault was detected, we must assume that it was due to one or more of the three following reasons:

--fault al the detectors responsible for supply the necessary information for the condition of transition to next stage S(i).
--Fault in the actuator or its related accessories for activity S(i).
--fault at detectors responsible for supply the data necessary for next condition of transition CT(i+1).

As showed at figure 3, the output delivered by expresion (1) is not deterministic, being necessary local analysis of the problem to define exactly the location of the fault. There is three deterministic outputs, described by expresions (2), (3) and (4), which are specific outputs to denote a fault in the stage S(i), CT(i) or CT(i+1). In the expresion (1), it is implicated CT(i+1) and S(i) as candidates for being included in the conclusion for fault finding. If the outputs to be generated as response to a supervision process belongs to expresions (2), (3) or (4), then two outputs for each response are needed. The first is to indicate the stage in which appeared a fault and the second is to express the component that fails as a deterministic description.

Such responses are given at expresions (5), (6) and (7) and (8) as,

$$A(i) = T(i).\underline{CT(i)}.\underline{CT(i+1)} \Longrightarrow \text{fault of } S(i) \text{ and/or } CT(i+1) \quad (5)$$

$$B(i) = T(i).\underline{CT}(i).CT(i+1) \Longrightarrow \text{slow actuation at stage (i)} \quad (6)$$

$$C(i) = T(i).CT(i).\underline{CT}(i+1) \Longrightarrow \text{fault of } S(i) \text{ at stage (i), (general actuactors supply)} \quad (7)$$

$$D(i) = T(i).T(i+1) \Longrightarrow \text{fault of } CT(i) \text{ at stage (i), (general detectors supply)} \quad (8)$$

The extended grafcet for control supervision as per expresions (2), (3) and (4), is showed at figure 4.

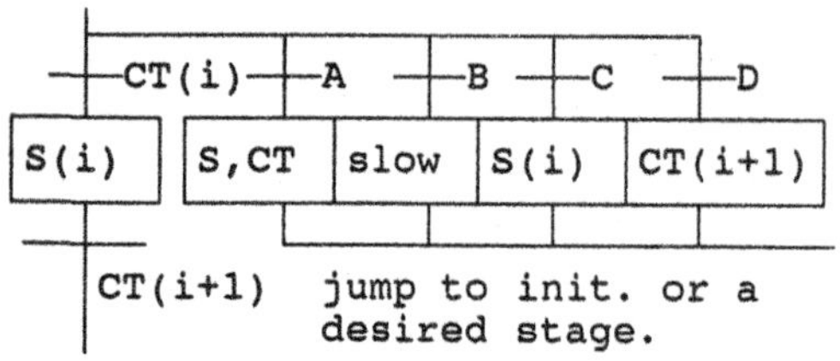

Fig. 4 Description of supervision task.

Under the presence of some faults, the options to be taken as outputs can be the following:
--the logic stage of the plant.
--the main characteristic of failures based on the transition generated.
If the fault invokes the transition A then ambiguity of response exist and a local analysis of the problem must be realised by an expert human oprrator.
If the fault invokes the transition B, then the actuator operation is too slow.
If the fault invokes the transition C, then some component of stage (i) fails, with high probability of failure at main supply to actuactors.
If the fault invokes the transition A then the most probably cause is the mains or supply to the detectors of CT(i).
At figure 5 it is showed the fragment of a block diagram for the proposed supervising algorithm by the proper extension of grafcet.

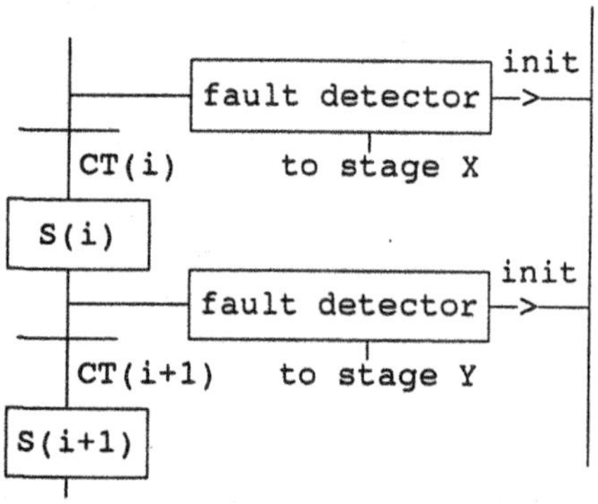

Fig. 5 Flow graph for a supervision algorithm

In order to reduce the number of outputs in the fault finding task, a minimum of one output must be used as watch-dog, based on the time necessary to perform an activity. So that, the time estimated to go from CT(i) to CT(i+1) is the reference time limit to advice about a malfunction. With two outputs, it can be described a system failure in the folowing terms: The first output describes the response of expresion (6). and the second output describes the response of expresions (5), (7) and (8) by expresion (9).

$$\underline{CT(i)} \,.\, \underline{CT(i+1)} + CT(i).\,\underline{CT(i+1)} + $$
$$+ CT(i).CT(i+1) \Longrightarrow CT(i) + \underline{CT(i+1)} \quad (9)$$

APPLICATION AND MAIN RESULTS

This application was implemented on a training board equipped with two pneumatic operated cylinders and the proper electro-neumatic solenoid valves each, whose signals are the outputs of a programable logic controller, (the TSX-17 of Telemecanique under the PL7-2 programming language). Two limit switches per cylinder are the inputs to the mentioned PLC. The case study to be tested under the described training board is a cycle to operate sequentially both cylinders according the grafcet of figure 6. According expresions (5), (6), (7) and (8), the implementation of extended grafcet for supervision of the proposed cycle is showed at figure 7. To check the function diagnosis some faults were injected. The first test was by reducing suply air pressure and the second, was by blocking the one cylinder limit switch.

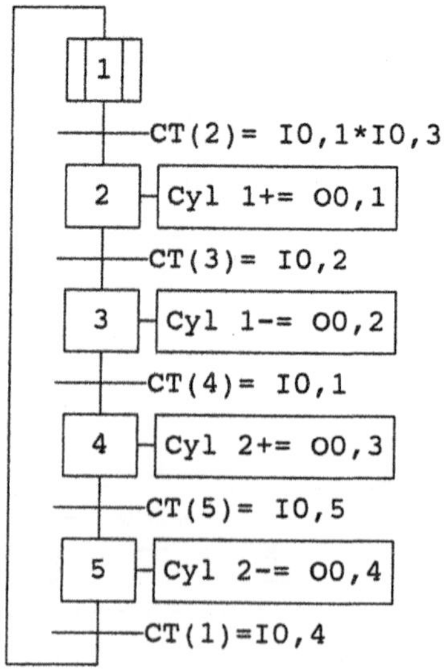

Fig. 6 Application grafcet

In figure 7 the transitions A, C, and D are indicated by E(i) and the interpretation is a fault in some components of CT(i), S(i) CT(i+1) or both. The fault due to slow actuation is indicated by B(i).

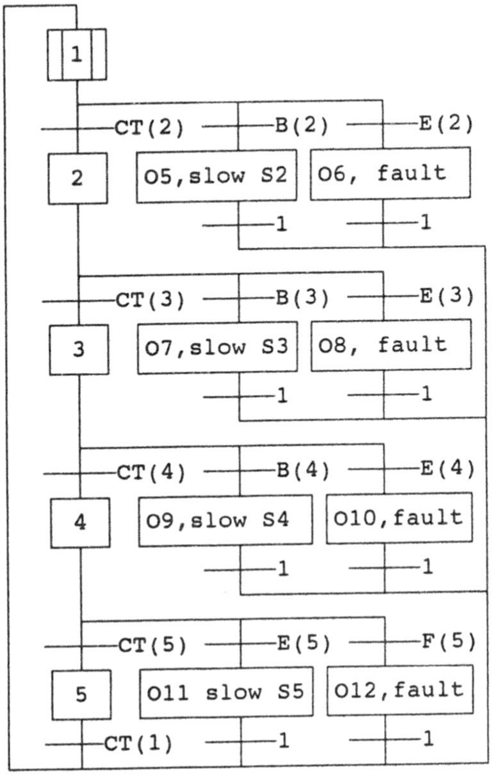

Fig. 7 Supervision by extended grafcet

From expresions (5) and (6), the transitions denoted in figure 7 as E(i) and F(i) are described at table 2.

TABLE 2 Description of conditions for transition in application example.

(i)	E(i)	F(i)
2	T2.I1.I3.I2	T2[I1.I3+I2]
3	T3.I2.I1	T3[I2+I1]
4	T4.I1.I5	T4[I1+I5]
5	T5.I5.I4	T5[I5+I4]

CONCLUSION

Logic and analog process control can be performed simultaneously in parallel with the fault diagnosis by means of the same PLC. The advantage of such procedure, is the increasing on reliability of the control system and system safety due to the avoidance of dangerous actuations after some fault detected. The indication of the stage in which the system fails, helps a lot in the minimisation of cost and time for repairing task.

REFERENCES

Silva, M. (1982) Hacia una nueva concepción del análisis y diseño de los sistemas lógicos secuenciales y concurrentes. Regulación y Mando Automático. Cetisa. (Ed.). Barcelona. Spain, No. 118, May, pp. 51-57.

Silva, M. (1985) Las redes de Petri: en la Automática y en la Informática, EC (Ed.), Victor de la Serna-46, Madrid. Spain.

Telemecanique, Service "Licences Logiciels" (1989). Modes operatoires en PL7-1 et PL7-2. Micro-Ordinateur IBM PC-PS/2 Language Boolean - Language Grafcet, 1989 P.I.A. Sophia Antipolis 06565 Valbonne Cedex, France.

Russo, Mark F. and Peskin R. L. (1987). Knowledge.Based Systems for the Engineer. Chemical Engineering Progress. September 1987. pp. 38-43

Shum S. K. et al. (1988). An expert system approach to malfunction diagnosis in chemical plants. Pergamon Journals Ltd, Grat Brtain, Comput. Chem. Engng., Vol.12, No. 1, pp. 27-36.

PROGRAMMABLE LOGIC CONTROL (PLC) FOR SMALL WATER DEMINERALIZATION PLANT

P. Góra*, E.M. Sroczan** and A. Urbaniak**

**Aqua-Dar Ltd., ul. Dabrowskiego 79, 60-900 Poznań, Poland*
***Technical University of Poznań, ul. Piotrowo 3a, 60-965 Poznań, Poland*

Abstract. Small water demineralization plant is described as an autonomous plant with full (complete) control system. The control system realizes all control functions for the production phase and for the renewal of the ionic filters' layers. The main part of the plant consists of two ionic columns (anion and cation column). High quality valves are used in this project what guarantees the reliability of the process. In the paper control devices based on a simple logic PLC controller are analyzed. Advantages and disadvantages solutions are discussed and on that basis some conclusions for future designers are drawn. The main features of developed solution, as well automatically work and low cost installation, is a possibility of application for the maitenace-free water demineralization plants.

Keywords. Programmable Logic Controll, water deminaralizing plant, chemical processes.

INTRODUCTION

Many technological processes demand high quality water. Chemical and biological treatment processes are often not sufficient, especially for water used in food production, for boiler water and air conditioning systems [Góra, Sroczan, Urbaniak, 1990, Mattiasson, 1987]. In these cases satisfactory parameters may be obtained by demineralization processes. For such applications, where small quantity of demineralized water is needed, the small water demineralization plant with the output of 1m^3/day has been developed by AQUA-DAR Ltd. [Góra, Stawicki 1991].

For control purposes the following parameters are measured: water level in tanks, flows in the given points of the installation and conductivity of the final product.

DESCRIPTION OF THE PROCESS

A general scheme of the water demineralizing plant - DEMI is presented in Fig.1. The developed technology enables reduction of mineral components with two ionit exchangers (cation exchanger K1 and anion exchanger K2). The following five different phases may be distinguished in the operation of the plant:

1. cation column (K1) regeneration,
2. rinsing of the column K1,
3. anion column (K2) regeneration,
4. rinsing of the column K2,
5. running demineralization.

The positions of valves and states of pumps corresponding to the above five phases are collected in Table 1.

The process itself is not complicated, but it demands strict discipline in control settings, as acidity increases may occur causing undesirable effects. Control system is equipped with many control elements, sensors and actuators, what may cause some difficulties at the beginning, but they allow of full automation of the process. According to the brief foredesign the plant is autonomous, enables continous operation with no supervising crew, except of temporary check.

This kind of water demineralizing plant can be used in many technological processes. Authors developed this project for replenishing of demineralized water in the heating system of a big food factory.

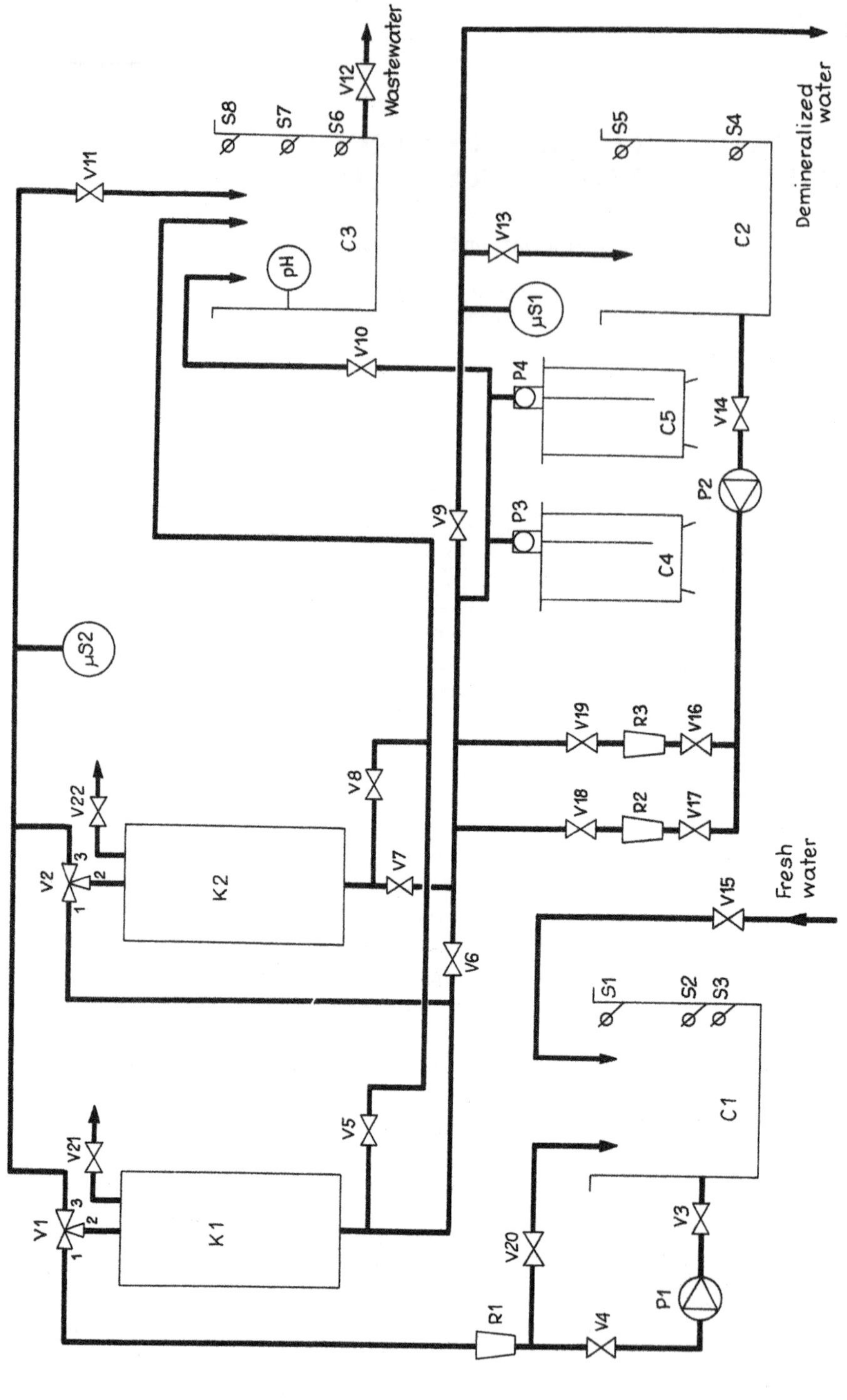

Fig.1. Scheme of water demineralizing plant

K1 - cation exchanger, K2 - anion exchanger, C1 - fresh water tank, C2 - tank of rinsing water,

C3 - washing tank, C4,C5 - solution tanks for regeneration, P1, P2 - water pumps; P3, P4 - dosing pumps,

V21, V22 - de-areating valves.

The plant can be used as a source of demineralized water in chemical laboratories and other installations where not too much clear water is needed.

The demineralizing process is based on a cyclic scheme given in Fig.3. A transition from the state "Running" to the states "Regeneration" and "Rinsing" should take place in case of the decline of treatment ability on ionit columns. This is continously controlled by the conductivity sensor S1. When increasing conductivity is detected, water from columns K1 and K2 is directed to the rinsing reservoir C3, and the regeneration process is started. It begins with the releasing of column contents to the reservoir C3 what requires a suitable action of valves and pumps (see Fig.1 and Table 1).

Table 1. Valves positions in accordance with work cycle.

Function	Valve Nº																						Pumps			
	1	2	3	4	5	6	7	8	9	10	11	12	13	14	15	16	17	18	19	20	21	22	P1	P2	P3	P4
Work/Run	1-2	1-2	O	O	C	C	C	C	O	Y	C	Y	Z	C	Y	C	C	C	C	Y	X	X	1	0	0	0
Regeneration K1	2-3	C	C	C	C	O	C	C	C	Y	O	Y	C	O	Y	C	O	O	C	Y	X	X	0	1	1	0
Rinsing K1	2-3	C	C	C	O	O	C	C	C	Y	O	Y	C	O	Y	O	O	O	O	C	X	X	0	1	0	0
Regeneration K2	C	2-3	C	C	C	C	O	C	C	Y	O	Y	C	O	Y	C	O	O	C	Y	X	X	0	1	0	1
Rinsing K2	C	2-3	C	C	C	C	O	O	C	Y	O	Y	C	O	Y	O	O	O	O	C	X	X	0	1	0	0

X - undefined position of de-areating valves, Y ,Z - undefined position of auxiliary valves

To provide proper operation the capacity of the reservoir C3 should allow of containing rinsings of one full cycle. High acidity of rinsings from the column K1, and a low one from the column K2 give opportunity to neutralize the rinsings when they are mixed together. Before the reservoir C3 is released to a sewage system its contents must be further neutralized with the solution of HCl or NaOH respectively. The capacity of clear (demineralized) water reservoir C2 is determined by the amount of clear water necessary for preparation of the regeneration solution and for columns rinsing. Preparation of the solutions for the regeneration processes requires precise measurement of their flow time. Capacity of reagent reservoirs C4 and C5 is calculated on a basis of brief foredesign for the amount of acid and lye needed for one full cycle minimum. Functional relations between regeneration and rinsing processes are shown in Fig.2.- as algorithms for both columns are similar, only operations connected with column K1 are shown. The neutralizing process which appears at the end of the algorithm is realized by adding a solution of HCl - when pH is greater than 7, and a solution of NaOH when it is smaller than 7 to the rinsings.

STRUCTURE OF THE CONTROL SYSTEM

As it was mentioned above the water demineralizing plant ought to be a fully automated and functionally autonomous item. For this reason we propose the use of a programmable logic controller for realization of all control functions. Programmable logic control enables designer to create flexible control structures. At the moment a large variety of these devices is at designers disposal [Technical catalogues].

The list of all system inputs and outputs is given below.

1. Digital binary inputs - 20
2. Digital counting inputs - 6
3. Analog inputs - 3
4. Digital outputs - 24
5. Timers - 2

All listed digital inputs are connected with valves, water level sensors and pumps (thermic fuse, solenoid control and Auto/Manual position switch).

For simple process balancing and effectiveness calculations flows of all substracts are measured. This is why we connected rotameters to the counting inputs of PLC.

The analog inputs are used to convert signals from two conductivity sensors and one pH sensor. The pH sensor is in some sense redundant as the designed capacity of the tank C3 and technology applied in the process ought finally produce neutral wastewater. For environmental security the tank C3 release may take place only after pH sensor acidity check. In fact, the level of conductivity, which is the measure of process quality, may be changed in respect to required parameters of the final product and other restrictions, for

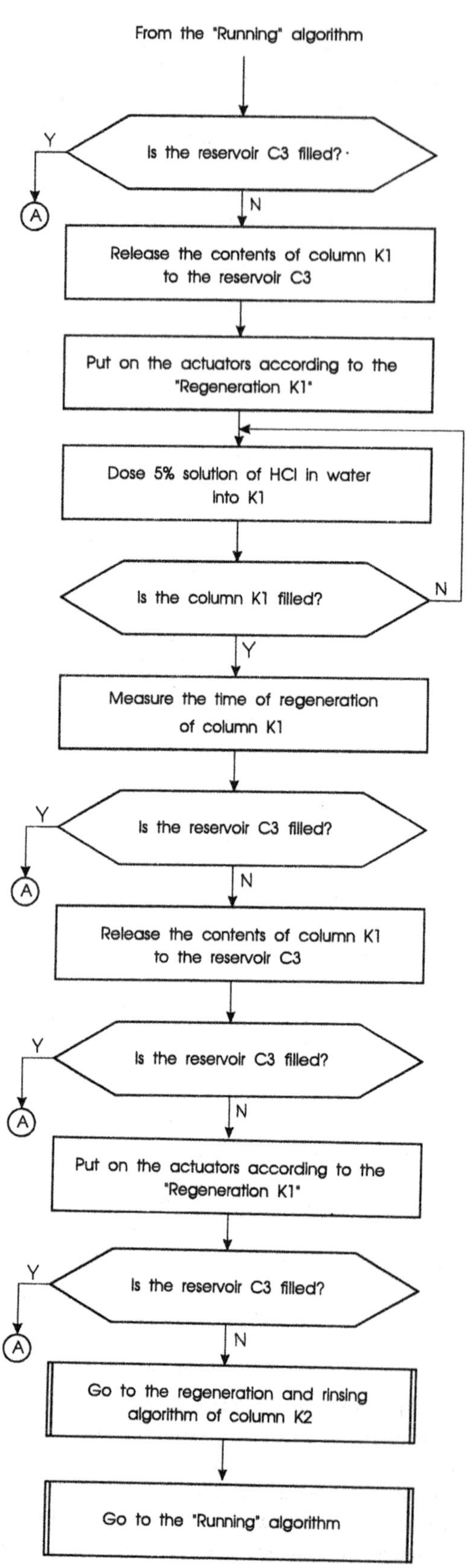

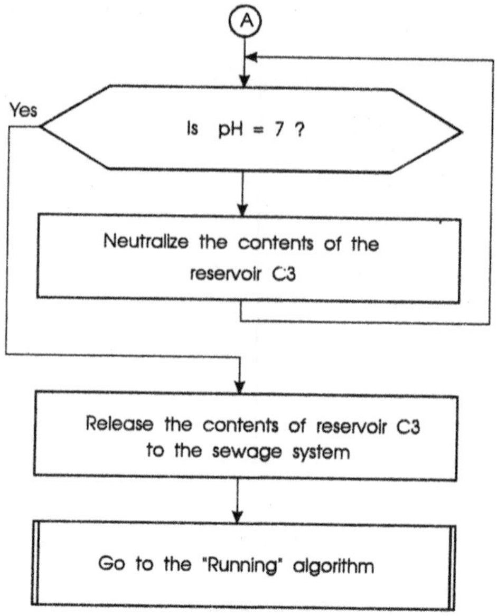

Fig.2. Regeneration and rinsing algorithms of column K1.

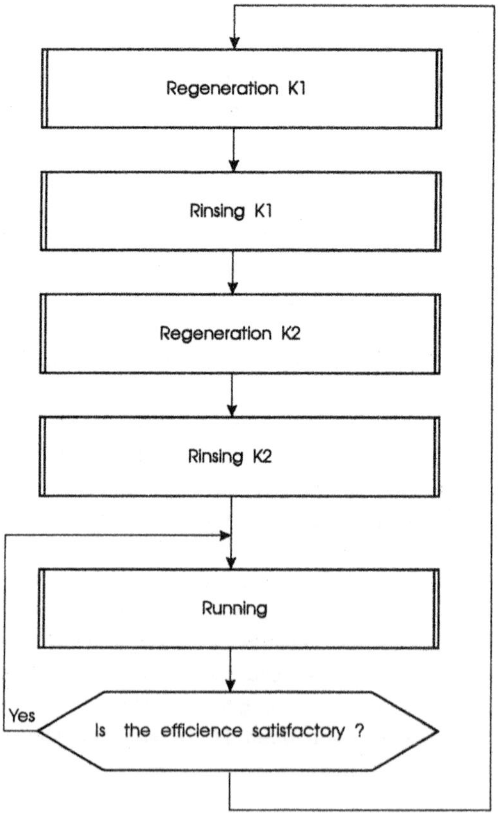

Fig.3. General diagram of work cycle.

example such as energy and reactants consumption. Decisions concerning these parameters are taken at the supervisory control level.
As a result 24 digital outputs were determined to control all valves actuators and pumps P1, P2.
The dosing pumps P3 and P4 are time-controlled (with the assumption of constant substracts flow). Apart from the measurement of time, system provides additional flow values given by rotameter counters.

The choice of the control unit which satisfies all requirements and enables effective processing should not cause difficulties.

As an option to conventional level sensors very accurate ultrasonic devices, which can be set by the program with respect to the state of the process, can be used to determined the liquid level in tanks.

Because the designed water demineralizing plant is one part of the technological chain we suggest to include the PLC system into the SuPervisory Control (SPC) structure (Fig.4). This structure is open and gives possibility for extension of the designed system. For example the proposed control system enables the optimization of

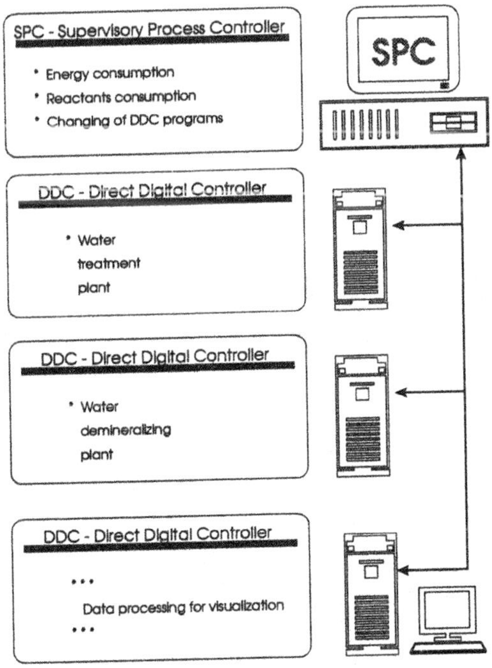

Fig.4. The structure of supervisory control.

energy consumption. In many cases the implemented PLC should provide some essential information about the process state and a level of demanded power. Energy management system (EMS) which supervises energy consumption in the process of water demineralization collects some chosen data about:
- active and reactive power,
- amount of reactants consumption,
- alarms and process state, etc.
Proposed algorithm of process control makes use of the PL controller flexibility. Any changes of the program - on request by the EMS system - and the optimal tuning of requested quality level are possible. It is realized by a remote setting of input values of the process, such as the conductivity of pure water.

We proposed the conception of the fully controlled water demineralization plant. The technical solution was based on two popular programmable logic controllers. First proposal is based on FESTO system and is given in Fig. 5a. It is the compact system constructed with the basically module of controllers FPC 101 AF. It is friendly programmable and contains the analog inputs and outputs. Additionally, we proposed to use the extension input/output module.

Another proposition is based on the SAIA system produced in Switzerland. The system has the modular construction. For our plant we must use modules listed in Fig. 5b.
The both proposals can be included in the local computer network (LAN). The network enables the expansion of control system and gives the possibility of extension of control functions.

FINAL REMARKS

Both of the presented solutions of automation of demineralizing water plant are compatible according to the range of control functions and was designed as a low-cost realization of measure and control circuits.

The results was achieved by omitting the analogue technique of measurement and control.

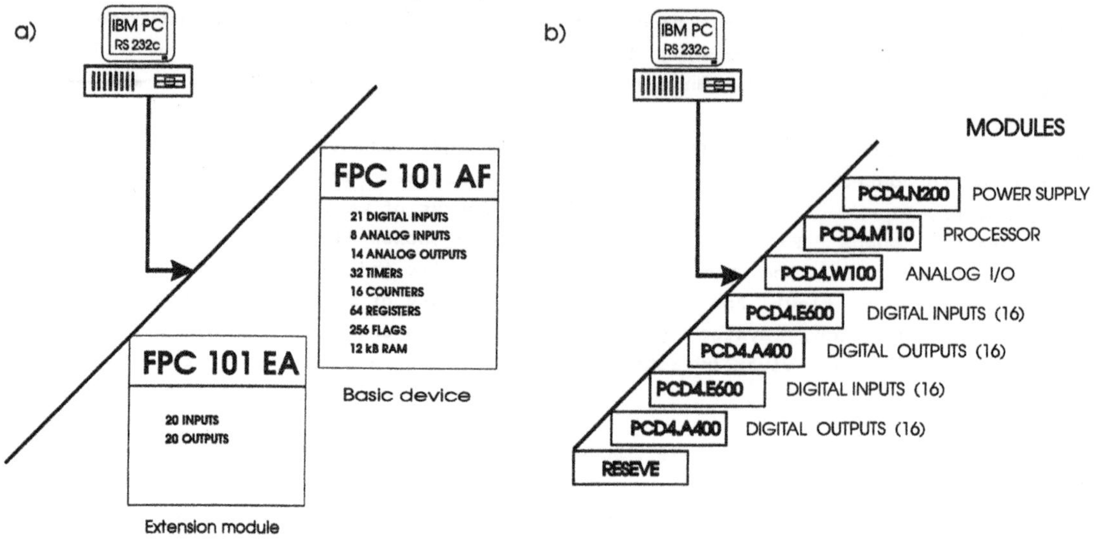

Fig.5. The structure of proposed control system based on: a) FESTO System , b) SAIA System

We suggest to add to the list of input/output signals several extra ones to allow future extent ion. Moreover, we must provide certain number of signals for visualization panel (operator's desk). The number of output signals used for this function strictly depends on the kind and accuracy of the process presentation and visualization.

If we take into account control signals only we can choose a small microcontroller with the A/C converter which is equipped with minimum 3 channals.

REFERENCES

Góra P.,E.M.Sroczan, A.Urbaniak, *Microcomputer logic control in the wastewater treatment plant*, IV Inter. Conf. on Integrated Problems of Industrial Control, Kiev, 1990

Mattiasson B., *Measurement and control*, Proc. 4th European Congress on Biotechnology, vol.3, 1987

Góra P., M.Stawicki, *Technological research of the water from the urban network*, AQUA-CLAR Ltd., Poznań, 1991 (in Polish)

Technical catalogues: FESTO, AEG, SIEMENS, TELEMECHANIQUE, SAIA

MICROPROGRAMMED ALU FOR A LOW COST LOGICAL CONTROLLER

Th. Borangiu

Department of Control and Computers, Polytechnical Institute of Bucharest, Romania

Abstract. The purpose of the paper is to present the basic design principles of a high-speed, low-cost microprogrammed ALU, integrated in the central processing unit of a Programmable Logical Controller (PLC) as a special LSI circuit. By extending the group of logical instructions of of the PLC with a minimum number of statements which act as logic separators such as "open bracket", "closed bracket" and "equal", a complete set of logic functions can be implemented using an Algorithmic State Machine (ASM) as ALU. For a defined set of logic instructions, four basic types of powerful general logic expressions are proposed, which cover practically all process control requirements in PLC applications. A detailed description of the hardware architecture of the ALU is given as well as of the data flow between the working registers; also, suggestions for one-chip VLSI implementation and integration in the central processing unit of the PLC are made.

Keywords. Microprogramming; programmable controller; algorithmic state machine; logic design; instrumentation; process control.

INTRODUCTION

Arithmetic Logic Units (ALU) based on accumulator techniques are built in integrated or discrete versions with a small number of components, but they impose certain constraints concerning the programming of logical expressions which are intensively used in the bit processing (test) section of the application software for Programmable Logic Controllers (PLC).

Thus, for the canonic disjunctive form used in logic bit processing functions, the minterms or the prime implicants must be temporarily saved, which requires an increased number of accesses to the working memory (RAM), leading therefore to a growth in size of the application programs

On the other hand, in writing an application program, the user has to permanently pay attention to the following aspects:

- to comply with the priority rules necessary for logic operations;

- to take into account the effects of complementing partial or final results as well as of separating logical expressions by means of brackets.

By extending the group of logic, bit processing instructions with a minimal number of statements acting as logic separators, a complete set of bit processing PLC instructions can be efficiently implemented by means of a microprogrammed ALU, acting upon an one bit DATA field (Borangiu, 1986).

When designed as a VLSI component, this microprogrammed ALU can also handle both the selection of the operand source: I/O field or Memory section (I, O, M), and the indexing of the operand address.

Such a powerful microprogrammed Address Control / ALU chip (AALU) can be used either in low cost bit processing PLCs, or as an integrated high speed bit processing unit in extended word processing PLCs.

INSTRUCTION SET AND MICROPROGRAMMING OF ALU SECTION

For bit processing in a PLC structure, let us consider the following instruction set, included in Table 1.

As can be seen, this set provides a complete functional system which allows the implementation of any logic function in a PLC configuration, with operands from the I, M fields.

In addition, there are also available conditional set/reset commands of a specified bit in one of the O, M fields.

TABLE 1 List of Instructions for Logic/Transfer Bit Processing

OPCODE	PARAM	SIGNIFICANCE
!	adr	Operand (adr) loading
!/	adr	Complemented operand (adr) loading
+	adr	OR operand (adr)
+/	adr	OR complemented operand (adr)
+(	adr	OR open bracket operand (adr)
+(/	adr	OR open bracket complemented operand (adr)
.	adr	AND operand (adr)
./	adr	AND complemented operand (adr)
(	adr	Open bracket (implicit AND operand (adr))
(/	adr	Open bracket (implicit AND complemented operand (adr))
=	adr	Equal: the value of the expression is assigned to the (adr) O, M field
=/	adr	Equal: the complemented value of the expression is assigned to the (adr) O, M field
)		Closed bracket
/		Complementing a logic expression
RTC	adr	Reset of the (adr) O, M field, if current result is 1
STC	adr	Set of the (adr) O, M field, if current result is 1

These instructions are executed by a microprogrammed ALU having the structure of an Algorithmic State Machine (ASM) in PLA version; it contains two basic registers:

- CIR: current instruction register;
- PIR: preceding instruction register,

the contents of which reflect respective the current and preceding transfer/logical bit instruction op codes.

The microprogrammed ALU contains also four working registers, which are used for the computation of the logic result:

- RR : current result register;
- RTP: parallel / series result (operand) temporarily saved;
- RSS: serial result temporarily saved;
- RPS: parallel result temporarily saved; this is the final result register.

Thus, the data transfers and the logic computation, as well as the sequence in which they take place depend for a current instruction fetched from the program memory, on the preceding instruction; at the same time, the previous partial result is taken into account.

The functional structure of the microprogrammed ALU section is presented in Fig. 1.

The instruction fetch cycle from the program memory consists from loading CIR, at the beginning of an instruction cycle, with the four bits of the current transfer / logic processing op code.

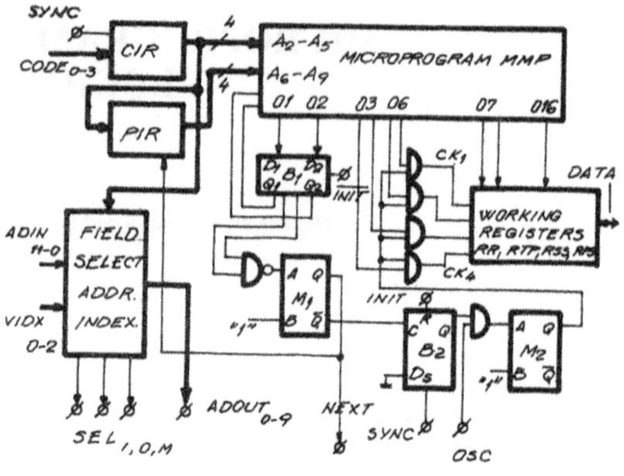

Fig. 1. Functional structure of the microprogrammed ALU

After executing the current instruction, the content of CIR is transferred into PIR, which will represent from this moment the op code of the preceding instruction, for the next instruction cycle.

The instruction execute cycle includes a number between one and four states, coded with two bits which give the least significant two address lines A1, A0 of the microprogram memory MMP; the most significant eight address lines A9 - A2 are given by the contents of CIR and PIR, which are fixed for a pair of current/preceding instruction.

MMP has a capacity of 1K x 16 bits. The first two output lines O1, O2 generate the next state for the execution cycle; the next four output lines O3 - O6 provide the load signals CK1, ..,CK4 of the four working registers RR, RTP, RSS and RPS, and the last ten output lines O7 - O16 represent the commands C5 - C14 which select the data paths for the four working registers of the ALU section.

The access cycles at the MMP and the load sequences for the working registers are generated by two flip-flops B1, B2 and two monostables M1, M2 under the external clock signal OSC.

For all the transfer and logic PLC bit processing instructions, a complete analysis of the necessary sequencing for all the allowed combinations (CIR) - (PIR) has been carried out, the result being the complete specification of the data transfers with or without logic computation required by a complete functional system (AND, OR, NOT, 1, 0) between the four registers RR, RTP, RSS and RPS (Das, 1973).

All the data transfers and logic computation which are declared on a single operational line in the following examples take place in an unique state of the current instruction execute cycle, involving a single access cycle to the MMP microprogram memory.

An example of the sequencing analysis is presented below.

The (adr Instruction

The CIR content is: (CIR) = (; the allowed (CIR) - (PIR) combinations generate three possible state sequencies for the execution of this instruction, depending on the preceding one:

I. (PIR) =) or / with the structural details:)(a or /(a.

1. (RSS) <-- (RSS)·(RR)
2. (RTP) <-- (DATA); (RR) <-- 0

II. (PIR) = · or ·/ or + or +/ or ! or !/ with the logic computation details: ·a(b or ·a(b or +a(b or +a(b or !a(b or !a(b.

The execution of the instruction is performed in this case in four microinstructions:

1. (RPS) <-- (RPS)+(RR)
2. (RR) <-- (RTP)
3. (RSS) <-- (RSS)·(RR)
4. (RR) <-- 0; (RTP) <-- (DATA)

III. (PIR) = any op code except those declared in the above mentioned I and II situations causes initialization of the registers in two states, at the beginninig of a logic expression.

1. (RPS) <-- 0; (RSS) <-- 1; (RR) <-- 0
2. (RTP) <-- (DATA)

BASIC LOGIC EXPRESSIONS COMPUTED BY MICROPROGRAMMING

The sequencing technique performed by the microprogram control unit which offers direct access to the hardware resources RR, RTP, RSS and RPS allows to implement four basic types of logic expressions E, depending on the variables $u_1, u_2, .., u_m$:

$$E = f(u_1, u_2, .., u_m)$$

The four basic types of complex logic expressions are:

A. $E = (e_{11})\cdot..\cdot(e_{1n_1}) + (e_{21})\cdot..\cdot(e_{2n_2}) + .. + (e_{p1})\cdot..\cdot(e_{pn_p})$,

where e_{ij} represents the disjunction of the prime implicants:

$$e_{ij} = \bigcup_{(\alpha_{K_1}, \alpha_{K_2}, .., \alpha_{K_m})} u_{K_1}^{\alpha_{K_1}} \cdot u_{K_2}^{\alpha_{K_2}} \cdot .. \cdot u_{K_\ell}^{\alpha_{K_\ell}},$$

and $K_1, K_2, .., K_m \in \{1, 2, .., m\}$; $m \geq 1$; $i = 1, 2, .., p$; $j = 1, 2, .., \max_i \{n_i\}$.

Also,

$$u_K^{\alpha_K} \underset{Def}{=} \begin{cases} u_K, & \text{for } \alpha_K = 1 \\ \bar{u}_K, & \text{for } \alpha_K = 0 \end{cases}$$

In the following example the related logic processing and data transfer mechanism which forms the type A result is suggested:

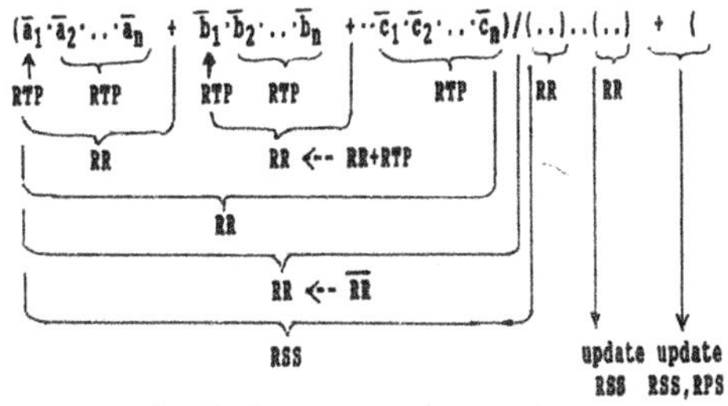

B. $E = [E_1]\cdot(e_{11})\cdot..\cdot(e_{1n_1}) + .. + [E_p]\cdot(e_{p1})\cdot..\cdot(e_{pn_p})$,

where e_{ij} has the same significance as in A, and

$$[E_i] = (((..(..)/..(..)/a_1^{\alpha_1}\cdot..\cdot a_n^{\alpha_n} + b_1^{\alpha_1}\cdot..\cdot b_n^{\alpha_n} + .. + c_1^{\alpha_1}\cdot..\cdot c_n^{\alpha_n})/\cdot d_1^{\alpha_1}\cdot..\cdot d_n^{\alpha_n} + e_1^{\alpha_1}\cdot..\cdot e_n^{\alpha_n} + f_1^{\alpha_1}\cdot..\cdot f_n^{\alpha_n})/..)..)$$

Computation of a type B expression is done like follows:

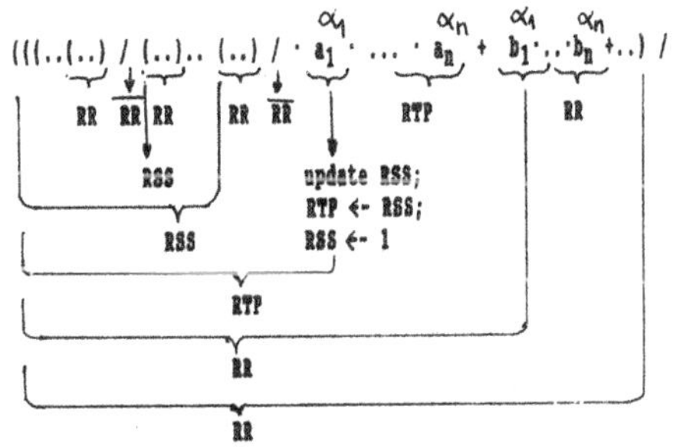

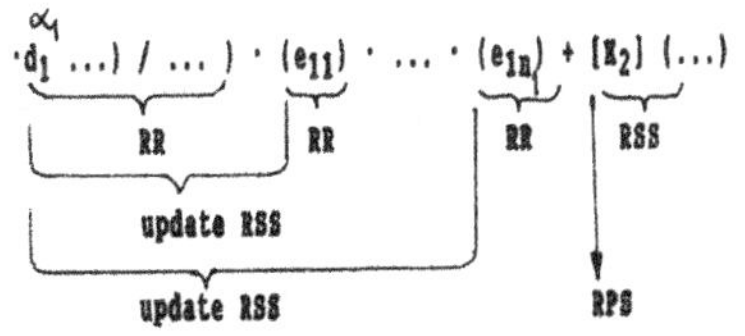

C. $E = [E_1] + [E_2] + .. + [E_n]$,

where $[E_i]$ has the logic structure given in type B expressions. The logic computation for a C type expression is based on the following mechanism:

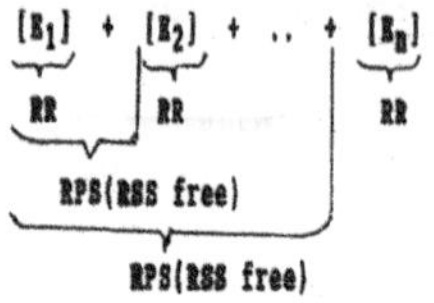

Concerning the type B and C expressions, the following specifications are necessary:

(a) A number of consecutive open brackets ((..(will be programmed with a single open bracket: (;

(b) An expression of the type $[E_k]$ uses temporarily the RSS register, but the final result of the logic computation will be saved in RR;

(c) It is not possible to implement expressions having the construction $\bigcup_{i,j} [E_i] \cdot [E_j]$, because although the final result of $[E_i]$ is saved in RR at the first open bracket of $[E_j]$, (RSS) <-- $[E_i]$ and for the computation of $[E_j]$ RSS will be further used, which means that the final result for $[E_i]$ will be destroyed.

(d) For similar reasons, expressions of the type $\bigcup_{i,j} e_{ij} \cdot [E_i]$ cannot be computed.

D. $E = e_{11} + I_1\ (e_{12})/\ ..\ (e_{1n_1})/ + ..$
$.. + I_p\ (e_{p1})/\ ..\ (e_{pn_p})$,

where $I_i = u_{i_1}^{\alpha_{i1}} \cdot \ldots \cdot u_{i_\ell}^{\alpha_{i\ell}}$ is a prime implicant of E. Such an expression is computed like follows:

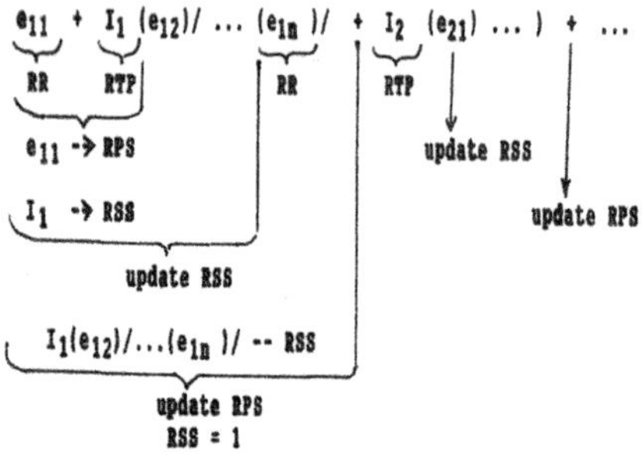

Taking into account the generality degree offered by the A - D type expressions, it is assumed that using the instruction set in Table 1 all the applications of logic computation are practically covered (***, 1990).

DATA FLOW AND INTERCONNECTION OF WORKING REGISTERS

The interconnection between RR, RTP, RSS and RPS and the linking of the microprogrammed ALU section with the DATA bus of the PLC depend on the informational paths which are created by the operational section of MMP in each microinstruction.

These paths can be grouped together like follows:

For RR:

(RR) <-- $(\overline{RR})$
(RR) <-- (RTP)
(RR) <-- (RR) + (RTP)
(RR) <-- (RPS)

The source of data for RR must be multiplexed 4 to 1; in addition, the current result register is subject to the initialization (RR) <-- 0.

For RTP:

(RTP) <-- (DATA)
(RTP) <-- (RTP)·(DATA)
(RTP) <-- (RSS)·(DATA)
(RTP) <-- $(\overline{DATA})$
(RTP) <-- (RTP)·$(\overline{DATA})$
(RTP) <-- (RSS)·$(\overline{DATA})$

The data transfer in RTP implies two levels of information multiplexing: a 3 to 1 level for the proper data source, and a second 2 to 1 level in order to take over the normal or complemented content of the one-bit DATA bus.

For RSS there is a unique information path

(RSS) <-- (RSS)·(RR),

in addition being required the initialization (RSS) <-- 1.

For RPS:

(RPS) <-- (RPS) + (RR)
(RPS) <-- (RPS) + (RSS)

These transfers are possible using a 2 to 1 multiplexing, with an additional initialization (RPS) <-- 0.

If in the input mode (I) the connection of the DATA bus to the ALU section is done by loading into RTP either directly the binary content (DATA), or $(\overline{DATA})$ in logic conjunction with the content of the RTP or RSS register, then in the output (O) mode, by means of 4 to 1 multiplexing the DATA bus receives:

(DATA) <-- (RR)
(DATA) <-- $(\overline{RR})$
(DATA) <-- 0
(DATA) <-- 1

Figure 2 presents the connection of the working registers in the microprogrammed ALU section.

The four working registers are implemented with D flip-flops which are loaded from the data source multiplexors under the control of the signals CK1, .., CK4.

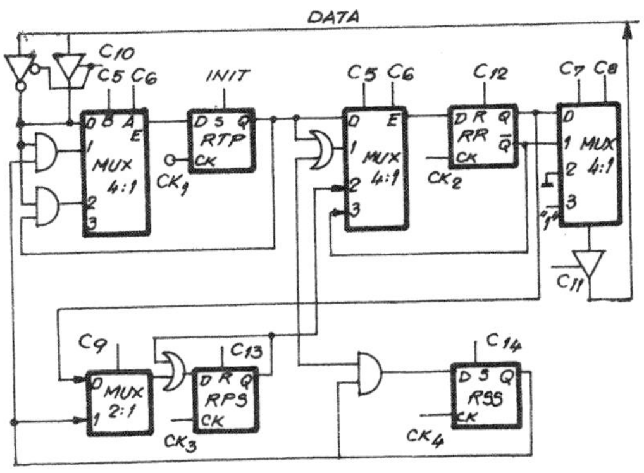

Fig. 2. Connection of the working registers

A certain time delay has to be provided for the active edge of the clock signals, with respect to the activation of the MUX controls C5-C9 in each state of the microprogram (Husson, 1970). Initialization of the working registers is synchronously controlled by C12-C14.

In the (I) mode, the content of the DATA bus is complemented under the control of C10; for all output operations (O), the ALU section is connected as data source to the DATA bus under the control of C11.

ADDRESS INDEXING AND AALU INTEGRATION IN LOW COST PLC STRUCTURES

The AALU concept allows the PLC to operate in the multiprogram mode with respect to transfers and logic processing of data to or from the three information fields: process (I/O fields) and temporary storage (M field).

The multiprogram mode allows the reduction of program memory by simplifying the application software. With a unique program section a number of different processes having identical control algorithms but distinct I/O channels can be simultaneously controlled.

In the multiprogram mode, the address of the operands is obtained as follows:

$$adr_I = adr_P + (index),$$

where adr_I is the indexed address of the data source or destination in the I, O, M fields; adr_P is the parameter field in the instruction (program address) and (index) is the content of the INDEX register, enabled.

The address indexing can be made:

- adjacent (fixed);
- nonadjacent (fixed or variable).

For the fixed indexing, the ordering rule of the parameters of the unique application program is given in the PLC's macroprogram memory; for the variable indexing the ordering sequence is configurated from the operator console.

In any case, the unique macrosubprogram will be written for the first loop or controlled subprocess, which corresponds to (index) = 0.

The AALU design assumes that for PLC applications the operand addresses are expressed on 10 bits, which leads to the following indexing format:

$$(index) = (NI) * (AI),$$

where * is the symbol for appending the nonadjacent index (NI) and the adjacent index (AI); NI gives the most significant 7 bits of the indexed address and AI the least significant 3 bits of the indexed address.

Thus, a PLC equipped with an AALU of this type can control at most 128 subprocesses having each at most 8 I/O parameters adjacently indexed, respectively at most 8 subprocesses having each at most 128 nonadjacently indexed I/O parameters.

Considering n subprocesses each of them having m I/O channels, the adjacent indexing creates for each loop the following operand addresses:

$$AI,\ n+AI,\ ..,\ (m-1)n+AI, \text{ with}$$

$$AI \in \{0,\ 1,\ ..,\ n-1\}.$$

The unique program will be run n times, the subprocesses being served in the following order: n-1, n-2, .., 0. After each each execution of the program AI is decremented.

The ordering of the control sequences for nonadjacent indexing will be defined in the application program, by assigning to each execution of the unique control program a NI. For the same (n, m) process configuration, the AALU will create the address indexing as follows:

$$IN_k m,\ IN_k m + 1,\ IN_k m + 2,\ ..,\ IN_k m + m-1,$$

with $IN_k \in \{0,\ 1,\ ..,\ n-1\}$, and the subprocesses are served in the order IN_1, IN_2, .., IN_n.

Unlike the fixed nonadjacent indexing, at the variable indexing IN is not loaded from the macroinstruction memory, but from a stack-type RAM memory, which can be configurated by the user.

The structure of the Address Indexing section of the AALU is given in Fig. 3.

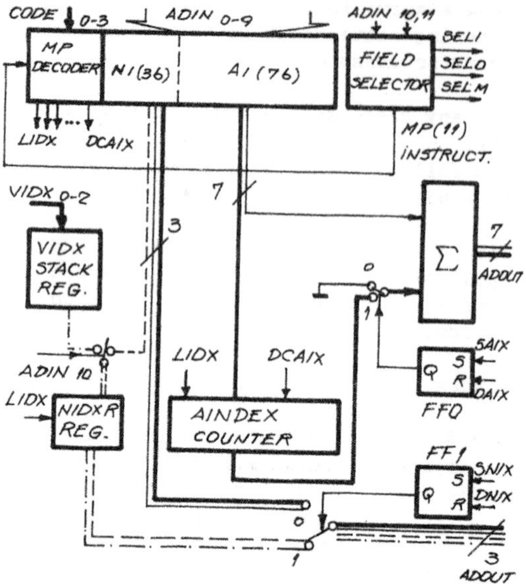

Fig. 3. Structure of the Address Indexing section in AALU.

The structure uses a 7 bit AINDEX counter for AI, a 3 bit NIDXR register for NI, a 7 bit ADDER, and two flip-flops FF0, FF1 which control data paths.

The list of multiprogram instructions which are processed in PLCs with address indexing AALUs are given in Table 2:

TABLE 2 Multiprogram Instructions

OPCODE	PARAM	SIGNIFICANCE
LIDX	-	Load index
SNIX	-	Start multiprogram with NI
SAIX	-	Start multiprogram with AI
DAIX	-	Disable AI
DNIX	-	Disable NI
DCAIX	-	Decrement AI
RRLH	-	(RR) -- 1, if (INDEX) = 0
RRLL	-	(RR) -- 1, if (INDEX) = 0

A remarkable facility offered by the multiprogram mode is the possibility to use both AI and NI in a PLC program cycle.

The integrated 40 pin one-chip AALU to be used either in low-cost PLCs or in bit- and word processing industrial control structures has the following pin assignment:

$ADIN_{11}$-$ADIN_0$ (input): macroprogram operand address; the two MSB lines $ADIN_{11}$,$ADIN_{10}$ identify the I, O, M fields for the states 00, 01, 10, but enable the multiprogram operations when 11.

$CODE_3$-$CODE_0$ (input): opcode for transfer / bit processing / multiprogram macroinstructions.

SYNC (input): synchronization line for microprogram start.

OSC: external supplied clock phase.

INIT (input): while this signal is activated, the contents of all registers and flip-flops are placed in a known state.

DATA (input/output three-state): data bus.

NEXT (output): signals the completion of a microprogram and returns the control in order to fetch the next instruction.

$ADOUT_9$-$ADOUT_0$ (output): indexed address bus.

SELI (output): select I field.

SELO (output): select O field.

SELM (output): select M field.

Vcc: +5 +5% Volts.

Vss: Ground Reference.

$VIDX_2$-$VIDX_0$ (input): external stack IN.

CONCLUSIONS

Microprogrammed structures can improve the speed quality of data processing and simplify the programming effort for PLC applications. Additional economic features of the PLC central processing unit can be provided by implementing multiprogram operations. The two functional objectives lead to the idea to build a large scale integrated Address Indexing / ALU chip, AALU, which can be integrated either in low cost, bit processing PLCs, or in bit and word oriented process control structures.

REFERENCES

Borangiu, Th., and R. Dobrescu (1986). Automate Programabile. Romanian Academic Press, Bucharest.

Das, S., D.K. Banerji,and A. Chattopadhyay (1973). On Control Memory Minimization in Microprogrammed Digital Computers. IEEE Trans. on Computers, C-22, 9, 47-50

Husson, S.S. (1970). Microprogramming Principles and Practice. Prentice Hall, Englewood Cliffs, New Jersey.

*** (1990). Modes Operatoires PL7-3 V3. Manuel 05-1990 (022), Telemecanique, TSX D22 004F.

SOFTWARE DESIGN METHOD FOR SMALL REAL TIME SYSTEMS USING MODIFIED PRO-NETS

B.R. Andrievsky, A.A. Vasiljev and V.N. Utkin

Department of Control Systems, Mechanical Institute, St. Petersburg, Russia

Abstract. The software design method for small real-time control systems is presented in this paper. The software is made as set of concurrent processes. Modified PRO-nets and interactive application independent language are used for description of these processes. These PRO-nets are used to develop concurrent programs such as flowchart for sequential. Each process is sequential. There are several dedicated languages for its programming. The design method is based on permanent transition from simulation of control system to half-natural modelling and then to working software. The implementation is based on FORTH language, so it has minimal differences between software for down and upper levels, technology and work software.

Keywords. Programming languages, real time computer system, software development,control system design, simulation, microcomputer-based control.

INTRODUCTION

When designing real-time control systems, software development for its microprocessor's control units takes very much time, because these control systems usually consists of many correlating components. They work asynchronous and parallel with each other.

Petri nets are widely used for concurrent process description. It is proposed to use Petri nets extension - PRO-nets (Noe,1980) for software design of small real -time control systems.

MODIFIED PRO-NETS

T-transition modification was made for simplification of its realization and semantics. The main element is shown below:

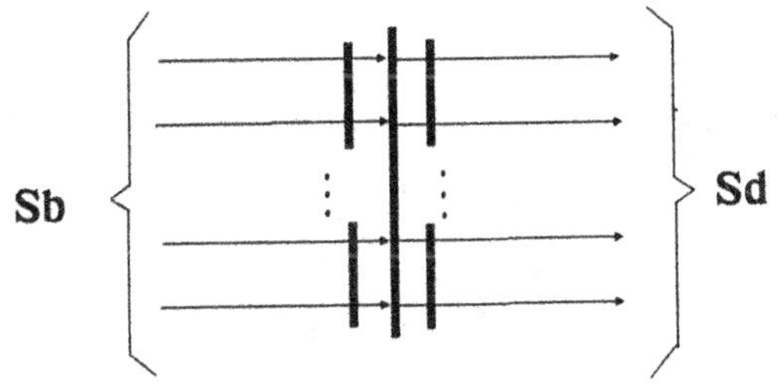

Subsets of input places Sb are named type "and" and "or" input places and subsets of output places Sd are named type "and" and "or" output places.

Sets Sb and Sd are divided on subsets of inputs and outputs. It is represented by the segment, that connects the arcs. Transition is on fire , if number of markers in each positions of some subset of Sb is greater than 1.

Selection functions of input subsets Sb (ro) and output subsets Sd (pi) are connected with each transition. After a firing time, markers momentary remove from selected by ro function places in Sb and transfer to places in selected by pi function subset in Sd. Ro and pi functions may be written on any algorithmic language.

It is important, that there are places of type "or" in modified PRO-nets. There are not one in original Petri nets.

Ro function is fixed: selected subset - the first type "or" places subset (from top to bottom), which has number of markers in each of its places greater than one.

If Sb or Sd has single subsets, then it is not necessary to connect arcs with segment.

At fire begining, subset number is being transferred to transition procedure. Fire time is being defined by the transition procedure, which gives output subset number from Sd. At fire end markers are transferring from input to output places subset. During fireing time, the transition does not react on marker's appearance in input places. The markers do not reserve, but conflicts are excluded, because each place may be on input for single transition only.

NET'S DESCRIPTION LANGUAGE

The next dedicated language is suggested for the model description in PRO-nets form:
Place description with some name:

PLACE name

n places description with any names:

n PLACES name_1 name_2 .. name_n

Transition description with some name:

TRAN name

n transitions description with any names:

n TRANS name_1 name_2 .. name_n

t transition inputs description

t INPS expression

t transition outputs description

t OUTS expression

Expression is made of places names (inputs or outputs), connected with "&" ("AND") and "!" ("OR") signs. The same place may be input and output. The same place may be in several subexpressions connected with "!" sign (subexpression places connected with "&" sign). It allows to simulate original Petri nets.
Start marking definition

((name_1 name_2 .. name_n))

The places that have one marker at start time, are in this sentence.
Building the transition procedure with name_2 from name_1 procedure

TRANSPROC name_2 name_1

This step consists of building of the procedure with the infinite cycle which contains the standard prologue, name_1 procedure and the standard epilogue. The name_2 transition procedure, which has been getting on before is connecting with the name_1 transition

name_1 LINK name_2

The processes, which may be in active or passive state, is connected with every transition. Active processes organize circular queue and each of them gets the single time-quantum. All processes have the equal priority.
At initial marking execution, markers are placed into the showed places and transitions which having these positions as inputs are going in the active state. (Each place has only one transition of this kind). When getting the time-quantum, the active process makes prologue. Prologue procedure verifies the transition execution condition (i.e. markers existence at all places connected by sign "&" of some subset from Sb). If this condition is not fulfilled, the transition is deactivated else input places subset's number remains on the stack for inner transition procedure. This transition procedure is executed and remains output positions subset's number on the stack. The epilogue procedure replaces the markers in these positions and deletes them from the input ones.

SOFTWARE DEVELOPMENT METHOD

Software design method consists of the following stages:

1. System's model description as a modified PRO-net.
2. Transition's models creation (objects and control laws).
3. Model description using PRO-nets' language and dedicated programming languages for transitions.
4. Simulation.
5. Step by step objects' models substitution to the real hardware (the semi-nature simulation). If necessary, control algorithm may be distributed among the different processors.
6. Working software version receiving.

PROGRAM SUPPORT

The instrumental software complex SCPD (Simulation and Control Program Development) has been created to support the described design method.
The control system can be distributed and usually has two levels: lower level is controllers and upper level is PC for user interacting. For user interface development on the high level, SCPD uses the standard interactive engineer's graphic tool AutoCAD and DBMS dBASE III.
The implementation is based on FORTH language, so it has minimal differences between software for low and upper levels, and software for development and work.
The special software complex includes the proposed programming language realization,

library for external description of concurrent processes and software tools for internal description asynchronous processes, which are the FORTH extension and/or TURBO-Pascal procedures. For example, there are means for description of continued and discrete time dynamical objects and standard control laws. SCPD is opened for language extension by users for other domain process description.
Program complex contains converter from AutoCAD Data eXchange Format to FORTH language, programs for getting initial parameters values files and parameters description files on FORTH or Pascal languages.
SCPD informative support includes the AutoCAD and dBASE user's guides, PRO-nets knowledge base, instructions for FORTH application to parallel process internal description and may be extended by HyperText possibilities.

EXAMPLE OF DESCRIPTION

For example, consider distillation column control system (Rey,1981). Mathematical process description includes six inertia units with mutual influence of outputs. Three proportional-plus integral (PI-) controllers are used for desired distillates concentrations level maintaining (every one controller for corresponding output flow concentration).
SCPD language program is shown below:

```
( 1) TRAN T1 TRAN T2 PLACE P1 PLACE P2
( 2) T1 INPS P1 T1 OUTS P1 T2 INPS P2 T2 OUTS P2
( 3) TRAN PROC PR1 control TRAN PROC PR2 object
( 4) T12 LINK PR1 T2 LINK PR2 (( P1 P2 ))
( 5) : REG ['] G ['] Y ['] U ['] integr PI-REG ;
( 6) : OBJinp 1 U B 2DUP 1 inrt -
( 7) 2DUP 2 inrt - 4 inrt -
( 8) 2 U B 2DUP 3 inrt -
( 9) 5 inrt - 3 U B 6 inrt - ;
(10) : OBJout 1 inrt - 1 Y B 2 inrt -
(11) 3 inrt - F+ 2 Y B 4 inrt -
(12) 5 inrt - F+ 6 inrt - F+ 3 Y B ;
(13) : control REG Tc integr BSTEP Tc WAIT ;
(14) : object OBJinp OBJout inrt []STEP To WAIT ;
```

PRO-nets description is at the lines 1-4. Net includes two independent parallel processes: object and controller, which realize object model and control law, respectively.
Transitions T1, T2 and places P1, P2 are described at line 1. Net's structure is given at line 2. Line 3 contains transition's procedures PR1, PR2 building. Line 4 defines the linkage between procedures and transitions and also the initial net's marking.

Individual processes descriptions are given at the lines 5-14. Line 5 includes procedure REG, where the required output values (array G), real object's outputs (array Y) are transmitted to the PI-regulators' inputs. Control influences are placed in the array U. PI-regulators' integrate operation is being done using the integrators' block with name "integr".
Lines 6-9 contains inertia units inputs definition procedure. Output flows concentrations' calculation with the mutual channels influence is given at the procedure OBJout (lines 10-12).
The individual processes are described by the procedures "control" (line 13) and "object" (line 14). "Control" contains procedure REG's call and makes PI-regulators' working step with the time interval Tc. By analogy, "object" includes procedures' OBJinp and OBJout calls and it makes object dynamic's simulation at the interval To. To reduce this program to real-time plant control is sufficient to change "object" procedure only. In that case "object" has to include the I/O operations with the external equipment data sets.

CONCLUSIONS

Proposed method may facilitate real time software development, because complex algorithm is divided on interacting parallel processes. Each process is sequential and modified PRO-nets supports process interacting design.
Realization of modified PRO-nets description language provides kernel software unchanging during transition from simulation to real system. Using FORTH language make easy development of dedicated languages for sequential processes programming by user.

REFERENCES

Noe, J.D. (1980). Nets in modelling and simulation. Lecture Notes in Computer Science, 84, 347-368.

Ray, W.H. (1981). Advanced process control. McGraw-Hill Book Company, New York.

MAN-MACHINE INTERFACE ORIENTED AND DYNAMIC SIMULATION BASED CONFIGURATION OF A LOW-COST PROCESS CONTROL SYSTEM

K.A. Reimann

Swiss Federal Institute of Technology (ETH) Zurich, Switzerland

Abstract. By means of an industrial application, it is shown that the use of dynamic simulation during software development, and a proper man-machine interface design, are no longer restricted to large systems like nuclear power plants, aircraft or petrochemical processing plants. If modern low-cost process control systems, designed for medium-size processing plants, are used to their full extent and in an intelligent manner, it is possible to achieve high quality, and at the same time to reduce the project costs.

Keywords. Industrial control, energy control, process control, man-machine systems, simulation, software development, program testing, operator training, education.

INTRODUCTION

In large control engineering projects, such as occur in the nuclear, petrochemical or aeronautic industries, it is common to build plant simulators long before the plant itself has been completed. Electricité de France, for example, have built their Simulator S3C even before the design of the corresponding nuclear plant type N4 was completed (Skull, 1988). They used it for verifying the man-machine interface design and to train the future operators.

For smaller applications in the process industries, this approach is often considered inappropriate. In this paper, however, it is shown that even when a typical low-cost automation system is used, state-of-the art methods of application software development and man-machine interface design can be used without extra cost, but with better results, as it has been demonstrated in a specific project.

AUTOMATION TASK

The plant and its control needs

In a plastics processing plant in India, two heat supply and waste fuel incineration units, each comprising a 10 MW burner, had to be automated. Thermal oil is heated in coils and then vaporized in a flash system. The vapor is used to heat various production units throughout the plant. The vapor pressure is the one variable that best indicates the balance of energy supply and demand. It is therefore used as the main control variable. The output variable of the pressure controller is used to generate setpoints of the air and fuel flows. A base fuel (either natural gas or Diesel oil) and two waste fuels from the production process are fired together.

The total flow of combustion air has to be set according to the actual mixture of fuels with their specific stoichiometric air demands, in order to keep the flue gas as clean as possible. Too much air would cause waste of energy and generation of NOx, whereas too low air flow would lead to unburnt hydrocarbons leaving the stack. During load transients, the air flow has to be on the safe side (i.e. rather too high than too low). In conventional firing control, this is achieved using a lead-lag (also called cross-over) scheme (e.g. Honeywell, 1980). In our case, due to the presence of three fuels, the air flow setpoint is calculated according to the flows and specific air demands of the individual fuels and the total burner load. A lead-lag algorithm is included.

Theoretically, this leads to an optimally adapted air flow at all times. However, to accommodate disturbance effects not covered by the calculations, a control loop for the flue gas oxygen content has been added. For this controller, the new method of dynamic filtering the reference input signal, according to Profos (1988) was implemented. This filter is in fact a model of the process.

In order to achieve smooth transients when fuels are swapped, a fuel heat flow balance has been implemented. Furthermore, a priority logic stops one or two of the three fuels in the case of low heat demand. If demand is even lower, the control will switch to on-off mode. Start-up of the burner itself has to be commanded by a certified sequence controller.

Automation functions

What follows is a list of the functions the automation system has to provide:
- process display and operation
- monitoring (alarms and automatic emergency shut-downs)
- logging
- burner start/stop
- base and waste fuels start/stop (manually and automatically)
- loop control of vapor pressure, fuel flows, air flow, flue gas oxygen content
- calculations: combined air demand, heat flow balance, burner load, burner efficiency
- communication with: certified burner sequence controller, parallel hard-wired safety logic, plant-wide data processing unit.

HARDWARE AND FIRMWARE

An automation system of type Siemens AS 215, along with a burner sequence controller by Durag and a parallel safety logic (for redundancy) consisting of relays were used to implement the automation functions.

The Siemens AS 215 is a typical and widely-used low-cost automation system. Its hardware is essentially that of the SIMATIC S5-115U programmable logic controller, whereas the system software (i. e. the firmware) is derived from that of the larger automation systems of the type TELEPERM. It can handle a maximum of about 20 control loops. The cycle time of the analog inputs and outputs of 0.5 s may seem high, yet it is sufficiently low for many applications in the process industries. The system memory of 128 kbytes does seem not much either; however, it is enough to accomodate the application software of half a man-year's design and development time. This seems reasonable, as the costs for hardware and application software will be of the same order of magnitude.

Of course, there is no window and mouse technology, neither for operation nor for programming (or structuring, according to Siemens terminology). Yet, for systems of this order, window technology would rather intrigue than help the operators, and the skilled development engineer will quickly get used to moving cursors or typing short commands. Drawing the software flowsheets using paper, pencil and eraser was even regarded an asset by a developer who did not want to spend eight hors a day in front of the computer monitor (software is usually developed on IBM compatible PCs and then loaded into the target system). One might argue that it is tiresome that there is no automatic update of the software flowsheet representation. On the other hand, doing some paper and pencil work means also more careful thought before touching the keyboard.

The firmware combines sequential control, interlocking, loop control, and man-machine interface in one configuration methodology. It is function block oriented und thus follows the way of thinking of the control engineer. With that, it differs from the more sequencial control oriented methodologies of programming PLCs such as ladder diagrams, which appeal more to electricians. The application software developer selects pre-configured function blocks from the firmware, such as timers, logic gates, arithmetic operators, PID-action controllers, bar graphs, toggle switch soft keys, etc. These blocks are linked and parametrized (Graf and Wettach, 1987).

APPLICATION SOFTWARE DEVELOPMENT

In contrast to conventional configuring of such systems, a new approach was used for this project in two respects.

Man-machine interface orientation

The development of the application software started with the design of the man-machine interface. The prototype screen pages were used as a specification for further software development. This follows a recent philosophy of automation software development which is also used for new development tools (Pomberger and Bischofsberger, 1990). The prototype pages were discussed with experts and customers, which led to successive improvements.

Following results of research at Twente University (Spenkelink, 1987), displaying process flowsheets on the monitor was renounced. Instead, the state of sequence control is shown by means of a tabular representation, whereas the state of the control loops is displayed by means of bar graphs. The whole set of bar graphs, comprising process variables, operator adjusted setpoints, calculated set points and other important analog values, is concentrated in one screen page. The operators get quickly used to recognising patterns on the tabular page and the bar graphs page, respectively. Thus, they obtain a fast overview of process "health" and operation stage by surveying the approximately 40 binary and 30 analog values shown on the two pages.

Besides that, there are two alarm pages relating alarms to plant tag numbers in a tabular way. Trend curves have also been implemented. They are rather for commissioning, tests and loop controller tuning than for serving the operator's hourly needs.

Dynamic process simulation

Once the man-machine interface was agreed upon, the application software "behind" was developed (i.e. configured) in a piecemeal manner. In order to validate each part of the software, a dynamic process simulation was implemented as well. Part of it was a discrete-event simulation (corresponding to the sequential and interlocking aspects of the process), and the rest was a continuous simulation based on a highly non-linear wide-range model of 20th order, taking mass and energy balances, physical properties and heat transfer correlations into consideration.

The simulation part of the software was built using the same function blocks that are available for the control part of the software. For the discrete-event simulation, logic gates and timers are the key elements, whereas arithmetic operators, integrators and first order delays (PT1) predominate in the continuous simulation. Heat transfer and physical property correlations for the whole range of temperatures and pressures that may occur had to be taken into account in a precise manner; table functions were used for that purpose.

The control part and the simulation part of the software are linked by using the simulation signal inputs of the analog and binary input function blocks. Thus, by setting a few bits, the software is switched from process control (i.e. on-line) operation to simulation operation. As simulation and control have to be carried out by the same processor, the simulation functions are deactivated for on-line operation. With that, the sequential and continuous control algorithms can run in shorter cycles. On the PC used for software development, the control and simulation software together runs three times as fast as real time.

The dynamic model was subject to extensive validation. Verification concentrated on steady-state correctness with respect to plant design calculations for all possible states of operation. Since the software was developed by an engineer who had been trained in modelling at university, the extra cost for developing the simulation was negligible. The benefits, on the other hand, were considerable, as is shown in the following.

BENEFITS

Concurrent development and validation of the control software. Each piece of software was validated after completion. With that, erroneous links, logic and calculatory errors were detected and corrected immediately.

Comprehensive validation of the control software. For all possible states of plant operation, including emergencies, the control software was thoroughly tested. Doing this in the office instead of the site, much time and expenses for travelling and fuel could be saved.

Training. Both commissioning engineers and operators could be trained in advance. There was neither need to wait for the slow reaction of the real process, nor to spend fuel, nor to risk damage to the plant.

Commissioning. The controllers were pre-tuned using the simulation. With that, there were sound starting values for fine tuning on site, which then took less time. Furthermore, during commissioning, the control engineer in the supplier's office could advise the commissioning engineer on site via telephone and facsimile in an effective way, as he had a simulation of the whole system at hand.

Sales promotion. The simulated system was shown to the management of the supplier. This led to a better appreciation of the work done by the control engineering group. It was also shown to prospective customers, thus making a successful instrument for sales promotion.

Education. The Siemens AS 215 system is particularly suited to education purposes and therefore found in many schools. The software development approach presented in this paper was taught during laboratory sessions with small groups of students in both a technical university and an engineering college. Groups of two to three students had to contribute project parts such as the process model, sequential control or loop control. For most of them, this was the first time they were exposed to a real-life process control system, and they learned how to apply their knowledge of control theory.

CONCLUSION

By means of an example it was shown that in medium-sized processing plants, where low-cost automation hardware and firmware is appropriate, application software can be developed in as an intelligent manner as in large scale automation projects. If a simulation model of the plant is included in the application software, the automation system, with its state of the art man-machine interface, becomes a plant simulator itself. With that, software validation and traning are carried out in a more effective manner.

In any case, a control engineer who develops the application software for the automation system has to understand the physics of the plant for which he designs the automation system. Now, this is already the first step of plant modeling. If he has already some experience in modeling, he can use the function repertoire of the firmware to implement a process simulation at only small extra cost. This is worthwile at any rate.

Creative control engineers are challenged to make use of low-cost automation systems as if they were dealing with large scale systems.

REFERENCES

Graf, O., and S. Wettach (1987). Zeitsparende Systemstrukturierung in der Prozeßleittechnik mit Automatisierungssystem AS 215. Energie & Automation, 9, Special «Hannover-Messe Industrie 1987».

Honeywell (1980). Oil/Gas Fired Boiler Control. Application Note D71-25-07-05. Honeywell Process Management Systems Division, Phoenix, Arizona.

Pomberger, G., and W. Bischofsberger (1990). The TOPOS Prototyping Environment. 12th ICSE Software Engineering Conference, Nice, France.

Profos, P. (1988). A new oxygen control concept in boiler furnaces. Gas wärme international, 37, 108-113.

Skull, G. (1988). Computergestützte Betriebsführung bei den französischen 1400-MW-Druckwassereinheiten N4. Computereinsatz im Kernkraftwerk. Schweizerische Vereinigung für Atomenergie, Bern, Switzerland.

Spenkelink, G.P.J. (1987). Structuring of process information on VDUs. Fourth year report II, Twente University, Ergonomics Working Group, Enschede, Netherlands.

INTELLIGENT FAULT DIAGNOSIS OF THE ICATS

Xu Lina and Deng Zhenglong

Department of Control Engineering, Harbin Institute of Technology, Harbin, Heilongjiang, PRC

Abstract. In this paper ,we introduce the structure and function of the inertial component automatic test system first .And then we analyse various methods which are used to dispose the system failure by the experts of the test system in detail . A fault diagnosis system of the inertial component automatic test system is founded by using the theories ,techniques and methods of knowledge engineering. The diagnosis system can improve the diagnosis rate by using other kinds of knowledge on the baiss of the hierarchy diagnosis model. The establishment of the protomodel system verifies the feasibility and correctness of the system .

Keywords.Knowledge engineering , gyroscope, automatic testing , fault diagnosis , problem–solving , hierarchy diagnosis model .

INTRODUCTION

Gyro is a precise inertial component which is the important part of the inertial system .The inertial component automatic test system (ICATS) is a kind of complex electro–mechanical system which is controled by the computer ,and it is the advanced test equipment of evaluating the performance of the gyro for the persons who are the gyro developers or the gyro producers .

In order to improve the effectiveness ,the maintenability and the level of the automatic test of the testing system ,the system must pocess the ability of selfdetection and selfdiagnosis .

As the developers of the test system ,we have been engaging in the research work on fault diagnosis of the ICATS by using the theories ,techniques and methods of knowledge engineering ,the knowledge of system design ,system structure and system function ,and the experience of debugging and gyro evaluating experience which has been accumulated for many years .

THE RESEARCH ON INTELLIGENT FAULT DIAGNOSIS

The purpose of the research work on fault diagnosis of the ICATS is to found a intelligent fault diagnosis system .What we will discuss is as followings :

1.The Analysis of the ICATS.

The automatic test system which is applied to evaluate the performance of the precise gyro is a kind of high precision electrmechanical system with the complex structure . Its performance is :to realize real time control for the test table and the measure equipment ; automatic acquisition of the test data and the the on–line identification of gyro error model by its computer .

In order to realize all the performance above

,the hardware of the system should consist of three subsystems which are comparatively independent ,but have certain connection .

Figure 1 is the block diagram of the ICATS.

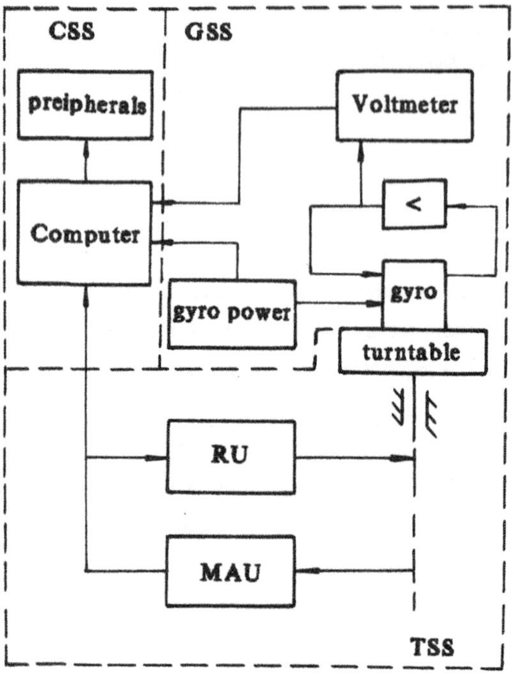

Fig.1. Block Diagram of The ICATS

The ICATS has three subsystems . The three subsystems of the ICATS are :

* the computer subsystem (CSS)
* the high precision test table subsystem(TSS) (The position accuracy is less than 1 second of arc ,it can measure gyro drift rate which is less than 0.01 ° / hr)
* the tested gyro subsystem (GSS)

While the test system operates in a normal state ,all of the subsystems can operate according to the design standard and realize their own performance ,they can also coordinate each other to realize the performance of the whole system .

Each subsystem can be divided into several subsubsystems for example ,the test table subsystem can be divided into the rate unit(RU) and measuring angle unit(MAU)),and the subsubsystem can also be divided till single unit or component .

If the ICATS is not in a normal atate ,that means some performance of the system has lost ,and it can be shown when the corresponding state parameters or characteristic quantities are abnormal. The procedure of fault diagnosis (or the procedure of failure problem –solving) is : to detect the abnormal information ,then to set failure , We call abnormal information failure information .

2.The Foundation of The Hierarchy Diagnosis Model (HDM)

The diagnosis model of the tree hierarchial structure can be founded on the basis of the structure and function knowledge of the test system (Figure 2).The highest level stands for the system failure ,and the first level stands for the subsystem failure ,etc .This model imitates the diagnosis behavior of the expert : it can search first from the highest level ,and then to the first levels ,that is ,it can search level by level .The features of this diagnosis problem –solving model ,are as followings :

1).The higher level failure can show the main function omen of the lower level failure;

2)Any lower level failure can be regarded as the reason of its higher level failure ,and its position is the failure source of its higher level.

So ,the relation between the lower level and its higher level is the causality.

Figure 2 is an example of HDM.

where :

I1: No datum is acqired by the computer

T1: The turntable vibrates with the amplidute of 0.5 ° .

T2: The turntable does not reach the required position .

G1: The digital voltmeter has no output datum .

M1: Angular value changes 1 ° .

M2: Coupling ciruit failure .

M3: Control circuit failure .

R1: There is no rate pulse .

C1: The error is more than the normal value .

CE1: The amplidute error is more than the normal value .

CE2: The phase error is more than the normal value .

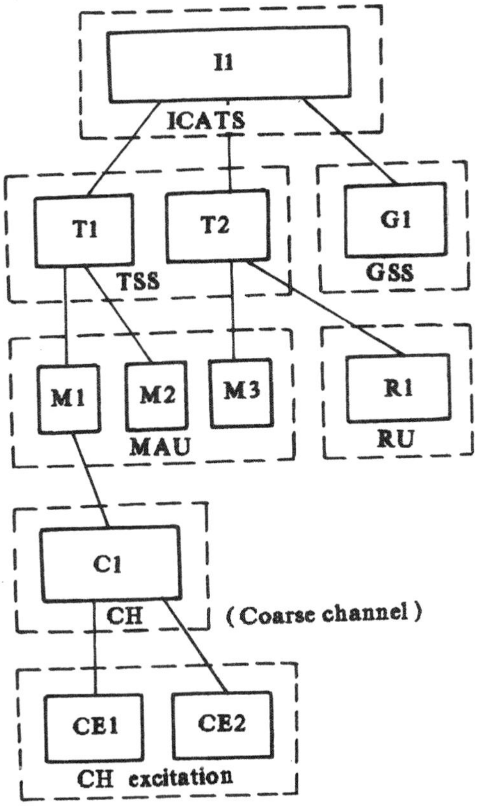

Fig.2. An exampe of HDM

3. The Methods of Improving the Diagnosis Rate .

On the basis of the hierarchy diagnosis model(HDM), we have summed up some methods of improving the diagnosis rate by analysing the methods which have been used by the developers (experts)of the test system when they dispose the failure :

1).To find the optimal path of the problem–solving from the experience knowledge .

By using the experience of the experts ,sometime we can find the optimal path of the problem–solving without considering some middle levels with causality .The feature of this procedure is that the uncertainty knowledge has been used .

For figure 2 , according to the experience ,we can reach the conclusion if the table has vibration with a amplidute of 0.5 ° at a given position ,it is more possible that the measuring angle excitation power has failure .(CH excitation)

2).To reduce the failure scope by using the various states of the ICATS .

If only rely on one failure omen of the higher level subsystem ,it is difficult to determine where the failure of the lower level subsystem is .The reason is that there is not only one lower level subsystem ,and all of the lower level subsystems have certain connection .So ,we should acquire various states of the high level subsystem to reduce its failure scope .

3)The optimal selection of the detection parameter and detection point.

The principle of selecting the detection parameter and detection point is : (a). To select import parameter or character value which defines the system operate state. (b). To select the location where the system is easy in trouble .

4). To diagnose the failure by using various singnal analysis methods (statistic analysis ,for example).

All the discussion above is the procedure of using knowledge .So ,founding an intelligent fault diagnosis system with high efficiency ,we must let the system pocess the knowledge which is essential for solving the failure problem .

THE ESTABLISHMENT OF THE IN TELLIGENT DIAGNOSIS SYSTEM

We can found the intelligent diagnosis system

of the ICATS, which is composed of the single–axis test table ,as its background(Figuer 3).

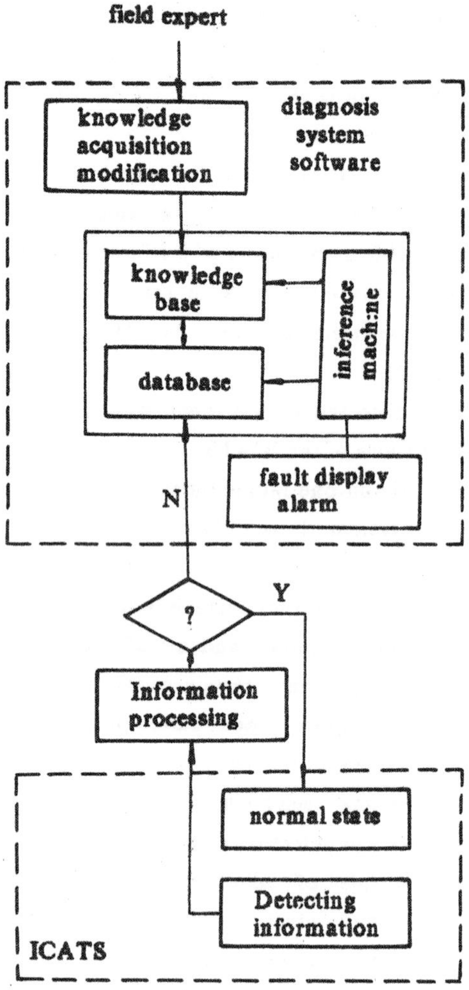

Fig.3. Intelligent Diagnosis Block Diagram

1.The diagnosis system is on the basis of HDM, and uses various knowledge to improve the diagnosis rate .

2.The diagnosis system can be realized by the production systems , and its import part is the knowledge base . The problem–solving knowledge is stored in the knowledge base . This knowledge expressed by the rules.
The rules can only be expressed by two forms (normalization)

if a then w (CF) (1)

if $a_1 \wedge a_2 \wedge \cdots \wedge a_n$ then w (CF) (2)

where :

$a, a_1 \wedge a_2 \wedge \cdots \wedge a_n$:premise

w : conclusion

CF :confidence factor,

$CF \in [0,1]$

For example :

if the turntalbe can't reach the given position,

and the rate system is in the normal state,

and the measuring angle data is unchangeable,

then there is failure in the measuring angle control circuit(CF= 0.8).

it can be expressd by the rules:

if $a_1 \wedge a_2 \wedge a_3$ then w(0.8)

IF the rule is :

if $a_1 \wedge (a_2 \vee a_3)$ then w (3)

it must be expressed by above two rules(normalization) (1) or (2):

if $a_1 \wedge a_2$ then w (4)

if $a_1 \wedge a_3$ then w (5)

It show that ,it is useful for the problem–solving .

3. The Uncertian Inference Algorithm

The inference procedure is drawing the conclusion from the premise .

1). The CF algorithm on the conclusion of rule(2)

If the confidence of the premise $a_1, a_2, \cdots, a_n$ is $CF(a_1), CF(a_2), \cdots, CF(a_n)$,

suppose $a = a_1 \wedge a_2 \wedge \cdots \wedge a_n$

the confidence CF (a) of the premise is

$$CF(a) = \min[CF(a_1), CF(a_2), \cdots, CF(a_n)]$$

the confidence CF of the conclusion is

$$CF = CF(a) \times CF(w)$$

2). The CF algorithm on the conclusion of

rule (3)

After having found out the CF of the conclusion of rule (4) and (5) respectively ,we select the bigger one .

4. The One-step Forword or Globle Backword Inference Control Strategy .

1). Starting the diagnosis system with failure information .

2). Searching all the possible failure sources in the lower level by using the one-step forword inference control strategy.

3). Verifying the failure source by the global backword inference .

4). Determining the failure source which is the most possible by the uncertaity inference.

5). Searching the system from the highest level ,and then level by level , till the solution —trouble-location problem .

CONCLUSION

The research work on founding the intellingent diagnosis system has achieved certain successes:

1). The diagnosis protomodel system with two subsystems has been founded;

2). Some diagnosis functions have realized (on-line)in a test system .

All of the research work we have been done above has verified the feasibility and correctness of designing the diagnosis system ,and laid a foundation on further engineering application .

REFERENCES

Nilsson ,N.J.(1980). Principles of Artificial Intelligence. Tioga PublishingCo.

Negoita ,C.V.(1985).Expert Systems and Fuzzy Systems. The Benjamin / Cummings Publishing Company . Inc. Menlo Park,California.

Fu.J.S, Cai Z.X and Xu G.Y. (1988). Artificial Intelligence and Its Appli cation .Qinghua Univetsity Press, Beijing .PP.277-300.

Xu,L.N.,and Sun, C.Y. (1990) .The Research of Fault Diagnostic System of The test-turntable angle Measurement System. Research Report of Harbin Institure of Technology,No.63. Harbin Institute of Technology, Hatbin.

Xu, L.N. ,and Wu, C.S. (1991) A Way of knowledge Formalization . Journal of Harbin Institute of Technology, Vol.23.6. Harbin Institute of Technology Press, Harbin.58-63

GENERATING ANALYTICAL REDUNDANCY FOR FAULT DETECTION SYSTEMS USING FLOWGRAPHS

Z. Kowalczuk* and J. Gertler**

**Faculty of Electronics, Technical University of Gdańsk, Majakowskiego 11/12, 80-952 Gdańsk, Poland*

***Electrical Engineering Department, George Mason University, Fairfax, VA 22030, USA*

Abstract. The paper concerns the analytical redundancy methods of diagnosis of monitored physical systems. Redundancy relations, describing how plant variables are nominally related to each other, are used to generate residuals. To address the problem of generating the redundancy relations that are unaffected by unmeasurable disturbances, we propose to apply the flow graph description that can reflect the internal structure of a given physical system, as a basis for the derivation of the primary redundancy relation. By reduction of the system's internal variables, a "concise" flow graph, having only the variables of interest (inputs, outputs, and disturbances), may be obtained. A resulting set of connected subgraphs containing only measurable variables provides the equations of the disturbance-decoupled redundancy relations. All the system variables can be identified as control inputs, measured inputs, sub-key variables (inputs and outputs), and key variables (sole outputs). The proposed general design methodology is suitable for instrument failure detection and isolation systems monitoring linear dynamic systems. It may also be useful for a class of nonlinear systems.

Keywords. Analytical redundancy, fault detection, diagnosis, parity equations, system description, system structure, flowgraphs.

INTRODUCTION

The paper concerns the analytical redunduncy methods of diagnosis that utilize the mathematical model of the monitored physical system. Checking this model for consistency with the actual system is the essence of analytical redundancy. Redundancy relations, describing how plant variables are nominally related to each other, are used to generate residuals. The presence of non-zero residuals merely indicates that something has gone wrong in the system (*fault detection*), and some further analysis is usually needed to decide upon the location of the failure (*failure isolation*).

There is a general uncertainty about an internal structure of a dynamical system when using the input-output system description (Kowalczuk, 1989a; 1989b). In the case of the redunduncy relations, the problem concerns the generation and selection of redundancy relations out of the set of all such relations that optimally detect certain changes in the dynamic system. Similarly, a wide variety of redundancy relations is possible when using state-space models (and the equivalent diagnostic observer problem is characterized by the choice of the innovation-feedback matrix with stability constraints). There is also a problem with unmeasurable system variables, which should not be included in the redundancy relations (undesirable variables or disturbances should not affect the design).

With the low-cost, limited-performance microprocessors in mind, the structured parity equation approach seems to be a reasonable methodological basis for many industrial projects, as compared to complex diagnostic algorithms based on observers and Kalman filters. A survey on model-based fault detection and diagnosis (Gertler, 1991) puts a new perspective on the structured parity equation technique. It shows that any diagnostic observer performance can be matched by an appropriate parity equation design, though similar results are not available with respect to multiple Kalman filter schemes.

It may then be concluded that fault detection theory is very much dependent on a theory for modelling of systems and it is worth trying to explore the modelling viewpoint. To address the above mentioned problems, we propose to apply the most specific system representation, the flowgraph description that can reflect the internal structure of the given physical system, as a basis for the derivation of a definite set of redundancy relations.

By reduction of the system's internal variables, we obtain a "concise" flowgraph having only the variables of interest (inputs, outputs, and disturbances). A resulting set of connected subgraphs containing only measurable variables provides the (primary and secondary) equations of the disturbance-orthogonal redundancy relations. Some of the nodes (physical system outputs) may turn out to behave as (formal) measured inputs and, finally, the system variables can be groupped as follows: regular/control inputs, measured inputs (physical outputs), sub-key variables (inputs and outputs), and key variables (sole outputs).

The proposed design methodology is illustrated using a linearized model of an automobile engine.

FAULT DETECTION AND ISOLATION

Fault detection and isolation/diagnosis methods usually utilize a mathematical model of the monitored plant. In spite of different motivation and origin, various approaches can be represented by a two-stage operation of generation and analysis of residuals. Residuals are quantities which should be of "zero" value in an ideal setting; some (not necessary all) nonzero values among them indicate the presence of faults, which have to be distinguishable from possible noise and modelling errors.

The presence of non-zero residuals merely indicates that something has gone wrong in the system (fault detection), and some further analysis is usually needed to decide upon the location of the failure (failure isolation). (A next level in this hierarchical decision making / pattern recognition scheme would be a failure value estimation). Conditions which have to be satisfied to assure isolation of all the respective system faults are called *isolability* conditions.

Analytical redundancy

Analytical redundancy (consistency relations) are used to generate residuals. A consistency relation is any mathematical equation describing how plant variables are nominally related to each other. A nominal system model may serve to provide a set of *primary* consistency relations. Since the primary system models do not usually satisfy isolability requirements, supplementary *secondary* relations, obtained by some transformations, are frequently used. Such a set of consistency relations is also referred to as balance or parity equations (Chow and Willsky, 1984; Gertler and Singer, 1990).

Dynamic parity equations are derived from a dynamic plant model. The model and the resulting parity equations will be assumed here to be linear in the continuous- or discrete-time domain.

Consider a system with inputs **u** and outputs **y**

$$\mathbf{u}(t) = [u_1(t),\dots,u_k(t)]^T$$

$$\mathbf{y}(t) = [y_1(t),\dots,y_m(t)]^T$$

Then the dynamic input-output relationship may be symbolically described as $\mathbf{Y}(\xi) = \mathbf{F}(\xi)\,\mathbf{U}(\xi)$ or a bit less formally as

$$\mathbf{y}(t) = \mathbf{F}(\xi)\,\mathbf{u}(t) \qquad (1)$$

where ξ is the complex frequency operator: s, z or δ (Kowalczuk 1989b). Here $\mathbf{F}(\xi)$ is a matrix of rational functions, $F_{ij}(\xi^{-1})$ of the (shift, digital-, or continuous-integration) operator ξ^{-1}

$$\mathbf{F}(\xi) = \left[F_{ij}(\xi^{-1})\right]$$

The maximum order of the component functions may be considered as the order of the system.

The input vector can be decomposed into three principal elements

$$\mathbf{u} = [\mathbf{u}_M^T\ \mathbf{u}_C^T\ \mathbf{u}_D^T]^T \qquad (2)$$

which are measured inputs $\mathbf{u}_M$, controlled inputs $\mathbf{u}_C$, and disturbance inputs $\mathbf{u}_D$. With these variables, (1) may be re-written as

$$\mathbf{y}(t) = \mathbf{F}_M(\xi)\,\mathbf{u}_M(t) + \mathbf{F}_C(\xi)\,\mathbf{u}_C(t) + \mathbf{F}_D(\xi)\,\mathbf{u}_D(t) \qquad (3)$$

where $\mathbf{F}_M$, $\mathbf{F}_C$ and $\mathbf{F}_D$ are the respective submatrices of **F**.
Because the measured inputs are subject to a sensor bias fault $\Delta u_M(t)$, the actual measurements $\tilde{u}_M(t)$ are

$$\tilde{u}_M(t) = u_M(t) + \Delta u_M(t) \qquad (4)$$

The controlled inputs can be distorted by an actuator bias fault $\Delta\tilde{u}_C(t)$, therefore the command value $\tilde{u}_C(t)$ is related to the actual input $u_C(t)$ as

$$\tilde{u}_C(t) = u_C(t) + \Delta u_C(t) \qquad (5)$$

The outputs are measured and, like the measured inputs, are subject to sensor bias

$$\tilde{y}(t) = y(t) + \Delta y(t) \qquad (6)$$

All the bias instrument faults are assumed to be deterministic signals of unknown constant amplitude, occuring at a random time instant. The measurement situation in the monitoring system is illustrated in Fig. 1.

Note that the measured input may be associated with any measurement point in the plant or its environment that serves as a signal source. The disturbance inputs are normally not directly measurable. Therefore we shall use $\mathbf{u} = [\mathbf{u}_M^T\ \mathbf{u}_C^T]^T$ rather than (2) as the input vector. And, consequently, a subset of equations for which the function $\mathbf{F}_D$ is identically zero will be the most desirable

$$\mathbf{y}(t) = \mathbf{F}_M(\xi)\,\mathbf{u}_M(t) + \mathbf{F}_C(\xi)\,\mathbf{u}_C(t) \qquad (7)$$

Certainly, this system description should not be obtained with any essential loss of information. Therefore, from the data availability viewpoint, some measured variables (which are, as a matter of fact, additional output points of the system) should contain the necessary information about some operating conditions, load, disturbances, *etc.*

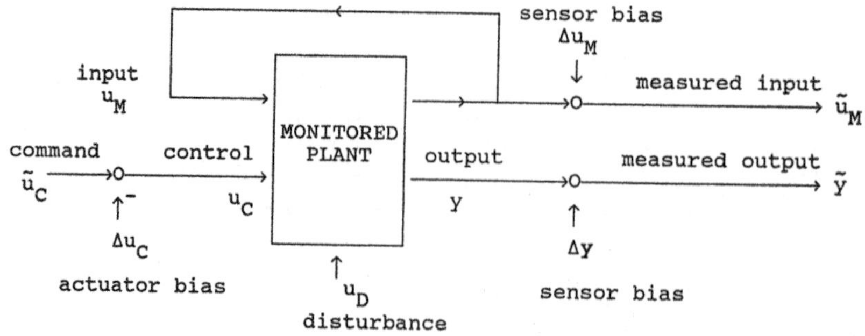

Fig. 1. Data collection scheme.

This may be possible, for instance, if the disturbance enters the system in such a way that it affects all the respective outputs only via certain variables and these latter variables are measurable. Then these measured system's outputs (or intermediate variables), which are not fully modeled (because of the lack of information about their input signals), will not be included in **y**. But because of carrying the relevant information about the disturbance-like signals, they can be treated as additional measured inputs (see Fig. 1) and used in respective redundancy relations. In conclusion, the set of the measured input variables can be specified as a minimum set of the measurable disturbance-dependent output variables and the regular input variables.

Using eqns. (4), (5) and (6) in the redundancy relations of eqn (7) results in the *computational* form of the residuals

$$\mathbf{e}(t) = \tilde{\mathbf{y}}(t) - \mathbf{F}_M(\xi)\, \tilde{\mathbf{u}}_M(t) - \mathbf{F}_C(\xi)\, \tilde{\mathbf{u}}_C(t) \qquad (8)$$

The remaining terms of eqn (7) constitute the *internal* form of the primary parity equations, demonstrating how the residuals depend on the faults

$$\mathbf{e}(t) = \Delta\mathbf{y}(t) - \mathbf{F}_M(\xi)\, \Delta\mathbf{u}_M(t) - \mathbf{F}_C(\xi)\, \Delta\mathbf{u}_C(t) \qquad (9)$$

Measurement, actuator and plant noises and modelling errors, which also affect the residuals, have been ignored above.

For the derivation of the necessary number of consistency relations (in order to satisfy the conditions for isolability) one is looking for different (*secondary*) analytical redundancy representations of the physical (*primary*) output variables included in **y**. Thus the above given system descriptions will usually be extended by redundant equations and the original output vector **y** will be extended by redundant output variables $\mathbf{y}_r$, *i.e.*

$$\mathbf{y} := [\mathbf{y}^T\ \mathbf{y}_r^T]^T$$

The redundant parity equations (or residuals) can simply be obtained by linear combinations (involving polynomial coefficients) of two or more parity equations. This can also be shown as a matrix transformation of the primary residual set **e**(t)

$$\mathbf{r}(t) = \mathbf{W}(\xi)\mathbf{e}(t) \qquad (10)$$

where $\mathbf{W}(\xi)$ is a matrix of rational functions.

With an approriate choice of the transformation $\mathbf{W}(\xi)$, this procedure makes it possible to create residual sets that exhibit the desired (structural) properties with respect to the faults.

Isolability requirements

It is important for diagnosis to have the set of residual equations constructed in such a way that only a fault-specific subset of the residuals becomes non-zero in response to a single fault. Then each residual subjected to a threshold test yields a single test bit ("fault detected"). The vector of the test bits is the fault signature (Gertler, 1988; Gertler and colleagues 1991).

The set of residual equations can be characterized by its "structure" indicating the faults by which the respective residuals are affected, say by 1's (*i.e.* ones symbolize the existance of a possible *cause-effect* relation and zeros indicate the lack of it). This structural format yields the *incidence matrix* of the diagnostic system, where each column makes the fault signature (the boolean code obtained in response to the given fault). For different faults to be distinguishable, their fault signatures must be different (and non-zero). A structure which has all distinct non-zero columns is "weakly isolating". (Note that disturbance decoupling can be interpreted as a special case of residual structuring.)

There is a problem with matching the threshold levels (triggering the diagnostic tests) to the actual faults. As a result of an intermediate-size fault, the fault code may have some excess zeroes. If there is another fault whose signature happens to be this pattern then the fault gets misdiagnosed. There are structures that have the so called, "statistical high-threshold isolability" property, that make such misdiagnosis not possible. One way of checking for and/or getting this kind of isolation (if possible at all) is by searching for a "column canonical" incidence matrix, in which each column has the same number of zeroes, each in a different configuration.

GENERATING LINEAR REDUNDANCY RELATIONS

The linear system description is the most developed field of engineering. Since many industrial plants operating in certain confined subspaces of the system variable space are liable for this treatment, the linear approach attracts considerable attention in engineering practice.

The following procedure is based on the assumption that the internal structure of the objective system is given. This structure can be most conveniently represented by a signal flow graph. At the first stage, a precise description (transfer functions) of the respective graph branches need not to be known. The procedure will indicate which subsystems are pertinent to the diagnostic algorithm and which next have to be analytically or experimentally derived/identified.

At this stage, the flow graph is a qualitative representation of the observed system that models mutual interrelations between the system variables. It reproduces the structure of the system and therefore it carries information relevent to the incidence matrix. The graph model indicates, for instance, how to deal with unmeasurable system disturbances: By chosing a set of reachable/measurable nodes of the graph a collection of all equations unaffected by such disturbance signals can easily be determined and used as the set of primary parity equations.

In this development we shall find the above given notions of the measured inputs (which can be physical outputs). All the "final" output nodes (and variables) that supply useful parity equations can be divided into two groups: *key nodes* (sole outputs) and *sub-key nodes* (that can serve both as outputs and inputs). The key nodes have a kind of "precedence" over the sub-key nodes in defining the parity equations.

Given a linear flow graph of the observed system the procedure for deriving a consistent and limited set of all possible disturbance-free parity equations is given below.

The procedure of obtaining a minimal subgraph

1. Eliminate all intermediate nodes to obtain a *minimal* flow graph composed of the measurable nodes (related to measurable inputs and outputs of the system) and of the unmeasurable disturbance nodes (related to pertinent disturbing variables);

2. Find a node that is entered by an unmeasurable

disturbance signal. (This measurable physical output node may then be used as a measurable pseudo-input only);

3. Construct a *reduced* flow graph by removing all the input branches entering this node and all the resulting disconnected nodes of the graph;

4. If there is still a disturbance node in the flow graph return to step 2;

The minimal flow graph description of the system without the disturbance variables, is suitable for deriving the (partial) parity equations (describing the subsystems of the observed system) completely unaffected by disturbances.

All that remains, is to write down the (primary) equations for the final (key and sub-key) output variables and then to combine them properly so as to obtain the secondary equations, and have, in such a way, the set of all possible analytical redundancy relations of the system.

MONITORING A LINEAR SYSTEM: AN EXAMPLE

As an examle, let us consider a linearized model of an automobile engine (Kowalczuk and Gertler, 1991) obtained based on the works of Dobner (1980, 1982) and Dobner and Fruechte (1983) (see also Gertler and colleagues, 1991) and shown in Fig. 2.

Appart from the essential (input U and output Y) variables necessary to describe the the system

$$[U^T \ Y^T] = [U_1 \ U_2 \ U_3 \ U_4 \ Y_1 \ Y_2 \ Y_3] \qquad (11)$$

where $U_4 = U_D$ is the unmeasurable disturbance input, the graph contains three auxiliary state variables

$$[X^T] = [\ X_1 \ X_2 \ X_3] \qquad (12)$$

Note that taking into account the origines of the model flowgraph from Figs 1 and 2 we can name the as follows (Kowalczuk and Gertler, 1991)

U_1 - the measured throttle position Thr

U_2 - the controlled exhaust gas recirculation Egr

U_3 - the controlled fuel injection Fuel

U_4 - the disturbances, namely the Load and the Spark advance

X_1 - the effective torque Trq

X_2 - the exhaust pressure P_{exh}

X_3 - the manifold model parameters Mam

Y_1 - the engine speed Rmp

Y_2 - the intake manifold pressure Map

Y_3 - the exhaust oxygen sensor voltage VO_2 or the air to fuel ratio Afr.

The flow graph can be described by a corresponding vector-matrix equation

$$\begin{bmatrix} X(\xi) \\ Y(\xi) \end{bmatrix} = H(\xi) \begin{bmatrix} U(\xi) \\ X(\xi) \\ Y(\xi) \end{bmatrix} \qquad (13)$$

For the sake of simplicity, instead of giving the full description of the system matrix $H(\xi)$ we may use only the corresponding incidence matrix Φ of H

$$\Phi = [\phi_{ij}] \qquad (14)$$

$$\text{with} \quad \phi_{ij} = \begin{cases} 1 & \text{if} \quad h_{ij}(\xi) \neq 0 \\ 0 & \text{if} \quad h_{ij}(\xi) = 0 \end{cases} \qquad (15)$$

where h_{ij} are elements of $H(\xi)$, $H(\xi) = [h_{ij}(\xi)]$.

The incidence matrix of the system from Figure 2 is

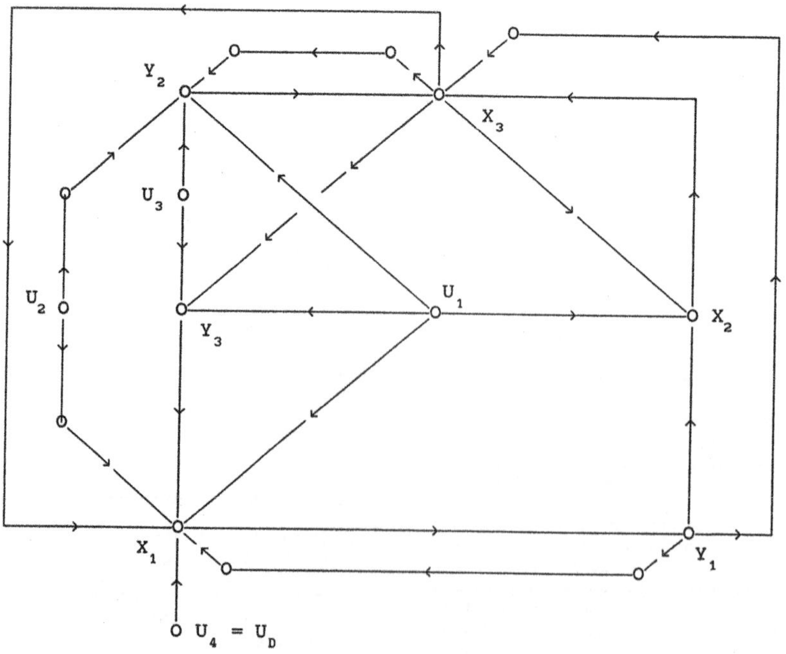

Fig. 2. Linear system flow graph.

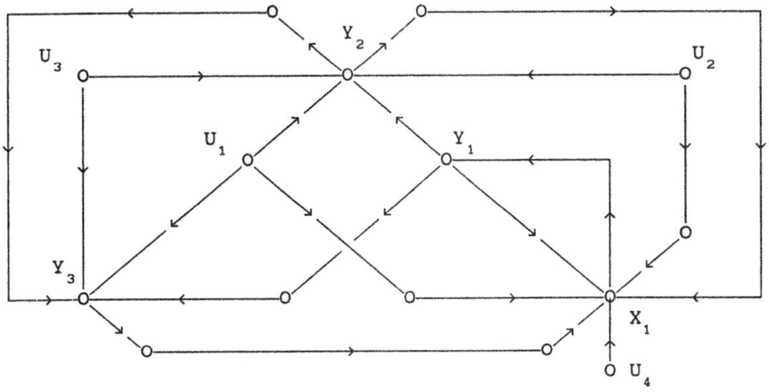

Fig. 3. Simplified (intermediate) system model.

$$\Phi = \left[\begin{array}{cccc:ccc:ccc} 1 & 1 & 0 & 1 & 0 & 0 & 1 & 1 & 0 & 1 \\ 1 & 0 & 0 & 0 & 0 & 0 & 1 & 1 & 0 & 0 \\ 0 & 0 & 0 & 0 & 0 & 1 & 0 & 1 & 1 & 0 \\ \hline 0 & 0 & 0 & 0 & 1 & 0 & 0 & 0 & 0 & 0 \\ 1 & 1 & 1 & 0 & 0 & 0 & 1 & 0 & 0 & 0 \\ 1 & 0 & 1 & 0 & 0 & 0 & 1 & 0 & 0 & 0 \end{array}\right] \qquad (16)$$

After reducing two auxiliary variables (X_2 and X_3) we have only one internal state X_1, which is entered by the disturbance U_4, as shown in Fig. 3. The structure (Φ) of the simplified systen is now given by

$$\begin{array}{c} \\ X_1 \\ Y_1 \\ Y_2 \\ Y_3 \end{array} \begin{array}{cccc:c:ccc} U_1 & U_2 & U_3 & U_4 & X_1 & Y_1 & Y_2 & Y_3 \\ 1 & 1 & 0 & 1 & 0 & 1 & 1 & 1 \\ \hline 0 & 0 & 0 & 0 & 1 & 0 & 0 & 0 \\ 1 & 1 & 1 & 0 & 0 & 1 & 0 & 0 \\ 1 & 0 & 1 & 0 & 0 & 1 & 1 & 0 \end{array} \qquad (17)$$

The minimal graph (according to Procedure, p.1) is depicted in Fig. 4

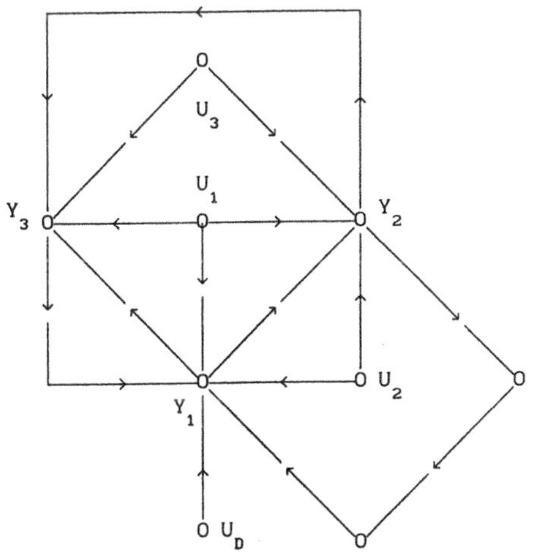

Fig. 4. Minimal flow graph of the system.

and may be represented by the structure Φ as follows

$$\begin{array}{c} \\ Y_1 \\ Y_2 \\ Y_3 \end{array} \begin{array}{ccc:c:ccc} U_1 & U_2 & U_3 & U_D & Y_1 & Y_2 & Y_3 \\ 1 & 1 & 0 & 1 & 0 & 1 & 1 \\ 1 & 1 & 1 & 0 & 1 & 0 & 0 \\ 1 & 0 & 1 & 0 & 1 & 1 & 0 \end{array} \qquad (18)$$

As seen from Fig. 4, the node Y_1 is entered by the disturbance U_4. Therefore, according to p. 2 of the procedure, this node cannot be used as an output and its respective variable (Y_1) may only serve as a measured input signal.

After removing the input branches of this node and the (disconnected) node of the disturbance (p.3), one comes to a final structure of the parity equations shown in Fig. 5 as a reduced flow graph.

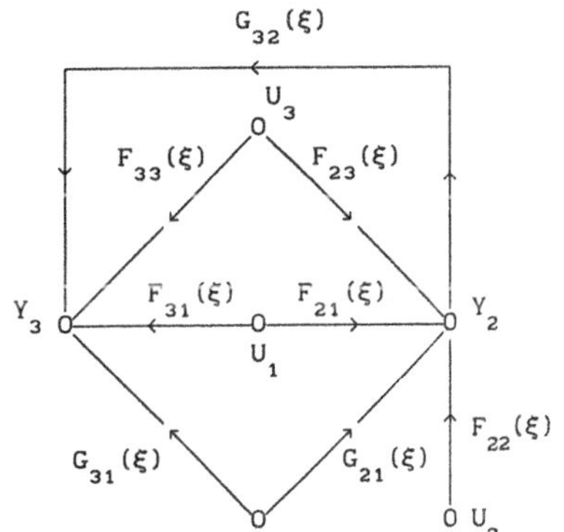

Fig. 5. Structure of the primary parity equations.

with the incidence matrix Φ

$$\begin{array}{c} \\ Y_2 \\ Y_3 \end{array} \begin{array}{ccc:c:cc} U_1 & U_2 & U_3 & Y_1 & Y_2 & Y_3 \\ 1 & 1 & 1 & 1 & 0 & 0 \\ 1 & 0 & 1 & 1 & 1 & 0 \end{array} \qquad (19)$$

Note that the parity equations obtained above are

not affected by the disturbance and the node Y_1 (a physical system output) plays the role of the (formal) measured input. The structure shows two "evident" outputs: Y_2 and Y_3. The variable Y_3 is the key (or pure output) variable, whereas the variable Y_2 is the sub-key variable since it serves both as an output variable (there is a node equation) and as a measured input variable (used in another node equation).

If we assign (by modelling and/or identification) transfer functions to the particular branches of the reduced signal flow graph as shown in Fig. 5, the corresponding primary parity equations can be expressed as follows

$$Y_2(\xi) = F_{21}(\xi)\,U_1(\xi) + F_{22}(\xi)\,U_2(\xi) + F_{23}(\xi)\,U_3(\xi) + G_{21}(\xi)\,Y_1(\xi)$$

$$Y_3(\xi) = F_{31}(\xi)\,U_1(\xi) + F_{33}(\xi)\,U_3(\xi) + G_{31}(\xi)\,Y_1(\xi) + G_{32}(\xi)\,Y_2(\xi) \tag{20}$$

The parity equations can be recognized by octal numbers representing the coincidence vector of the measured variables used in each of the equations, respectively, with a "natural" arrangement from (11)

$$[U^T\ Y^T] = [U_1\ U_2\ U_3\ Y_1\ Y_2\ Y_3] \tag{21}$$

The equation for Y_3 may then be referred to as equation #57, and the Y_2 equation as #76.

Note that the system variables in (19) and (21) are arranged in a special sequence: the regular (or control) inputs (U_1, U_2, U_3), the measured inputs (the physical output: Y_1), the sub-key variables (the input and output: Y_2), and the key variables (the pure output: Y_3).

By simple algebraical combination of these two (primary) equations a set of new linear parity equations (all concerning the key variable, Y_3) can be derived. Combining equations makes it possible to eliminate a chosen variable, *i.e.* to obtain a new equation unaffected by this variable (and by its associated fault). The four new equations obtained in this way can be marked with #37, #67, #73, and #75. The whole set of six parity equations has a nice property of statistical high-threshold isolability shown in Table 1.

TABLE 1 Structures of Parity Equations with High-Threshold Isolability

no	U_1	U_2	U_3	Y_1	Y_2	Y_3
37	0	1	1	1	1	1
57	1	0	1	1	1	1
67	1	1	0	1	1	1
73	1	1	1	0	1	1
75	1	1	1	1	0	1
76	1	1	1	1	1	0

CONCLUSIONS

In the paper a method for obtaining a definite, consistent and limited set of all possible disturbance-free parity equations is given based on a linear flow graph of the observed system.

By reduction of the system's internal variables, a minimal flow graph, having only the variables of interest (inputs, outputs, and disturbances), may be obtained. A reduced flow graph containing only measurable variables provides parity equations that are disturbance-decoupled.

Some of the nodes (physical system outputs) turn out to behave as (formal) measured inputs. All the system variables can be groupped as follows: regular/control inputs, measured inputs (physical outputs), sub-key variables (inputs and outputs), and key variables (sole outputs).

The proposed general design methodology is suitable for instrument failure detection and isolation systems monitoring linear dynamic systems. It may also be useful for a class of nonlinear systems (Kowalczuk and Gertler, 1991).

REFERENCES

Chow, E.Y. and A.S. Willsky (1984). Analytical redundancy and the design of robust failure detection systems. IEEE Trans. Aut. Control., **AC-29**, 603-614.

Dobner, D.J. (1980). A mathematical engine model for development of dynamic engine control. SAE Paper No. 800054.

Dobner, D.J. (1982). Dynamic engine models for control development. Part I: Nonlinear and linear model formulation". Int. J. Vehicle Design, (GM Report GMR-3783).

Dobner, D.J. and R.D. Fruechte (1983). An engine model for dynamic engine control development. Proc. American Control Conf., San Francisco (CA), June, 22-24, 1983, pp. 1-5 (WA4). (GM Report GMR-4364)

Gertler, J. (1988). Survey of model-based failure detection and isolation in complex plants. IEEE Control Systems Magazine, **8**, (6), 3-11.

Gertler, J. (1991). Analytical redunduncy methods in Fault detection and isolation. Prepr. IFAC/IMACS Symposium Fault Detection, Supervision and Safety for Technical Processes. Baden-Baden (Germany), Sept., 10-13, 1991, vol.1, pp. 9-22.

Gertler, J. and D. Singer (1990). A new structural framework for parity equation - based failure dtection and isolation. Automatica, **26**, 381-388.

Gertler, J., M. Costin, X-W. Fang, R. Hira, Z. Kowalczuk, and Q. Luo (1991). Model-based onboard fault detection and diagnosis for automotive engines. Prepr. IFAC/IMACS Symposium Fault Detection, Supervision and Safety for Technical Processes. Baden-Baden (Germany), Sept., 10-13, 1991, vol.4, pp. 241-246.

Kowalczuk, Z. (1989a). Finite register length issue in the digital implementation of discrete PID algorithms. Automatica, **25**, (3), 393-405.

Kowalczuk, Z. (1989b). On a dynamical linear-flowgraph description. IEE European Conf. Circuit Theory and Design, Brighton, pp.547-551.

Kowalczuk, Z. and J. Gertler (1991). Instrument failure detection and isolation using stationary nonlinear redundancy relations of a car engine. Research Report, G. Mason Univ., Fairfax (U.S.A.), no. GM/GMU/ZFK002/10/91, 42 pp.

DAML - A LANGUAGE FOR MODELLING AND SIMULATION OF DIGITAL CONTROL SYSTEMS

Z. Kowalczuk

Department of Automatic Control, Faculty of Electronics, Technical University of Gdańsk, Majakowskiego 11/12, 80-952 Gdańsk, Poland

Abstract. The hybrid characteristics of digital control processes, which have to be analysed both in the discrete and the continuous time domains, makes the problem of optimal control design especially difficult. A prior theoretical design is usually carried out in one domain and does not include all the nonlinear constraints resulting from finite precision implementation. In the paper a computer-aided simulation system, suitable for solving these problems, is presented. A general concept of an associated programming language, the digital and analogue modelling language - DAML, is given. The language supplies facilities for programming analog plants and digital controllers, various types of arithmetic, and different levels of precision. The block-diagram representation of the plant structure, including non-linear and user-defined functions, and a number of discrete controller descriptions, including the z-transfer function, the state-space representation, extensions of the linear-flowgraph notation, as well as a high-level programming language, can be applied. In order to model typical restrictions of implementation, the fixed-, floating- and block-floating-point digital-controller mechanizations can be used. A discrete PID regulator for a stationary plant and an adaptive algorithm for a robotic manipulator illustrate applicability of DAML.

Keywords. Computer-aided design, system analysis, computer control, DDC, modelling, simulation, finite precision implementation.

INTRODUCTION

Optimal design of control strategies is connected with various types of scientific computations (Troch, 1988). Simulation is a type of complementary numerical support that provides an effective tool for rigorous examination of analytically derived solutions and for related sensisivity studies. It may also be an efficient means of direct parameter optimization, provided that the designer has an experience of using the "proper" structure of the designed control system.

Solving engineering problems via simulation has always been one of the most important means in the design. Development in theory and technology imposes new demands upon the simulation tools. Special modelling languages (Orłowski and Hawryluk, 1971; Kruger and Mylius, 1988; Funke, 1988) are usually created in order to ficilitate model formulation and carying out simulation experiments.

The hybrid characteristics of digital control processes which are given both in the discrete- and the continuous-time domain make the design problem especially difficult, and the prior (or *proper*) theoretical design and/or optimization is usually carried out in one, either *pure* discrete- or continuous-time domain (Kowalczuk, 1985).

What is more, it is demanding to include all nonlinear constraints resulting from finite precision implementation in the proper design, and only different deterministic or stochastic approximate techniques (Kowalczuk, 1989a) are usually applied with the aim of attaining crude estimates of deviations from optimality in view of certain measures. Hence, in most cases simulation is the best way of studying nontrivial control problems (Isermann, 1989) before putting them into practice.

The objective of this paper is to present a versatile language along with an integrated system for digital control simulation purposes.

The power of the simulation tool is illustrated in an implemantation study of typical analog-plant control problems, where a digital PID regulator for a stationary plant, and a discrete adaptive algorithm for a robot arm are applied.

THE LANGUAGE AND ITS BACKGROUND

The digital and analog modelling language (DAML) is a formal tool facilitating simulation of complete hybrid (DDC) control loops. It has been used at TUG since 1985. The associated simulation system DAML was originally implemented on a mainframe computer and its new modification (Kowalczuk, 1988) is suitable for personal computers.

DAML distinguishes two kinds of systems, discrete- and continuous-time. Special language expressions are designated to control the simulation process.

In order to be close to the origines of real analog systems, we apply the block-diagram representation which enables constructing the plant model to the best of our knowledge about its internal structure. A suitable set of operations (including different non-linear and user-defined functions) is used.

Even though a preference for linear time-invariant systems has been observed, advanced adaptive computer control algorithms or self-tuning controllers based on recursive identification and using on-line controller re-design (*e.g.*, Niedźwiecki and Kowalczuk, 1988; Kowalczuk, 1989c, 1991, 1992), demand some extention of this

description, at least into time-variant systems.

Therefore, we use the following collection of discrete controller descriptions: the z-transfer function, the state-space representation, and a cascaded combination of them, a dynamical linear flowgraph description (Kowalczuk, 1989b), and an extension of the linear-flowgraph notation (including signal multiplication/division). The possibilities for using a general high-level programming language are also provided.

The finite precision of digitally implemented control systems as compared with idealized mathematical design, leads to degradation due to quantization noise, coefficient inaccuracy, and limit cycle oscillations. These issues are well known in pure digital signal processing. They have also to be taken into account in the application of digital techniques to the discrete-time controllers especially when implementing digital controllers on microprocessor-based system with short word length (Kowalczuk, 1989a).

For implementation it is necessary to inspect several issues of the design regarding the digital network, the type of arithmetic, and the word length, and consequently optimize the possible solutions taking into account combined performance indices (Kowalczuk, 1989a; 1989d).

In order to model the implemental restrictions the fixed-, floating- and block-floating-point digital - controller representations (with truncation or roundoff) are considered.

ANALOGUE PLANT DESCRIPTION

It is a well-known fact that apart from the simple case of linear continuous-time objects, analogue systems (or environmental signals) subject to digital control (or signal processing) may be of multiple types and may incorporate a number of non-linearities and implicit relations. Therefore, similarly to the analog-process modelling language CEMMA (Orłowski and Hawryluk, 1971), we use one of the most general (continuous) system description, the block-diagram (structural) representation, to construct the objective model according to its presummed internal structure and based on a suitable set of elementary operations (blocks).

The semantics of the language DAML includes the following six groups of the elementary operations:

1) basic arithmetical operations (+,-,*,/),
2) integrating performed by different numerical methods (rectangular, trapezoidal, and the 4th order Runge-Kutta scheme),
3) standard mathematical functions (modulus, square root, exponential, logarithmic and trigonometrical functions),
4) nonlinear characteristics (delay, hysteresis, "diode", "limiters", "relays", "dead-band"/ "neutral zone", "clearance"/ "shake"),
5) signal generators (deterministic jumps and linear functions as well as stochastic uniform and normal distribution functions),
6) user programmed functions (piece-wise linear and higher-order Lagrange interpolation).

Note that all the dynamical operations characterized by a memory of the past (such as integration and hysteresis, for instance) do certainly need initialization.

DIGITAL CONTROLLER DESCRIPTION

The structures of digital controllers used in the design, optimization, and implementation of control, is chosen by the designer. Therefore in most cases, classical linear time-invariant discrete-time models (having a fixed structure) are applied (Iserman, 1989; Kowalczuk, 1985; 1989a; 1989d). It may be noted that the observed preference for linear time-invariant systems has been a reflection of the up-to-date state of the art of describing and analysing the real world rather than a reflection of a deep believe in that type of representation (Troch, 1988).

Nevertheless, since also nonlinear and adaptive algorithms are often considered (Niedźwiecki and Kowalczuk, 1988; Kowalczuk, 1989c, 1991, 1992; Gertler and co-workers, 1991; Kowalczuk and Gertler 1991), structures with time-varying parameters and freely-programmed options can be useful.

Taking into account the above considerations we focus our attention on the following collection of discrete-time controller descriptions:

I. Formal, time-invariant models:

1) the z-transfer function (of SISO type)

$$F(z) = \frac{Y(z)}{U(z)} = \frac{a_0 + a_1 z^{-1} + \ldots + a_N z^{-N}}{1 + b_1 z^{-1} + \ldots + b_N z^{-N}} \qquad (1)$$

2) the state-space representation (of MIMO type)

$$\begin{aligned} z\,\mathbb{X}(z) &= \mathbb{A}\,\mathbb{X}(z) + \mathbb{B}\,\mathbb{U}(z) \\ \mathbb{Y}(z) &= \mathbb{C}\,\mathbb{X}(z) + \mathbb{D}\,\mathbb{U}(z) \end{aligned} \qquad (2)$$

3) the cascade of subsystems,

$$F(z) = F_1(z) * F_2(z) * \ldots * F_K(z) \qquad (3)$$

where all the $F_k(z)$, for k = 1, 2,...K, are of the form either of the above defined SISO system (1) or the MIMO one (2). In the latter case each subystem must be of appropriate dimension, of course.

4) the dynamical linear flowgraph description, which is one of possible general system representations,

$$\begin{aligned} \mathbb{X}_p(z) &= \mathbb{F}_p\,\mathbb{X}_d(z) + \mathbb{G}_p\,\mathbb{U}(z) \\ \mathbb{X}_d(z) &= \mathbb{F}_d(z)\,\mathbb{X}_p(z) \\ \mathbb{Y}(z) &= \mathbb{H}_p\,\mathbb{X}_d(z) \end{aligned} \qquad (4)$$

where $\mathbb{X}_p$ and $\mathbb{X}_d$ are vectors of *pseudo-dynamical* and *dynamical* (Kowalczuk, 1989b) node (or state) variables, respectively. The matrices $\mathbb{F}_p$, $\mathbb{G}_p$, and $\mathbb{H}_p$ are of constant coefficients, and only the *dynamical* state transition matrix $\mathbb{F}_d(z)$ is composed of z-transfer functions which need not to be strictly proper (or delayed[1]) functions of z^{-1}.

II. User-defined structure (graph):

[1]The necessary delay is assigned to the digital-to-analog converter (DAC).

5) an extension of the linear-flowgraph notation comprising

   - standard summing nodes,
   - unit delay arcs representing the shift operator z^{-1} (">"),
   - standard scaling arcs ("-") multiplying by a constant ("=" if with quantization),
   - extra signal-multiplying branches ("*"), and
   - extra signal-dividing branches ("/").

III. User-programmed subroutine:

6) If none of the above formal descriptions meets the actual requirements of the modeller, he or she has the possibility for using a general high-level programming language (Fortran or Pascal, depending on the computer environment).

QUANTIZATION ISSUE

The finite word length effects of digitally implemented discrete controllers (*i.e.* quantization noise or limit cycle oscillations) are even more difficult to analyse as compared to open-loop digital signal processing (Kowalczuk, 1989a).

On the other hand, the finite precision simulation of digital systems is as simple as their implementation.

The proposed language provides the opportunity of inspecting several issues of the design (regarding, for example, the digital controller structure, some types of arithmetic, and the word length) and of optimizing the possible solutions taking into account proper performance indices. It may also be an important support in finding proper scaling coefficients in the procedure of *normalization*[2] of the controller algorithm (Kowalczuk, 1989d).

In order to offer the system engineer the possibilities for modelling the mechanization restrictions the following three digital controller variables' representations are considered (see APPENDIX):

1) the fixed-point format,
2) the floating-point representation, and
3) the block-floating-point (with a single common exponent) format.

Each of the numerical representations assume a specific form on decreeing the length of the mantissa and the length of the exponent (in bits).

Similarly, both the input (ADC) and output (DAC) converters are accommodated with the same as above but individual parametrization.

In each of the foregoing cases two types of quantization, *i.e.* either truncation or roundoff, may be applied.

THE LANGUAGE SYNTAX

The modelling language **DAML** facilitates simulation of a digital controller and an analog environment or plant as a whole even though the two parts of the DDC system are defined separately and are characterized by their specific attributes.

Both types of models have to be described by means of the DAML **terms** or **identifiers** and **records** or **expressions**. Additional language expressions are designated to control the simulation process.

A suitable collection of the language identifiers is given in APPENDIX. They are used to construct the programming **structural records**, a sequence of which formulate the programming **segments**, and finally, the **simulation program**.

Each programming record consists of a special DAML identifier (**name**) preceded by the sign "#" (*i.e.* # *name*) and/or **parameter fields** ending in the semicolon separator (";"), which can be followed by a **commentary**. The parameter field is composed of numerical parameters and some *separators*.

In the case of the analog part description, each structural record must be labelled with a **description-unit identifier** (*natural number*) preceded by the sign "*" (*i.e.* * *natural number*).

In the *parameter field*, prefix separators are used to indicate an **input description unit** (< *natural number*) and a parameter value (: *number*). The star ("*") following the input-description-unit number denotes an external input (*i.e.* the number of the digital controller output).

Programming the first four of the digital controller's models simply needs an approriate DAML identifier showing the type of description (the above formulas (1)÷(4)) and a list of separated parameters given in a predetermined arrangement.

The fifth the flowgraph description announced by the graph identifier (**#GRAPH**) is given in terms of input and output node numbers coupled by the **connection symbol** (- or = for transmission and > for shifting). Premultiplication is admissible in connection with the standard (unit transmission) arcs.

Programming the control algorithm in a high-level language (as subroutines of FORTRAN or PASCAL) is facilitated by predefined system-interfacing variables and a library of quantization utilities.

The whole simulatiom program is composed of the following segments:

1 - HEADING/COMMENTARY,

2 - ANALOG STRUCTURE
and related parameters,

3 - DIGITAL STRUCTURE,

4 - DIGITAL CONTROLLER PARAMETERS
(including the parameters of the sampling and quantization processes),

5 - SIMULATION PARAMETERS
(the total time, the integration technique, the type of display facilities),

6 - initialization (**#ENTER**),

7 - program termination (**#LEAVE**),

The USER-PROGRAMMED SUBROUTINE, which can be defined in an external file of a predetermined name, stands for the segment of DIGITAL CONTROLLER PARAMETERS. The digital controller structure and its parameters are given in two separate segments. This is to simplify modification of the controller since segments 4, 5, and 6 may be programmed repeatedly (it will not need extra compilation).

---

[2] That is, *structurally scaling* - in order to diminish the probability of overflows.

Owing to its characteristics, the program-compiler works in an interactive mode and enables, in such a way, the modeller to define the last four segments of the simulation program in an interactive manner.

A SIMPLE PID CONTROL

Let us consider a simple analog plant,

$$\mathfrak{G}(s) = \frac{h\, e^{-sT_0}}{1 + sT_1} = \frac{10\, e^{-s}}{1 + 10s} \qquad (5)$$

and a simple constant digital PID controller (Kowalczuk, 1989a):

$$R(z) = K + K \frac{1}{1 - z} + K \frac{z - 1}{\gamma z + 1 - \gamma} \qquad (6)$$

tuned by the typical analog Ziegler- Nichols settings and working with the sampling frequency of 5 Hz.

The whole, closed loop system has been modelled according to the diagram shown in Fig. 1. The digital controller has been mechanized by the most simple and common-in-use canonical structure 1D (Kowalczuk, 1989a), modified for the scaling purposes.

A maximum-type of variable scaling (Kowalczuk, 1989d) has been used to avoid overflow in 16-bit controller registers with a fixed-point arithmetic applied. Aditional numerical matching has been obtained by choosing different weighting in binary exponents both of the internal registers and of the input/output converters.

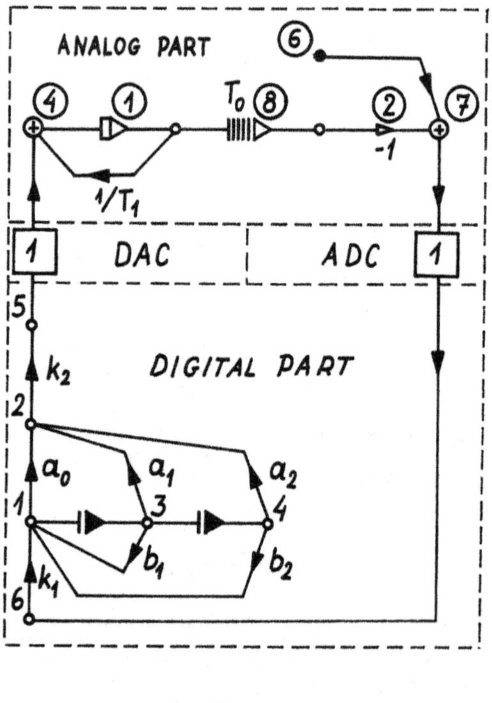

Fig. 1. Modelling simple DDC loop.

A possible implementation of the modelled control system using the graph description is given by the following DAML program:

```
;A digital PID regulator  controlling  a plant
;                         [10*exp(-0.1s)/(s+1)]
; begin of the analog description
* 1  #INTGR  <4 : 0;
* 8 #DELAY  <1  : 1.0 ;              (5 * 0.2 s)
* 2   #INVERT < 8  ;
* 06  # CONST  : 1. ;
*7 #SUMM1<2<6;
* 4  #SUMM2  < 1*  <1  :1  :-0.1 ;
                              ; (1* - means DAC No 1
# DIGIT         ;             begin of digital part
# SINGLDIM      ;        single dimension (SISO)
#ARCS: 13;
#INPUT:7;          analog unit No 7 to be an input
# CLOSE ;            end of the controller structure

#CDATA          ;             controller data
#GRAPH;
6 = 1, 0.90171;         = means quantizing after
                ;            multiplication/scaling
1 > 3 ;              shift between nodes No 1 and 3
3 > 4 ;
3 = 1, 0.888888;    there will be quantization
4 = 1, 0.111111;              as above
1 = 2, 1.109;                        .
3 = 2, -1.8626;                      .
4 = 2, 0.7861;                       .
2 - 5, 4.533; transfer to DAC and final scaling
6*6 - 7, 1;            quadratic error gained by
           ;          signal multiplication
#INPGRAPH :6; input node (connected with
           ;                             ADC No 1)
#OUPGRAPH :5; output node (to DAC No1)
# SAMPLED : 0.2;             the sampling time T
# QUANT #FIX ;           fixed-point quantization
#ADC #TRUNC #EXP:1 #BITS:16 ; truncation
                            ;             in ADC
#DDC #ROUND #EXP:4 #BITS:16 ; rounding in
                            ; the controller
#DAC #TRUNC #EXP:1 #BITS:16; truncation
 #SDATA                ;      simulation data
# SIMUTIME : 16       ;
# INTGSTEP : 2.0 e -03 ;        integration step
#LAGERROR : 0.05;             precision in delays
# RUKU4         ;        integration method
# SCREEN: 6, 8 ;          display the 6th and 8th
                ;              analog-unit outputs
# RANGE : 0, 2.5;  range of the ordinate of
                ;              the displayed graph
# ENTER
```

The step response obtained in the simulation run is shown in Fig. 2.

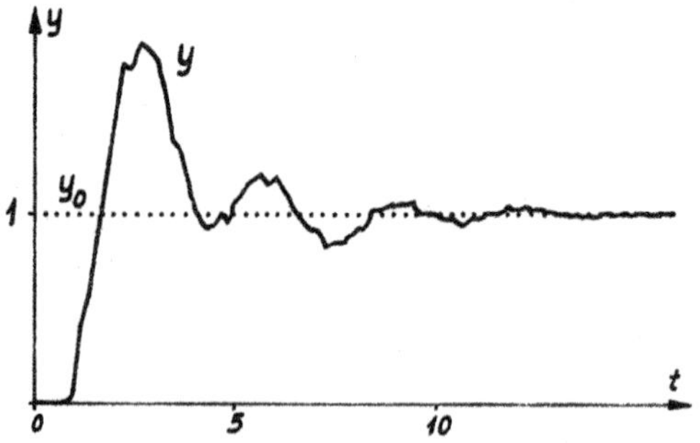

Fig. 2. Output of the controlled system.

ROBOTIC MANIPULATION

Consider now the control problem of an robotic manipulator arm, which, as opposed to the previous example, is a strongly nonlinear object.

Let us take a three-degree of freedom model consisting of one rotational joint (ϕ) and one translational joint (r) in the (x,y) plane as well as one lifting joint (z), and given by the following kinematic equations (Freund, 1983; Dwyer, Lee, and Chen, 1986):

$$\ddot{r} = r\,\dot{\phi}^2 - \frac{m_R\, l}{2m}\,\dot{\phi}^2 + \frac{1}{m}\,u_r$$
$$\ddot{\phi} = \frac{-2mr + m_R\, l}{\kappa - m_R l r + mr^2}\,\dot{r}\,\dot{\phi} - \frac{1}{\kappa - m_R l r + mr^2}\,u_\phi \qquad (7)$$
$$\ddot{z} = -\,g + \frac{1}{m}\,u_z\,,$$

where $m = m_R + m_L$; m_R and m_L are the masses of the arm and load, respectively; κ is an (equivalent) inertia constant; g is the acceleration of gravity; and l is the length of the arm.

Different output- and state-linearization techniques (Freund, 1983; Dwyer, Lee, and Chen, 1986) were proposed for the control syntheses purposes. The linearization of the state description by an appropriate choice of nonlinear feedback, seems to be especially valuable. Following this approach, the system of (7) can be rewritten into a linear framework:

$$\ddot{x} = K_x\,(u_x + \upsilon_x) \qquad (8)$$

where, for $x \in \{r,\ \phi,\ z\}$ we have the gains

$$K_r = K_z = 1/m, \qquad K_\phi = 1/(k - m_R l r + mr^2),$$

and the interactive force/torque perturbations:

$$\upsilon_r = \dot{\phi}^2\ [2mr - m_R l]/2$$
$$\upsilon_\phi = -\,\dot{r}\,\dot{\phi}\ [2mr - m_R l] \qquad (9)$$
$$\upsilon_z = -\,m\,g$$

Note that in the steady-state conditions only the gravitation force has to be compensated. It may be done for a nominal mass load, *e.g.* $m_L = m_{L0} + \Delta m_L/2$.

Having the simple decoupled double-integrator model of the manipulator, one can easily design a linear single-variable control algorithm, which may then be augmented by taking into account corresponding correcting terms.

We propose to use a simple PD regulator in feedback, and a proportional one (k_x) in the feedforward path of each control channel:

$$u_x = [x_0 - (x + T_d\,\dot{x})]k_x \ - \ \upsilon_x \qquad (10)$$

where υ_x is the controller-output correcting term, calculated on-line based on measurements of coordinates $\{r,\ \phi,\ z\}$.

The PD regulator gain is constant $(k_x = k)$ in all but one channels. Namely, the ϕ-controller additionally tunes its coefficient k_ϕ so as to compensate the radius-dependent model gain K_ϕ (with the nominal load).

The simulated digital controller is illustrated in Fig. 3, where only the structure of one most complex part (ϕ) of the algorithm is indicated. A discrete PD equivalent similar to that of (6) is applied.

Appart from its main task this type of algorithm is also able to supply the derivatives necessary for the output recalculation. That is why the digital PD regulator has a non-canonical form.

The pertinent parameters were chosen according to Dwyer, Lee, and Chen (1986). The digital controller (without quntization) worked with the sampling time of 0.05 sec. The full analog model of the robot manipulator was simulated with trapezoidal integration and the integration step of 0.003 sec.

Fig. 4 illustrates the transient behaviour of the robotic system in the (x,y) plane for jump-type changes in all the setpoints and a load mass variation Δm_L. The response in the z-direction was similar but for a small steady-state offset (0.1% in the extreme), which can be remedied by nonlinear filtering.

CONCLUSIONS

A digital and analog modelling language DAML, suitable for simulation of DDC systems has been

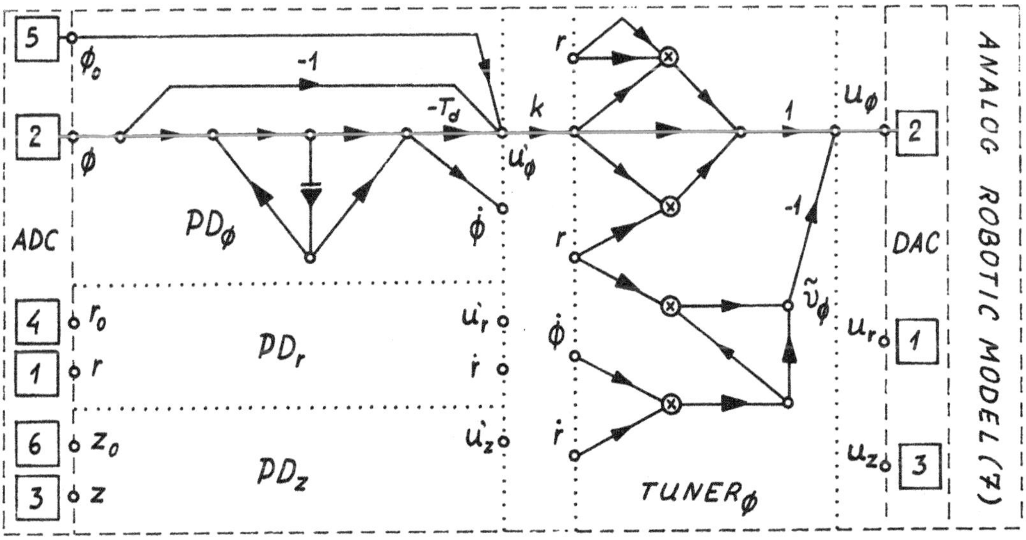

Fig. 3. Structure of the digital algorithm controlling manipulator.

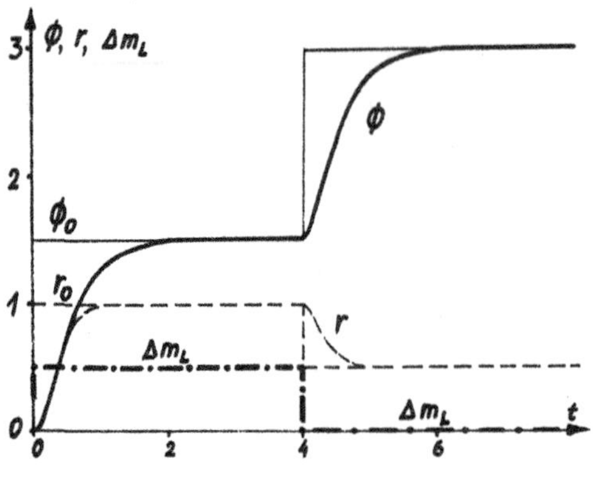

Fig. 4. Robotic manipulator trajectories.

presented and exemplified taking into account two control problems.

The programming segments allows for different structures of the analog plant and the digital controller. The parameters of the controller and the simulation process are declared separately making it possible to perform repetitive modifications and runs. A suitable set of system descriptive tools and implementation utilities facilitates the design and examination of DDC systems as well as considering its finite precision.

The proposed language provides the opportunity of inspecting several issues of design regarding, for example, a digital controller structure, a type of arithmetic, and a word length, and of optimizing the possible solutions taking into account different performance indices. It may also be an effective means of finding proper scaling coefficients in order to avoid overflows in the registers of the digital controller.

REFERENCES

Dwyer, T.A.W., G.K.F. Lee, and N. Chen (1986). A terminal controller for a robotic manipulator arm with corrections for perturbations. Intern. J. of Robotics and Automation, 1, 16-22.

Freund, E. (1983). Direct design methods for the control of industrial robots. Computers in Mechanical Engineering (ASME), pp. 71-79.

Funke, R. (1988). CANDYS/CM - A dialog system for modelling continuous dynamical systems with chain structure by differential equations. System Analysis and Simulation, Akademie-Verlag, Berlin, 1988, pp. 169-171.

Gertler, J., M. Costin, X-W. Fang, R. Hira, Z. Kowalczuk, and Q. Luo (1991). Model-based on-board fault detection and diagnosis for automotive engines. Prepr. IFAC/IMACS Symposium Fault Detection, Supervision and Safety for Technical Processes. Baden-Baden, vol.4, pp. 241-246.

Isermann, R. (1989). Digital Control Systems. Springer, Berlin.

Kowalczuk, Z. (1985). Synthesis and microprocessor implementation of digital systems for analog plant control application. Ph. D. Dissert., Techn. University of Gdańsk. Gdańsk.

Kowalczuk, Z. (1988). A system for modelling analog and digital processes - DAML. Report Inst. of Computer Sci., Techn. University of Gdańsk, Gdańsk. 88 pp.

Kowalczuk, Z. (1989a). Finite register length issue in the digital implementation of discrete PID algorithms. Automatica, 25, (3), 393-405.

Kowalczuk, Z. (1989b). On a dynamical linear-flowgraph description. IEE European Conf. Circuit Theory and Design, Brighton, pp.547-551.

Kowalczuk, Z. (1989c). Cooperative parallel estimators in tracking of nonstationary stochastic systems. URSI Int. Symp. Signals, Systems and Electronics, Erlangen, pp.225-228.

Kowalczuk, Z. (1989d). Improving the performance of fixed-point PID controllers by structural optimization and scaling. IFAC Symp. Low Cost Automation, Milano, pp.W203-208.

Kowalczuk, Z. (1991). Simple self-tuning control of a nonstationary stochastic analogue system based on competitive identification: an implementation issue. System Science, 17,(2), 63-78.

Kowalczuk, Z. and J. Gertler (1991). Instrument failure detection and isolation using stationary nonlinear redundancy relations of a car engine. Research Report, G. Mason Univ., Fairfax (U.S.A.), no. GM/GMU/ZFK002/10/91, 42 pp.

Kowalczuk, Z. (1992). Competitive identification for self-tuning control: robust estimation design and simulation experiments. Automatica, 28, (1), 193-201.

Kruger, S. and W. Mylius (1988). A modular computer-aided modelling and simulation system in chemical engineering. System Analysis and Simulation, Akademie-Verlag, Berlin, 1988, pp. 165-168.

Orłowski H., and J. Hawryluk (1971). Modelowanie Cyfrowe (Digital Modelling). WNT, Warsaw

Niedźwiecki, M., and Z. Kowalczuk (1988). Improving parameter tracking properties of finite-memory adaptive filters. 4th European Signal Processing Conference, Grenoble, vol. 2, pp. 799-802.

Troch, I. (1988). Optimization and simulation in control design. System Analysis and Simulation, Akademie-Verlag, Berlin, 1988, pp.226-231.

APPENDIX : THE DAML IDENTIFIERS

ANALOG STRUCTURE

Arithmetic : DIVDE, INVRT, MULT1, MULT2, MULT3, SUMM1, SUMM2.

Integration : INTGR.

Standard Functions : ABSLT, ATAN1, ATAN2, COSIN, EXPNT, LGRTH, SINUS, SQROT.

Nonlinearities : DELAY, DIODE, FZONE, HYST2, HYST3, LIMIT, RELAY, RLAY3, SHAKE.

Signal Generators : CONST, GTIME, LIGEN, HEAVS, UNOIS, GNOIS.

User Pathes : FUNC1, FUNC2, NODES, POINT.

DIGITAL STRUCTURE

Begin : DIGIT.

Dimensions : MULTIDIM, OWNECALL, SINGLDIM, STATEDIM.

Structure : ARCS, CASCADE, DYNGRAPH, INPUT,

End : CLOSE.

DIGITAL CONTROLLER PARAMETERS

Begin : CDATA.

Single & Cascaded : ACOEFFS, BCOEFFS.

Multiple & Cascaded : AMATRIX, BMATRIX, CMATRIX, DMATRIX.

Dynamical Graph : FPMATRIX, FDMATRIX, GPMATRIX, HPMATRIX,

Graph : GRAPH, INPGRAPH, OUPGRAPH, : (graph layout symbols : -, =, >, *, /).

Sampling Time : SAMPLED.

Quantization : ADC, DAC, DDC, ROUND, TRUNC, QUANT, BITS, EXP, FIX, FLO, BFP.

SIMULATION PARAMETERS

Begin : SDATA.

Time : SIMULTIME.

Accuracy : DIVPARAM, LAGERROR.

Integration : EULER, INTGSTEP, RUKU4, TRAPZ.

Display : PRNTSTEP, SCREEN, RANGE, TABLE.

INITIALIZATION : ENTER.

TERMINATION : LEAVE.

APPROXIMATE QUADTREE AND OCTREE REPRESENTATIONS FOR MANUFACTURING TASKS

J. Vörös

Nabrezna 91, CS-940 73 Nove Zamky, Czechoslovakia

Abstract. This paper deals with approximate quadtree and octree representations of two-dimensional and three-dimensional objects for manufacturing tasks and presents some useful algorithms for robot grasp choice and further applications. These are based on a memory saving "top-down" form of tree nodes descriptions, enabling inner as well as outer approximations of regions and solid objects. A simple approach is presented for generation of gripping positions on solid object dealing with its octree model. Application of approximate representations to robot path-planning and to investigation of robot moveability is sketched.

Keywords: Solid models; hierarchical representation; robots; CAE; manufacturing processes.

INTRODUCTION

Modelling and representation of two-dimensional (2D) and three-dimensional (3D) objects based on computer systems have become very useful in many fields that deal with regional and/or spatial phenomena. However, there exist many tasks in a robot environment modelling and in the object-oriented simulation of manufacturing processes, where we do not need necessarily the complete representations of given objects. Proper approximations of scenes, objects and areas may be sufficient for quick decision processes in many low cost applications as machining, assembly, manipulation and so on.

It is appropriate to choose such approach to modelling and representation of 2D objects which coicides with that of 3D objects. The so-called tree representations seem to be a very good choice in such situations. Recently a new method was developed for quadtree generation in "top-down" manner where a minimal size and memory saving form quadtree exists right after processing of binary image data.

This method enables creation of majorizing or minorizing approximate representations. It was also extended to the top-down generation and description of octree representation using the silhouettes of object, which are given in the form of quadtrees.

This paper deals with approximate quadtree and octree representations of 2D and 3D objects for manufacturing tasks and presents some useful algorithms for robot grasp choice and further applications. A simple approach is presented for the generation of gripping positions on solid object dealing with its octree model. Finally, application of approximate representations to robot path-planning and to investigation of robot moveability is sketched.

APPROXIMATE REPRESENTATIONS

A quadtree representation of binary picture is a quadtree whose leaves represent square areas or quadrants of the picture and are labeled with the color of corresponding area, i.e. BLACK, WHITE or GRAY. A new memory saving method of quadtree description (Vörös, 1989) is based on the fact that the GRAY or mixed nodes of quadtree are the most relevant ones from informational point of view. Hence we can uniquely characterize a quadtree as a list or an ordered array of all the GRAY nodes' descriptions.

The record describing a GRAY node consists of four entries and for the i-th node at k-level of quadtree representing $(2^k \times 2^k)$-dimensional region it can be written as an ordered quadruple

$$P_i = (Q_{i1}, Q_{i2}, Q_{i3}, Q_{i4}) , \qquad (1)$$

where P_i denotes the parent node and Q_{im}, m=1,...,4, denote its sons at (k-1)-level. In the case that Q_{im} is a terminal node, we substitute the corresponsponding color into (1) as follows:

$$Q_{im} = \begin{cases} 1 & \text{for BLACK quadrant ,} \\ 0 & \text{for WHITE quadrant .} \end{cases} \qquad (2)$$

If Q_{im} is a nonterminal node, it is identified with a new parent node being at (k-1)-level, i.e.

$$Q_{im} = P_j \qquad \text{for GRAY quadrant ,} \qquad (3)$$

j > 1, and it means that for P_j a new quadruple of form (1) must be created in the same way as for P_i.

Now the quadtree A corresponding to a binary picture of $(2^n \times 2^n)$-dimension can be characterized as an ordered array of quadruples

$$A = \{P_1, P_2, P_3, \dots \} . \qquad (4)$$

The proposed description of quadtree is a top-down one (P_1 is the root node), but not the known depth-first one. It means that the subdivision of GRAY quadrants is performed step-wise always for all the nonterminal nodes at the same level simultaneously. Hence during the generation of quadtree we have a complete and minimum size tree up to recent level with (possibly) unexamined nonterminal nodes.

This enables to receive different approximations of given quadtree representation at any level. Inserting BLACK nodes for nonterminal ones at the desired level and cancelling all the lower level nodes, we obtain the outer approximation, majorizing the original picture. In the same way but inserting WHITE nodes we obtain the inner approximation, minorizing the original picture.

As in the case of quadtrees we introduce a new memory saving method of octree description. The record describing a GRAY node consists of eight entries and for the i-th node at k-level of octree associated with $(2^k \times 2^k \times 2^k)$-dimensional volume it can be written as an ordered eight-tuple

$$F_i=(S_{i1},S_{i2},S_{i3},\dots,S_{i8}) \ , \tag{5}$$

where F_i denotes the parent node and S_{im}, m=1,...,8, denote its sons at (k-1)-level. In the case S_{im} is a terminal node, we substitute the corresponding assignment into (5) as

$$S_{im} = \begin{cases} 1 , & \text{for FULL octant ,} \\ 0 , & \text{for VOID octant .} \end{cases} \tag{6}$$

If S_{im} is a nonterminal node, it is identified with a new parent node being at (k-1)-level, i.e.

$$S_{im} = F_j \ , \qquad \text{for GRAY octant ,} \tag{7}$$

and it means that for F_j a new eight-tuple must be created further in the same way as for F_i.

Now the octree corresponding to an object space is characterized by an ordered array of eight-tuples

$$V = \{F_1,\ F_2,\ F_3,\ \dots\ \} \ . \tag{8}$$

The eight-tuples in (8) are ordered in the same way as the quadruples of quadtree in (4).

The eight-tuple (5) is the shortest form of octree node description. However, it may be important to know the exact position and dimension of an octant. Hence we can augment the description (5) as

$$F_i=(S_{i1},S_{i2},S_{i3},\dots,S_{i8},x_i,y_i,z_i,d_i), \tag{9}$$

where x_i, y_i, z_i denote the coordinates of corresponding octant (a chosen corner of cube) and d_i is the octant size. This extended form of node description and the whole octree representation can be easily generated from its basic form (8).

The proposed description of octree is again a top-down one. It means that the subdivision of GRAY octants is performed step-wise always for all the nonterminal nodes at the same level. Hence during the generation of octree we have a complete and minimum size octree up to the recent level with (maybe) unexamined nonterminal nodes.

This enables to receive different approximations of given octree representation. Inserting FULL nodes for nonterminal ones at the desired level and cancelling all the lower level nodes, we obtain the outer approximation, majorizing the original object space. In the same way but inserting VOID nodes we obtain the inner approximation, minorizing the original object space.

SPECIAL OPERATIONS WITH OBJECTS

The quadtree and octree representations of 2D and 3D objects, respectively, given as ordered arrays (4) and (8), are appropriate for realization of basic boolean operations (union, intersection, difference and complement) with 2D/3D objects. Two further operations are described bellow.

Cross sections of solid object

The form of space decomposition in the octree modelling enables very quick realization of object cross sections by planes which are orthogonal to a coordinate axis or parallel with the three main planes (x-y), (y-z), and (z-x). In these cases the planes are determined only with one parameter, i.e. the distance of this plane from the corresponding main plane, and we can speak about section in given distance from the corresponding main plane.

The fact that the plane lies in a certain integer distance from the main plane or from the origin of coordinate system implies that only certain octants of 3D object are meeting this plane. It means that not each node of octree is needed to be investigated by searching the cross section.

As the distance plays fundamental role by the section of objects, it will be reasonable to include dimensions into the octree description. That is why we will us the extended form of octant description (9). This enables to determine the coordinates of all suboctants S_{ij} as follows:

$$x_{ij} = x_i + c(1,j).d_i \ , \tag{10}$$
$$y_{ij} = y_i + c(2,j).d_i \ , \tag{11}$$
$$z_{ij} = z_i + c(3,j).d_i \ , \tag{12}$$

where the coefficients c(k,j) are given by the following Table 1 and result from the chosen order of suboctants in the octant description (5).

k/j	1	2	3	4	5	6	7	8
1	0	1	0	1	0	1	0	1
2	0	0	1	1	0	0	1	1
3	0	0	0	0	1	1	1	1

TABLE 1

Now by means of the origin and dimensions of octant F_i we are able to determine whether this octant is in contact with the given section plane or not. We can also determine which "layer" of four suboctants is involved in the section plane, i.e. the layer which is either "near" to or "far" from the main plane. This "active" layer is of fundamental importance for the creation of resulting quadtree representation of cross section. Namely, the types of suboctants in active layer (i.e. FULL, VOID, GRAY) directly determine the types of corresponding subquadrants of cross section quadtree (i.e. BLACK, WHITE, GRAY). It is evident that suboctants being included in the near and far layers differ according to the given section plane. Keeping the chosen order of suboctants given by Table 1, the active suboctants for three mentioned types of object section and their relation to the corresponding subquadrants of cross section quadtree for the near layer are given in the Table 2 and for the far layer in the Table 3.

	subquadrants			
type	1	2	3	4
x-y	1	2	3	4
y-z	1	3	5	7
z-x	1	5	2	6

TABLE 2

subquadrants			
1	2	3	4
5	6	7	8
2	4	6	8
3	7	4	8

TABLE 3

Determination of octants, which is involved in the given section plane, is based on its distance from the main plane. This must be always less than the maximal dimension of scene and because of chosen octree definition it may be equal to zero.

Assign as h_z, h_x, and h_y the parameters of section planes, which are parallel with the (x-y), (y-z), and (z-x) planes. According to the chosen description of octant F_i we can state conditions for the octant to be involved in a section plane. An octant with coordinates $(x_i,\ y_i,\ z_i)$ is intersected by the plane, which is parallel with

1) (x-y)-plane, if

$$z_i \le h_z < z_i + 2d_i \ , \quad (13)$$

2) (y-z)-plane, if

$$x_i \le h_x < x_i + 2d_i \ , \quad (14)$$

3) (z-x)-plane, if

$$y_i \le h_y < y_i + 2d_i \ . \quad (15)$$

After we have found out that an octant is intersected by the section plane, still we have to determine the active layer of suboctants. For these types of cross section we define the following values:

$$m_z = h_z - z_i - d_i \ , \quad (16)$$
$$m_x = h_x - x_i - d_i \ , \quad (17)$$
$$m_y = h_y - y_i - d_i \ . \quad (18)$$

Now the active layer determination depends on the sign of (16)- (18). If m_z is negative, then for the section parallel with (x-y)-plane the active layer is the near one according to the data in Table 2. If m_z is zero or positive, the active layer is the far one according to the Table 3. The same is valid for the values m_x and m_y, corresponding to the sections parallel with (y-z) and (z-x) planes.

On the basis of preceding ideas an algorithm for cross section of arbitrary 3D scene with one or more solid objects was proposed. The resulting quadtree representation of section area is given in the similar manner as that of octree, i.e. as an ordered array of nonterminal nodes (4). A nonterminal node Uj at the k-level is written as

$$U_j = (K_{j1}, K_{j2}, K_{j3}, K_{j4}) \ , \quad (19)$$

where

K_{ij} = 0, if the active layer suboctant is VOID,
K_{ij} = 1, if the active layer suboctant is FULL,
K_{ij} = u, if the active layer suboctant is GRAY,

and in the last case the number "u" points again to the description of mixed subquadrant which is in the form of quadruple but at (k-1)-level.

Now let the octree representation of ($2^N x 2^N x 2^N$)-dimensional object scene be given as the ordered array (8), but the nonterminal nodes are described in extended form (9). If the section plane is given by its type and parameter, then the resulting ($2^N x 2^N$)-dimensional cross section can be represented by a quadtree as an ordered array

$$C = \{U_1, U_2, U_3, \dots\} \ . \quad (20)$$

This is generated in the top-down way beginning with the root node U_1 at the N-th level. The entries K_{1j}, j=1,2,3,4, are substituted from the entries of octree root node F_1 in accordance with the Table 2 or Table 3 after determination of the sign of corresponding value of (16)-(18). The following nodes at lower levels are treated in the same way always using the correct octants determined by preceding level nodes.

Note that the resulting quadtree may contain reducible nodes. However, these can be eliminated by a simple reduction algorithm to receive a minimum size quadtree.

<u>Solid over surface</u>

This operation is again based on the quadtree representation of considered area in 2D plane, which is given as an ordered array (4) of nonterminal nodes in the form (1). As the aim of this operation is to receive a 3D object with the height equal to that of object space, we shall consider its representation in the form of octree described by an ordered array (8) of nonterminal nodes (5). Hence the procedure can be demonstrated on (2^k x 2^k)-dimensional quadrant corresponding to a nonterminal node at k-th level of quadtree.

It is evident that the full volume will be only over BLACK subquadrants of given quadrant. If we cover a square surface corresponding to a BLACK subquadrant K_{ij} of quadrant U_i, i.e. K_{ij} = 1, with a cube of same dimensions, we receive a kind of "bottom" layer of created volume. The whole volume will be received by putting an equal cube on the preceding one. This cube will be in the "top" layer of created volume. However, these cubes represent suboctants of that octant which is created over the quadrant, containing the considered BLACK subquadrant. The space over a WHITE subquadrant will be void. A GRAY subquadrant will demand going to the lower level node for both bottom and top layers and repeating twice the covering of BLACK squares with cubes of corresponding dimensions.

Generally speaking the types of quadrants in the description of area determine the types of octants and always in two layers of the volume octree representation. This is the basis for the proposed method of creation of solid over an area placed in some of the main planes of 3D space. Keeping the fix order of subquadrants in the description of quadtree as well as the fix order of suboctants in the created octree we can easy find the relationship between the subquadrants and suboctants. This is given in the Table 4.

	suboctants							
type	1	2	3	4	5	6	7	8
x-y	1	2	3	4	1	2	3	4
y-z	1	1	2	2	3	3	4	4
z-x	1	3	1	3	2	4	2	4

<u>TABLE 4</u>

According to this table the type of given suboctant will be equal to the type of corresponding subquadrant. Because of the identity of corresponding suboctants in both layers of each octant, all pointers to lower level nodes appear in respective octant description twice. Note that this is an interesting property of proposed top-down form description of octree hierarchical structure, namely the possibility to use only one eight-tuple (5) for more identical nodes being at the same level of tree. However, this is not possible if we are using the extended form of octant description (9), because the coordinates will not be equal.

Now the generation of volume will agree with the form of above mentioned octree creation. Let the (2^N x 2^N)-dimensional area be given by the ordered array (4) of quadruples (1). Then the resulting ($2^N x 2^N x 2^N$)-dimensional object space can be written as the ordered array

$$W = \{T_1, T_2, T_3, \dots\} \ . \quad (21)$$

where

$$T_i = (R_{i1}, R_{i2}, R_{i3}, \dots, R_{i8}) \ . \quad (22)$$

We begin with the first item in (21) representing the octree root node. The entries R_{1j}, j=1,2,...,8, of T_1 will be received by substitution of corresponding items of P_1 according to the Table 4. The following nodes at the (N-1)-th level of octree will result from T_1. They will be created using the corresponding nodes P_2, P_3, ... , of (4) applying the Table 4. The resulting octree will be of minimal size.

ROBOT GRASPING

Numerous approaches (analytic, knowledge-based) have been proposed for the characterization of robot grasping and the modelling of manipulation processes (Bekey et al, 1991), (Cutkosky, 1989), (Wolter et al, 1985). However, the proposed approaches seem to be still very expensive, although they are built on simplifying assumptions and carefully structured experiments. Proper low-cost applications for manufacturing tasks as machining, assembly, handling, and fixturing parts and tools, need further simplification and approximate approaches.

Recently a low-cost method of gripping position determination was proposed using hierarchical tree representations. This is based on the assumption that in many cases it is sufficient to search contact areas on a manipulated solid object which are parallel with one of the main planes of coordinate system. It means that a gripper with two parallel jaws is considered, what is the case in many robot installations. This assumption greatly simplifies the analysis and the results may be appropriate for many low cost applications.

Assume the analyzed solid object is given by its octree representation in the form of array (8). As this representation is generally associated with the orthogonal coordinate system with origin in (0, 0, 0), we can find the real coordinates of an octant as

$$X_i = x_i + x_r , \quad (23)$$
$$Y_i = y_i + y_r , \quad (24)$$
$$Z_i = z_i + z_r , \quad (25)$$

where (x_r, y_r, z_r) are the coordinates of the octree origin in the coordinate system of robot or a reference space considered. Then the extended form of octant description will be

$$F_i=(S_{i1},S_{i2},S_{i3},\dots,S_{i8},X_i,Y_i,Z_i,d_i), \quad (26)$$

and the ordered array

$$V = \{ F_i \} , \qquad i = 1,2,3, \dots \quad (27)$$

will determine the position and the shape of analyzed solid object from the robot point of view.

From (27) we can also determine the maximal and minimal values of object coordinates. It is evident that the planes which are in this maximal and minimal distance, respectively, will contain the possible contact areas for gripping in corresponding direction. To find them we proceed as follows.

We make two orthogonal/normal sections of object V by planes which are perpendicular to the chosen direction of gripping. The first section will be in the distance "H", which is equal to the maximal dimension of object in the corresponding direction. Using the proposed algorithm we receive the quadtree representation of section surface as the following ordered array

$$C_H = \{ {}^H U_j \} , \qquad j = 1,2,3, \dots \quad (28)$$

where the nodes descriptions are in the form of (19). The second normal section will be in the distance "h", which is equal to the minimal dimension of object and the quadtree of corresponding section surface will be

$$C_h = \{ {}^h U_j \} , \qquad j = 1,2,3, \dots \quad (29)$$

The quadtrees (28) and (29) represent two binary images. They contain regions which are surely accessible for the corresponding jaws of gripper.

Then these free surfaces are subject to proper grasping criteria. For the chosen case of two parallel jaws standing opposite each other, those regions of C_H and C_h are of interest which are also opposite each other. They can be found as the intersection of binary images

$$C_c = C_H \cap C_h = \{ {}^c U_j \}, \qquad j=1,2,3,\dots \quad (30)$$

implemented by means of their quadtree representations. Their common covering C_c, if nonempty, can be further investigated according to the conditions of chosen criteria to find a feasible gripping position.

The use of two normal sections to find a gripping positions need not be sufficient for complex shape 3D objects because possible contact areas may be also on further parts of the object surface. It implies that we must analyze all the parts of accessible surfaces perpendicular to the gripping direction. To make it possible we have to introduce a system of properly defined subobjects.

A subobject V_k must differ in at least one of its limiting sizes H_k and h_k in given direction from that of subobject V_{k-1} to be able to apply the method of two sections. By creating the subobject V_1 we start from the two sections of object V given by (28) and (29). The projection of the section with the smaller area into the corresponding main plane will be used to create a solid object D_1 over its surface according to the algorithm mentioned previously. The resulting octree can be used for realization of difference operation

$$V_1 = V - D_1 , \quad (31)$$

discarding the already analyzed part of V and giving the subobject V_1, which has at least one of its limiting sizes different from that of object V.

The same process can be carried out with the subobject V_1 as with the object V. We can make two normal sections in distances H_1 and h_1, respectively, which are equal to the limiting sizes of subobject. Then we analyze the received regions and search for gripping positions. Creating of D_2 and its discarding from V_1 yields the subobject V_2.

We can proceed in the creation of further subobjects V_k by means of

$$V_k = V_{k-1} - D_k , \qquad k = 2,3, \dots \quad (32)$$

using the solids D_k generated over corresponding section surfaces while the subobject is nonempty.

The set of all the gripping positions received in the before described process fulfil the basic accessability requirement. If this set is nonempty it can be further analyzed according to proper criteria, which can be again geometric oriented, to find an optimal gripping position.

The proposed method of searching free regions on the surface of solid can work in the described way only if the intersection (30) of two sections consists of one connected region. In the case of more regions included, the analysis is a little complex and may even lead to a failure. An approach how to overcome this problem is the use of centered moments of the first order (Vörös, 1990).

ANOTHER APPLICATIONS

Planning a collision-free path is one of the fundamental requirements for a mobile robot to execute its tasks (Garcia, 1989), (Fujimura, 1989). The objective is to navigate a robot from its start position to a goal position in the presence of a given set of obstacles on a 2D plane and the principal requirement is not to collide with any of them.

The path can be represented as a sequence of specific points in 2D. Hence the path planning process requires proper geometric model for the moving area of robot. A quadtree can represent obstacles and their allocation in the workspace determining the free regions (WHITE quadrants) and forbidden regions (BLACK quadrants) for robot paths.

For the purpose of simplicity it is often considered by the path-planning that the robot is a point in the 2D plane and the obstacles are expanded by the size of the robot. While shrinking the robot to a point causes no problems, this is not the case by expanding the obstacles. When using quadtree representation we have two possibilities to cope with

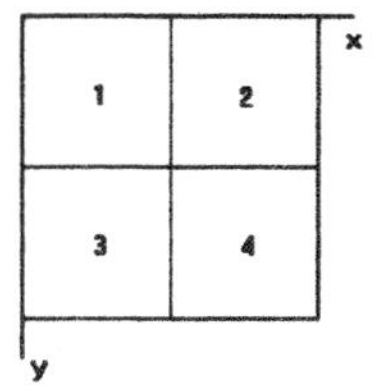

Quadtree decomposition

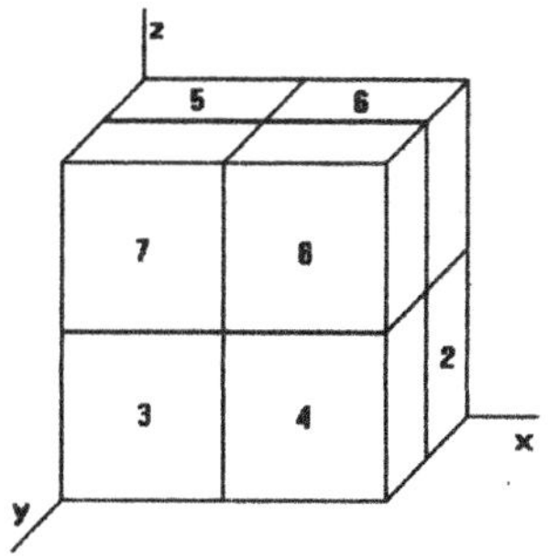

Octree decomposition

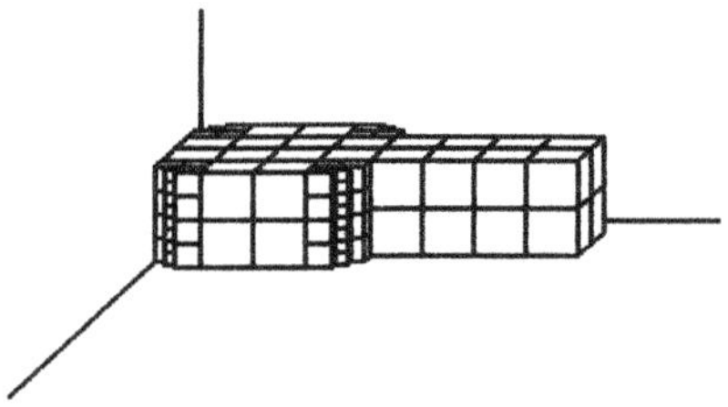

The analyzed object V

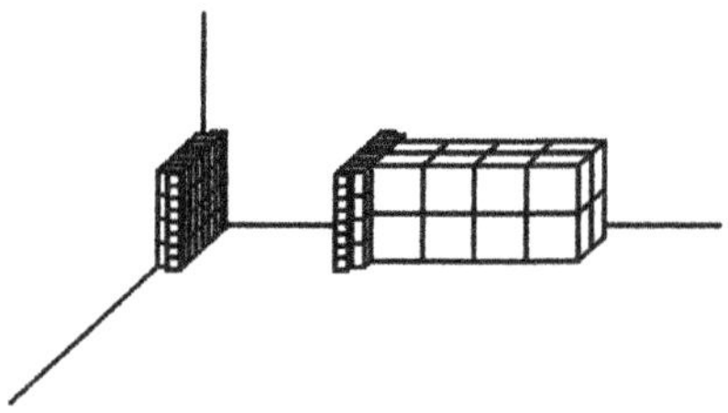

The subobject V_1 generated from V

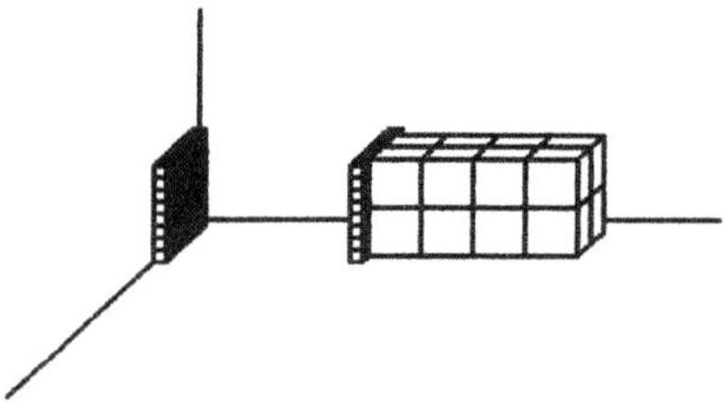

The subobject V_2 generated from V_1

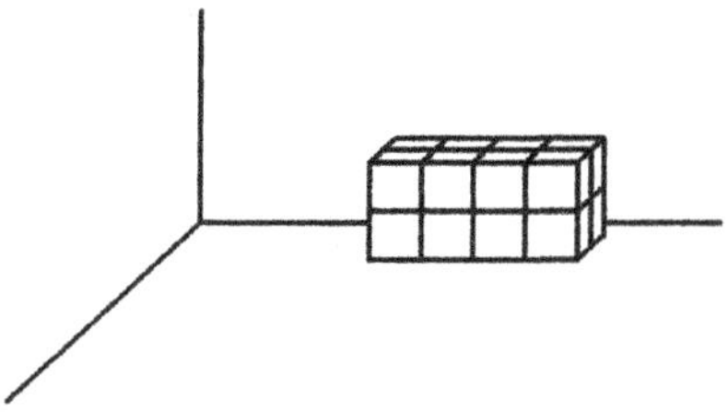

The subobject V_3 generated from V_2

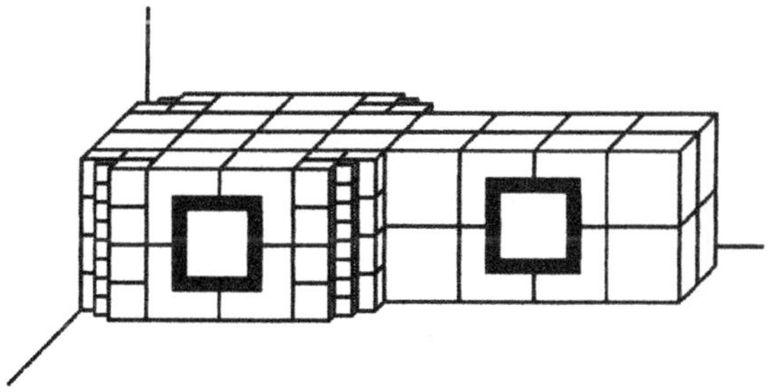

Gripping positions on the analyzed object V

this simplification. First, we can replace the robot by a point, but the expansion of obstacles is solved by the majorization of workspace model. It means that we search the path in an approximate quadtree where all the GRAY nodes of a given level are made BLACK and all the lower level nodes are cancelled. Naturally, the choice of this level depends on the size of robot.

Second, we can identify the robot with a quadrant or a square of dimensions, which corresponds to the robot size. Then we need not expand the obstacles, however, we work again with the majorized quadtree approximation of robot workspace, where the least level nodes correspond to quadrants of the same or greater dimensions than those of the square that represents the robot. Then the robot path will be a sequence of adjacent WHITE quadrants of possible different sizes.

The description of collision-free workspace of a robot moving in an encumbered environment is very useful when verifying the robot's ability to reach a set of points describing the task to be achieved. However, mapping the workspace of a robot, according to the joint limits and the various obstacles in the environment, is not sufficient to ensure the robot's global mobility. The free workspace may be connected while the end effector of robot arm is not able to move between two different points within it. This is the problem of "moveability" of a robot or its arm among obstacles.

The representations of workspace cartesian and robot configuration collision-free spaces can be given in the form of octrees. Then the moveability areas are computed by means of boolean operations applied to these spaces and following connectivity analysis.

Low cost solutions can be reached applying approximate octree representations for one or both of mentioned spaces. Octrees minorizing these spaces may significantly simplify the evaluation of required boolean operations as well as the whole analysis. The resulting moveability areas may be sufficient for the given task realization.

However, there are further aspects of using tree representations. First, we may use the previously described method of a solid creation over surface for generation of workspace or even robot configuration spaces. The generation of spatial obstacles on the base of their 2D projections will lead to approximate but memory saving representations, however, the forbidden areas of workspace may be extended. Second, besides the boolean functions we can also use the above mentioned special operations with quadtrees and octrees in the analysis of moveability. And finally, the presented technique for gripping position searching can be properly modified for the determination of moveability areas.

CONCLUSIONS

The processing of 2D and 3D complex shape objects is based on compatible forms of geometric models. Quadtree and octree representations generated in proposed top-down manner enable similar approaches in dealing with regions and solids. At the same time we can create different majorized or minorized representations for a given object, what is not a trivial problem especially in another solid modelling methods (Boundary Representation, Constructive Solid Geometry).

A new method for searching of gripping positions on objects to be grasped by robot is described. It is based on the repeated use of two normal sections applied to the object and the set of generated subobjects. The resulting possible more gripping positions can be further investigated to find optimal grasp according to given constraints. Naturally, the presented method of two normal sections is also applicable in another areas, where the shape analysis of complex 3D objects is required. It can be used for solids with concave surfaces, too.

The use of approximate representations leads in many cases to appropriate solutions of geometric oriented problems, considerably reducing the computational efforts and memory requirements and provides a powerful and sophisticated tool for low cost automation of manufacturing tasks.

REFERENCES

Bekey,G.A., T.Iberall, R.Tomovic and H.Liu (1991). Knowledge based models of human and robot grasping. Preprints 8th IFAC Symp. Identification and Sys. Parameter Estimation, Budapest, pp 1050-1055.

Cutkosky,M.R. (1989). On grasp choice, grasp models and the design of hands for manufacturing tasks. IEEE Trans. Robotics and Automation, Vol.5, No.3, pp 269-279.

Fujimura,K. and H.Samet (1989). A hierarchical strategy for path planning among moving obstacles. IEEE Trans. Robotics and Automation, Vol.5, No.1, pp 61-69.

Garcia,G., Ph.Wenger and P.Chedmail (1989). Comput ing moveability areas of a robot among obstacles using octrees. Proc. 4th Int. Conf. Advanced Robotics, Columbus, Ohio, (K.J.Waldon ed.), Springer-Verlag, pp 385-396.

Noborio,H., T.Naniwa and S.Arimoto (1990). A quad tree-based path-planning algorithm for a mobile robot. Journal of Robotic Systems, 7(4), pp 555-574.

Vörös,J. (1989). Representation of three-dimensio nal objects. Research Report III-8-2/13, Slovak Academy of Sciences, Bratislava, (in Slovak).

Vörös,J. (1990). Automatic gripping of complex shape solid objects: algorithms and programs. Research Report III-8-2/13, Slovak Academy of Sciences, Bratislava, (in Slovak).

Wolter,J.D., R.A.Volz and A.C.Woo (1985). Automatic generation of gripping positions. IEEE Trans. System, Man, and Cybernetics, Vol. SMC-15, No. 2, pp 204-213.

MINIMUM LOSSES CONTROL OF THE DC MOTOR DRIVE INCLUDING SUPPLY SYSTEM LOSSES

P. Halamski and R. Muszyński

Technical University of Poznań, Poland

Abstract. Power losses in the elements of the supply system (transformer, line, compensator) at the minimum losses control of the DC motor drive are discussed. The separate components of the losses are described. It is examplified that the supply system losses can be notable in the group drive with the 18 kW motors. They have effect on the course of the optimum motor field current as a function of operating point (torque or armature current, rotational speed). This dependence is determined for the investigated case, and then it is reduced to the form of approximating polynomial by means of the mathematical apparatus of the experiment design theory. Finally, the possibility of the inclusion of the supply system losses in the known minimum losses controllers is being suggested.

Keywords. Optimization; electric drives; extremum control; dc motor; minimum losses control.

INTRODUCTION

Electric drives of all types are heavy consumers of energy. Green and Boys (1982) inform that in the United States about 60% of the generated electricity goes to drive systems. Economic operation of drives can save much energy and reduce costs in many fields of activity.

Generally, the drive system loss minimization consists in applying a control algorithm that selects the best from among infinite number of value combinations of two or several variables, securing the required torque and speed. Minimum of power losses in the system is a choice criterion.

As the energy costs rises the drive system is more and more treated as a whole including elements of power connection to the motor. In approximate losses minimization (Kusko and Galler, 1983 and Tsuchiya and Egami, 1987) only motor losses are taken into account. In more precise efficiency optimization (Kirschen, Novotny and Lipo, 1985 and Abgrall, 1986) converters are included. This approach also is not correct because some losses related to operation of the drive system arise outside it, in the supply system. It refers especially to the thyristor drives which apart from active load can absorb reactive power of the power network and generate current harmonics to it. The reactive power and harmonics produce additional losses. The losses can be imited in the elements of the power supply system and in electric devices connected to the system. If the drive is equipped with reactive power compensation and/or harmonic filters then the losses related to reactive power and harmonic arise mostly in the compensation devices. It appears that in some cases the losses on the supply system side can be relatively large. The aim of the paper is to demonstrate influence of these losses on the optimum drive control.

THE TOTAL TREATMENT OF THE DC MOTOR DRIVE LOSSES

The system chosen for the analysis is presented in Fig.1 (Halamski, 1991). It is assumed that 10 identical drives are supplied through the common line from the same transformer and they have the common compensator (harmonic filters and/or adjustable condenser bank). Each converter has separate commutation inductors on the AC side.

The losses in all devices of the system are included in the analysis. Only field converter losses are neglected. For the common elements, the losses are calculated due to one drive operation. The appendix contains particular components of losses. For the analysis, it is suitable to express all loss components using the operating point coordinates (torque T or armature current I_a and rotational speed Ω).

Motor

A motor loss model contains armature circuit windings loss (P_1), field circuit loss (P_2) armature iron loss (P_3), brush contact and brush friction losses (P_4), friction and windage losses (P_5) and stray loss (P_6). The expression P_3 (see appendix) includes together the armature core and teeth, and P_6-all the different small losses which are usually neglected as the separate ones, but together they have influence on the efficiency of the machine. Since many of construction data of the analysed motor were known, it was possible to determine the coefficients for all loss formulas P_1 to P_6.

Armature circuit converter and inductors

A converter loss model contains conduction loss in the valves (P_7), reverse-blocking mode loss (P_8), snubber circuit loss (P_9) and converter controller loss (P_{10}). It is assumed that the losses in the smooth (P_{11}) and commutation (P_{12}) inductors are dependent on the square of the rms value of their currents (see appendix).

The supply system

In order to calculate the supply line loss due to one drive operation, the line is replaced with ten equivalent lines. Each of them feeds its drive, and it has resistance ten times greater than a real line supplying a group of ten drives. The transformer and filter resistances are treated in the same way. According to this, the constant losses of common transformer are also divided into ten parts. This approach allows to count the losses due to one drive operation in the conditions when the whole drive group is operating. It is assumed that the compensator gives the whole reactive power for each converter, and it filters fifth and seventh harmonics so that only 33% of these harmonics (plus the other harmonics and active component of fundamental frequency) are flowing into the supply line. In the supply system there are the following loss components (see appendix): compensator losses (P_{13}), loss in the line (P_{14}) connecting transformer with converter and transformer losses (P_{15}). The compensator losses have two components; that is the capacitor loss which is proportional to the reactive power and loss in the filter inductors which is proportional to the square of the rms value of filter currents. Since there is also located the reactive power in the filters, through each of filter branches is flowing the current of filtered harmonic and fundamental frequency.

MINIMUM LOSSES CONTROL

In the analysed case, the optimization problem is one-dimensional and it consists on a search for such a value of a field current I_f for which there are the minimum loses. For each operating point with given

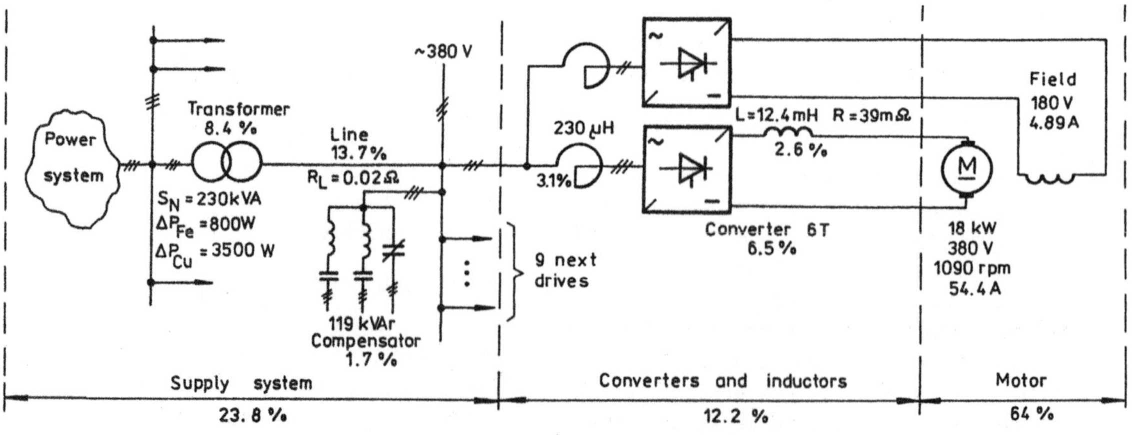

Fig.1. The diagram of analysed system with losses distribution

load torque T and speed Ω, the algorithm starts with the rated field current and reduces it in succesive steps so long as the minimum losses are reached. The following quantities are calculated in each step on the basis of T, Ω and I_f:

- motor flux using the magnetization curve in the form of approximation

$$\phi = (0{,}603+8{,}56I_f-1{,}43I_f^2+0{,}0871I_f^3)\ 10^{-3} \quad [V \cdot s] \qquad (1)$$

- electro-magnetic torque (torque T plus torque of mechanical and iron losses),
- average value of the armature current I_a,
- voltage and the delay angle α of the converter,
- the form factor k_i of the armature current,
- all the losses from P_1 to P_{15} (see appendix).

In this analysis, only the continuous current range is taken into consideration. In order to determine the influence of the supply system on the losses and optimal control the following three cases are compared

a - drive system without the supply system and commutation inductors,

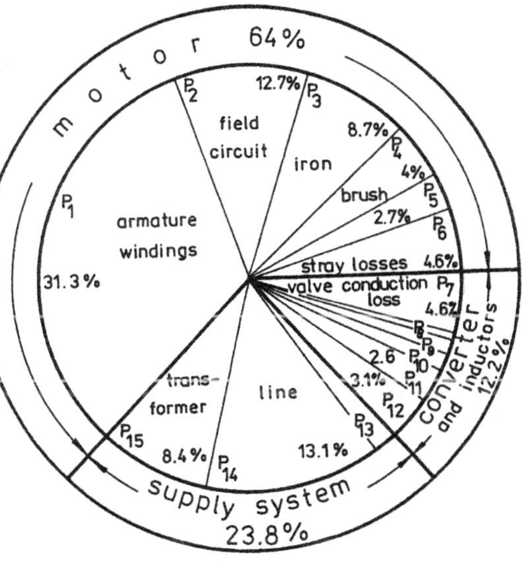

P_5 - friction and windage losses
P_8 - reverse-blocking mode loss 0,1%
P_9 - snubber circuit loss 0,7%
P_{10} - controller loss 1,1%
P_{11} - smooth inductor loss
P_{12} - commutation inductor losses
P_{13} - compensator losses 1,7%

Fig.2. The loss distribution for case b and minimum loss control at rated speed and load

b - whole system with the line of 80 m and 0,02 Ω (shorter line),

c - whole system with the line of 200 m and 0,05 Ω (longer line).

Fig.2 shows the loss distribution for case b, at the optimum control. The percentage of the loss values for this case is given in Fig.1. The losses in the supply system (transformer, line and compensator) constitute 23,8% of all the losses. This part increases to 36,6% in the case with a longer line (case c). The optimum control reduces the losses mostly in a low load range with respect to the constant field current operation (Fig.3). The economy decreases when the torque increases, and practically, there is none in the rated operating point.

In Fig.4 and 5, there is shown the course of the optimum field current as a function of operating point parameters. In these dependencies, instead of torque T the armature current I_a is used with regard for ease of the current measurement. Fig.4 shows that the field current I_f varies in the notable range approximatelly according to a square root function when current I_a increases.The comparison of the cases in Fig.4 shows that if there are more devices on the supply side, the optimum field current is higher. It is comprehensible as the losses in the supply circuit depend mainly on the armature current. For a given torque the current and losses can be smaller when the field current is higher. Fig.5 shows that the influence of speed on the optimum field current is not great. With the increase of speed, the field current should insignificantly decrease.

In order to realize optimum control, the dependence $I_f=f(I_a, \Omega)$ should be obtained for the whole operation range of the drive. The function can be given in the form of EPROM table or in the form of analytical approximation. The mathematical apparatus of experimental design theory (Box and Hunter,1957) allowed to obtain, on the basis of numerical data for the cause with shorter line, the following polynomial

$$I_f = -0{,}6585+0{,}6694\sqrt{I_a} -0{,}009549\,\Omega + +0{,}001164\sqrt{I_a}\,\Omega\,. \qquad (2)$$

The polynomial approximates the optimum field current in an operation area which is determined by the variation range of current

$I_a = (0{,}28 \div 1)\ I_{aN} = (15 \div 54{,}4)\ [A]$
and speed

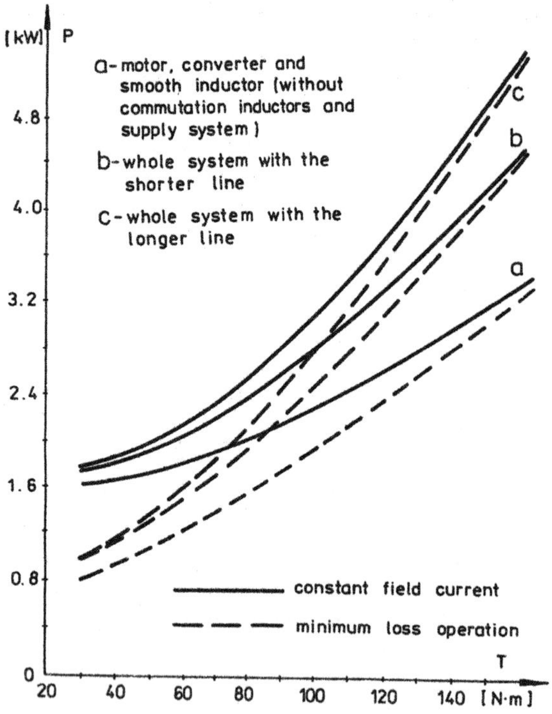

Fig.3. Power loss as a function of load torque at the rated speed

$\Omega = (0,18 \div 1)\,\Omega_N = (20,95 \div 114,2)\ [\frac{rad}{s}]$.

The rms error of the approximation (2) for this whole area constitutes 1,3% of the means value of the field current, which is 3,01 A. The controller realization according to the equation (2) needs two non-linear operations (extraction of root and multiplication). If the interaction of current and speed is omitted during the creation of the approximation, the field current is

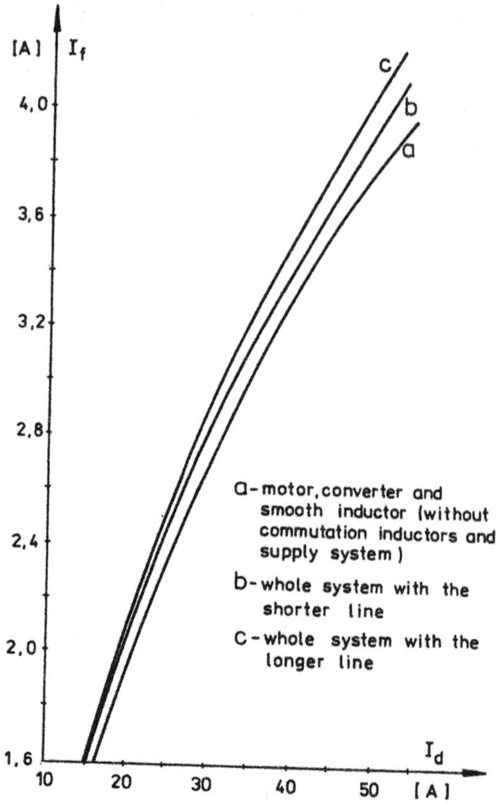

Fig.4. The optimum field current as a function of armature current at the rated rotational speed

$$I_f = -1,111 + 0,7481\sqrt{I_a} - 0,003003\,\Omega\ , \qquad (3)$$

and the rms error is 2,3%. Generally, the speed signal is of low significance, and the controller can be realized without it. Then, the equation takes the following form

$$I_f = -1,304 + 0,7481\sqrt{I_a} \qquad (4)$$

and the error is 4%.

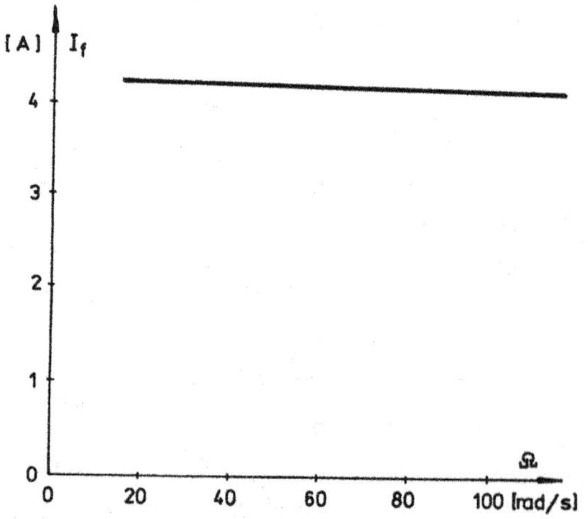

Fig.5. The optimum field current as a function of rotational speed at the rated armature current for case b

Obviously it is possible to include the supply system losses in the known optimizing controllers (Kusko and Galler, 1983). In the optimizing controller with a loss model (open-loop controller), it will do if the loss components of the supply system will be introduced to the model. The formulas (2), (3) or (4) are the solutions of the minimum loss equations and they can be a base for operating such a controller. In the case of testing controller (closed-loop controller) which searches for the minimum of input power, it is sufficient to measure the power in the point as close to a supply source as possible. If some of the supply circuit devices (for instance transformer) can not be included in the power measurement, the losses in these devices should be calculated analytically and added to the measurement result. Only the sum of these signals is a minimized quantity for the testing controller.

CONCLUSION

The losses in the devices of the supply system (transformer, line, compensator) can be the notable part of the total loss in the drive system and they should be included at maximum efficiency control. In thyristor DC shunt motor drives, the optimum field current is higher when the supply system losses are taken into account than without them. The inclusion of these losses in the known optimizing controllers does not require additional costs. For this purpose, it can be used existing input power measurement or current and speed signals which are available in the control system.

REFERENCE

Abgrall, J.P. (1986). High power modules and optimization of efficiency in inverter drives for induction machines. Proc.International Conf. on Evolution and Modern Aspects of Induction Machines, Turin, pp.614-618.

Box, G.E.P., Hunter J.S (1957). Multifactor experimental designs for exploring response surfaces. Annals of Mathematical Statistics, 28, No 1, pp.195-208

Green, R.M, Boys J.T. (1982). Inverter AC-drives efficiency. IEE Proc, vol.129, pt. B, pp.75-81.

Halamski, P. (1991). Minimum losses control of the thyristor DC drive. (Polish), diploma work No IEp 3/90-1483, Technical University of Poznań,

Kirschen, D.S., Novotny D.W., Lipo T.A. (1985). On-line efficiency optimization of a variable frequency induction motor drive. IEEE Trans.on Industry Appl. Vol. IA-21, No 4, pp.610-616.

Kusko, A., Galler A. (1983). Control means for minimization of losses in AC and DC motor drives. IEEE Trans.of Industry Appl., Vol. IA-19, No 4, pp.561-570.

Tsuchiya, T.T., Egami T. (1987). Efficiency optimized speed control - application of optimal control and adaptive control. 10th World Congress on Automatic Control, IFAC, Munich, pp.343-348.

APPENDIX

Loss components

The armature circuit copper losses are

$$P_1=(R_a+R_{cm}+R_{cp})(k_i\ I_a)^2$$

where

$R_a=0,23\ \Omega$ is an armature winding resistance,

$R_{cm}+R_{cp}=0,248\ \Omega$ is a sum of commutating and compensating winding resistances,

$k_i=\frac{I}{I_a}\sqrt{1+\left(\frac{U_{do}}{2\pi f\ I_a L_E}\ D\right)^2}$ is a form factor of the armature current,

I [A] is an effective (rms) armature current,

I_a [A] is an average armature current,

$U_{do}=\frac{6}{\pi}\sqrt{2}\ U \sin\frac{\pi}{6} = 512$ V is an average DC voltage of converter at $\alpha = 0^o$,

U=380 V is an effective line-to-line voltage on the AC side of converter,

f=50 Hz is a network frequency,

$L_E=L_M+L_s+2L_c=21,2$ mH is an equivalent inductance,

$L_M=8,3$ mH is a motor inductance,

$L_s=12,4$ mH is an inductance of smoothing reactor

$L_c=0,23$ mH is an inductance of commutation reactors,

$D=0,048-0,0203\ \cos\alpha+0,0291\ \cos^2\alpha + -0,0275\ \cos^3\alpha - 0,0194\ \cos^4\alpha$ is a damping factor which depends on delay angle α and number of pulses (the approximation obliges for six thyristor bridge).

The shunt field copper loss is

$$P_2 = R_f I_f^2$$

where

$R_f=33,5\ \Omega$ is a shunt field winding resistance,

I_f [A] is a field current.

The armature iron loss is

$$P_3 = (H\,\Omega + E\,\Omega^2)\ \phi^2$$

where

Ω [rad/s] is rotational speed,

H=5690 W/V·rad is a hysteresis loss coefficient,

E=44,7 Ws/Vrad^2 is a eddy-current loss coefficient,

ϕ [V·s] is a motor flux according to Eq.1.

The brush contact and brush friction losses are

$$P_4 = 2U_B I_a + F_B\Omega$$

where

$U_B=1$ V is a voltage drop at the brush contact,

I_a and Ω are explained above,

F_B=0,631 Ws/rad is a brush friction coefficient.

The mechanical losses are

$P_5 = F\Omega + V\Omega^3$, where

F=1,01 Ws/rad is a bearning friction and windage losses coefficient,

$V=3,75\cdot10^{-6}$ Ws^3/rad^3 is a ventilation loss coefficient,

Ω is explained above.

The stray loss is

$P_6 = S(k_i\ I_a)^2$, where

S=0,07 W/A^2 is a stray loss coefficient,

k_i and I_a are explained above.

The conduction loss in the valves is

$P_7 = 2U_{TO}I_a + 2r_T(k_iI_a)^2$, where

U_{TO}=0,65 V is a reverse connected voltage in the thyristor conduction characteristic model,

I_a and k_i are explained above,

r_T=0,0227 Ω is a dynamical resistance of the thyristor in conducting direction.

The reverse-blocking mode loss is

$P_8=4Q_1U_{RM}I_R+4Q_2U_{RM}/R_R=5,2$ W , where

Q_1=0,478 and Q_2=0,403 V are coefficients which depend on converter configuration,

U_{RM}=539 V is a peak value of reverse voltage,

I_R=0,005 A is a reverse current of thyristor,

R_R=56 kΩ is a dynamical reverse resistance.

The snubber circuit loss is

$P_9 = k_{RC}\ \sin^2\alpha$, where

k_{RC}=68,4 W is a coefficient which depends on configuration and parameters of converter and supplying network and on parameters of valves and snubber circuit,

α is a delay angle.

The converter controller loss is

P_{10} = 50 W

The smoothing reactor loos is

$P_{11} = (k_i\ I_a)^2R_S$, where

k_i and I_a are explained above,

R_S=0,039 Ω is a smoothing reactor resistance.

The commutation reactor losses are

$P_{12} = 3I_{AC}^2\ R_C$, where

$I_{AC} = \sqrt{\frac{2}{3}}\ I_a$ is an effective current on AC side of converter,

I_a is explained above,

R_C=0,0241 Ω is a commutation reactor resistance.

The losses in the compensator are

$P_{13} = P_C + P_I$, where

$P_C=l_CU_{do}I_a\sin\alpha$ is a power loss in the capacitors,

l_C=3 W/kVAr is a specific power loss in the compensating capacitors,

$P_I=P_{I5}+P_{I7}$ is a power loss in the filter inductors,

$P_{I5}=3(10R_5)I_S^2$ and $P_{I7}=3(10R_7)I_7^2$ are losses in the fifth and seventh harmonic filters respectively,

R_5=0,082 Ω and R_7=0,045 Ω are resistances of filter inductors,

$I_5=\sqrt{I_{55}^2+I_{51}^2}$ and $I_7=\sqrt{I_{77}^2+I_{71}^2}$ are effective currents of fifth and seventh harmonic filter respectively,

$I_{55}=\frac{2}{3}\frac{I_1}{5}$ is a fifth harmonic current of fifth harmonic filter,

$I_{77}=\frac{2}{3}\frac{I_1}{7}$ is a seventh harmonic current of seventh harmonic filter,

$I_1=0,955\ I_{AC}$ is a converter current of fundamental frequency,

$I_{51}=\frac{w_5Q}{\sqrt{3}\,U}$ and $I_{71}=\frac{w_7Q}{\sqrt{3}\,U}$ are currents of fundamental frequency in the fifth and seventh harmonic filter respectively,

w_5=0,624 and w_7=0,376 are coefficients of reactive power distribution for fifth and seventh harmonic filter,

$Q=U_{do}I_a\sin\alpha$ is a reactive power of converter,

$U_{do}, I_a, \alpha, I_{AC}$ and U are explained above.

The line loss is

$P_{14} = 3(10R_L)^2I_L^2$, where

R_L=0,02 Ω or 0,05 Ω is resistance for line of 80 m or 200 m respectively,

$$I_L=\sqrt{(I_1\cos\alpha)^2+(\frac{I_1}{3\cdot5})^2+(\frac{I_1}{3\cdot7})^2+(\frac{I_1}{11})^2+(\frac{I_1}{13})^2+\ldots+(\frac{I_1}{25})^2}$$

is an effective line current,

I_1 and α are explained above.

The transformer losses are

$P_{15} = \frac{P_{Fe}}{10} + \frac{P_{Cu}}{10}\ (\frac{10\ I_L}{I_{2N}})^2$, where

P_{Fe}=800 W is a constant (iron) loss in transformer,

P_{Cu}=3500 W is a rated copper loss in transformer,

I_{2N} =349 A is a rated secondary current of transformer,

I_L is explained above.

COMPARATIVE STUDY OF DEDICATED MOTION-CONTROL PROCESSORS

E. Bertran*, A. Guimerá** and G. Montoro*

*Department of Signal Theory and Communications, Universitat Politècnica de Catalunya, P.O. Box 30.002, 08080 Barcelona, Spain

**Oteman, S.A., P.O. Box 205, 0877 Igualada, Spain

Abstract:

A comparison of dedicated motion-control microcomputers is presented. Following an overview of different hardware alternatives and devices to D.C. motor control, two usual motion-control processors are compared in terms of manufacturer's characteristics and development experience. In addition, the behaviour of both processors in an specific application, emphasizing dynamical aspects and control law programming facility, is shown.

Keywords: *Microprocessor Based Systems. Motion Control. Control Algorithms. D.C. Motors.*

1.- Introduction:

Motion digital control of d.c. motors is a usual problem in Control Engineering, for which there exist many low cost solutions. Early ones were based on the development and application of general purpose microcomputers. Later integrated circuits dedicated to resolve some sub-functions, as was the case of universal pulse processors, or pulse width modulators, appeared. In other level, same circuits intended to design a D.C. motor closed loop control, as motor drivers, specialized converters and speed transducers have been widely used. Performances and design facilities of this kind of integrated circuits are, generally speaking, very reduced.

In last years, in parallel with microcontrollers capable to acquire analog signals, or pulse counting, and to generate outputs oriented to facilitate the excitation of the motor, dedicated microprocessors with capacity to ease the programming of the control law, to generate position or velocity profiles, to change set-points, to interconnect incremental shaft encoders and to communicate with another processor operating as a host, have appeared in the market. In spite there are few microprocessors dedicated to implement the overall motion-control loop, they are different in conception, facility to use them and performances.

In this paper, a comparison of two dedicated microprocessors is presented. One of them is based on Z-domain design, allowing to adjust a zero, a pole, the regulator gain and the sampling rate. The other is based on a digital implementation of a PID regulator, being transparent to the user this digital conception (it is no required to design in the Z-plane). In this case, sampling period can be adjusted independently for data acquisition and for derivative (noised) stages implementation.

Though different, both processors are capable to read optical encoders, to generate motion profiles and to interconnect with a host computer in order to perform a high level command set. They are very useful and represent an economic option to develop CAD - CAM systems where there are more than one axis to control the motion (as cutting machines or high velocity plotters), or in one axis applications, as in rolling machines. The performance versus cost ratio is very attractive in both processors.

After manufacturer's (Hewlett-Packard and National Semiconductor) characteristics comparison, design difficulty and additional hardware needs are compared. Moreover, the behaviour of both processors, experimented in the same process (plotting machine), is presented and discussed.

2.- Partial solutions.

The implementation of different subsytems in an analog or digital control loop has a wide range of possibilities. Apart from more classical elements (drivers, transducers, ...), some partial tasks specialized IC's have appeared in the market several years ago. It is over the aim of this paper to review this kind of IC's, but, as an example, incremental encoders interfaces (i.e., THCT 2000) or pulse processors (i.e, HD63140) reduce the interface cost from 40 $ for classical TTL or CMOS devices to 30 $ (approximately).

In motor control (speed, current or torque), different manufacturers offer a lot of solutions to drive the motor or, even, to implement a single analog loop (usually with current or velocity feedback). This is the case of IC's type L292, TL377, UAA4003, etc. Some of these IC's can be included in a digital control loop (DAC-type) by using a digital to analog converter.

The control signal is, normally, a pulse width modulated (PWM) one, so some PWM regulators initially oriented to switched-mode power supply design are useful in motor control. PWM signals reduce power amplifier costs, but, if the possibilities of the chosen IC are not enough, it is difficult to include internal control loops.

3.- DDC global solutions.

Except for mechanical elements, all the blocks of a direct digital control loop can be supported by commercial one-board or one-chip solutions (figure 1).

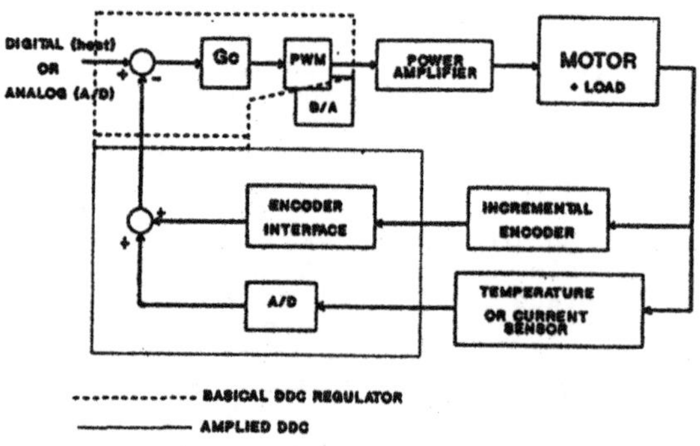

Figure 1.- Direct Digital Control.

The more classical way uses a microcomputer-based (8 or 16 bits) single board, with restrictions in computing time (sampling period), transducers interface, input/output capabilities and host communications. These restrictions are solved by more complex circuitry, with the consequent increase in development time, cost, board surface and fault problems.

A low-cost alternative is the use of microcontrollers (or single-chip microcomputers). With list prices, component cost could be reduced from 60 $ to 20 $ if the tasks supported by a single-board microcomputer are resolved with a microcontroller.

Microcontrollers offer internal RAM (around 128 or 256 KBytes), ROM (4 - 8 KBytes) or external EPROM, input / output lines (24 - 32), timer and serial communications facilities. Moreover, some microcontrollers offer hardware interrupts expansion, PWM output, watchdog system, full-duplex communications (in order to facilitate an RS-232 connection to a PC), and A/D converters.

Recently some 16-bits microcontrollers have appeared, so precision and resolution are not problem. Main restrictions are computing time and memory expansion capacity. Memory problems are not present in a DDC control loop, anyhow this problem corresponds to set-point calculation level that can be supported by another microprocessor working as a host.

If system bandwidth is high (around 50 - 100 KHz), a solution for computing time is the use of digital signal processors (DSP). If plant bandwidth is lower but high speed and precision are required, as is the case of a high speed plotter with a maximum speed of 2 $m.s^{-1}$ and a precision under 1 mm, DSP products are also a solution. Third case of application of DSP are advanced control laws, where state observation, parameter identification, optimal filtering or adaptive strategies requires significant computational efforts.

At present, the cost of DSP-based designs advises to reserve this solution to the implementation of complex control algorithms. Typical cost of a system board with memory expansion, A/D and D/A converters, interfaces, and serial communications could be between 600 and 1000 $. On the other hand, a classical DSP of the first generation, with fixed-point, 16 bits, and a cycle time of 160 ns (new DSPs can reduce this time to 50 ns), can perform a PID controller in a time lower than 3 μs, and adaptive controller (indirect adaptation, with 6 parameter estimation) in a time lower than 50 μs. PID controller, implemented with a new generation DSP, is carried out in 0,5 μs. Floating-point DSP are devoted to signal processing applications; in control applications fixed-point is recommended.

A solution between the use of microcontrollers and the use of DSPs are dedicated motion-control processors. These products can implement all the digital control loop (encoder interfaces, control algorithm, profile computation, emergency survey, output signal to power amplifiers generation and host communications). The global structure is supervisory-type, based on a host computer, (from a single 8-bit microcomputer to a PC) devoted to high level algorithms, and one or same dedicated processors devoted to motion control from set-points sent by the host (figure 2).

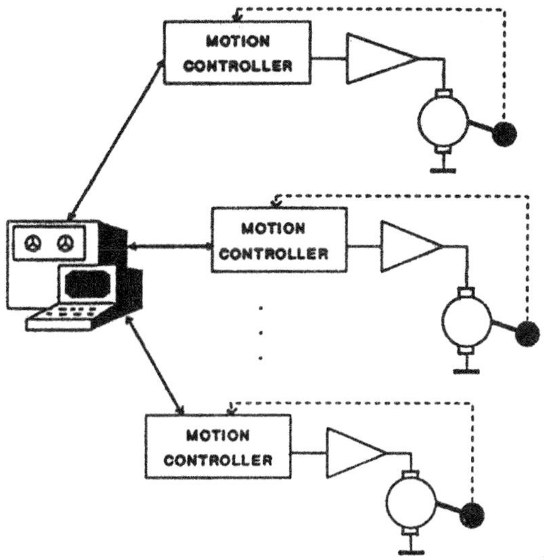

Figure 2.- Supervisory (DAC) control.

Sampling period could be around 60 µs, so they are useful in usual d.c. motor applications. This order of sampling period is 20 times larger than the period obtained using a DSP.

Compared with 16-bit microcontrollers, dedicated motion-control processors are faster in an order that varies between 2 and 40. On the other hand, microcontrollers have an open control algorithm (user-made) that can be modified according to the special characteristics of the application.

4.- Motion-control processors.

Two similar price (around 50 $) motion-control processors are considered: HCTL 1000 (recently an alternative, the HCTL 1100, has appeared), from Hewlett-Packard and LM 628 / LM 629 from National Semiconductor. Although both processors are different in host interface, resolution, motion control possibilities, interrupts and outputs, could be described by an structure similar to one shown in figure 3.

From manufacturer's information and development experience, the comparative table shown at the end of the paper has been obtained.

General hardware comparison shows lower power dissipation and clock frequency in the HCTL 1000. On the the other hand, LM 628/629 can have higher clock frequency (6 or 8 MHz according to IC type), with the obvious increased difficulty in

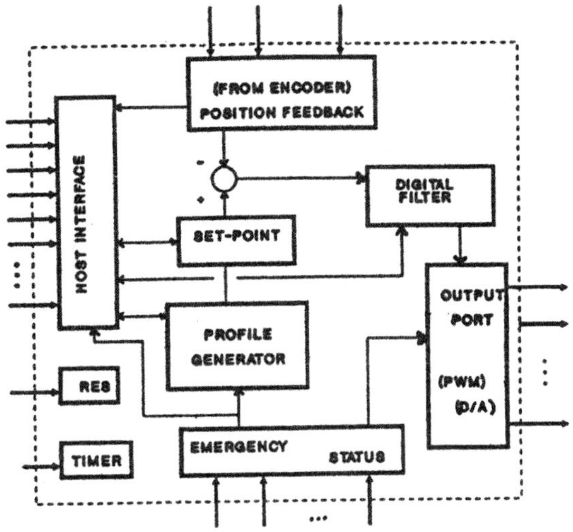

Figure 3.- General structure.

printed circuit design. Curiously, sampling period can be faster in HCTL 1000.

The multiplexed bus of HCTL 1000 implies scarcely additional hardware as compared to the asincronous bus of LM 628/629. But LM 628/629 could require also additional hardware if microprocessor data bus is read or written immediately after R/W signal changes: LM 628/629 doesn't considers valid the bus data immediately after its WR signal goes down. This problem can be solved with an one-shoot circuit.

In HCTL 1000 commands can be changed and parameters can be read during motion; in LM 628/629 it is only possible if a "busy bit" is correct. Reset, interrupts, filter parameters variation and data reporting are executable during motion, but any trajectory control commands may be changed during motion.

HCTL 1000 filter is a single zero-pole with gain. The pole must be in the left side of unit circle (Z-transform), and the zero in the right. LM 628/629 filter is a digital version of classical PID controller, with the possibility to define a different sampling period in derivative stages in order to avoid noise problems. If a constant signal is corrupted by noise, derivative stages can magnify the noise if T is low in the discrete derivative relation:

$$S \rightarrow \frac{f[kT] - f[(k-1)T]}{T}$$

As response to a reset input both processors outputs 0 volts to power amplifier. LM 628/629 clears its filter parameters, so motor turns free (open loop), whereas HCTL 1000 forces a controller pole to the left and a zero to right. This residue distribution is a good choice: loop remains closed with a stabilizing distribution (figure 4).

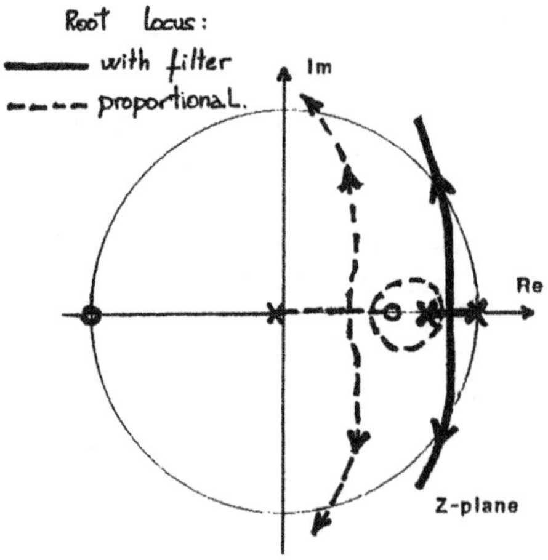

Figure 4.- Effect of controller pole and zero after reset.

Precision in control commands, measurements and filter parameters (studied later) are better in LM 628 / 629, but HCTL 1000 offers more kinds of profiles (see table).

5.- Dynamics comparison. Effects of filters structure and precision.

Figure 5 shows a general structure for a dc motor control using incremental encoder feedback. The factor 4N corresponds to encoder resolution (quadrature), Ka is the voltage source amplifier gain, Kmv the relation speed / voltage in motor, and τe and τm the electrical and mechanical time constants. Filter algorithm is modelled as Gc(z) plus an one step delay in order to consider that sampling process is carried out at the beginning of the sampling period and D/A signal is sent at the end of this period.

This structure has been used in a high speed plotter design, with values near to:

$4N / 2\pi = 318,3$ counts / rad
$K_{mv} = 26$ rad.$s^{-1}.V^{-1}$
$K_a = 0.2$ V / count.
$\tau_m = 50$ msec, 66 msec, 100 msec.
τ_e = negligible.

In order to have a speed of 2 m/s with a resolution close 1 mm, a sampling period of 330 µs has been selected. Lower sampling periods derivates to acoustical noise problems (low PWA frequencies). With these parameters, the following open loop transfer functions have been obtained (using invariant Z-transform):

(τ_m = 100 msec)

$$Gc \frac{1655,16}{z} \; \frac{5,43\ E\text{-}6\ z + 5,43\ E\text{-}6}{100\ (z - 1)\ (z - 0,9967)}$$

and, with τ_m = 50 msec,

$$Gc \frac{1655,16}{z} \; \frac{2,174\ E\text{-}5\ z + 2,168\ E\text{-}5}{400\ (z - 1)\ (z - 0,9934)}$$

Rational variation of sampling rate does not vary significantly resultant zero and poles. If a sampling period of T = 1 msec is selected, then the open loop transfer function becomes (for τ_m = 50 msec):

$$Gc \frac{1655,16}{z} \; \frac{1,987\ E\text{-}4\ z + 1,973\ E\text{-}4}{400\ (z - 1)\ (z - 0,9802)}$$

From these transfer functions, we observe that a single pole is at the Z-plane point z = 0, and single zero is near the point z = -1. Other poles are at frequency origin (z = 1) and close to this origin. With a proportional control law, system becomes easily unstable.

Using the HCTL 1000 in Gc(z) implementation,

$$Gc(z) = \frac{K}{4} \; \frac{z - A/256}{z + B/256}$$

the desired solution is to place the regulator zero over a plant pole (figure 6). This solution improves stability and time response (faster).

Since HCTL 1000 has 8 bit registers, possible zero positions are A = 254 and A = 255. In the first case, 254/256 = 0,9922 and in the second case 255/256 = 0,9961. Comparing this ratios with transfer functions obtained above, it is not possible to have a perfect cancellation. So, result could be that shown in figure 7a or the one shown in figure 7b. In the case of figure 7a an undesired slow response mode is present, whereas in the case of figure 7b second order dynamics can apperar.

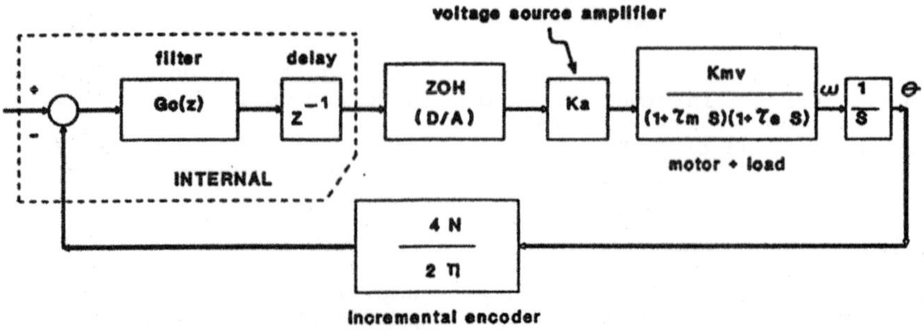

Figure 5: *DC motor control loop.*

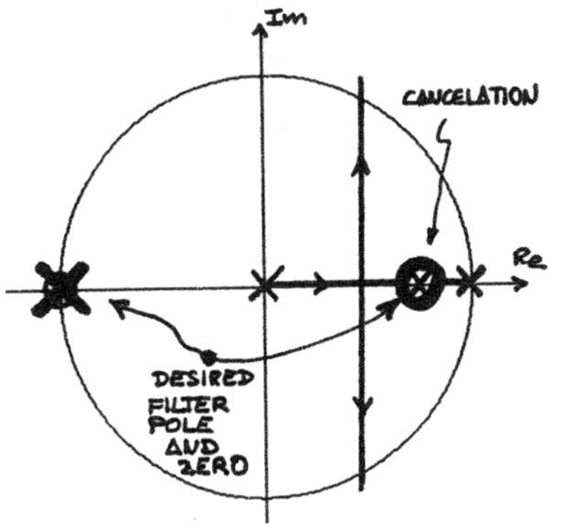

Figure 6: *Desired pole-zero location*

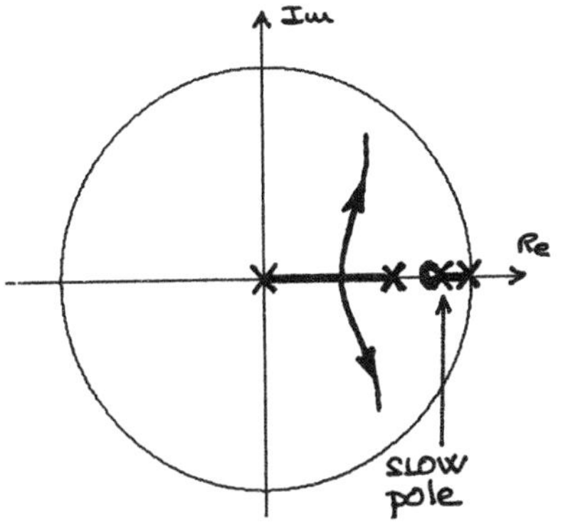

(a)

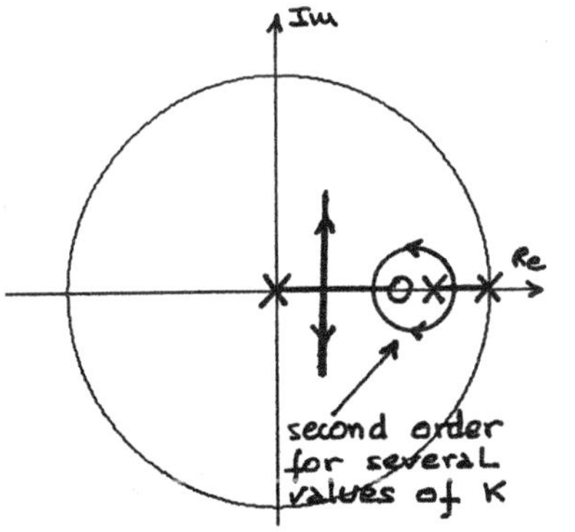

(b)

Figure 7: *Imperfect cancellations.*

These problems have been found in a high speed plotter laboratory development. Figure 8 shows a pattern drawn with filters of figures 7a and 7b.

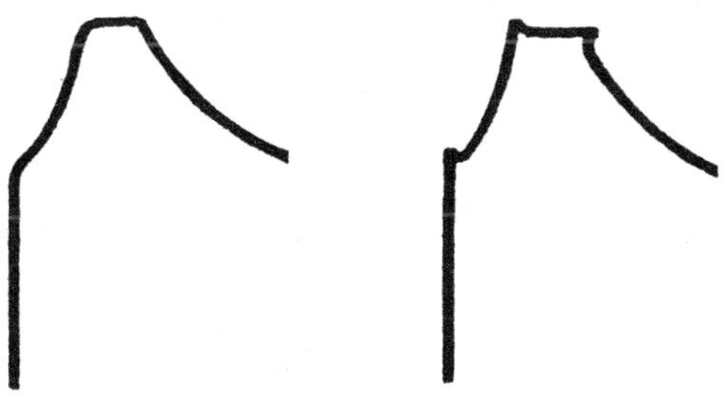

Figure 8: *Pattern drawing.*

The LM 628 filter algorithm can be expressed in Z-domain as:

$$\frac{Kp\ (z-1)\ z + Ki\ Ts\ z + (Kd\ /\ T'_d)(z-1)^2}{z\ (z-1)}$$

Ts being the sampling time and T'_d the derivative sampling time. The integral action is not possible due to a double pole in z = 1 that unstabilizes the system, but a PD regulator is enough. In this case filter parameters are 16 bits long. Difference between successive pole positions is 65534 / 65536 = 0,99996 and 65535 / 65536 = 0,99998. It is possible to adjust until the 5th decimal, so, cancellation is possible.

Another problem is acceleration error. During profiles that accelerates the system until obtaining a desired velocity, the system must have a double pole in z=1 in order to have precision in acceleration ramps. Open loop transfer function (figure 5) in acceleration ramp is:

$$Gc(z)\ .\ \frac{K_t\ \ (z-1)}{(\ z\ -\ e^{-pT})\ z}$$

p being the motor pole and T the sampling period.

The HCTL 1000 filter structure can't locate a pole over the zero placed on z=1. So acceleration error is present in the system. This problem is reduced if PID filter supported by LM 628 is used; in this case it is possible to place a pole (not two) in z = 1. In practice this problem is not significant in both processors if normal applications are developed: profiles generators of both ICs compensates this error if position or velocity modes are selected. However, if precise acceleration profile is desired, it is important to consider these filter aspects. A solution is the use of internal control loops, only possible if motion processor output is D/A type; PWM outputs difficults the behaviour modification of the set power amplifier + motor.

6.- References:

- *"LM 628 / LM 629 Precision Motor Controller"*. National Semiconductor, 1989.
- *"Design of the HCTL - 1000's Digital Filter Parameters by the Combination Method"*. Hewlett-Packard, Application Note 1032
- Jacob, Tal. *"Motion Control by Microprocessors"*. Galil Motion Control, 1984.
- Kuo, B.C. *"Digital Control Systems"* Hold-Saunders Int. 1981.
- *"General Purpose motion Control IC"*. Hewlett Packard. Technical Data.
- *"Motion Control Handbook"*. SGS - Thompson.
- *"HD 63140 UPP Application Note"*. Hitachi.

GENERAL HARDWARE	HCTL 1000	LM 628 / 629
NUMBER OF PINS	40	28
MAXIMUM POWER	950 mW (NMOS)	550 mW (NMOS)
CLOCK FREQUENCY (Fc)	1 MHz - 2 MHz	1 MHz - 6 MHz 1 MHz - 8 MHz
Vcc	5 volts	5 volts
MAX. OPERATING TEMP.	70 °	85 °

HOST INTERFACE	HCTL 1000	LM 628 / 629
BUS TYPE	multiplexed addres/data data : 8 bits	asynchronous data : 8 bits
COMMANDS (READ OR WRITE)	possible allways, even during motion.	possible only if bussy bit = 0. Filter parameters variation, data reporting, reset and interrupts possible during motion.
RESET ACTIONS	D/A output = 00 h A = 229 K = B = 64 Position counter = 0 t sampling = 656 / Fc	D/A output = 80 h Ki, Kp, Kd, derivative period and position counter = 0 t sampling = 2048 / Fc

CONTROL	HCTL 1000	LM 628 / 629
CONTROL LAW (FILTER)	$\frac{K}{4} \quad \frac{Z - A/256}{Z + B/256}$	$Kp.e(n) + Ki \sum_{N=0}^{n} e(n) +$ $Kd.[e(n') - e(n'-1)]$ n'= derivative sampling rate
FILTER PARAMETERS (BITS)	8	16
SAMPLING PERIOD	64 μs - 2.048 ms (2 MHz)	256 μs (8 MHz)
PROFILE GENERATION	POSITION CONTROL PROPORTIONAL VELOCITY INTEGRAL VELOCITY TRAPEZOIDAL VELOCITY	POSITION MODE VELOCITY MODE
COMMANDS SIZE (BITS)	POSITION: 24 VELOCITY.(PROPORTIONAL) AND ACCELER.: 16	POSITION : 32 VELOCITY AND ACCELERAT.: 32 (16+16)
COUNTERS SIZE (BITS)	POSITION: 24 VELOCITY (PROP), ACCEL:16	POSITION, VELOCITY AND ACCELERATION: 32
POSITION	ABSOLUTE	ABSOLUTE OR RELATIVE
ENCODER INDEX	NO	YES

OUTPUT SIGNAL	HCTL 1000	LM 628 / 629
D/A RESOLUTION	8 BITS	8 OR 12 BITS (628)
PWM RESOLUTION	1/100 (SAME CHIP)	1/128 (629)
FREQUENCY PWM	20 KHz (2 MHz)	15 KHz (8MHz)

INTEGRAL PROGRAMMING FOR MECHATRONIC SYSTEM

K. Belov and V. Geortchev

Institute of Mechatronics, Bulgarian Academy of Sciences, Acad. G. Bontchev Str., Bl. Nr.2, 1113 Sofia, Bulgaria

Abstract.

Mechatronic systems integrate different, autonomous sybsystems for execution of common activities. They require appropriate system's organization and integral programming. An approach to such programming is described in the paper. A system structure which allows it is described. The problem of structure development is discussed and a block called "system organizer" is proposed for its control and evaluation. The main guidelines and restrictions for mechatronic system and its integral programming building are given.

Keywords. Mechatronic systems, flexible manufacturing, control systems, programming, drives, intelligent machines, artificial intelligence.

INTRODUCTION

Mechatronic systems integrate different subsystems for performance of common activities. This leads to complexity in controlling and programming such systems and raises difficulties in their modification.

This problem of variety and system's complexity could be treated in different ways. Usually the systems are provided with advanced means, intelligent structures, soft module's identification and so on. An approach for intelligent systems building was described by Saridis and Valvanis (1988).

The inherent development logic of mechatronic systems as well as our own experience led us to the idea that the programming should reflect the abilities of the mechatronic approach. The system's organization has to explore this potential using intelligent means.

When developing flexible systems, a question about the system structure arises. Numerous approaches exist in this field. It is obvious that the flexible system structure should allow multi-functionality and development. The criteria of optimal organization could not be entirely objective, independent from the system. They should reflect the essence of the system, the goals before it etc. The flexible system should possess means for structuring and evaluation. It is especially important for the low- cost approach where constant upgrading of the system is very appropriate.

In the paper short description of mechatronic system is given. The main guidelines and restrictions for building mechatronic systems are listed.

The operating system, or more precisely, meta-operating system of the mechatronic system is described. As follows from the mainstream of the outlined notions, the organization of the system and especially of its programming should be arranged much like the organization in the living systems cells.

Description of module called "System's organizer" is given in the paper. It plays central role in the realization of the developed principles of mechatronic system organization.

MECHATRONIC SYSTEM

The notion of mechatronics was formulated in recent years. It reflects the levels of development in different fields - theory of mechanisms, robotics, computer applications. In fact it is more an approach, a technology than a class of entirely new systems. A technology which emphasizes on the flexibility of the system.

The mechatronic approach gives new abilities to create flexible systems. Changes, modifications could be made in different subsystems (mechanical, drive, control, organization, sensory etc.) at any time. So a new need arises: the process of structural changes has to be monitored and controlled constantly. Key role in this process of structural organization plays the programming system.

The main guidelines of mechatronic systems building are, as follows:

(1). Cellular, modular structure with autonomy of different cells, transparency of each module, language homogenity of all cells and the whole system;
(2). Non-traditional way of the whole system organization, especially when sensory subsystems and adaptive control is used;
(3). Common standarts for most modules in all levels. Provisions should be made for automatic module recognition;
(4). Special means for automatic configuration performance (automatic module search, recognition, compiling, checking etc.);
(5). Hot helps provided in every point should facilitate system use;
(6). Common approach to different problems. Common libraries on disposal to all modules.

These guidelines form the main demands for the programming of the mechatronic system.

INTEGRAL PROGRAMMING FOR MECHATRONIC SYSTEM

Because of the mechatronic system's properties, its programming has to be integral. It has different aspects: structural (organizational), functional, evolutional and others which exist in one common system for given time interval. The flexibility demands that the control structure should not be rigidly connected with any block in the system. One and same block of the system can play the role of execution level at one instance and the role of coordinating level at the next instance. These roles are determined from the needs of the system as a whole, from the resources available etc.

Programming in the Mechatronic System.

Usually the production systems (FMS, CIM and similar) are organized in Cell, Area and Plant subdivisions, or levels. Of course, this is more or less conditional interpretation. For example, the three levels could build the control system of one robot. Or the whole area system could be organized on one level. It depends on the complexity of the goals and the abilities of the system.

Cell structure. Figure 1 shows the basic cell sketch. Because of the low- cost automation subject, a flexible manufacturing cell is discussed here, but any other interpretation is possible.

There are processing modules - machines, which perform operations on workpieces; material handling (servicing) modules - robots, conveyors, etc.; energy monitoring modules; storage modules, programmable logic controllers and so on.

A module called "System organizer" is shown on the figure. It performs most of the functions conserning system's structure: set-up, optimization, modelling, evolution and similar. In more detail this module is described below. It works closely with the cell's Knowledge (data) base.

As it is shown on the figure, all blocks in the system are interconnected through system bus. The bus communication approach was chosen because it is the most flexible means for communication. Practically each scheme of connection

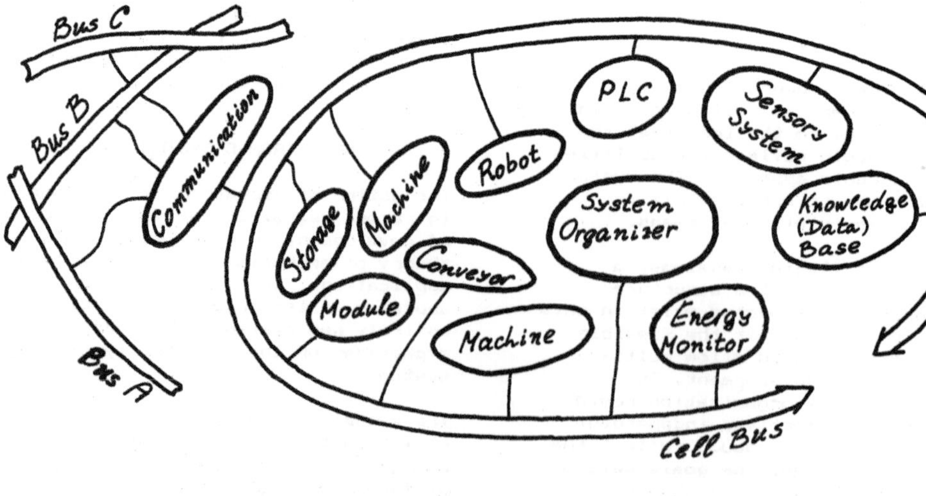

Fig. 1. Basic cell organization.

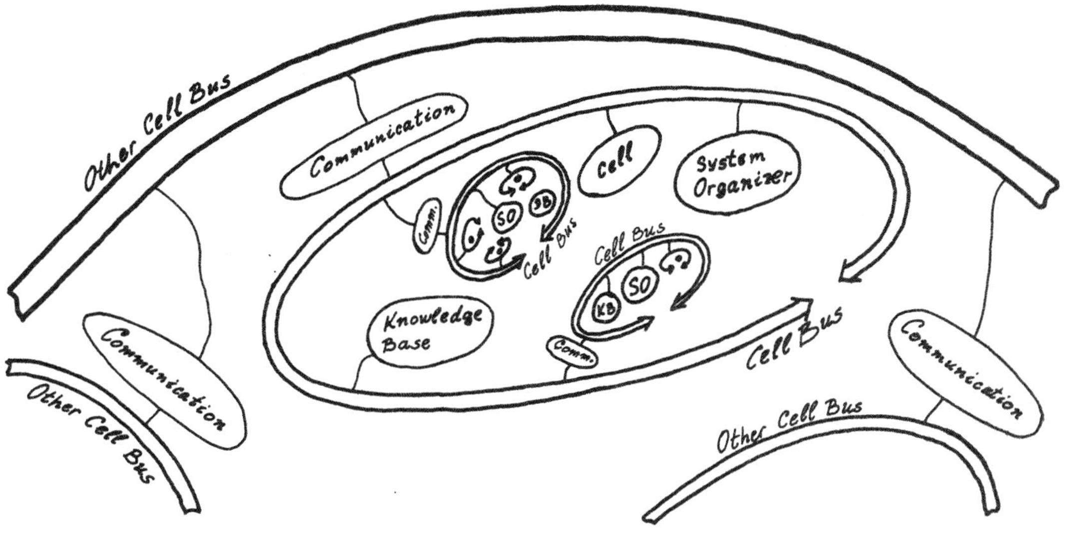

Fig. 2. Cells integration.

(point- to- point, star, chain etc.) could be realized on the bus only by reprogramming it (changing protocols etc.). These system bus abilities create very flexible environment for appropriate restructuring of the system when it is required.

The communication with the external world is performed through the special <u>communication module</u> capable to use different protocols.

<u>Integrating the cells</u>. Figure 2 shows an example of cells' integration in a system. The only elements which remain in all the cells are the system organizers, data (knowledge) bases and communication blocks. These elements create the basis for the whole system's structure building. Through communication blocks the information flows between different cells. Thus different control structures can be built. For instance, a "tissue" can be formed when the cells are all on the same level, as is shown on Fig. 3. Or classical three-level structure, if the hierarchical approach is more adequate to the needs of the moment.

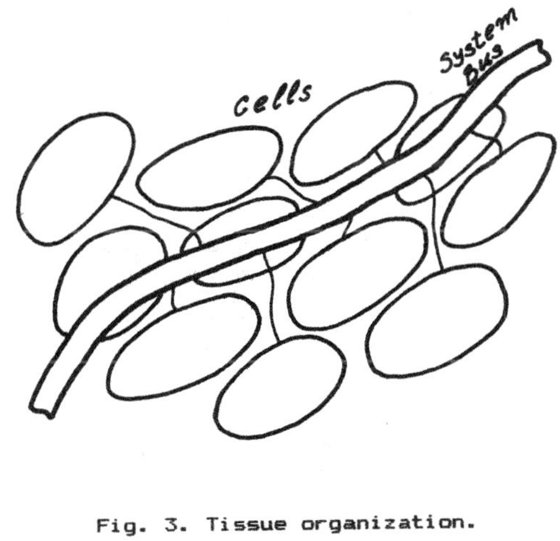

Fig. 3. Tissue organization.

On Fig. 4 a driver cell is shown. It integrates motors, other actuators, encoders, sensors. Here the intelligence is implemented on a very low level, which gives big flexibility and autonomy opportunities using simple or more elaborated sensor blocks and corresponding integral controls.

Following this approach different systems could be created consisting of cells and capable of development, change of structure and evolution.

FLEXIBLE SYSTEM STRUCTURE DEVELOPMENT.

<u>The Problem of Purposeful Organization.</u>

In the process of flexible system's construction a question about its purposeful structure arises. Usually the experience of experts- constructors gives the criteria of structure optimality. Then the behaviour of the system is observed, and the constructor makes

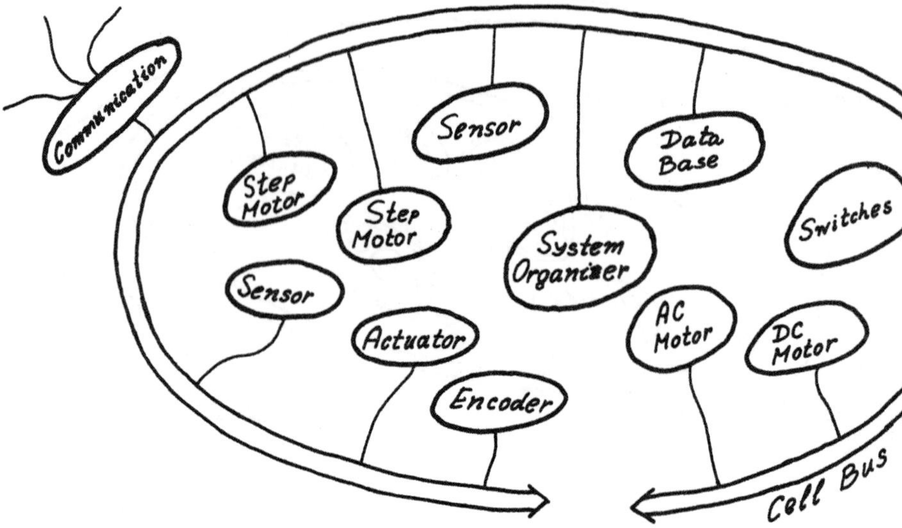

Fig. 4. Driver cell structure.

changes in the structure in order to continue the optimization and to perform the evolution of the system.

This designer's approach suffers, by our opinion, at least from two shortcomings. First, it is expensive when modifications are to be made in the existing complex system. Often it is more easy to replace the entire system with new one though it could have blocks which are still "alive", like driver modules, power supplies etc. Drastic changes happen mainly in the computing blocks and especially in the programming. Secondly, the decision about system's competence is taken by the outside observer - by the constructor. Its knowledge is incomplete, it rests on assumptions which could not be proven reliably.

Such were some of the reasons for us to try to undertake another approach in flexible systems development. It rests on the following assumptions:

1). The system is hold integral by its internal properties, organization (by its non- additive properties, in first place).
2). The structure optimality criteria could be found in the relation between external, environmental factors and the internal system's factors.

There is no exhaustive knowledge about internal systematic factors or about their relationship with the external world (also taking into account the relativity of these notions). And the near future flexible systems will be so complicated that it will be difficult to describe the processes in them with pure physical laws. Relying on euristics and human experts only will also be no more satisfying because they use their own values and give recommendations which could be far from the system's internal needs.

We think that the exit from this situation could be found by investigating the living nature organization. Some its features could be used for development of principles of organization of artificial flexible systems.

Structural Problem Formulation.

The system's structure could be presented as follows:

1). Set of elements, E:

$$E = \{e_1, e_2, \ldots, e_m\}$$

2). Each element e_i has its own aspects, sides, which could be denoted as:

e_{ij} - j-th aspect of the i-th element

where $i = 1, 2, \ldots, m$

$j = 1, 2, \ldots, k$. This k could be different for each element e_i, which should be taken into account.

3). Set of relations between elements:

$$R = \{r_{ij}\}, \quad i = 1, 2, \ldots, m \quad j = 1, 2, \ldots, m.$$

For instance,

$$e_i \; r_{ij} \; e_j$$

is the relation between i-th and j-th element of the system.

4). The structure S of the system could be denoted as subset of elements and relations betwen elements in given time period:

$$S \subset T \times E \times R$$

A space of all possible structures could be defined for given system. The structures could be classified in sets, families, classes etc.

5). The goal before the system could be presented (as in Kalman and colleagues, 1969):

$$G \subset T \times X \times Y$$

Given goal is actual for a period of time.

Now the problem of optimal configuration could be formulated as follows:

A structure, subset of S has to be found which satisfies the goal G in the best way for the time of its actuality for the system.

This problem could be hardly solved by pure analitical approach. So we could use more open approach and convert this optimization problem into well known decision making problem. Now it will look as follows:

A structure from the set of available structures has to be chosen in regard to a set of criteria, and from the point of view of their effectiveness for realization of the goal before the system for the time of its actuality.

As is shown in Belova and Belov (1984), the decision could be made through measuring or evaluation. For evaluation of the alternative's effectiveness for realization of the goal before the system, a value position is used, which is the set of goal- dependent patterns (standards) defining the way of valuation in the control system.

This means that in our case, in order to be capable to take decisions about proper organization, a set of goal- dependent patterns, standards, for each goal before the system has to exist and to be on disposal of the decision maker in the moment when the organization is evaluated.

These functions are performed by the system organizer - a block in the flexible system which holds the value positions for system structures.

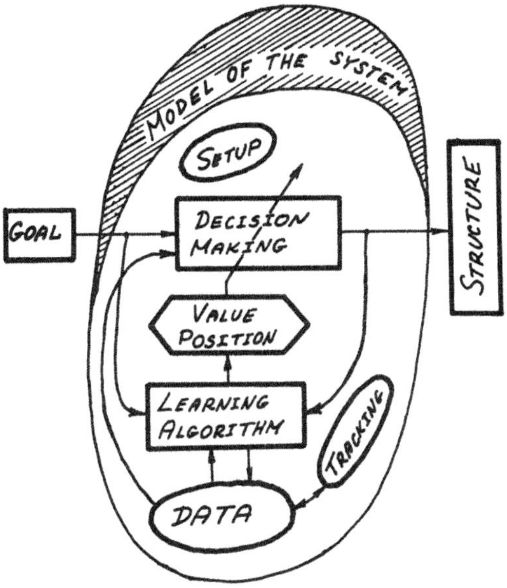

Fig. 5. System organizer.

System Organizer.

The system organizer is present in all cells. It is an organization center around which the cell is structured. The expedience of the system's structure is realized by means of the system organizer. Its view is shown on Fig. 5.

The most important features of the system organizer are the following:

Model of the system. The system organizer is built around some model describing elements of the system with their properties, and relations between them.

Set-up, configuring the system. This is routine function. Automatic module recognition is used here. For those cases when it is not possible, an approach with automatic discrimination of modules' features is used.

Ensuring the expediency of system's organization. This is the most specific function of the system organizer. It is performed by holding the set of goal-dependent patterns, standards. When given goal is actual, a search for corresponding structures is activated. The effectiveness of the structure is evaluated in the process of its functioning, and this learned experience is used in further development of system's structure.

Modification tracking. Holds tracks of all changes in the system's structure for the history of its development.

System's learning. Experience gathering about different structures and their performances. Different learning algorithms to change the structure in appropriate way.

The system organizer structures the system resources in an adequate manner for the task before the system. For instance, when the mechatronic system is used for palletization, a configuration is proposed which matches this task. When the same system is used for welding - another configuration is proposed. When performing given task, system's behaviour is monitored and changes in control structure and parameters are performed in order to optimize the structure.

CONCLUSION

An approach to integral programming of mechatronic system was outlined in the paper. It relies on the flexible system's structure described and gives opportunities for its development, evolution. The main features of the integral programming are:

1). No particular control structure is permanent. The structure can be changed in any moment.
2). The system organization is based maximally on software, in order to give maximum flexibility.
3). The preferred control organization is heterarchical. Almost every block has the potential for controlling the whole system.
4). The structure is prone to development. The first step in this direction is the system organizer's ability to evaluate the system's behaviour.

REFERENCES

Belova, N., and K. Belov (1984). Valuation and Control. In R.Trappl (Ed.). Cybernetics and Systems Research 2. Elsevier Sci. Publ. B.V. (North- Holland). pp. 93-96.

Kalman, R., P. Falb, and M. Arbib (1969). Topics in Mathematical System Theory. McGraw Hill.

Saridis, G.N., and K. Valvanis (1988). Analytical Design of Intelligent Machines. Automatica, V24, No.22, pp. 123-133.

LOW COST AUTOMATIC CONTROL APPLIED TO AN EXPERIMENTAL GREENHOUSE

G.M. Claudia, H.R. Gilberto and P.G. Guillermo

Computer Sciences Department, Monterrey Institute of Technology, Monterrey, México

Abstract. The main goal in controlling greenhouses environment is to have a high level of intensity in plant production with a low cost investment in automation. This paper describes the control system developed for an experimental greenhouse where several crops were produced. The system is based on a 68000 Motorola microprocessor. Three kind of sensors are used for monitoring of temperature, relative humidity and CO_2. These variables, also as water pumps and fans are controlled through the intelligent system and a power module.

Keywords. Microprocessor control; temperature control; humidity control; warehouse automation.

INTRODUCTION

The investment costs for greenhouses as well as labor and energy costs are much higher than in conventional plant production. Therefore research and development for a better utilization of the yielding potential of plants and a high labor productivity and energy efficiency of greenhouses is necessary.

In addition to the greenhouse temperature control the control of humidity, CO_2, light intensity, and water supply are also required. Some of these parameters are used in commercial greenhouses including plant nutrition. In several countries plant factories have been developed and they are expected as the plant cultivation system for the future, for them, as well as for plant production research through the use of greenhouses, the control systems become very important because they force the environment variables to follow some pre-established trajectories in spite of the external climate changes or any other disturbance.

Greenhouses and plant factories have reached the stage of dynamic control by means of computers, and automation systems that are required for the purpose of the dynamic control. A minimal microprocessor system can be a solution when low cost automation becomes an extremely important requirement for development, specially in those places where economic resources can be very limited.

This paper describes a greenhouse control system based on a 68000 Motorola microprocessor. The main signals to be controlled are sensed by temperature integrated circuit sensors, relative humidity and CO_2 transducers. Through the real time clock of the microprocessor it is possible to activate a set of switches which turn on or off the lights and the fans of the greenhouse area.

The control system was applied to an experimental greenhouse of the Monterrey Institute of Technology in México. The development of the project which refers also to the kind of variables to be controlled, was designed and implemented according to the Agronomy Department (DCAM) requirements, and its research activities in the field of Agriculture.

SYSTEM DESCRIPTION

The whole control system consists of five modules: the master control module, the data acquisition module, the power module, the communication module, and the remote monitoring module (see Fig. 1).

Temperature is monitored in the environment through integrated circuit sensors. Relative humidity (from which absolute humidity can be derived), and CO_2 are sensed by special transducers. The physical location of such sensors can be seen in Fig. 2, which represents the greenhouse area. The same figure was programmed in a remote PC computer as a user's interface. The system was

developed according to the next criteria (Gauthier and Gray, 1990):
- The system must respond in real time to the sensor's perception of a somewhat unpredictable and changing environment.
- The data from sensors must be interpreted and reasoned about. The system must identify sensor or time-based events by matching them with patterns stored in the computer.
- The system must react safely and reasonable to inaccurate and/or uncertain data inputs.
- The system must handle multiple zones within the greenhouse(s) having different configurations and possibly different crops or crop stages.
- The system must allow the transfer of information between controllers installed in a greenhouse complex and a central computer.
- The system should log and maintain histories describing the evolution and performance of crops, climate, equipment, and resources.

I The master control module
This module deals with the control of the whole system in the greenhouse. It gets the data from the user about the several variables to be controlled, specifying limits and oscillation ranges. It also transfers the information to a remote PC where the greenhouse environment parameters can be programmed. The module controls the whole greenhouse system according to the user's program.

The master control module consists of a minimum system based on the 68000 Motorola microprocessor, with 64 Kbytes of nonvolatile memory RAM, 64 Kbytes of memory ROM, and a real time clock. It has a keyboard, several expansion slots and one video card. This module is assembled together with the communication and data acquisition modules in one cabinet.

The software of the system allows the data acquisition module to sense the temperature every 3 minutes and send a signal to the power system in order to power on or off the fans or the air condition device. It also allows the power system to activate the water pumps according to the user's program and the light intensity which becomes crucial during the night.

II The data acquisition module
The sensing task of the greenhouse system is part of the data acquisition module, which makes the electronic signal conditioning in order to provide data for the control module. Eight temperature integrated circuit sensors are distributed along the greenhouse area together with 2 relative humidity, and 2 CO_2 transducers.

The data acquisition task is performed by a card which was designed for this purpose. It is based on an analog-to-digital converter of 8 bits. The card can handle 8 simultaneous input signals. The electronic conditioning for temperature, relative humidity and CO_2 is implemented in the data acquisition card, also as the circuitry to interface the microprocessor system.

The data acquisition card can measure temperature in the range between -10 to +45 centigrades with a maximum error of 0.25 centigrades along the whole range. The relative humidity parameter can also be measured in the range from 18 to 100% of relative humidity and a maximum error of 2.8%. An hysteresis loop of $\pm$ 2.5% of relative humidity can be found in this transducer due to the fact that the absorption is faster than the adsorption process. The transducer is a non-linear device which changes impedance according to the changes in relative humidity.

III The power module
The power module is a cabinet with 8 relays of 250 VAC and 25 amperes each. The module handles the on and off actions for every device involved in the control system which are 4 water pumps, 3 fans, 1 air conditioned mechanism and a set of 20 lamps for artificial light.
The power module receives the control signals from the parallel port of the master control module, activating the several devices mentioned above and turning on a light indicator for the user advisement.

IV The communication module
This module consists of the serial card which was designed for the communication purpose. It transmits the information of the master control module to a remote PS 50 system which remains in a research room 100 meter away from the greenhouse. The asynchronous communication is based on the RS-232 standard configured at 9600 baud, with one bit for start, seven bits of information and one more for stop.

The card consists of an integrated circuit for communication, a real time clock, and a parallel port with digital outputs. These devices make possible the information transfer and control functions between the several modules.

V The remote monitoring module
This module asks the information from the master control module about temperature, relative humidity and CO_2, which corresponds to the data stored during one week and that is used for statistical purposes.

The remote monitoring module is built up of a PS/50 computer and some software developed in "C" language. As frequently as the researcher desires, the information concerning the sensors and the situation of the water pumps can be asked through this module. It is not necessary to be at the greenhouse place to know about the performance of the control system and the parameters sensed by it on the room. Through the remote monitoring module it is possible also to activate from the research room the lights and the water pumps of the greenhouse. After one week, the data can be stored in a diskette according to the researcher desire.

PROGRAMMING OF THE GREENHOUSE PARAMETERS

The control system is programmed by the user concerning the sensor's calibration, water supply, and light exposure times.

Once the system has been turned on, it will ask the user if he or she want to calibrate the sensors; we must answer Y (yes) or N (not). After this procedure, the system will display the temperature, the relative humidity, and the CO_2 concentration inside the greenhouse.

When the sensors have been calibrated, the water supplies should be programmed; we must specify the frequency of every of them in one day with a value between 0 and 8. It is also necessary to enter the duration time of the water supplies between 1 and 59 minutes; the system will distribute in automatic form these frequency and duration of the water supplies avoiding the fact of having the pumps in the "on" state for long periods of time.

In the same way the light exposure can be programmed only by entering the times when the system should turn on or off the lamps.

After the programming of the greenhouse parameters, the screen will show the user the data to verify the temperature which should be in accordance with the desired value; it will also allow the user to verify the real time of the system. This procedure is repeated every 15 minutes.

The data from the temperature, humidity and CO_2 sensors is sent to the computer every hour. This time the system can not be programmed unless the data has been stored in the auxiliary memory.

The information of the greenhouse performance can be read in the remote PS/50 computer at the research room. The screen will show the Fig.1 where a window will display the next options:

Humidity sensor
CO_2 sensor
Temperature sensor
Summary of the week
Update data
Air conditioned
Fans
Water Pumps
End

The mouse of the computer sill help the user in selecting any of the options above. When data are required every week about the state of the greenhouse it should be typed the number of the file (FILE.DAT) where these data should be stored, and immediately the system will start the information transfer between the controller and the remote monitoring modules. When selecting the option END the system will return to DOS system.

RESULTS

A microprocessor based control system was developed for a low cost automation application in a greenhouse. The system can be easily programmed according to the requirements of any specific type of plant production. After three months of uninterrupted testing a raise in productivity was observed in an experimental greenhouse of the Monterrey Institute of Technology in México, where optimization in tomato, cucumber, strawberries and flowers production was obtained for research purposes. The development resulted in a low cost control system based on a minimal microprocessor master card without the necessity of having a dedicated expensive computer for the control task. Data could be transferred to a remote station through RS-232 connection where statistical analysis helps the research work.

CONCLUSIONS

A minimal system based controller can be a solution in plant production where the investment for automation requires low cost components. Once the sensors are adequately chosen, the system can be easily programmed to control the greenhouse with a wide range of reliability. A remote monitoring system can help in obtaining the information from the controller for statistic purposes, facilitating the research work where a computer can be shared for several tasks and becomes at the same time a good interface between the greenhouse control system and the user. This interface management can be programmed in such a friendly way that the researcher can easily access it and select the different tasks to perform concerning his/her work.

ACKNOWLEDGMENTS

The authors are pleased to acknowledge the assistance of Dr. Donald Vega and Professor Garcia Blanca from the DCAM Department with the financial and technical assistance.

REFERENCES

Analog Digital Conversion Handbook. (1990). Analog Devices. Prentice Hall, USA.
Linear Products Databook.(1990). Analog Devices.
Frederick, E. S.(1991). Techniques for Automation Systems in the Agriculture Industry. Control and Dynamic Systems, Vol. 49, pp. 99-128.
Gauthier L. and R. Guay (1990). An Object-Oriented Design for a Greenhouse Climate Control System. Trans. of the ASAE 36, pp. 999-1004.
Tantau, H. J.(1990). Automatic Control Application in Greenhouse. Annals of the CIRP,66, pp 302-305.
Jacob, M. (1983). Digital and Analog Circuits and Systems.

USA.
Alan, C. (1987). Microprocessors Systems Design 68000. Hardware, Software and Interfacing. PWM, USA.
Robert, G. S. (1983). Transducers, Sensors, and Detectors. Prentice Hall, USA.

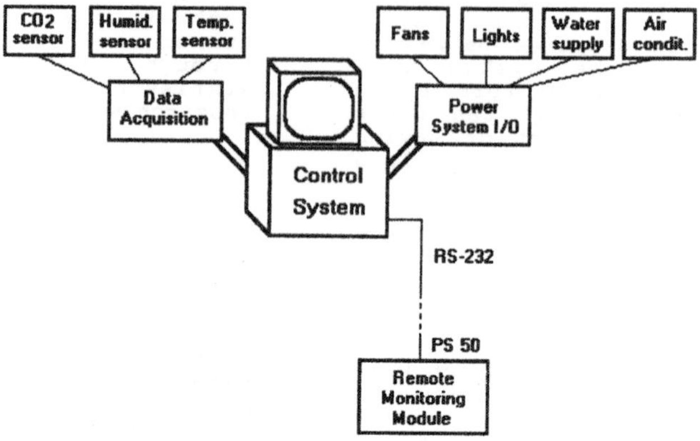

Fig. 1 Diagram representation of the Control System for a greenhouse

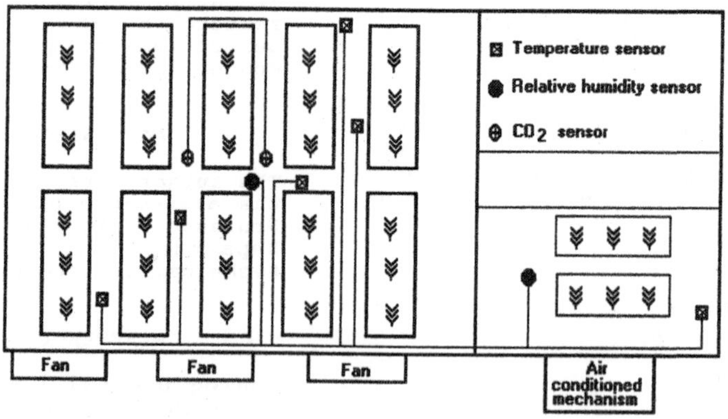

Fig. 2 Greenhouse area and location of the several sensors in each room. The upper right side room is only considered for future plans

SUPERVISION ISSUES IN SEMIAUTOGENOUS GRINDING

A. Cipriano*, M. Guarini*, C. Muñoz*, J. Cáceres** and R. Collado**

*Faculty of Engineering, Catholic University of Chile, Chile
**Colón Concentrator, Division EL Teniente, Codelco-Chile, Chile

Abstract. This paper presents an application of parameter estimation and signal prediction to the supervision of semiautogenous grinding plants. Firstly, simple dynamic models for mill power and mill load prediction are stated. Then the models are compared using experimental data. The predicted signals show very good agreement with the corresponding measurements.

Keywords. Mining industry; mineral grinding; SAG mill supervision; parameter estimation; signal prediction.

INTRODUCTION

Mining is one the most important economical activities in our country. As mineral grade declines and the international competition increases, major efforts are directed to more efficient operation of mineral processing plants.

During the last twenty years much attention has been focused on the control of grinding circuits. Many successful applications of ball-mill computer-based PID or more advanced control strategies have been reported. Autogenous and semiautogenous (SAG) grinding circuits are more difficult to control, because the mill performance changes strongly with the size of the feeding ore particles. To decrease this effects, more sophisticated controllers and supervision algorithms. are required. Therefore, many differents supervision schems and control strategies have been proposed and tested in semiautogenous grinding plants of Australian (Mount Isa), Canada (Lornex), Chile (Los Bronces, El Soldado, Chuquicamata), United States (Chino Mines), Finland (Enonkoski), Philippines (Benguet Dizon) and South-Africa (Kinross). Two of these strategies are described in Páris and Cipriano (1991) and Cipriano et al (1992).

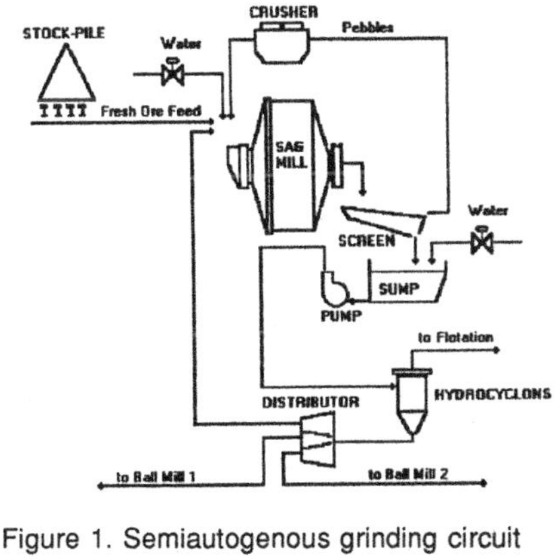

Figure 1. Semiautogenous grinding circuit

GRINDING CIRCUIT

Figure 1 shows the flowsheet of a typical semiautogenous grinding plant, which is comprised of a primary section, two secondary sections and a flow distributor. Primary grinding includes SAG mill, vibrating mill discharge screen, crusher, sump, pump and a battery of hydrocyclones as classifier. A secondary grinding section consists of ball mill, sump, pump and classifier.

Figure 2 shows SAG mill circuit instrumentation. SAG mill control variables are feeders speed, fresh ore feed rate to the mill, feed rate of water to the mill and mill speed. The regulatory SAG mill control loops are solids feed rate controlled through variable speed feeders and feed water ratioed to the solids feed rate.

Major controlled variables are mill power and mill load. Feed rate is sometimes cascade-controlled by mill power. When the ore is soft, the mill is allowed to draw as much power as possible and the feed rate is increased. If the set-point for mill power controller remains constant, this power-feed rate cascade loop could not differentiate between decreasing power draw due to mill overload or to softer ore. Mill overload occurs when the power significantly decreases at the same time that the mill load increases.

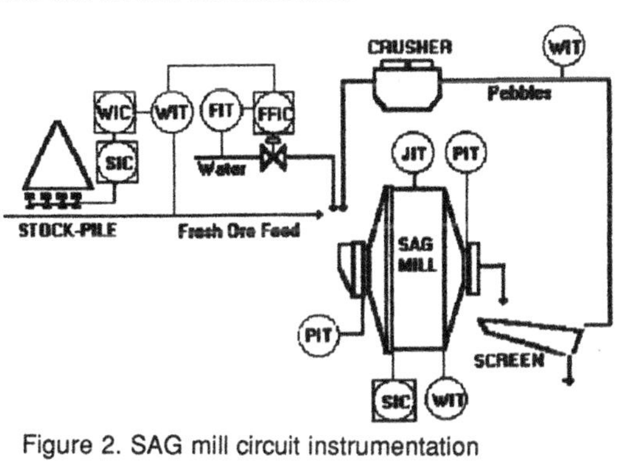

Figure 2. SAG mill circuit instrumentation

In many SAG plants, in order to maximize the throughput without overloading the mill, both mill power and mill load (or bearing back pressure) are used to determine the feed rate set-point. For example, the setpoint of the solids feed rate control loop is obtained by checking the setpoints of the mill power and of the mill load controllers, and selecting the lower of the two values.

Standard methods for overload detection are based on the J-M mill power - mill load curve (see figure 3). Some methods require to determine the derivates of both signals. In these cases it is said that a mill overload is in progress when $dJ/dt < -\alpha$ and $dM/dt > \beta$, with α and β positive values. For example,

$\alpha = 15$ KW/min., $\beta = 10$ Ton./min.

Other overload detection methods are based directly on the analysis of mill power and mill load trends. For these purposes it is necessary to predict the evolution of both signals. This work describes design and experimental evaluation of dynamic models for mill power and mill load prediction.

MODELS FOR MILL POWER PREDICTION

Experimental data for this study was obtained at the Colón Concentrator SAG plant of CODELCO-Chile using a PC connected to a distributed control system Honeywell TDC 3000 (see Varela and Cipriano, 1991). Operation records of the most important variables were obtained by filtering and averaging every 1 minute each one of the 0.3 Hz sampled signals.

Figures 4 and 5 show mill power and mill load evolution. Figures 6 and 7 show two important variables for modeling: new solids feed rate and mass flow of pebbles. Pebbles form the coarse fraction of the mill discharge that is removed by a vibrating screen and returned to the crusher.

It is possible to see that interaction between these four variables is very strong. SAG mill operates without overloads, so that mill power and mill load present similar trends. When the new solids feed rate is cut, mill load decreases notoriously. Mass flow of pebbles also changes, but not so much.

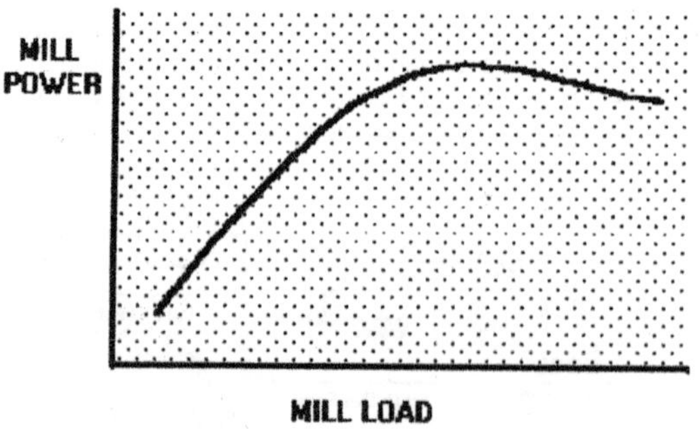

Figure 3. Mill power - mill load curve

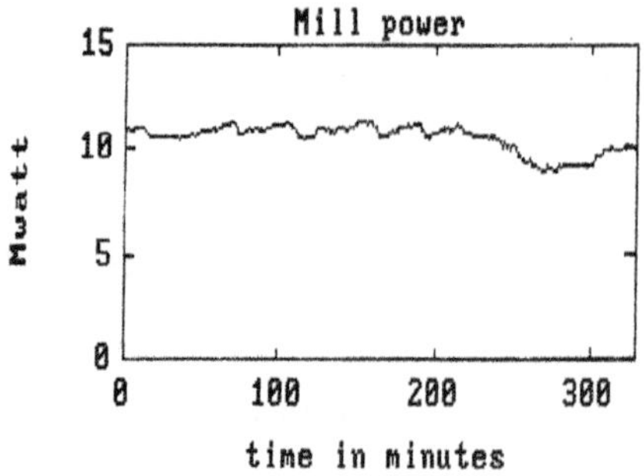

Figure 4. Mill power

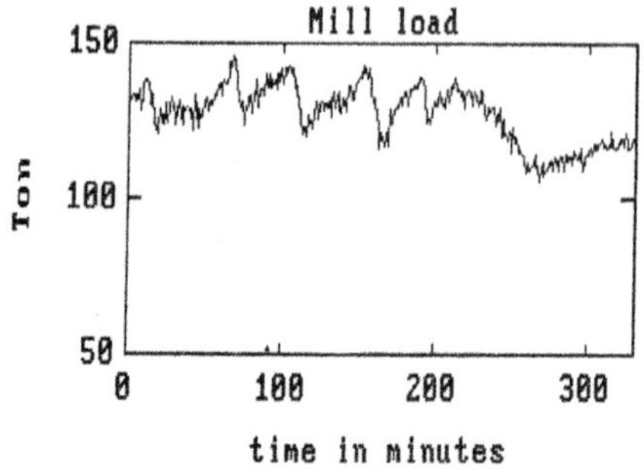

Figure 5. Mill load

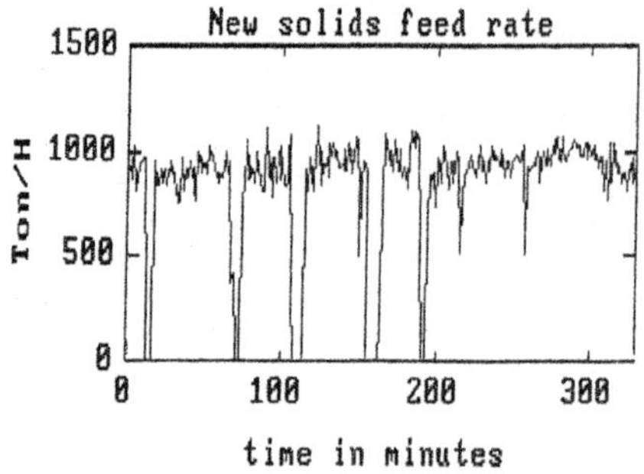

Figure 6. New solids feed rate

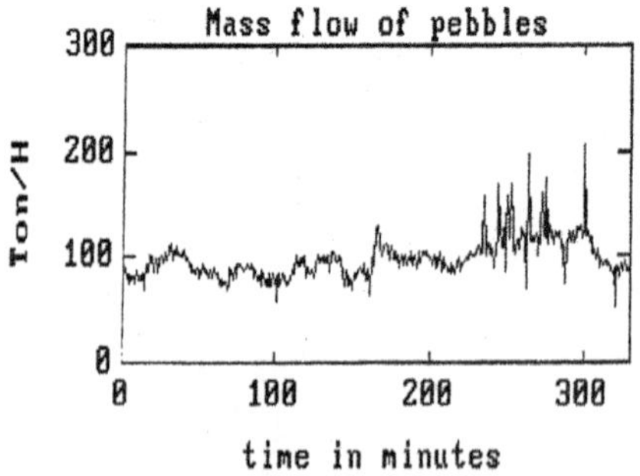

Figure 7. Mass flow of pebbles

It is possible to state several models that verify these conditions. The study is constrained to simple linear and nonlinear ARMAX models, therefore we have examined only four:

M1 : $J'(k+d) = C_0 + C_1 J(k) + C_2 M(k)$

M2 : $J'(k+d) = C_0 + C_1 J(k) + C_2 M(k) + C_3 W_m(k)$

M3 : $J'(k+d) = C_0 + C_1 J(k) + C_2 M(k) + C_3 W_m(k) + C_4 W_p(k)$

M4 : $J'(k+d) = C_0 + C_1 J(k) + C_2 M(k) + C_3 M^2(k) + C_4 W_m(k) + C_5 W_p(k)$

where J'(k+d) is the mill power prediction with time horizon d = 3, 4, 5 or 6, and J(k), M(k), $W_m(k)$ and $W_p(k)$ are mill power, mill load, new solids feed rate and mass flow of pebbles, respectively.

The parameters C_0-C_5 were calculated using recursive identification. Typical parameter convergence graphs obtained with model M4 and d=3 min. are shown in figures 8 to 13. Parameters C_1, C_2 and C_4 are positive and parameters C_3 and C_5 are negative. Therefore, the sign of all these parameters agree with the interaction existing between the four variables.

Table 1 presents the root mean square of prediction error obtained with each model for differents time horizons. In all cases model M4 achieves better performance. With this model the root mean square of the prediction error for 6 minutes of time horizon is 0.2 MW, which means 1.9% of the mean mill power.

Figures 14 and 15 confirm the good results obtained with model M4. Power prediction follows rather closely the measured variable.

TABLE 1. Root mean square - power prediction error

d	M1	M2	M3	M4
3	0.1534	0.1334	0.1312	0.1308
4	0.1853	0.1620	0.1590	0.1568
5	0.2326	0.1886	0.1856	0.1814
6	0.2355	0.2132	0.2105	0.2041

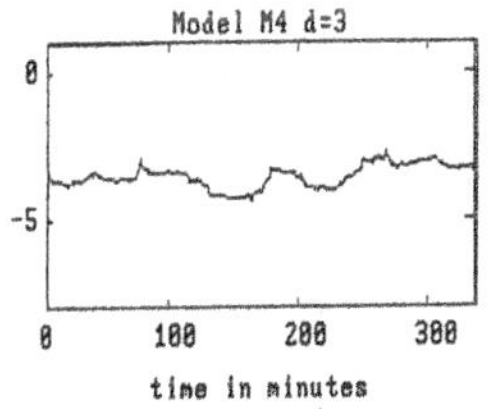

Figure 8. Parameter C_0

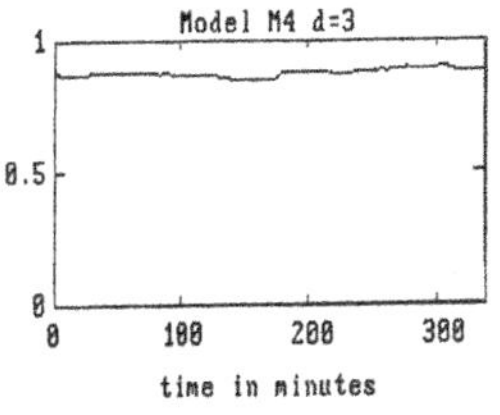

Figure 9. Parameter C_1

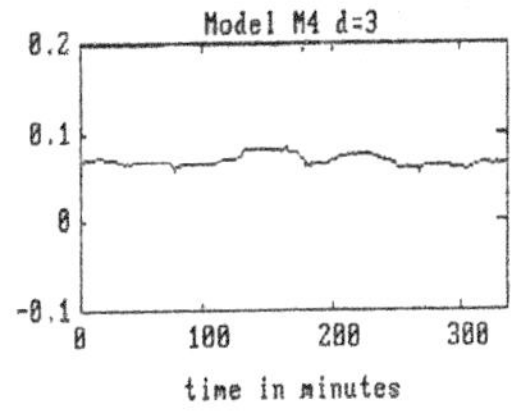

Figure 10. Parameter C_2

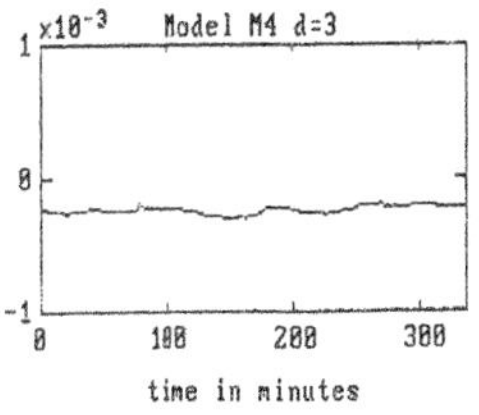

Figure 11. Parameter C_3

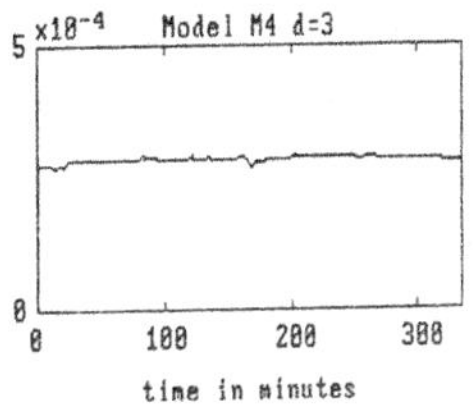

Figure 12. Parameter C_4

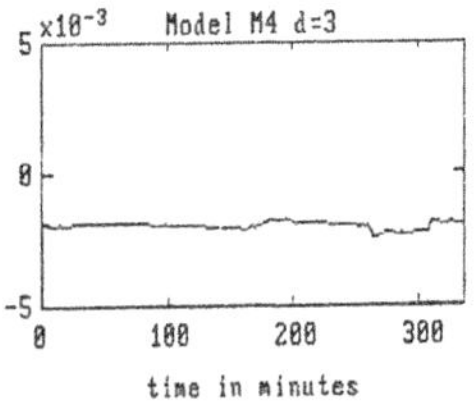

Figure 13. Parameter C_5

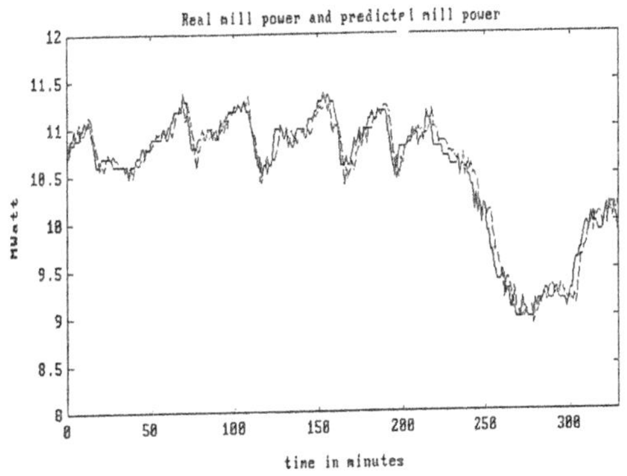

Figure 14. Measurement (_) and power prediction (...)

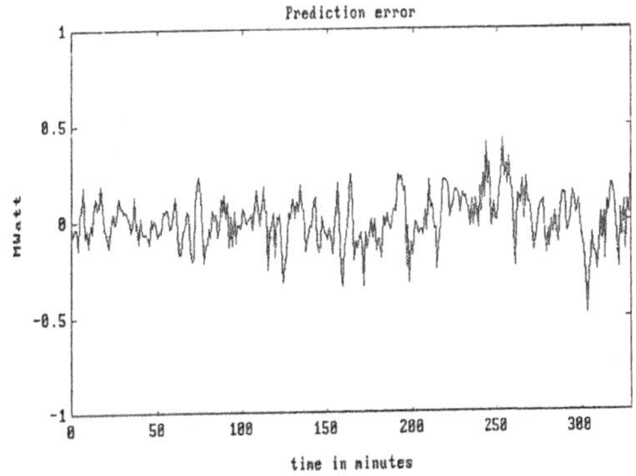

Figure 15. Power prediction error with model M4, d=3.

MODELS FOR MILL LOAD PREDICTION

In this case we have analyzed the following three linear ARMAX models :

M1 : $M'(k+d) = C_0 + C_1\,M(k) + C_2\,J(k)$

M2 : $M'(k+d) = C_0 + C_1\,M(k) + C_2\,J(k) + C_3\,W_m(k)$

M3 : $M'(k+d) = C_0 + C_1\,M(k) + C_2\,J(k) + C_3\,W_m(k) + C_4\,W_p(k)$

where $M'(k+d)$ is the mill load prediction with time horizon d = 3, 4, 5 or 6.

Figures 16 to 20 show the parameter convergence for model M3, d=3 min. Parameters C_1, C_2 and C_3 are positive and parameter C_4 is negative, which agree with the interelation observed between the variables.

Table 2 presents the root mean square of mill load prediction error obtained with each model for differents time horizons. In all cases the minimum predictor error is achieved with model M3. The minimum root mean square prediction error for 6 minutes of time horizon is 4.2 Ton., which means 3.4 % of the mean mill load.

The performance of the prediction algorithm with model M3 is shown in figures 21 and 22. It can be seen that a very good tracking of the mill load is achieved.

TABLE 2. Root mean square - load prediction error

d	M1	M2	M3
3	3.4283	2.9746	2.9680
4	3.8166	3.3060	3.2671
5	4.4411	3.8167	3.8031
6	4.9780	4.2195	4.2139

CONCLUSIONS

We have presented an application of parameter estimation that is being used to detect overloads in semiautogenous grinding. The detection algorithms is based on mill power and mill load predicted signals.

If the SAG mill speed changes, overload detection algorithm doesn' t work. The reason is that mill speed strongly affects the mill power and the mill load. Therefore, we are developing new prediction models that considere mill speed as input variable.

Parallel to this study, a control strategy based on PID controllers and logic rules is currently under implementation on the Honeywell TDC 3000 distributed control system of Colón Concentrator in El Teniente. This strategy includes mill power and mill load controllers. According to our results, both variables are significantly coupled, so that a multivariable controller may be necessary. In this case the models that we have developed will be the basis for the multivariable controller.

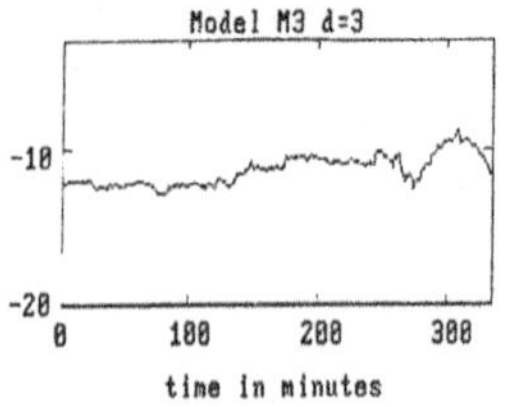

Figure 16. Parameter C_0

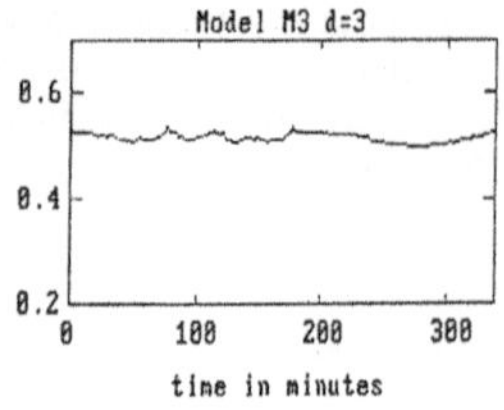

Figure 17. Parameter C_1

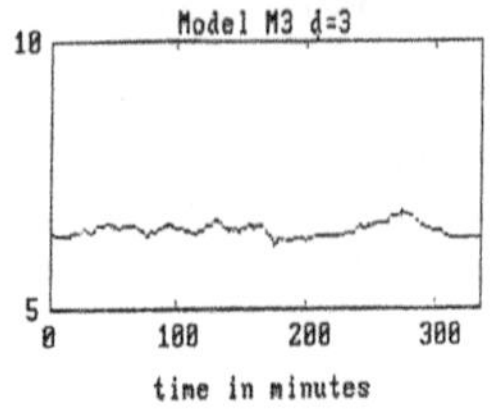

Figure 18. Parameter C_2

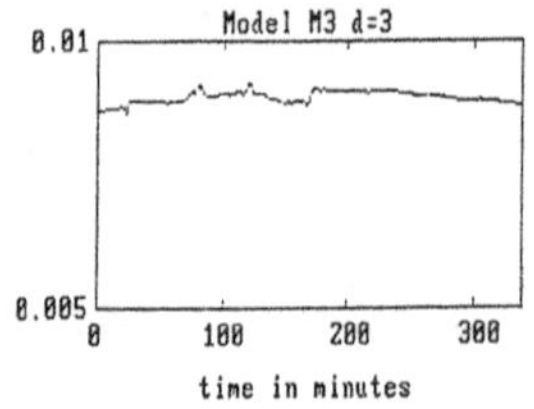

Figure 19. Parameter C_3

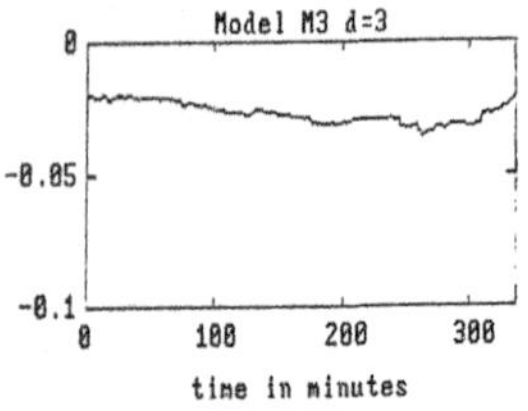

Figure 20. Parameter C_4

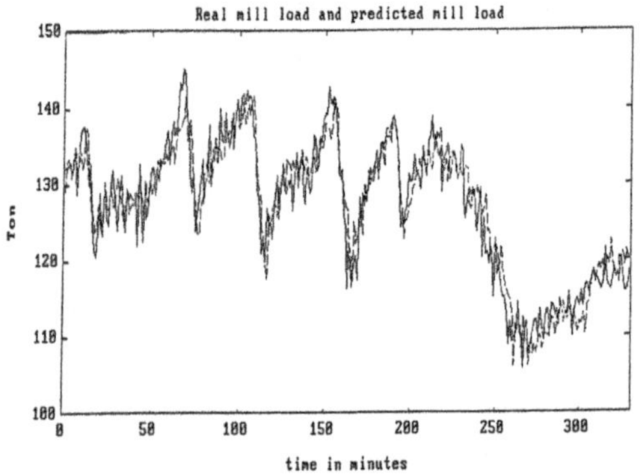

Figure 21. Measurement (_) and load prediction (...)

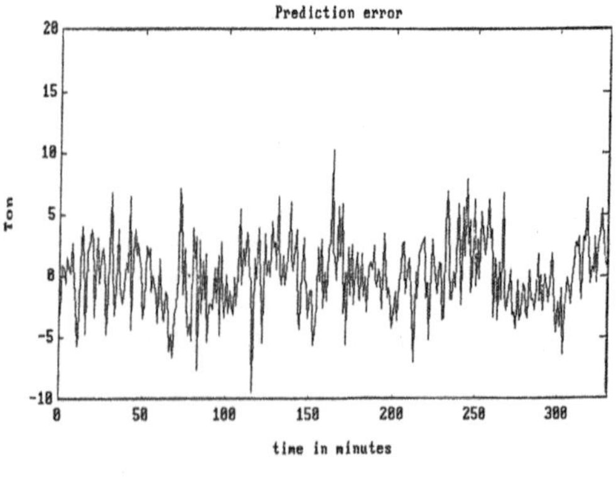

Figure 22. Load prediction error with model M3, d=3.

REFERENCES

Cipriano, A., Labarca, E., Varela, V., Jeréz, J. (1992). Expert supervision and robust control strategy for a semiautogenous grinding plant. Preprints of the IFAC Symposium on Mining, Minerals and Metal Processing, August 26-28, Beijing, P. R. China (in press).

Herbst, J. A., Hales, L. B., Pate, W. T., Sepúlveda, J. E. (1989). Report on actual benefits arising from the application of expert control systems in industrial semiautogenous grinding circuits. Preprints of the IFAC Symposium on Mining, Minerals and Metal Processing, September 4-8, Buenos Aires, Argentina, Vol. 1, 56-63.

Páris, A., Cipriano, A. (1991). Supervisory expert control of a semiautogenous grinding circuit. IFAC Workshop on Expert Systems in Mineral and Metal Processing, August 26-28, Espoo, Finland.

Varela, V., Cipriano, A., Guarini, M. (1991). An ONSPEC based advanced supervisory system for grinding plant. Preprints of the IFAC/IMACS Symposium on Fault Detection, Supervision and Safety for Technical Processes, September 10-13, Baden-Baden, Germany, Vol. 2, 295-298.

ACKNOWLEDGMENTS

This work was supported by CODELCO-Chile Division El Teniente and by grants FONDECYT 0743/90 and FONDECYT 0834/92.

A CONTROL SYSTEM USING COMPUTER VISION FOR MARKING CARBIDE CUTTING TILES

J. Honec and F. Solc

Department of Control Engineering, Technical University of Brno, Bozetechova 2 Str., 612 66 Brno, Czechoslovakia

Abstract. Cemented carbide cutting tiles which are produced for cutting tools must bear inscription data which specifies the kind of tile, batch number etc. The inscription must be placed on a relatively narrow wall of the tile with a laser beam. the tiles of size 15x15x4 mm approx. are fixed in columns in an open box on a turning table of the marking machine. Due to production tolerances and palletization methods it is not possible to specify the position of individual tiles by mechanical means. that is why it was decided to use a computer vision system which enabled the control system to define the position of each tile with required precision and speed. The paper describes main features of the control system. The control system removed a bottleneck in cutting tile production and highly improved quality of tile marking.

Keywords. Computer control; Computer applications; Kalman filters; Metallurgical industries; Optimal filtering; Picture processing;

INTRODUCTION

For a newly designed automatic machine a method of marking carbide cutting tiles (placing inscription at a required place) has been developed. The marking is performed by means of laser with the galvanoelectric head. A simplified layout / design of this machine can be seen in Fig. 1. Carbide cutting tiles (or sintered carbide cutting tiles) are placed on a pallet. The size of the open box in the pallet is 100 x 100 mm and tiles are marked on the narrower wall. The size of the tiles depends on their type, e. g. if the size is 15 x 15 x 4 mm, there are 144 tiles placed in the pallet in 6 columns, each containing 24 tiles. The positions of tiles in columns must be measured as they cannot be defined a priori with sufficient precision because of manufacturing tolerances and air gaps. The positions of columns are defined by the pallet and need not be measured.

DESCRIPTION OF THE SYSTEM

A new method evaluating the position of tiles in a pallet from the digitized picture has been developed. Tiles display / exhibit different properties and some reflections may even occur. As an example copies of digitized pictures of two different types of tiles are shown in Fig. 2. It is not possible to simply define the edges from brightness evaluation because scattered optical properties do not guarantee proper lighting conditions that would enable location of edges. In addition , it was found that placing the inscription on a tile requires precise measurements which can hardly be taken with a common television camera resolution of which is limited by a number of TV signal lines.

It was necessary to develop a reliable and not very complicated picture processing method which would not be a weak point in technological process. The two-phase picture processing method proved to meet these requirements. During the first stage the signal is processed by filtering through a special filter with a mask illustrated in Fig. 4a. The shape of the mask was set so as to highlight boundaries between the tiles. The mask is defined in discrete points with weights a_1 or a_2 so as the sum of all weighted points is zero.

$$\sum_{i=1}^{Y-1} \sum_{j=1}^{N-1} a_{ij} \qquad (1)$$

This is the usual way of relating the results of filtering to zero value (Horn, 1986). The mask of dimensions X and Y corresponds to the size of tiles. The filter with mask is very much like compliance filters known from one-dimensional signal processing theory and its properties are similar. Moreover in y direction it behaves like the second-order derivative approximation of brightness function $\partial^2 b(x,y)/\partial y^2$. From the above it follows that filter is not very sensitive to overall brightness level of the pallet and to non-uniform illumination. Relatively big sizes of the mask may partially eliminate local differences in brightness along the identified edges. The results of filtration were so good that where the edges had been detected, interpolation used in astronomy for detecting positions of fixed stars by means of CCD area image sensor could be applied, Fig. 4 b. Thus several times more precise detection of edges could be attained than with a CCD image sensor precision of which is given by size its elements and by distances between them.

It would be sufficient to use the third-order regression polynomial to interpolate the position of edges.

$$f(y)=a_0+a_1y+a_2y^2+a_3y^3 \qquad (2)$$

Coefficients a_i have been calculated from five points in the vicinity of the assumed maximum. The interpolated position of maximum can be defined according to the following conditions

$$\frac{df(y)}{dy} = 0 \;;\quad \frac{df^2(y)}{dy^2} < 0 \qquad (3)$$

Points $z(k)$, at which the extremes have been detected, need not correspond to the positions of edges. For example reflections can be misinterpreted as edges. In other cases the boundary lines may not be detected at all because of small contrast. At the second processing stage the results of filtration were based on a priori data on the positions and dimensions of the tiles in a pallet. The positions of boundaries between the tiles in an arbitrary column can be described by a model see Fig. 3.

$$y(k+1) = y(k) + d + w(k) \qquad (4)$$

where d is the mean interval between position of the tiles, and $w(k)$ is random error of placing the tiles. For our purpose it can be assumed that $w(k)$ is uncorrelated, random process with a zero mean value and dispersion W. A more complicated model did not bring much better results.

The aim of further processing of obtained values is to estimate the coordinates of boundary lines

$$\bar{y}(k/Z) \qquad (5)$$

where $Z = [z(1),z(2)\ldots,z(N)]^T$ is the vector of all measured variables (the result of the first stage of picture processing), the measured variable $z(k)$

$$z(k) = y(k) + v(k) \qquad (6)$$

can be modelled as an additive mixture of signal (the real position of a tile) and noise.

To optimize the edge position estimation (values $y(k)$), a criterion function can be constructed

$$J = \sum_{k=1}^{N} \text{var}\,\{y(k) - \bar{y}(k)\} \qquad (7)$$

with coupling condition (4). The criterion function (7) with coupling (4) is a marginal problem and can be solved using the methods of the calculus of variations (Sage and Melsa, 1971b). For a linear coupling condition and for the criterion of mean square error the problem splits into Kalman filter (8) and optimum interpolator (9).

$$y^*(k) = y^*(k-1)+d+K(k)[z(k)-y^*(k-1)-d] \qquad (8)$$

$$\bar{y}(k) = y^*(k)+A(k)[\bar{y}(k+1)-y^*(k)-d] \qquad (9)$$

Equation (8) is solved with the initial condition $y(0)$, which is given by positioning the pallet on the machine turning table. Equation (9) is solved with the terminal condition $\bar{y}(n) = y^*(n)$.
The methods of solving these problems, the computation of Kalman gain $K(k)$ and matrix $A(k)$ are described in detail in Sage and Melsa (1971a). Statistic variables for designing algorithms (8) and (9) must be sought by experiment carried out for each type of produced tiles. The calculation procedure is shown in Fig. 4. The dispersion of assesment error which is calculated as conjugate equation in filter (8) enables construction of a "window" for elimination from the processing of ambiguous measurements.This procedure is used in radiolocation during secondary processing.

CONCLUSION

The above mentioned procedure specifies the position of individual tiles with sufficient precision. All calculations are done on a computer. The computer at the same time controls the laser beam head movement and the turning table movement. The movement itself is driven by stepping motors. Using the machine, the productivity and quality of marking cutting tiles with inscriptions are significantly increased.

REFERENCES

Horn, B.K.P (1986). Robot Vision. McGraw-Hill,NY.

Sage,A.P., Melsa,J.L.(1971a) Estimation Theory with Application to Communication and Control. McGraw-Hill,NY

Sage, A.P., Melsa, J.L. (1971b) System Identification. Academic Press, NY.

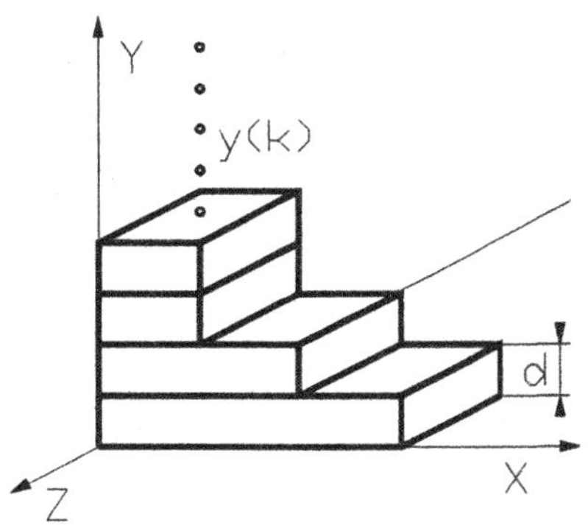

Figure 3. The position of tiles in a pallet. Axes x and y are in horizontal plane.

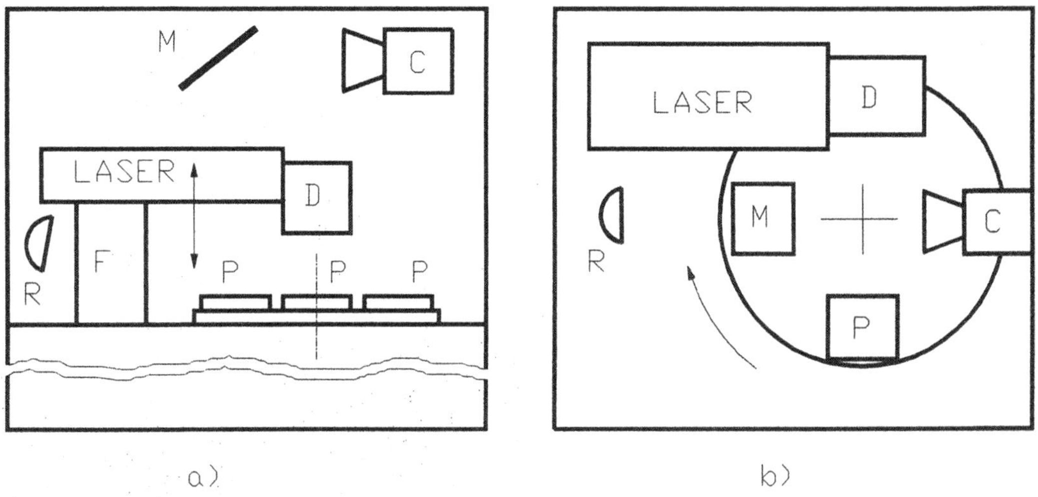

Fig.1. Front view (a) and the ground plan (b) of the marking machine.
F is laser focusing mechanism, R is reflector, D is deflection system with mirrors C is CCD camera, P is pallet with sintered carbide cutting tiles, M is a mirror.

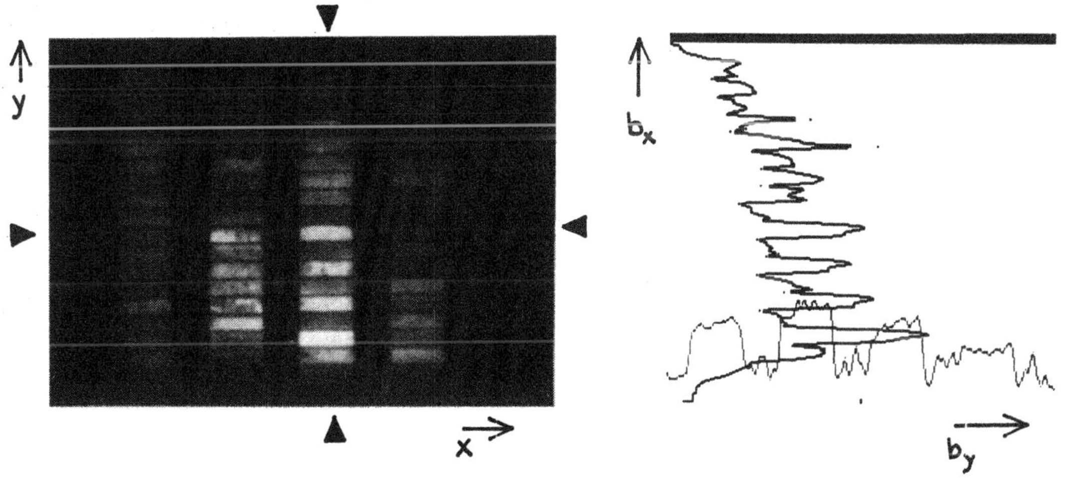

Fig.2a. A digitized picture of various types of tiles in pallets and in xy cross-section through brightness matrix in the direction of arrows. The picture of pallet with tiles was scanned with a CCD camera on 256 x 256 raster and digitized to the 256 grey levels.

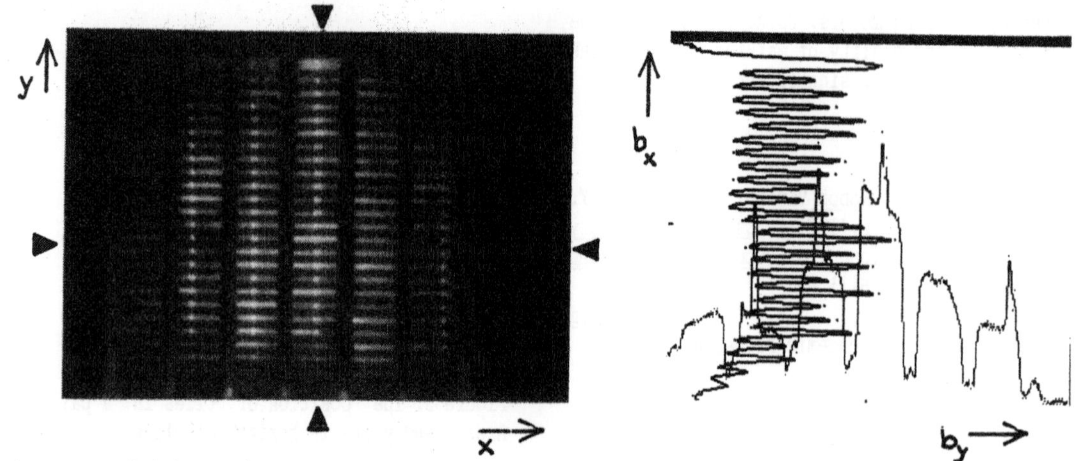

Fig.2b. A digitized picture of various types of tiles in pallets and in xy cross-section through brightness matrix in the direction of arrows. The picture of pallet with tiles was scanned with a CCD camera on 256 x 256 raster and digitized to the 256 grey levels.

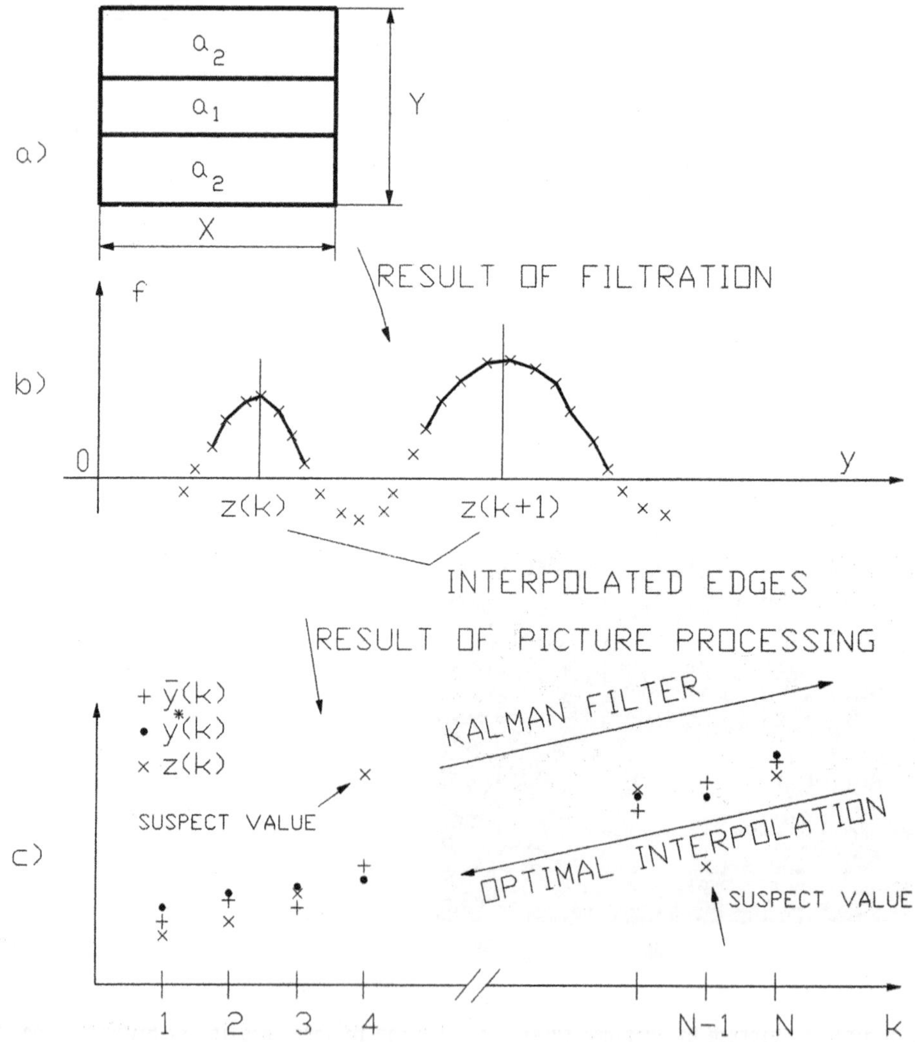

Fig.4. Sequence of image processing. (a) the mask of the filter.(b) The result of filtration in axis of a tile column. (c) The final result of processing by Kalman filter and optimal interpolator.

AUTHOR INDEX

KEYWORD INDEX

SYMPOSIA VOLUMES

PONOMARYOV: Artificial Intelligence
PUENTE & NEMES: Information Control Problems in Manufacturing Technology (1989)
RAMAMOORTY: Automation and Instrumentation for Power Plants
RANTA: Analysis, Design and Evaluation of Man-Machine Systems (1988)
RAUCH: Applications of Nonlinear Programming to Optimization and Control*
RAUCH: Control of Distributed Parameter Systems (1986)*
REINISCH & THOMA: Large Scale Systems: Theory and Applications (1989)
REMBOLD: Robot Control (SYROCO'88)
RIJNSDORP: Case Studies in Automation Related to Humanization of Work
RIJNSDORP *et al*: Dynamics and Control of Chemical Reactors (DYCORD'89)
RIJNSDORP, PLOMP & MÖLLER: Training for Tomorrow-Educational Aspects of Computerized Automation
ROOS: Economics and Artificial Intelligence
SANCHEZ: Fuzzy Information, Knowledge Representation and Decision Analysis*
SAWARAGI & AKASHI: Environmental Systems Planning, Design and Control
SINHA & TELKSNYS: Stochastic Control
SMEDEMA: Real Time Programming (1977)*
STASSEN: Analysis, Design and Evaluation of Man-Machine Systems (1992)
STRASZAK: Large Scale Systems: Theory and Applications (1983)
SUBRAMANYAM: Computer Applications in Large Scale Power Systems
TAL': Information Control Problems in Manufacturing Technology (1986)
TITLI & SINGH: Large Scale Systems: Theory and Applications (1980)*
TROCH, DESOYER & KOPACEK: Robot Control (1991)
TROCH, KOPACEK & BREITENECKER: Simulation of Control Systems
TU XUYAN: Modelling and Control of National Economies
UHI AHN: Power Systems and Power Plant Control (1989)
VALADARES TAVARES & EVARISTO DA SILVA: Systems Analysis Applied to Water and Related Land Resources
Van WOERKOM: Automatic Control in Space (1982)
VERBRUGGEN & RODD: Artificial Intelligence in Real-Time Control (1992)
WANG PINGYANG: Power Systems and Power Plant Control
WELFONDER *et al*: Control of Power Plants and Power Systems
WESTERLUND: Automation in Mining, Mineral and Metal Processing (1983)
YANG JIACHI: Control Science and Technology for Development
YOSHITANI: Automation in Mining, Mineral and Metal Processing (1986)
ZAREMBA: Information Control Problems in Manufacturing Technology (1992)
ZWICKY: Control in Power Electronics and Electrical Drives (1983)

WORKSHOP VOLUMES

ASTROM & WITTENMARK: Adaptive Systems in Control and Signal Processing
BOULLART *et al*: Industrial Process Control Systems
BRODNER: Skill Based Automated Manufacturing
BULL: Real Time Programming (1983)*
BULL & WILLIAMS: Real Time Programming (1985)
CAMPBELL: Control Aspects of Prosthetics and Orthotics*
CHESTNUT: Contributions of Technology to International Conflict Resolution (SWIIS)
CHESTNUT *et al*: International Conflict Resolution using Systems Engineering (SWIIS)
CHESTNUT *et al*: Supplemental Ways for Improving International Stability*
CICHOCKI & STRASZAK: Systems Analysis Applications to Complex Programs
CRESPO & DE LA PUENTE: Real Time Programming (1988)
CRONHJORT: Real Time Programming (1978)*
DI PILLO: Control Applications of Nonlinear Programming and Optimization
ELZER: Experience with the Management of Software Projects
FLEMING & JONES: Algorithms and Architectures for Real-Time Control
GELLIE & TAVAST: Distributed Computer Control Systems (1982)*
GENSER *et al*: Safety of Computer Control Systems (SAFECOMP'89)
GOODWIN: Robust Adaptive Control
HAASE: Real Time Programming (1980)
HALANG & RAMAMRITHAM: Real Time Programming (1991)
HALME: Modelling and Control of Biotechnical Processes*
HARRISON: Distributed Computer Control Systems (1979)
HASEGAWA: Real Time Programming (1981)*
HASEGAWA & INOUE: Urban, Regional and National Planning - Environmental Aspects
JAMSA-JOUNELA & NIEMI: Expert Systems in Mineral and Metal Processing
JANSEN & BOULLART: Reliability of Instrumentation Systems for Safeguarding and Control
KOPETZ & RODD: Distributed Computer Control Systems (1991)
KOTOB: Automatic Control in Petroleum, Petrochemical and Desalination Industries
LANDAU, TOMIZUKA & AUSLANDER: Adaptive Systems in Control and Signal Processing
LAUBER: Safety of Computer Control Systems (1979)
LOTOTSKY: Evaluation of Adaptive Control Strategies in Industrial Applications
MAFFEZZONI: Modelling and Control of Electric Power Plants (1984)*
MARTIN: Design of Work in Automated Manufacturing Systems*
McAVOY: Model Based Process Control
MEYER: Real Time Programming (1989)*
MILLER: Distributed Computer Control Systems (1981)
MILOVANOVIC & ELZER: Experience with the Management of Software Projects (1988)
MOWLE & ELZER: Experience with the Management of Software Projects (1989)
NARITA & MOTUS: Distributed Computer Control Systems (1989)
OLLUS: Digital Image Processing in Industrial Applications - Vision Control
QUIRK: Safety of Computer Control Systems (1985)(1986)
RAUCH: Control Applications of Nonlinear Programming*
REMBOLD: Information Control Problems in Manufacturing Technology (1979)
RODD: Artificial Intelligence in Real Time Control (1989)
RODD: Distributed Computer Control Systems (1983)
RODD: Distributed Databases in Real Time Control
RODD & LALIVE D'EPINAY: Distributed Computer Control Systems (1988)
RODD & MULLER: Distributed Computer Control Systems (1986)
RODD & SUSKI: Artificial Intelligence in Real Time Control
RODD & SUSKI: Artificial Intelligence in Real-Time Control (1991)
SIGUERDIDJANE & BERNHARD: Control Applications of Nonlinear Programming and Optimization
SINGH & TITLI: Control and Management of Integrated Industrial Complexes*
SKELTON & OWENS: Model Error Concepts and Compensation
SOMMER: Applied Measurements in Mineral and Metallurgical Processing
SUSKI: Distributed Computer Control Systems (1985)
SZLANKO: Real Time Programming (1986)
TAKAMATSU & O'SHIMA: Production Control in Process Industry
UNBEHAUEN: Adaptive Control of Chemical Processes
VILLA & MURARI: Decisional Structures in Automated Manufacturing

Out of stock • High quality repro print available. Details of prices sent on request from the IFAC Publisher.

IFAC related titles

BROADBENT & MASUBUCHI: Multilingual Glossary of Automatic Control Technology
EYKHOFF: Trends and Progress in System Identification
NALECZ: Control Aspects of Biomedical Engineering

Printed and bound by CPI Group (UK) Ltd, Croydon, CR0 4YY
03/10/2024
01040320-0017